———————

Start where you are,
use what you have,
do what you can.

Arthur Ashe

———————

———————
This quote by Author Ashe was borrowed from an article by ASCE Past President Robert W. Bein in *ASCE News*, May 2001, on ASCE's infrastructure report card and local needs assessment.

Contents

Preface

Probabilistic methods allow input variables and mathematical model parameters to have ranges instead of single numbers. This in turn provides results in the form of ranges and outcomes. Suppose that benefits and costs of a contemplated action, each with ranges of possible values, under some probability distribution, are entered into the analysis. The evaluation of the difference of the variables leads to a net benefit that also has a range with some probability distribution. This result allows determination of not only the average value of the net benefit, but also of the probability of critical threshold values, for example, the probability of the net benefits exceeding or not exceeding zero, in other words the probability of this activity being an economic success of failure. This gives the decision maker additional information that allows him or her to select an alternative that not only is cost-effective on average, but also has a reasonable probability of success or an acceptable risk or failure. Conventional deterministic methods, in contrast, gloss over the fact that there may be a range of outcomes that represents the probability of failure. Many projects that have become known as failures might have indicated a rather high and possibly unacceptable probability of failure if a wide range of possible results had been evaluated probabilistically. Instead, crucial decisions may have been based on a special favorable result without paying attention to many other possible favorable and unfavorable results.

Two important extensions of traditional engineering methods that are provided by probabilistic methods are discussed in connection with dealing with risk. The first is the interpretation of the expected value as the insurance premium of a risk-neutral risk taker or insurer, and the second is the inclusion of personal attitude toward risk into the expected value and thus into the insurance premium of a risk-averse decision maker or insurer. Insurance is an important aspect when dealing with risk and this aspect is given some room because risk control, risk taking, and risk transfer are the essential options a decision maker is confronted with. The increased risk awareness that flows from dealing with probabilistic methods should lead to the selection of economical and risk-averse alternatives. Probabilistic methods can explain the sometimes seemingly puzzling cost overruns or even economic failures of apparently sound projects and the often encountered divergence among parties over the results of risk assessment.

Opponents of probabilistic methods consider them unusable because of the usual lack of probability information. Also, there is an aversion to "making up" information to fill data gaps. The old adage "garbage in, garbage out" certainly applies to all methods. It does not exempt the traditional deterministic relationship between input and output. The probabilistic approach is not immune to poor information but surprisingly may be better off because it does not put all eggs in one basket, so to say. In practice, engineers often must make decisions with "imprecise" information. If such decisions also lack analytical support by a structures approach, they are known as seat-of-the-pants decisions. They usually are justified by those who make them by substituting "experience" and "gut-feeling" for missing information. Such undocumented and unsupported decisions are said to come out of a smoke-filled room meaning that few people if anybody except the decision-maker know how the decision was arrived at. This kind of decision-making suffers from lack of repeatability and cannot respond to data and parameter variation. Probabilistic methods can help to bridge information gaps by making up information and analyzing its impact on the final result. In other words, sensitivity analysis can be performed on uncertain but crucial information and the final decision takes these uncertainties into account. It is the purpose of this text to convince engineers and decision makers that probabilistic methods are the right tools to use when information is incomplete and imprecise.

Acknowledgments

The idea of using probabilistic methods in civil engineering is not new. A landmark publication in the field, Benjamin and Cornell's *Probability, Statistics, and Decision for Civil Engineers* was published in 1970. As happens with new approaches, they don't catch on easily. The computational burden that probabilistic methods carry with them was undoubtedly a barrier in the past. In the 30 years since then, more pioneering work of civil engineers, including Harr, Ang, Tang, Schnitter, Idel, Rissler, Vrijling, Yen, and others has become known, and with the desktop computer, the computational burden as a barrier has disappeared.

The roots of this book go back to the American Society of Civil Engineers (ASCE) Task Committee on Probabilistic Approaches to Maintenance of Hydraulic Structures (TC-PAMHS). This activity was approved in 1989 by the parent committee, the Committee on Probabilistic Approaches to Hydraulics of the Water Resources Engineering Division of ASCE. The general purpose of the parent committee was the exploration of application areas for the probabilistic methods in the field of hydraulic engineering. The committee, under the direction of the author, conducted its work from October 1, 1989 through September 30, 1992, the usual life span of 2 to 3 years of such a committee. The activities included (1) a survey of how field personnel perceive the probabilistic nature of maintenance, and (2) an introduction to probabilistic methods for practicing maintenance engineers. A survey of operators and/or owners of hydraulic structures was conducted in the fall of 1990. A summary of the survey was published in the ASCE Hydraulics Division conference proceedings by the author in 1991.

This book is an outgrowth of the original idea that practicing engineers could profit in many ways by the use of probabilistic concepts and a probabilistic way of thinking. It thus can be considered the concluding work of the task committee's original plan. The compilation of this introduction to probabilistic methods got bogged down after the disbanding of the task committee in 1992 for various reasons. Through the encouragement of the late Professor Ben C. Yen, University of Illinois, Urbana-Champign, and Dr. Steve Melching, then with the USGS at Urbana, Illinois and now a professor at Marquette University, Milwaukee, the author carried on the work.

Both Dr. Yen and Dr. Melching were members of the parent committee in 1996 when the effort to complete the book was revitalized. Dr. Melching took on the task of a principle reviewer and advisor. He looked over the material, provided suggestions, and reviewed the revised material again. His encouragement and help were most important in keeping the effort going. Dr. William LaPay of Westinghouse, Pittsburgh, reviewed all chapters. The author is also grateful to J. R. Bowman, formerly with Harza Engineering Co., Chicago. When he heard about this project, upon his retirement, he left his collection of papers on hydroprojects to the author. The papers provided interesting information on the years of the hydraulic structures building boom from the 1920s through the 1950s.

The author is very grateful all named and also many unnamed contributors, supporters, and constructive reviewers. The author never stopped learning, revising, and correcting as he worked his way deeper and deeper into this sometimes difficult but ever more fascinating subject matter. He gratefully acknowledges the help and encouragement of the ASCE publishing staff, especially Suzanne Coladonato. The author finally had to concede that he was facing a probabilistic problem that allowed him to reduce the probability of errors but not to eliminate it so that in the end he felt subdued by the spirits he had called.

July 2004

Walter O. Wunderlich, P.E.
Consulting Civil Engineer
Knoxville, Tennessee

Summary

Chapter 1 defines maintenance and gives an overview of maintenance activities. It describes the state of the hydraulic infrastructure, followed by an overview of regulatory and agency approaches to maintenance. It concludes with a discussion of selected questions form a field survey of maintenance personnel undertaken by the ASCE Task Committee on Probabilistic Approaches to Maintenance of Hydraulic Structures (TC-PAMHS) in 1990.

Chapter 2 is an introduction to the basics of probabilistic methods. The chapter begins with sets, events, and probabilities. The various kinds of sets, their properties, and their probabilities are discussed. The characteristics of random variables are described and some simple probabilistic processes and models in which they are occur are discussed. The most important parameters of random number populations, such as the mean and the variance, are introduced. Comparisons between deterministic and probabilistic calculations are demonstrated. Probability density functions (pdfs) and cumulative distribution functions (cdfs) are introduced. The most important and also most widely used of all distributions, the Gaussian pdf, is discussed in some detail. An overview of eight other distributions concludes the chapter.

Chapter 3 begins with the prediction and data analysis methods that attempt to extract information from random processes. Then transformations of random variables are described. The log-normal transformation is described that uses the logarithms of data and exploits the amazing property of their normal distribution, a result that would otherwise be rather difficult to find.

This is followed by a section on arithmetic with random variables, such as addition and subtraction. Section 3.4 describes one of the most amazing theorems on probability theory, the central limit theorem (CLT). According to this theorem, whatever the original distribution of random variables, if only enough of them are sampled, the sample sums or means approach the normal distribution. The practical examples of this phenomenon abound and some will be discussed. For example, the distributions of quarterly mean flows of the Tennessee River at Chattanooga are different for each quarter and are especially skewed toward low flows in the summer and fall quarters. But when they are all added up, the annual mean flow shows a distribution that is remarkably close to a normal distribution; this tendency, while theoretically guarantees only for very large samples, practically already happens for the sum of elements from only four different distributions. In other words, the convergence to the normal distribution is very fast and robust, which makes the CLT a very practical and widely applicable tool for dealing with random samples.

The next section is devoted to reliability. The probabilistic approach to the capacity-load problem, or resistance-stress problem has many similar applications in very different fields, including civil engineering, power engineering, industry, economics, and medicine, and is one of the most powerful and convincing applications of probabilistic calculus. This approach can be used to give probabilistic meaning to arbitrarily selected safety factors and safety margins.

Section 3.6 deals with the derivation and use of the hazard function and related functions. It describes the conversion of mortality rates into the pdf of life length, and the derivation of the function of remaining life. This application demonstrates how probability data can be converted from one function into another, as required to meet the needs of probabilistic analysis and the availability of data. Section 3.7 deals with the conversion of probability into a return period. This conversion provides a scaling that tends to straighten out the cdf so that it can be more easily used for interpolation and extrapolation in the critical areas.

Chapter 4 explains hydraulic structures as systems composed of subsystems and components with associated probabilities of functioning and not functioning. Systems usually consist of subsystems and components in serial, parallel, or mixed serial-parallel arrangements. Disaggregation of such systems into subsystems and further disaggregation into components, analysis of the components, and aggregation or synthesis of the system are the steps for performing a reliability analysis of a complex system. The system structure function and the system reliability function are described as tools of a probabilistic approach to the reliability analysis of system performance. Systems that must achieve high reliability may require redundancy, which may be provided in the form of system redundancy and component redundancy. Reliability and economic factors may determine the adoption of one or the other. Reliability analysis by the cut set and tie set method is described followed by a discussion of standby as a reliability measure.

Chapter 5 deals with probabilistic maintenance concepts. Maintenance options, such as corrective and preventive maintenance, are described. The criteria for their selection are usually based on economics, project safety, and reliability. An important aspect of maintenance planning is downtime. It is crucial to bring the system back up after an outage has occurred to minimize disruptions of service caused by an unexpected outage or a scheduled downtime. Loss of service and possibly the required purchases of replacement service can add substantially to outage costs and can cause unexpected cost overruns.

Other sections deal with availability, maintainability, and maintenance strategies. In this latter context, maintenance is viewed as interference with a random deterioration process of a system. Section 5.8 is devoted to economic concepts. It deals with benefit-cost analysis, cost effectiveness analysis, and life-cycle cost analysis. These calculations deal with estimated costs and estimated future benefit streams. Life-cycle cost analysis is a typical application area for probabilistic methods because many ingredients to the analysis are uncertain. An interesting probabilistic problem is the internal rate of return calculation that matches probabilistic life-cycle benefit and cost streams with the front-end investment. The probabilistic input in the form of benefit and cost streams produces a probabilistic internal rate of return. The problem is, however, left to the reader to solve.

The section concludes with a discussion of the economy of premature demise. The last section briefly describes the probabilistic aspects of project planning methods, such as by critical paths and project evaluation and review technique.

Chapter 6 is devoted to the concept of risk and its role in engineering and maintenance. Risk is a problem that is all around us. Yet methods to quantify it are not well developed. At the beginning of the chapter, a definition of risk is given, followed by an overview of types of risk. Hazards are defined as conditions that increase risk and, thus, must be identified and controlled.

Also discussed is the attitude toward risk and risk perception, which underpin the subjective aspect of risk. An important aspect is dealing with risk. A section is devoted to the transfer of risk from a weary risk owner to a willing risk taker through insurance. There are many types and aspects of insurance, and there is insurable risk and uninsurable risk. Risk control often is a prerequisite for insurance as well as self-insurance, re-insurance, liability insurance, and surety. Risk management is discussed as a strategy that aims at reducing or eliminating risk. Maintenance decisions often must be made with incomplete information. They have the potential of being right or wrong, sometimes with considerable social and economic consequences. This requires approaches that prepare the decision-maker for risky decisions.

Two sections deal with methods of quantifying and measuring risk. Objective risk measures are discussed in Section 6.4. The expectation is risk-neutral and must be supplemented by measures that indicate the extent to which real outcomes can depart from the expectation. The standard deviation is a measure of dispersion to judge the risk content of an alternative. Hypothesis testing is also discussed as a hybrid between objective and subjective risk measurement. It uses a test statistic, the sample mean, and its normal distribution guaranteed by the central limit theorem.

A subjective significance level is used for rejecting or not rejecting the predefined null hypothesis. Subjective risk measures are discussed in Section 6.5. By the use of a utility function that reflects the decision maker's attitude toward risk, the risk-neutral expectation is transformed via a risk-averse expected utility into a risk-sensitive certainty equivalent or insurance premium. This monetary value provides a risk-sensitive criterion. Its minimization is in the interest of the risk owner whether he intends to transfer the risk to an insurer or whether he has to use self-insurance. An example that applies objective and subjective risk measures concludes the chapter.

In all chapters, theory and examples are kept simple to help the reader make rapid progress in gaining an understanding of the capabilities of probabilistic methods. Formula derivations and numerical integrations are used for the same purpose and to demystify advanced mathematical concepts, such as integrals, mathematical operators, and convolution integrals that may look forbidding but are actually quite easy to handle in spreadsheets where they are reduced to elementary operations. But some of the elegant solutions of classical calculus are also demonstrated where appropriate.

The desire for improved and more state-of-the-art techniques in maintenance planning was expressed in several comments by responses to the questionnaire discussed in Section 1.6.3. The collection of methods presented here is expected to pique the interest of analytically minded practicing engineers and computer-oriented younger engineers, many of whom may spend their entire career in the ever more important infrastructure maintenance and rehabilitation sector.

The work presented here may serve as a bridge between engineers who know the methods, but are not necessarily experienced in maintenance on the one hand, and maintenance engineers and personnel, who know the maintenance problems, but are not necessarily familiar with probabilistic methods on the other hand. The material presented here may also stimulate various aspects of maintenance work, such as measuring, monitoring, and data analysis, and research in areas such as the quantification of uncertainty and risk.

The presented material is not intended as an exhaustive treatment of probabilistic methods applicable to maintenance. Many more advanced concepts can be brought to bear on practical problems; some of them are touched in Sections 2.2.3 and 2.2.4. This text is thought of primarily as a first stepping-stone into the realm of probabilistic methods. Once this step is done, the reader should have developed enough enthusiasm that will make it relatively easy to continue to more advanced levels.

1. Maintenance of the Infrastructure

1.1 The Role of Maintenance

Maintenance literally means "hand-holding"(L.: *manu tenere* → Fr: *maintenir* → hold by the hand). A production facility, while in operation, even if automated to some extent, usually can not be left entirely to itself, at least not for prolonged periods. It needs supervision, adjustments, and occasional repairs to continue operating as expected. In a modern industrial environment, operation as expected means safe, reliable, and efficient operation. Maintenance is often used in one breath with "operation." Operation usually would not last long if it were not supported by maintenance. Maintenance literally holds hands with operation. Most major industrial production processes have associated with them an "operation and maintenance" (O&M) department devoted to keeping the process up and running.

The most extensive investigation of hydro infrastructure maintenance ever undertaken probably is the U.S. Army Corps of Engineers Repair, Evaluation, Maintenance, and Rehabilitation (REMR) research program, which began in 1984 and was completed in 1998 (Scanlon et al., 1983; ASCE, 2000, p. 27). REMR defines *maintenance as action that prevents or delays damage or deterioration, or corrects deficiencies* that would otherwise lead to early repair or the need for rehabilitation. *Repair* restores a deteriorated, damaged, or failed component to service. *Rehabilitation* is a major modification or reconstruction of an existing structure to bring it up to prevailing operation requirements and standards. *Replacement* is the exchange of defective, inadequate, or obsolete components or systems for new ones. In all maintenance-related activities, similar methods are used, such as work planning and scheduling, benefit-cost analysis, risk and reliability assessments, decision analysis, prediction, and so on. Therefore, no distinction is made here among these activities, and "maintenance" is used as a collective term.

The REMR Overview and Guide summarizing 14 years of studies became available in the fall of 2000 on the U.S. Army Corps of Engineers USACE web site (www.wes.army.mil/ REMR/reports.html). The study was done at a cost of $66 million. Five years after the first phase ended, the corps estimated savings from the new methods at $200 million. Additional similar savings are expected from the second phase, which ended in 1998. Given the very far-flung operations of the corps, the multibillion dollars worth of investment in all sorts of hydraulic structures, and the multibillion dollars of new investments through their annual civil works budget for new construction, maintenance, and rehabilitation, even a 1 % savings annually over 10 years could produce savings of this order of magnitude. Hence, enormous sums of money undoubtedly can be saved by making project operation and maintenance as efficient as the state of the art allows.

Maintenance has always been a part of the management of well-run projects. The machine halls of power plants and pumping stations have traditionally been shining examples of a spick-and-span operation environment. But there is more than meets the eye. Age, environmental effects, operational stress, and material erosion and decay take their toll by creating weaknesses that sooner or later lead to unexpected outage, loss of efficiency, or failure. When the number of outages increases, O & M costs also increase, output and its reliability decrease, credibility with clients is lost, and failure may bring the operation to a halt, for extended periods or forever.

In principle, all hydraulic structures are addressed here. They may serve various purposes, such as flood control, power supply, energy production, municipal and agricultural water supply, navigation, recreation, fish and wildlife habitats, and so on. For water projects with dams, the dam also is the keystone of the development. The dam creates a head for waterpower, the required depth for a navigation channel, the storage that transforms a random flow into a reliable water supply, a controlled discharge in a flood, an impoundment that adds a scenic accent to the landscape and creates a recreational water surface. The impoundment in turn provides fish and wildlife habitats, wetlands, and so on. But a dam's necessary interference with a water course modifies the natural flow and water quality regime, existing wildlife habitats and land uses, and thus creates environmental impacts. Some impacts are unavoidable, others can be mitigated, or alternatives can be provided by the new environment created by the dam as many of the larger dams are equipped to regulate flow and impoundment levels by a predesigned operation plan. Dams by the water mass they impound also create a potentially harmful exposure to life and property within the influence area of a potential dam failure. Contrary to many other killer events, such as hurricanes and tornadoes, this area is predictable. Nevertheless, even though dam failures are rare, dams are a source of risk to life and property that must be minimized. Besides, they are primary attention getters for the media and thus have the potential of giving the entire civil engineering profession a bad name.

1.2 Probabilistic Methods in Maintenance

Many people are more familiar with statistics than with probability theory and therefore give preference to the attribute "statistical" over "probabilistic." Most people have encountered statistics in their daily lives whereas the term *probabilistic* sounds rather theoretical. The population census taken every 10 years makes every citizen an element in a statistical population of hundreds of millions of elements. The reader may wonder what the difference is between statistical methods and probabilistic approaches. When speaking of statistical methods, one thinks of processing large amounts of data. Probabilistic methods evoke more the aspect of lack of precise knowledge and uncertainty, calculating the outcomes of a system that acts like a lottery, for example. Methods used on census data are statistical methods that extract means, dispersions measures, and trends from large data populations. Probabilities are defined in statistics as the ratio of the number of elements of a certain kind to the total number of elements in the data population when the total number of elements goes to infinity. So they are averages derived from very large populations. Statistical methods are intertwined with probabilistic methods but still there is a difference. Consider a mathematical analysis that attempts to determine the pros and cons of two possible construction alternatives of a rehabilitation project, each of which may produce different consequences when exposed to possible natural phenomena, such as floods and earthquakes. The solution of such a problem also makes use of statistical concepts, such as means, standard deviations, probabilities, and so on, not with emphasis on processing a large mass of data, but with emphasis on how to weigh even very little information when making a decision among a few alternatives in the face of uncertainty.

Statistical methods have discovered and incorporated in themselves the basics of probability theory. If 100 concrete samples are taken from a concrete production process, after organizing the results into groups of about equal concrete strength, it is almost impossible to overlook the emergence of a histogram, most likely bell-shaped in this case, of the number of items in each group over the total range of sample strengths. The associated data-based mean, standard deviation, probability distribution function, correlations, and so on, are objective results that sum up the information content of the data sample. The purpose of such sampling is to make inferences on the true strength of the concrete produced by the production process. The same result should be obtained regardless of who performs the analysis. This is why statistical methods are called objective methods. Probabilistic methods use statistical methods, especially their results, and deal with concrete strength, in this case, as not just one value but as a variable with a range and a frequency distribution over it. This means that none of the strengths within the variable's range has certainty of occurring, but only some measurable probability. The mean of the variable may be the point that has the highest probability of occurrence, but this value may be reached and exceeded by the rather high probability of 50 %. Probabilistic methods incorporate this uncertainty of variables or model parameters that are part of a complex process. There may be uncertainty, not only in numbers but also in relations between numbers, due to lack of experience, paucity of data, or simply ignorance of the process being dealt with. In many situations, large databases are often neither available nor may they ever be obtainable by the very nature of the process. One cannot delay action for years until a sizable data sample is collected to extract needed information and then perhaps discover that the process is not stationary after all, making the data a questionable sample for the prediction of the future. Also, one cannot produce extreme conditions for a real system to find out how it will react under extraordinary circumstances. Probabilistic methods can be used to analyze a large number of possible scenarios in the form of a wide array of combinations of variables with ranges and probabilities, and then process the many outcomes by statistical methods to arrive at ranges of results and their probabilities which provide an extended information base for decision making that includes a spectrum ranging from undesirable outcomes to desirable outcomes.

Probabilistic approaches also have to deal with the fact that statistically derived probabilities usually are just limiting values, or averages, that emerge from many trials. For example, throwing a coin produces a head or a tail, which is given here the value $+1$ and -1, respectively. Throwing the coin 1,000 times may produce 495 times a $+1$ and 505 times a -1. The decision maker may come to the conclusion that the probability of either outcome is about 0.5. Hence, the probability-weighted sum is $0.5 \cdot (+1) + 0.5 \cdot (-1) = 0$. This is known as the expectation resulting from many trials. If the decision maker is allowed only one trial, he cannot count on the statistical average. He faces the real outcomes, $+1$ or -1. This adds another aspect to the probabilistic approach, namely, the presence of *risk*. The following definition of probabilistic methods and approaches is proposed: *Probabilistic methods incorporate the uncertainty of variables and parameters in the analysis of problems. Probabilistic approaches are applications of probabilistic methods to solve engineering problems.*

To put this definition in perspective, the uses of the words *probabilistic*, *statistical*, and *stochastic* in a selection of references are discussed next. Moan (1982) calls the chapter he contributed to the *Reliability Handbook* (Ireson, 1982) "Applications of Mathematics and Statistics to Reliability and Life Studies." The chapter deals with sets (events), the probability of sets, probability density functions, probability distributions, and probability functions including reliability functions. The chapter also deals with predominantly statistical subjects, such as sampling techniques, sample analysis, and regression and correlation analysis. All these subjects are treated under "statistics and mathematics" but they are all elements of probabilistic methods.

In *Stochastic Processes*, Parzen (1965) deals with the theory of sequences of random variables. Any process that generates random realizations is a stochastic process. A major subcategory consists of Markov chains that deal with probabilistic transitions between states, state occupation times, and first passage times. Stochastic processes include various Poisson processes, interarrival times and waiting times, birth and death processes, renewal processes, queuing processes, time series, random variables, conditional probabilities, and others. These are subjects that already by their names indicate applicability to maintenance planning problems. For example, the arrival of repair jobs during a month can be construed as a Poisson arrival process. Given each job has a random repair cost and random repair time, monthly totals of cost and time can be construed as monthly realizations of an annual stochastic repair cost and repair time process (see Section 2.7.7). Parzen (1965, p. v) calls the theory of stochastic processes "the 'dynamic' part of probability theory." According to Parzen (1965, p. 2), the theory of stochastic processes is the mathematical foundation of both *statistical physics* and *statistical mechanics*. Considering the frequent use of probabilities in statistical mechanics, it could also have been called probabilistic mechanics or stochastic mechanics.

In *Principles of Statistical Mechanics*, Katz (1967, p. 28) deals with the probability of the system being in a specified region which is described by a "statistical function," and the probability-weighted average is called "statistical average or expectation."

In *Hydrosystem Engineering and Management*, Mays and Tung (1992, p. 164) define probability as a measure of likelihood of an event and uncertainty as a characteristic of an unknown and uncontrollable event-producing process whose random outcomes or random variables are unpredictable but its average behavior can be characterized by statistical properties such as expectation, variance, and other statistical parameters that can be calculated from an observed sample. In principle, all probabilistic (or stochastic) processes are uncontrollable but some can be influenced by a policy in the way a decay process can be influenced by maintenance.

In *Probability, Statistics, and Decision for Civil Engineers,* Benjamin and Cornell (1970, p. vii) state: "Although virtually all forward looking civil engineers see the rationality and utility of probabilistic models of phenomena of interest to the (civil engineering) profession, the number of civil engineers trained in probability theory has been limited." They continue "there has not been sufficient development ... to permit a unified probabilistic approach to the many aspects of the strength of materials, soil mechanics, construction planning, water-resource design, and many other subjects where the methods could clearly be useful." The present text is in good company when it adds maintenance of hydraulic structures to the list of subjects to be treated by probabilistic approaches and by making another attempt to promote the understanding and use of "probabilistic models" in engineering.

Benjamin (1974, p. 357) distinguishes three branches of probability theory and applications. The first branch is basic probability theory and stochastic processes. The second branch includes statistical analysis and inferences from data, hypothesis tests, confidence intervals, data-based probabilities, and data summarizing procedures, such as regression analysis. The third branch is Bayesian statistical decision theory.

In *Probability and Statistics in Hydrology,* Yevjevich (1972) considers *chance, random, probabilistic,* and *stochastic* as synonyms. He defines probability theory as "that branch of mathematics that investigates random phenomena," and mathematical statistics and stochastic processes as two offshoots of probability theory. Together, probability theory, mathematical statistics, and stochastic processes provide the tools for dealing with the random phenomena of hydrology (Yevjevich, 1972, p. 4). Commenting on the controversy between advocates of deterministic modeling and advocates of probabilistic modeling, Yevjevich (1972, p. 71) states "that the deterministic solution is nothing else than a particular solution of the general probabilistic or statistical solutions." By using a deterministic approach on a "fundamentally probabilistic phenomenon three errors may be committed: a part of the total information is lost, misleading results are obtained, and incorrect solutions are produced." He also points out that the use of means instead of random variables is only correct when variations about them are negligible. These statements are still fully valid and will continue to provide motivation for using probabilistic methods.

In *Statistics for Modern Business Decisions*, Lapin (1978) deals with populations and samples; frequency distributions; statistical parameters, such as mean and variance; probability; hypothesis testing; regression and correlation; time series; price indexes; inferences; decision theory; Bayesian analysis; and judgmental probabilities and utilities, including the making up of probability distributions by judgment, and attitude toward risk, all under the subject of "statistics."

In *Stochastic and Risk Analysis in Hydraulic Engineering*, Yen (1986) proposes a definition of *stochastic hydraulics*, a term he dates to the early 1970s, as "analysis and application of hydraulics involving temporal and/or spatial random processes. Diffusion processes, turbulence, sediment transport, and random forces due to waves or wind are examples of stochastic hydraulics." In *Reliability and Uncertainty Analyses in Hydraulic Design*, Yen and Tung (1993) address the application of probability principles that has long been limited to hydrologic frequency analysis to a wider variety of hydraulic engineering applications, including safety and reliability analysis of hydraulic structures, and to planning, constructing, and operating hydraulic projects and systems.

It seems that any of the words—statistical, stochastic, and probabilistic—could have been used as a qualifier of *approach* in the title of the present text. The word *probabilistic* was chosen because it best describes the basic idea behind the proposed approach: the use of variables with ranges that are spanned by frequency distributions as a way of quantifying the likelihood of occurrence of a value in the range. The use of probabilistic methods is more than just a different calculation technique. Aside from using computer-assisted numerical algorithms, which carry the substantial computational burden, the user acquires greater circumspection, and greater awareness of the usually existing multiple process pathways and the associated multisolution aspects of engineering problems. This expanded way of thinking should have a beneficial effect on planning for project safety, reliability, and economy.

In the maintenance sector, areas that can benefit from probabilistic approaches are conducting inspections, scheduling surveillance activities, collecting and analyzing data, making diagnoses, conceiving and selecting repair alternatives, evaluating alternatives, and selecting alternatives for implementation. Also persons who do not get involved in the computational aspects of probabilistic approaches should find a basic knowledge of probabilistic principles quite useful. Thinking in probabilistic terms makes it natural to look at the entire spectrum of inputs and outputs of a problem. It is a more comprehensive approach than the still widely used conventional deterministic approach. In the years to come, there will be more and more applications of probabilistic methods in areas where uncertainty affects analysis and decisions. Examples are engineering, arts and sciences, business, health sciences, human resources, and military sciences. Usually the results obtained by a probabilistic approach are used as decision support. Given that all pertinent information is being used, it is hoped that the probability of a correct decision is increased compared to the deterministic approach that relies on a few preselected special cases, or even worse, the seat-of-the-pants approach that does not use analysis at all. Swets et al. (2000) give an example of an application to medical diagnostics. The benefits cited are a reduction of unnecessary or harmful treatment of patients, a reduction of false alarms that trigger costly measures, and an increased probability of a correct diagnosis that is followed by a most likely correct treatment.

The purpose of the material presented here is to familiarize the reader with the basics of probabilistic concepts and methods. Mathematical derivations are given in some detail, not just final formulas, because it is thought that the derivations give additional insight into probabilistic concepts. Numerous application examples are given, all with solutions, because this will save time, facilitate the understanding of probabilistic concepts, and minimize frustration. It is hoped that the text leads readers to a level of understanding that facilitates recognizing applications and formulating problem solutions. It should also prepare the reader for the study of more advanced methods.

1.3 Areas of Maintenance Activity

1.3.1 General Maintenance Objectives

The objectives of maintenance and the approach to maintenance are rather similar in the various organizations, federal and nonfederal, active in the hydro sector. They are even similar in many sectors of diverse industries that have nothing to do with hydro operations. Each sector, however, has its own special problems to which the rather general methods must be adapted. Sometimes it may amount to nothing more than recognizing similarities in problem formulations for transferring solution methods from one sector to another. The term *hydraulic structures* is used here in a wider sense and includes structures and equipment, as needed, to operate systems that make use of or deal with water in some way, such as dams, hydropower plants, locks, irrigation structures, weirs, diversions, flood control structures, canals, intakes and outfall structures, river and coastal protection structures, cooling water systems, municipal water supply and drainage systems, and others. Basically, maintenance contributes to three areas:

1. Protecting the integrity of materials of which the system and its operational equipment are built. Such materials include concrete, steel, wood, rubber, plastics, and others. Damage may result either from the production process itself or from the environment in which the process operates.
2. Guarding the functioning as expected of all system components, subsystems, and the system as a whole.
3. Keeping the system prepared to withstand disruptive external events, such as seepage, foundation failures, floods, earthquakes, vandalism, and other natural and human events.

Successful maintenance in these areas can avoid damage to and loss of infrastructure, loss of property, and loss of life. The probabilistic aspects of these areas are discussed in Sections 1.3.2 through 1.3.4.

1.3.2 Maintenance of Structural Integrity

All or at least most hydraulic structures are built in the field by temporary fabrication facilities. The large amounts of materials that are required for dams or other large structures are produced as close to the construction site as possible. This applies mainly to concrete, earth, and rock materials for dams. The material is also put in place under field conditions. Steel and plastics are produced in manufacturing facilities off-site, but still must be assembled in the field. This requires that materials production, assembly, and placement be done under strict observation of high-quality standards. Mistakes or defects embodied in the structure in the construction phase may carry over into operation and maintenance of the completed project and may become a liability of maintenance by requiring more frequent and more expensive maintenance (see Section 1.5.2). In anticipation of shortfalls of reaching needed quality levels by probabilistic processes such as concrete manufacture, specifications may be imposed on the builders by the owner or operator by demanding a sufficiently high average quality level. For example, the Bureau of Reclamation (BUREC) requires from its construction contractors that 80 % of the concrete test specimens exceed the design strength. Depending on the spread of the sample distribution, which is a function of the coefficient of variation, a mean concrete strength is defined that the contractor must maintain (BUREC, 1975, p. 173).

1.3.3 Maintenance of System Operation

The two major types of maintenance are operational maintenance and repair-oriented maintenance. Operational maintenance or *routine maintenance* is the continuing surveillance and upkeep of the running system. It is intrinsically tied to operation and usually is part of the same department, the O&M department. It consists of "in-operation" support work such as checking sensors, keeping a clean operating environment, keeping bearings greased and oiled, keeping out rust. It is the kind of maintenance that comes to mind when the layman thinks of maintenance. From a probabilistic point of view, also routine maintenance has a probabilistic aspect as it reduces the probability of failure and thus increases equipment survival and reliability.

Hydroprojects are relatively simple production systems when compared with coal-fired or nuclear power plants, but they are still rather complex undertakings with major safety and environmental concerns. The creation of a substantial operation head, the impoundment of large amounts of water, the water pressure in pipes, on gates and valves, the speeding mechanical and electrical machinery, the potential environmental effects, the transport of electric energy at high voltage over long distances from the often remote project sites to demand centers requires sensitive project management. Hydropower generation is adaptable to rapid load variations. This may result in increased equipment wear and tear and reduce the life expectancy of equipment. There is increasing environmental pressure to eliminate or at least mitigate quick discharge changes at hydroplants which would reduce the economy of hydro operations.

Repair-oriented maintenance or *preventive maintenance* looks for incipient major problems by occasional in-depth inspection and diagnosis of the findings. As a result of such investigations, various maintenance measures may be considered and a remedial action plan developed if hazardous conditions are recognized that could lead to the development of major damage, unscheduled shutdown, or failure.

Planning for system shutdowns, repairs, rehabilitations, and renewals are typical activities where probabilistic methods apply, specifically for downtime and repair cost estimates. Many systems have a serial lineup of components so that the whole system is as reliable as its weakest link. It is the task of maintenance to watch over the functioning of all components so that the need for maintenance is recognized with sufficient lead time to accomplish repair and/or replacement during low load periods and the shortest possible time slots. Identifying and focusing maintenance on the weakest link effectively increases the reliability of the whole system.

1.3.4 Maintenance Aimed at Incipient Failure Mechanisms

A process system can be in one of two fundamental states: operating or shut down. If the system is in the operating state, there are two possibilities: the operation will continue as scheduled or it will terminate in a shutdown. A system without maintenance sooner or later will fail with certainty, that is with probability 1. The uncertainty about final failure is associated with *when* and *how*. Relatively few structures have come down to us from antiquity. Most of the survivors have been continually used and maintained. The demise of systems can be attributed to a number of factors such as climatic effects in the form of alternating heat and cold, freeze and thaw, takeover by vegetation, wind and water erosion, obliteration by sediment deposits, vandalism, abandonment due to obsolescence, collapse due to structural defects, and failure of the maintaining entities in the form of governments, communities, and owners (Jansen,1980; Wunderlich and Prins, 1987, pp. 3–189).

Apart from the continuous chipping away caused by the usual operational wear and tear, and environmental effects, there are extraordinary events that may overcome the resistance of a system. These events can be of natural or human origin. Natural events include floods and earthquakes. Human events include major operational error, such as creating a destructive water hammer or causing a dam to be overtopped, and acts of war. Maintenance usually is not able to prevent extraordinary events, but it may be able to keep a system prepared so that there is an increased probability that it will withstand such events, as long as they are within the design limits of the project. Next to foundation failure, the most frequent cause of dam failure has been "overtopping" in floods. Overtopping occurs because release equipment is inoperable or the spillway capacity is insufficient. In both cases, maintenance and rehabilitation could have prevented the failure. Earthquakes have destroyed a few dams. Others have escaped destruction by the fact that joint corroborating events usually are required to bring on a disaster. For example, in some cases, earthquake damage to the dam was severe enough so that it would have failed had it been under full water load. Since the earthquake occurred at low water level, as a result of either defensive operation policy, seasonal circumstance, or a combination of both, dam failure and associated consequences did not occur.

The *fair-weather dam failure* is the most hazardous kind of dam failure because it is aggravated by the surprise effect, which means minimum warning time and thus maximum exposure to harm of human life and property. Such dam failures can be caused by sudden structural collapse, usually foundation failures, or embankment slides due to earthquake shaking. A major preventive rehabilitation project is that of Saluda Dam in South Carolina, a dam about 60 m high, which impounds a huge reservoir with 202 km^2 surface and exposes some 120,000 people to possible harm immediately downstream. The dam of the hydraulic fill type has been found unsafe because of its liquefaction potential in a Charleston-type earthquake (7.1 to 7.3 on the Richter scale with a 1,000 year recurrence). An emergency dry dam constructed at the foot of the existing dam is expected to impound the breach should the existing dam fail (Rizzo et al., 2002). Acts of war have succeeded in producing major dam disasters when the delivery of the destructive device was combined with deficient protection of the target (Jansen, 1980).

The failures of St. Francis, Vaiont, and Malpasset Dams all occurred under full load with associated large loss of life and property (Jansen, 1980). Lack of maintenance was not a factor in any of these disasters, but they hammer home the presence of the enormous amount of potential energy behind dams that maintenance must help to keep under control. Impending structural failure is sometimes indicated in advance by subtle symptoms, such as cracks or leaks. These symptoms must be discovered and correctly interpreted in time, and necessary action must be taken to avoid an accident or failure. If one link in the sequential maintenance decision process fails, the entire process fails and the accident takes its course (see Section 5.7.1).

The Bureau of Reclamation (BUREC) requires an Emergency Preparedness Plan (EPP) for all dams whose failure would endanger human life or cause substantial property damage, which is defined as death of at least one permanent resident and damage on the order of $1 million (BUREC, 1986, p. 85). Maintenance can contribute to the implementation of an EPP in many ways, all aimed at reducing, to the extent possible, the probability of an emergency occurring, and at reducing the exposure of lives and property to harm. Examples of what can be done are as follows:

- Maintain adequate spillway capability.
- Ensure operation of spillway gates under adverse weather and flow conditions.
- Be prepared for unusual flow conditions during natural events, such as debris flows, that could make spillway gates inoperable or ineffective.
- Be prepared for and capable of lowering the water level as soon as a potential threat emerges.
- Inspect for hidden damages after an external event has occurred

without obvious effect on the structure.
- Check dam, its lateral groins, the dam slopes, and the riverbed downstream for seepage flow, wet areas, and slumps.
- Check seepage flow for suspended and dissolved materials.
- Check for cracks in concrete, steel, and embankment materials; have them diagnosed; and take corrective action without undue delay.

The findings of inspections are analyzed and causes of damage are determined. The true cause may evade detection. The suggested action and repair method may, therefore, be based on a wrong assessment of the true condition and thus may be wrong or ineffective, thereby allowing a hidden problem to persist. Consequently, a repeat of the condition or more serious damage may occur in the future. Such uncertainties can be assessed in a probabilistic damage assessment and repair plan. The decision based on the probabilistic assessment may still be wrong, but the probability of this being the case is also assessed and the decision maker is aware of it. Hence, the ultimately chosen repair method should be one that remedies all or at least the most likely causes. A repair may be apparently successful without ever remedying the true cause of the problem.

Because of the usually high damage potential, especially of dams, major hydroprojects are or should be operated with a preventive maintenance policy in place. A maintenance schedule for such a policy is based on a continuous surveillance program that tries to detect incipient damage or failure mechanism. Maintenance action takes place before serious damage can develop. Such an approach is or should be mandatory for all major hydroprojects. A probabilistic corollary of Murphy's Law (Section 2.1.6) should be heeded: *if there is a possibility of failure, then there is a probability of it to happen*. The fact that relatively few and mostly small dams fail out of the big number of some 78,000 dams indicates that it takes one or more unlikely conditions to occur for a serious accident to happen. Examples of such conditions are:

- Structural weakness (concrete swelling, corrosion, lack of compaction).
- Lack of construction quality control (poor filter construction, poor seepage control).
- Lack of water level control (too small, inoperable, or blocked outlets).
- Improperly prepared foundations and abutments (internal erosion, leakage, rock joint or bedding plane pressurization).
- Improper embankment drainage (high pore pressure, liquefaction).
- Insufficient foundation stability and drainage (buoyancy).
- Misjudgment of early symptoms (leakage, erosion, leaching).
- Vulnerability to external events (cracking, shearing, liquefaction).
- Reservoir seiches (wind, slope slides).

- Tunnel and penstock failures.
- Machinery incidents (gate and valve vibrations, jamming, bearing failures).

There are many ways in which such conditions can develop into critical failure mechanisms that lead to repairable or non-repairable accidents. From a probabilistic point of view, failure mechanisms and accidents can develop or occur as a consequence of accidental activation of latent defects, independent external random events, joint external events, and conditional events:

- *Latent defects*: A hidden defect, a hazardous condition, that has been present in an inactive state becomes activated by one or more accidental events leading to a failure mechanism. Example: seepage through a poorly consolidated core finds an outlet for erodible material in a defective filter area and begins a piping process.
- *Random events*: An external event that occurs unprovoked and unexpectedly, or an unexpected equipment failure. Examples: earthquake, fire, explosion, slide, generator failure.
- *Joint events*: Several independent events may occur simultaneously and result in an evolving failure mechanism. Example: the retention space of a reservoir is filled during the normal filling season. A landslide produces a seiche that overtops the dam and causes severe damage of the downstream embankment (of an earthfill dam) or undermines the foundations of some monoliths (of an arch or gravity dam).
- *Conditional events*: A cause event triggers follow-up events. Follow-up events are conditioned on simultaneously or previously occurring events leading to a chain of events. The events in this case depend on each other and are therefore dependent joint events. Example: an embankment dam made of fines (hydraulic fill) is subject to high pore water pressure. Under earthquake shaking of sufficient strength and length, such an embankment has a high probability of liquefaction which in turn with high probability leads to an embankment slide that may shear away the dam crest leading to overtopping and breaching.

The probability of such failure mechanisms can be evaluated in a manner similar to the evaluation of the probability of functioning of a multicomponent system (see Chapter 4) with possible events taking the place of components.

1.3.5 Hazard Ratings

A hazard is defined here as a condition that enhances the probability of occurrence of a harmful event or of its magnitude and consequences. For example, any condition that increases the probability of a dam failure or the consequences in the form of property damage or loss of life is a hazard. The hazard rating of a dam is based on the exposure to possible harm of downstream life and property that the mere presence of the dam creates. Similar hazard ratings are used by U.S. agencies with some differences in wordings. An example of such a rating used by the Federal Energy Regulatory Commission (FERC) is given in Table 1-1. The ratings refer to the exposure of downstream life and property to harm in a dam failure, regardless of the structural state of the dam or its likelihood of failing. A dam may thus be classified "high hazard" without having a known structural or operational defect.

Table 1-1: Hazard Ratings for Dams (according to FERC Engineering Guidelines)

Hazard Category	Definition
High hazard	*High hazard potential structures* are located where failure *may* cause serious damage to homes; agricultural, industrial, and commercial facilities; important public utilities; main highways; or railroads, and where there would be danger to human life.
Significant hazard	*Significant hazard potential structures* are located in predominantly rural or agricultural areas where failure *may* damage isolated homes, secondary highways, or minor railroads; cause interruption of use of service of relatively important public utilities; or cause some incremental flooding of structures with possible danger to human life.
Low hazard	*Low hazard potential structures* are located in rural or agricultural areas where failure *may* damage farm buildings, limited agricultural land, or township, or country roads; low potential hazard dams have small storage so that their sudden release would be confined to the river channel in the event of a failure and therefore would represent no danger to human life.

During dam inspections (see Section 1.4.3), a significant number of dams have been classified "high hazard" and at the same time many of these dams have been judged unsafe. This latter group of structures must be given priority by maintenance or rehabilitation. Fortunately, most of these dams are small, but even small high-hazard dams can wreak havoc on downstream areas, as has been demonstrated by historical examples (Jansen, 1980). From the definition of hazard given previously, it is clear that if hazard is measured by its enhancing the probability of a harmful event, then the combined hazards of a severe exposure to harm and a faulty dam produce a higher probability of harm than a relatively safe dam that creates the same exposure.

1.4 State of the Infrastructure

1.4.1 Historical Perspective of the Water Infrastructure

The major building period for hydroprojects in the United States lasted from approximately 1930 to 1975. According to a survey by the International Commission on Large Dams (ICOLD) during this period, 75 % of 14,700 dams 5 m and higher were built in 33 countries (participating in the ICOLD survey). Approximately 1,100 additional dams were built in non-participating countries excluding China (ICOLD, 1984, p. 26, Table 1.3.2). In China some 20,000 dams were counted in 1998, but only a small percentage of them fall into the large dam category (Britannica, 1999, p. 144). By the year 2000, many hydraulic structures had been in operation for 50 years or more. Fifty years is not an extraordinary age for the water resource infrastructure whose functioning will continue to be needed for the foreseeable future. Many facilities have, however, outlived their original objectives and priorities and are in need of adaptation to changed societal expectations. Such adaptations include fish passages, fish-friendly and oxygen-inducing turbine wheels, temperature-adjustable turbine intakes, the release of minimum flows at dams, and truly multipurpose power plant and lake operations.

Nature's water supply is, in general, not geared in any way to society's desires and needs. Water management has to step in to bridge the discrepancy between supply and demand. There is no way to meet these demands without water management structures and these structures cannot be managed without impact on the natural flow regime. Incompatible interests will continue to fuel controversy over water allocation. Reasonable water management and physically well maintained hydraulic structures will combat a nascent trend aimed at destroying or at least diminishing the Nation's hydraulic infrastructure.

Throughout history, hydraulic engineers have served society by meeting needs through hydraulic structures, such as dams, weirs, canals, locks, reservoirs, water power installations, aqueducts, tunnels, cisterns, drinking water supply systems, wastewater drainage systems, mine drainage systems, water lifting systems, irrigation systems, land amelioration systems, dikes, levees, military structures, decorative structures, and others. Those that were well constructed and properly maintained served for centuries. Some dam and reservoir designs, usually for drinking and irrigation water, date back a thousand years and more. The reason for this longevity was their vital importance in semiarid countries, such as Egypt, Syria, Arabia,

Mesopotamia, Persia, India, Sri Lanka, Spain, Italy, and Greece, where civilization development and sustenance were fundamentally connected with the supply of water (Jansen, 1980, pp. 1–20; Wunderlich and Prins, 1987, pp. 3–176).The dam and lake of Homs, on the Orontes River in Syria, is thought to have been built around 1300 B.C. by Egyptian engineers. It continues to provide water for today's population (De Camp, 1993, p. 50). Obviously, in this case, the water infrastructure was successfully handed down from generation to generation through the millennia. Most of the ancient hydraulic structures succumbed because of primitive building techniques, lack of hydrologic and hydraulic knowledge, lack of maintenance, and failure of the supporting institutions. Whenever cultures failed, the water infrastructure usually failed with them. Unprecedented technology advancements in the twentieth century have allowed the development of a water infrastructure beyond all previously possible scales. However, it was not only the available technology, but also its combination with a unique window of opportunity that has permitted this water infrastructure to be put in place within the relatively short time of about 45 years. It falls onto this generation and future generations to adequately operate and maintain it and to derive sustainable benefits from it.

1.4.2 Dam Inventory

The methods discussed here deal with hydraulic structures, not just dams. The creation of a storage impoundment removes a good part of the natural uncertainty associated with water control and water supply activities. This makes dams the keystones of many water projects. Dams also are the prime attention getters for the hydraulic structures sector because they expose people and property to possible harm. If dams do not command the attention of maintenance then which part of the water infrastructure will? Therefore, the inventory of dams and its findings throw some light on the state of the water infrastructure as a whole.

After a flurry of dam failures with heavy loss of life in the 1960s and early 1970s, the Congress of the United States passed the Dam Inspection Act (PL 92-367 of August 8, 1972). It authorized the USACE to undertake a national program of dam inspection. The USACE received authority to inspect all dams except those under the supervision of other federal agencies, including the BUREC, the Tennessee Valley Authority (TVA), the International Boundary and Water Commission (IBWC), and dams licensed by the Federal Power Commission (now the Federal Energy Regulatory Commission - FERC). Also, the secretary of the army had authority to exempt dams as he saw fit. No inspections were made for several years because of lack of funds, but initial steps were taken by developing a dam inventory. Additionally, the federal government wanted to have the states perform the inspections of nonfederal dams, but not all states had a dam inspection program. The activities undertaken by USACE were the following (Miles, 1977, p. 3; Morris, 1974, p. 18; BUREC, 1980, p. 2):

1. Compile an inventory of federal and nonfederal dams.
2. Conduct a survey of each state and federal agency's capabilities, practices, and regulations regarding the design, construction, and O & M of dams.
3. Develop guidelines for safety inspections and evaluations of dams.
4. Formulate recommendations for a comprehensive national dam safety program.

Table 1-2: United States Army Corps of Engineers Dam Inventory of October 1974 by Dam Types and Height (after Miles, 1977, p. 4)

Type		Height	
Number	Type	Number	Height
45,459	Earthfill and rockfill	43,143	7.5 m–15 m
3,334	Gravity	5,107	15 m–30 m
246	Buttress	737	30 m–60 m
244	Arch	304	60 m–150 m
39	Multiarch	31	over 150 m
49, 322	All types	49,322	over 7.5 m

The dam inventory required by the 1972 Dam Inspection Act was completed by May 16, 1975. In November 1976, the corps presented the results to Congress. Nationwide some 49,300 dams had been inventoried, 7.6 m (25 ft) high or higher, or with an impounded volume of 60,000 m^3 (50 acre-feet) or more. An overview of the types and heights of the dams is given in Table 1-2. Approximately 7,900 of these dams had never been inspected, and about 20,000 were located in areas where their failure could cause loss of life and significant property damage, including damage to homes, buildings, public utilities, highways, and railroads. Table 1-2 presents two population classifications, by type and by height. By far the largest group in the type category are earthfill and rockfill dams, and most dams are in the 7.5 m to 15 m height category, smaller dams not being included. Table 1-2 also shows that there are about 3,900 concrete dams of different types, and some 6,200 dams 15 m and higher. Since each of the two largest groups of dams almost make up the entire population, there is a large overlap between earthfill and rockfill categories and the smallest height category, in other words, most dams belong to both.

In the years immediately following the passage of the Dam Safety Act, from 1972 until about 1975, the corps "responded to 10 requests for assistance in developing or strengthening state dam safety programs and to more than 130 requests for technical assistance and advice regarding measures to eliminate or mitigate hazardous conditions" (Willis, 1976, p. 6). These are rather modest numbers of requests for assistance considering the large number of deficient dams that were later found when the inspections started in the 1980s.

Many of today's large dams were built from 1930 on. Most of the federal dams, including those of the USACE, the BUREC, and the TVA, among them the largest structures of the time, date from that period. Table 1-3 gives the total number of various types of dams in operation by 1973 and the percentage of this total that was built from 1930 through 1973.

Table 1-3: Total Dams in Operation by 1973 and Dams Added from 1930 through 1973 (after ASCE/USCOLD, 1975, Table I, p. 12)

Period of Completion	Earth	Earth Core Rock-fill[2]	Rock-fill	Gravity	Arch	Un-known	Total
Total dams in operation by 1973	3,604	151	141	487	276	255	4,914
Dams added from 1930 to 1973	3,095	123	91	239	123	136	3,807
% of total[1]	86	81	65	49	45	54	77

Notes: 1. dams added (2. row) in percent of total (1. row). 2. rockfill dams with earth core as impermeable element; also referred to as earthfill-rockfill.

Table 1-3 shows that half or more dams of all types were added to the total 1973 inventory between 1930 and the end of 1972, and 77 % of all dams in existence in the United States in 1973 date from that period. According to ASCE/USCOLD (1975, Fig. 1A, p. 9), the dam building rate for both earth dams and other types was highest from 1944 to 1968.

Table 1-4: USACE Inventory of Nonfederal U.S. Dams (defined by PL 92-367) by the End of Decades (after Tschantz, 1977, p. 22) and Federal Dams (after Dolcimascolo, 1980)

End of Decade[1]	Number of Nonfederal Dams	Annual Rate of Addition of Nonfederal Dams[2]	Number of Federal Dams	Annual Rate of Addition of Federal Dams[2]
Prior to 1900	1,586	–	80	–
1900 - 1909	3,787	220	–	–
1910 - 1919	5,551	176	160	–
1920 - 1929	7,730	218	230	7
1930 - 1939	11,354	362	–	–
1940 - 1949	14,895	354	920	–
1950 - 1959	24,855	996	1,330	41
1960 - 1969	42,004	1,715	1,960	63
1970 - 1977	47,418	773	2,078	17

1. On the occasion of the completion of the second millennium of the Christian time count, it was learned with some difficulty by many that the time count begins with instant t = 0, whereas the period that follows is period 1. Hence, periods like a decade are counted from year 1 to year 10, and not from year 0 to year 9, as there is only an instant time zero but not a year zero. The counts in the table being from two authors may not cover exactly the same periods.
2. Average annual rate calculated by dividing the number of dams added per decade by 10. Last period is assumed to cover 7 years.

Most dams in the United States are nonfederal dams owned by private owners, local governments, states, and public utilities. The rest are owned by federal agencies, such as the USACE, the BUREC, the TVA, and services of the Department of Agriculture. Table 1-4 gives an overview of nonfederal and federal dams and their growth in numbers by decades of the twentieth century. More than 6,000 nonfederal dams and very few federal dams were built in the first three decades until about 1930. Then the rate picked up from about 205 per year before 1930 to about 360 per year until 1950. Then the rate jumped to about 1,000 per year in the 1950s and to 1,700 in the 1960s before it plummeted in the 1970s. The addition of federal dams reached its maximum of about 50 per year during the two decades from 1950 to 1970, but the annual rate remained below 4 % or less of the nonfederal rate. The real number of all dams has never been established. Dam surveys counted from 47,344 to 47,418 nonfederal dams, large and small, and 2,078 federal dams in 1977. The total number of dams is only of marginal interest here as many are small and may not qualify for the application of advanced maintenance concepts.

Table 1-5: Nonfederal Dams by Types and Primary Purposes in Percent of Total—1974 (after Tschantz, 1977, Fig. 6, p. 29)

Primary Purpose	All Dams	Earth-fill	Rock-fill	Gravity	Buttres s	Arch	MA[2]	Other
	Percent of Total							
(1)	(2)	(3)	(4)	(5)	(6)	(7)	(8)	(9)
Recreation	34.7	31.45	0.34	2.18	0.24	0.07	0.01	0.46
Water supply	15.0	13.12	0.30	1.00	0.10	0.15	0.02	0.34
Navigation	0.1	0.01	0	0.03	0	0	0	0.03
Flood control	15.2	14.50	0.12	0.27	0.01	0.01	0.01	0.28
Power	2.8	0.59	0.18	1.60	0.03	0.14	0.03	0.12
Irrigation	12.6	12.04	0.14	0.24	0.04	0.05	0.01	0.11
Farm and livestock	9.6	9.15	0.01	0.01	0	0	0	0.01
Debris and silt control	0.7	0.63	0.01	0.01	0	0.01	0.00	0.05
Other	9.3	7.55	0.15	1.18	0.06	0.01	0.00	0.35
Total[1]	100.0	89.04	1.25	6.52	0.48	0.44	0.08	1.75

1. The inventoried total of nonfederal dams is 47,418. The percentages of each row, from column 3 to 9, add to column 2 with some minor discrepancies due to rounding. The last row represents the percentage of dam types of the inventoried total. 2. Multi-arch.

The primary purposes of nonfederal dams are shown in Table 1-5. Column 2 shows the total contribution of all nonfederal dams to eight identified purposes. Columns 3 though 8 break these contributions up among six identified types of dams. Recreation is the most frequent purpose with 34.7 % of all 47,418 dams, and it is to 31.45 % out of 34.7 % addressed by earthfill dams. The next most frequent purpose is flood control. Here again earthfill dams contribute 14.5 % out of the total of 15.2 %. With the exception of navigation and power, where concrete dams are the most frequent contributors, earthfill dams make the largest contribution to all other purposes.

The last row of Table 1-5 shows the distribution of nonfederal dams by types. The second largest category next to some 42,200 earthfill dams are about 3,000 concrete gravity dams followed by still smaller contingents of other dams. Not all of these dams are of interest to the application of probabilistic approaches to maintenance. Table 1-2 shows that earthfill and rockfill dam categories and small height categories are very large sets compared to the total dam population so that there is considerable overlap of these categories, in other words, many of these dams are also in the relatively low height category. Only the more complex and larger projects that require advanced planning and decision making, perhaps a few thousand, are of interest.

An inventory commissioned by ICOLD counted 4,974 "major dams" in the United States (ASCE/USCOLD, 1975, p. 8). This inventory refers to dams 15 m high and higher, which ICOLD considers "large dams" (in the U.S. 13.7 m and higher). In 1980, the 35 highest dams in the world ranged from 183 m to 330 m high. In this group are 5 earth dams, 25 concrete dams, and 5 rock fill dams. Of the 5 earth dams, 2 dams occupied ranks 1 and 2, with 330 m and 317 m, respectively. Twenty-five were gravity, gravity/arch, or arch dams. A massive gravity dam, 285 m high, occupied rank 3, and a thin arch, 272 m high, occupied rank 4. Of the five rock fill dams, one occupied rank 5 with a height of 264 m (Jansen, 1980, p. 87). A listing of the world's major dams and hydroplants from 1989 contains 58 U.S. projects that meet at least one of the following criteria (after Mermel, 1989, pp. 35 - 42):

	Number of U.S. dams
Dam height of at least 150 m above deepest foundation	20
Dam volume of at least $15 \cdot 10^6$ m^3	22
Reservoir volume at least $25,000 \cdot 10^6$ m^3	5
Hydroplant capacity of at least 1,000 MW—installed	16
Hydroplant capacity of at least 1,000 MW—planned	23

The numbers on the right-hand side indicate the number of U.S. projects that meet the particular criterion. There are very few projects that meet more than one criterion; among them are Grand Coulee (dam height and capacity), Glenn Canyon (dam height, reservoir volume, and capacity), and Hoover (dam height, reservoir volume, and capacity). Seven of the projects listed without power are mine tailing dams. With the exception of one, these mine tailing dams owe their inclusion on the list to their massive earth dam volumes. The largest of them has about $210 \cdot 10^6$ m^3. In comparison, the 230 m high Oroville Dam has a volume of $61 \cdot 10^6$ m^3. Of the 58 projects, 46 are located west, and 12 east of the Mississippi River.

For 36 major U.S. projects, Mermel (1989) gives the year of initial operation. The distribution by decades is given in Table 1-6. The rapid decline of these numbers in the 1980s and 1990s shows that hydro development, after a steep rise, rapidly lost steam in the last decades of the twentieth century.

Table 1-6: Initial Operation of 36 Major U.S. Dams (Mermel, 1989, pp. 35-42)

Year of Initial Operation	1930s	1940s	1950s	1960s	1970s	1980s	1990s
Number	2	3	7	8	8	6	2

In developed countries, the intense competition for land and water resources has now limited new construction to a few projects related to water control and water supply in areas where water management is absolutely vital. Three such projects were scheduled for completion in California in 1999 (Britannica, 1999, p. 143). The existing water infrastructure can be expected to gain in importance as demand will keep growing. Even if another source of water becomes available in some areas, such as ocean water osmosis or distillation, the sun-powered water cycle and its multipurpose management remains a relatively cheap, efficient, and sustainable source for meeting water-related needs.

Hydropower and other water developments still hold great promise for developing countries, as the number of projects under construction shows. In 1997, 1,500 dams were under construction worldwide, and 242 dams were completed. In 1998, 1,738 dams were under construction worldwide: 625 in India, 302 in China, 236 in Turkey, and 145 in South Korea (Britannica, 1998, p. 139; Britannica, 1999, pp. 142-144). Building and managing water projects can be done right or wrong, which makes all the difference between the creation of an infrastructure that can provide a sustainable source of water, hydropower, and other benefits for future generations and an ecological and/or social disaster. The decision is at the discretion of the politicians, financiers, and builders.

1.4.3 Dam Inspection Findings

The Dam Inspection Act of 1972 had exempted federal agencies from operational dam inspection as well as from (preoperational) dam construction inspection. Such inspections of all dam construction and operation activities had been proposed by the USCOLD Model Law (Willis, 1976, p. 9). The failure of the BUREC's Teton Dam during first impoundment, on June 5, 1976, caused $1 billion[1] in damages and 11 fatalities, as if to demonstrate that there was no good reason for exempting any organization from independent regulatory oversight.

The Teton Dam failure of 1976 triggered the establishment of the Federal Coordinating Council for Science, Engineering and Technology (FCCSET) to review federal dam inspection practices. While this group was preparing its report, Kelly Barnes Dam broke near Toccoa, Georgia, killing 39 persons. In 1977, the failure of Laurel Run Dam in Pennsylvania killed 40 persons. A recommendation of the council was that inspections be started of the 47,300 inventoried nonfederal dams, of which only 3,400 received some kind of federal regulation. Inspections began in 1978 with some 9,000 high-hazard nonfederal dams (Dolcimascolo, 1980). Of the 1,400 dams inspected after the first year, about 25 % were found unsafe. Of those, 90 % had inadequate spillway capacity. Other deficiencies included seepage (4.3 %), structural weakness (2.5 %), and embankment instability (2.5 %) (Dolcimascolo, 1980). By 1981, the Corps of Engineers had inspected 8,800 nonfederal dams that were classified as high hazard (see Section 1.3.5). Of the dams classified as high hazard, about one-third also were judged unsafe — a very critical group of dams. Since the first inspections in 1978, the poor state of at least a part of the water infrastructure (mostly nonfederal dams) has been known. There has been some action in the legal and engineering sectors aimed at improving the water infrastructure, but decisive improvements did not happen.

1.4.4 National Infrastructure Assessment

The National Council on Public Works Improvement (NCPWI) conducted a nationwide survey and published its report on America's public works in 1988 (NCPWI, 1988). The report summed up the general state of the infrastructure in all sectors: "We have worn through the cushion of excess capacity built into earlier investments. In effect we are now drawing down past investments without making commensurate investments of our own." The NCPWI study further stated: "The total public spending on infrastructure has dropped from 3.6 percent of the gross national

---

[1]Billion throughout this text means the American billion or 1,000 million = 10^9.

product (GNP) in 1960 to 2.6 percent in 1985. While spending on operations and maintenance has remained a constant share of GNP, capital spending has dropped from 2.3 percent of GNP in 1960 to 1.1 percent in 1985. The relative share of public works spending at all levels of government has declined drastically from nearly 20 percent of total expenditures in 1950 to less than 7 percent in 1984" (NCPWI, 1988, p. 8). Numbers supporting this statement are given in Table 1-7.

Table 1-7: Absolute Dollar Amounts of Spending for Three Selected Years and Percentages of Total (NCPWI, 1988, p. 8)

Year	Total	Public Works		Public Welfare and Education		Defense		Interest on Debt		Other Non defense
	$ billion[1]	$ billion	%	$ billion	%	$ billion	%	$ billion	%	%
1950	239	46	19.1	25	10.4	77	32.4	21	8.6	29.6
1960	491	60	12.3	135	27.4	154	31.4	30	6.2	22.8
1984	1428	97	6.8	581	40.7	248	17.4	144	10.1	25.0
factor	6.0	2.1		23.2		3.2		6.9		5.0

1. The costs are all in 1984 $ billion. The consumer price index rose from $1 in 1950 to $4.31 in 1984, and from $1 in 1984 to $1.73 in 2002. From 1950 to 2002, the price index rose from $1 to $7.44 (minneapolisfed.org/research/data/us/calc/).

The relative amount allocated to public works shrank from about 19 % in 1950 to 6.8 % in 1984, but the absolute amount doubled from $46 billion to $97 billion (both at 1984 price level). The dramatic percentage reduction reflects a reduction in priority ranking of public works. The change in priority ranking also becomes apparent when one compares the growth factors of the spending sectors in the bottom line of Table 1-7. From 1950 to 1984, total government spending grew by a factor of 6. This factor is exceeded in two sectors, public welfare and education (23.2) and the interest on national debt (6.9). The other sectors showed factors below 6.0. Public works had the lowest factor (2.1).

The NCPWI study (1987 and 1988) found that the neglect of maintenance in many sectors of the infrastructure had reached national notoriety and was expected to be a continuing problem. It quantified its findings by issuing a *report card of the national infrastructure*. An excerpt is shown in column 2 of Table 1-8. Ten years after the NCPWI study, the ASCE issued America's infrastructure report card in 1998. The ASCE cautioned against a direct comparison of its report card with the NCPWI results. With this caveat in mind, the ASCE results are given in column 3 of Table 1-8. The grade given to dams by the ASCE was D, as was the overall grade for all sectors.

Table 1-8: Report Card on the Nation's Public Works (after NCPWI, 1988; ASCE 1998, and ASCE, 2001)

Infrastructure Sector	NCPWI Grade[1] 1988	ASCE Report Card 1998	ASCE Report Card 2001[2]
(1)	(2)	(3)	(4)
Highways Roads Bridges	C+	D- C-	D+ C
Mass transit	C-	C	C-
Aviation	B-	C-	D
Water resources	B	D (dams)	D(dams)
Navigable waterways	B		D+
Drinking water supply	B-	D	D
Wastewater	C	D+	D
Solid waste	C-	C-	C+
Hazardous waste	D	D-	D+
Energy			D+

1.grades: A (excellent); B good; C satisfactory; D deficient; F fail.
2.www.asce.org/reportcard—March 2001.

In the face of growing infrastructure repair needs in all sectors, priority for infrastructure renewal funding is likely to go to the sectors with the greatest direct benefit to the public, such as roads, bridges, airports, municipal water supply, sewer systems, sewage treatment and disposal, and industrial waste disposal. These sectors affect more people across the land than dams and hydroprojects and thus command more urgency and public interest. Also, their malfunction or failures affect more people than those of the hydro sector even though dam failures attract more media attention.

Some salient points of the NCPWI (1988) report are summarized here as the probabilistic approach can respond to the problems identified in the report:

● Governmental investments should be operated "to get the maximum value out of each public dollar spent." In the water sector, rate payers are captives of the service providers, be it water supply, wastewater collection and treatment, recreation, hydroelectric energy supply, flood control, and navigation. The service providers should make it their policy to *use state-of-the-art maintenance technology* to minimize operation costs and to enhance safety, reliability, and efficiency.

● Design, construction, operation, and maintenance should be viewed as integral parts of a functioning project. Depending on the approach chosen, design and construction costs can be either small or large compared to lifetime operation and maintenance costs. Only by considering *all* costs and benefits over the lifetime of a project upfront, and by making them commensurate by discounting them to present, can the most cost-effective alternative be identified. Such an approach, known as *life cycle benefit-cost analysis* , can be greatly aided by the probabilistic approach, because benefits as well as costs over the life cycle are probabilistic quantities.

● Preventive maintenance keeps maintenance planning and decision making under the control of the maintenance department with the added benefits of increased project reliability and safety, and more satisfied maintenance personnel who feel that they are in charge and not driven by random events.

● Maintenance is an activity that sometimes is skipped because of lack of funds or because of an intentional deferral in the hope of avoiding maintenance costs. In such a case, a deliberate choice is made between a scheduled shutdown and all its ramifications including shutdown costs, repair costs, service replacement costs, and so on, and the possible costs of an unscheduled, forced shutdown that is possibly compounded by a more extended outage than the planned outage, by damage costs and failure penalties. The probabilistic approach can evaluate alternative actions, their costs, and the associated risks.

The ASCE published an updated infrastructure report card in March 2001 with the grades shown in column 4 of Table 1-8. The water resources infrastructure including dams, navigable waterways, drinking water supply, and wastewater again earned Ds. It is like repeatedly testing a student who has not understood the subject matter and giving him Ds time after time. The summary of the updated 1997 USACE dam inventory is given in Table 1-9 (the full table is available on the ASCE web site at www.asce.org).

Many of the 2,099 state-regulated dams that were found to be unsafe (Column 5 of Table 1-9) are also in high-hazard locations. Alabama had 150, Colorado 189, Maryland 278, Ohio 450, and Texas 452 state- regulated, unsafe dams. Definitions vary from state to state of the deficiencies that classify a dam as unsafe. At the time of the Laurel Run failure in 1977, Pennsylvania had 209 hazardous, unsafe dams. After the Laurel Run failure, the Pennsylvania state legislature passed a law that made owners responsible for remedial action. From 1981 through the 1980s, unsafe, high-hazard dams in Pennsylvania were reduced by 80 % (ENR, 1989). The aforementioned ASCE report reflecting the inventory of 1997, lists Pennsylvania with 7 state-regulated unsafe dams.

Table 1-9: Summary of 1998 Dam Inventory with Hazard Categories (after 1997 State Dam Inventory according to U.S. Army Corps of Engineers (Copyright 1996-2000 ASCE); the detailed list by states is found on the ASCE web page, www.asce.org/reportcard, ASCE, 2001)

State	Total National	Total State Regulated	State High Hazard	State Regulated Unsafe	Government Ownership
(1)	(2)	(3)	(4)	(5)	(6)
All States	75,299	94,517	9,133	2,099	19,367

Notes: Column 2: includes federal and nonfederal dams over 7.5 m high, and/or with about 60,000 m^3 reservoir volume.
Column 3: all dams under state regulatory control.
Column 4: high hazard by state definition; subset of column 3.
Column 5: dams with identified deficiencies by state definition; subset of column 3.
Column 6: subset of the national inventory, column 2.

The number of unsafe, high-hazard dams is increasing rapidly in some states. Population growth and urban sprawl make dam accidents increasingly costly in terms of both loss of life and loss of property. Small dams, like wildlife are becoming an endangered species, as their need for a free flood way is encroached upon by urban sprawl, which crowds dams out of their habitat. According to the ASCE 1998 report card, 9,281 dams were in the high-hazard category. By April 2001, the number had increased by 640 to 9,921. Among these dams were 2,100 dams with deficiencies that leave them highly susceptible to failure and put them into the high-hazard, unsafe category (ASCE, 2001). For many dams this means that they must be upgraded to satisfy higher safety standards, or they will be breached or removed. This may result in a loss of valuable water infrastructure.

The ASCE 1998 report gives the average cost of rehabilitating an unsafe dam as $500,000. Hence, the rehabilitation of only the 2,100 unsafe dams would cost about $1.1 billion. The main hurdle to fixing substandard dams is lack of funds. The Association of State Dam Safety Officials (ASDSO) estimates the cost of bringing all 75,193 USACE-inventoried dams into safety compliance at $40 billion, with $31.5 billion going to dams 7 m and higher (ENR, 2000). In comparison, the federal FY99 allocated $1.5 million (ASCE, 1998; ASCE, 2001).

In 1997, the National Dam Safety Program Act was passed. It was a five-year program to fund individual state dam safety programs. The states are overseeing 95 % of the nonfederal dams, the total being estimated by the ASCE 2001 report card at 78,000. The states have primary responsibility of protecting the public from dam failures and from terrorist acts toward dams. The program was up for reauthorization in 2002 and passed the House of Representatives with a vote of 401– 2 for a four-year extension. The added dimension of dam security increased funding from $4 million to $6 million for the state safety programs. In addition, $1.5 million for research and $1.1 million for training and FEMA staff were included in the total of $8.6 million (ASCE, 2002/9, p. 12). These federal funds share the financial burden with the states. It is still an insignificant amount considering the multibillion dollar bill that it will take to upgrade the dam infrastructure.

1.4.5 Dam Incidents and Failures

The ASCE issue brief on dams (ASCE, 2000) reports that "in the past 10 years, there have been more than 200 documented dam failures across the nation which caused millions of dollars in property damage and repair costs." Dam incidents are events in connection with floods, earthquakes, and inspections that alert dam safety personnel to deficiencies that threaten the safety of a dam. A failure is the most severe dam incident. Before 1900, there were approximately 38 significant failures, and from 1900 to 1965, there were about 164 failures. In a total of 202 major failures in the last 100 years or so, some 8,000 people lost their lives (Jansen, 1980, p. 94).

In 1964, ICOLD through its Committee on Failures and Accidents to Large Dams collected information through its national committees on all incidents, including failures, accidents, and major repairs to large dams. A large dam was defined as one being about 13.7 m (15 m in international parlance) high. The cut-off date for reported incidents was the end of the year 1972. Only about 11 % of the 2,000 questionnaires were returned. It is not known if nonrespondents had no incidents or simply did not respond. Probably only medium to minor incidents remained unreported as major incidents would have become known through other channels.

The dam population of the ICOLD survey is shown in Table 1-3. It consisted of 4,914 dams that were built between the 1850s and the end of 1972. Building techniques and engineering knowledge improved enormously over the 120 years that this sample covers and the failure rate diminished substantially during this time. It is estimated from this survey that it went from about 2.2 % for the dam population before the 1930s to about 0.3 % after that. Hydroprojects built after 1930 can still be considered modern structures today. Therefore, the analysis of failure mechanisms that follows was limited to structures built from 1930 to the end of 1972. Table 1-3 shows that about 50 % to 86 % of all types of dams were built during that 42-year period.

The ICOLD survey's purpose was to learn about the incidents of various types of dams (ASCE/USCOLD, 1975). This is also of interest for maintenance purposes. From the mid-1800s to 1973, 39 failures led to the abandonment of the dam out of a total of 349 incidents of varying severity, and 70 % of all dam failures occurred in the first decade of a dam's life, and especially in the first year (IWPDC, 1995).

Table 1-10: Incidents for Dams in Operation at the Time of the Incident (ASCE/USCOLD, 1975, p. 7)

Incident Category	Code	Dam Incidents from 1930 to 1973	
		Built Before 1930	Built After 1930
(1)	(2)	(3)	(4)
Major failure of an operating dam leading to abandonment of the dam	F1	10	3
Failure that may have been severe but permitted restoration of the dam	F2	8	7
Accident to an operating dam that may have been severe but was prevented from becoming a failure by remedial work or a specific operation (drawdown)	A1	12 (USC) 15 (IC)	25 (USC) 19 (IC)
Extensive repair because of deterioration of structures (concrete damage) or need for updating of certain features	MR		69

Notes: Columns 3 and 4: extracted from data by (ASCE/USCOLD, 1975, Figures 3, 4, 5A, and 5B); IC means "ICOLD count" and USC means "USCOLD count;" the MR category represents all incidences of major repairs from 1930 to 1973 without separation into old and modern dams (ASCE/USCOLD, 1975, Table II); MR represents major repairs and modernization (after ASCE/USCOLD, 1975, p. 70, and Figures 30 through 35).

The ICOLD study distinguishes eight types of incidents, of which four are of interest in connection with maintenance. A description of these four types of incidents and the recorded numbers of incidents are summarized in Table 1-10. They include unrepairable (F1) and repairable (F2) major failures, accidents during operation (A1), and extensive repair incidents (MR). Incidents not included in Table 1-10 refer to accidents that occurred during the placement of the dam into operation; accidents during dam construction before impoundment; accidents related to the reservoir; and damage to the partially constructed dam or temporary structures, for example, due to flooding of the construction site.

Table 1-10 shows that three F1 failures of dams built in 1930 or later occurred between 1930 and 1973. Details on dam and accident types are given in Table 1-11. All three F1 failures involved earth dams, two were caused by piping and one by spillway failure. Seven F2 failures occurred, which also involved only earth dams; five were caused by piping, one by spillway gate failure, and one by downstream slope slide. Twenty-five type A1 incidents occurred, which involved one earthfill-rockfill dam, one gravity dam, and the rest earth dams.

Table 1-11 shows that *piping* is the most serious incident leading to failure from which the project may (F2) or may not (F1) recover. In two cases, foundation leaking and piping could be repaired before becoming a severe accident (A1). Overtopping does not occur among the included incident types in the available data. A data item not included in the table, but of interest to maintenance, consists of 11 dams built from 1906 to 1921 that were modernized around 1970 to meet new design criteria, most likely spillway extensions.

Table 1-11: Incidents from 1930 to 1973 for Dams Built After 1930
(data taken from ASCE/USCOLD, 1975, Figures 12 to 24 and Figures 30 to 35)

Incident			Dam Type	Description
Type	Code	Number		
Embankment leakage, piping	F1	1	E	Piping
"	F2	2	E	"
Foundation leakage, piping	F1	1	E	"
"	F2	3	E	"
Flow, discharge	F1	1	E	Spillway destroyed
"	F2	1	E	Spillway gate failure
Sliding	F2	1	E	Downstream slope slide

Embankment leakage, piping	A1	1	E	Embankment leakage
Foundation leakage, piping	A1	10	E	Foundation leakage
"	A1	2	E	Foundation leak and piping
Flow discharge damage	A1	3	E	Damaged spillway
"	A1	2	E	Spillway destroyed
"	A1	1	E	Slide gate failed and damaged
"	A1	1	E	Butterfly valve malfunction
"	A1	1	R	Trashrack failed
Embankment, abutment sliding	A1	1	E	Downstream slope slide
"	A1	1	E	Abutment slope slide
Slope protection	A1	3	E	Too small riprap
"	A1	1	E	Concrete eroded
Deformation differential	A1	3	E	Transverse embankment cracks
Flow discharge damage	MR	1	MV	Spillway repair
"	MR	2	G	"
"	MR	4	E	Outlet and stilling basin repair
"	MR	1	E-R	Slide gate repair
Foundation sliding, leakage	MR	1	A	Sliding abutment
"	MR	1	E	"
"	MR	1	E	Leakage, foundation
"	MR	1	E-R	"
Slope protection	MR	2	E	Wave height

"	MR	3	E	Deteriorated riprap
"	MR	1	E-R	"
"	MR	2	E	Riprap too small
Concrete deterioration	MR	1	A	Severe climate
"	MR	1	G-E	Concrete deterioration

Notes: The abbreviations in column 4 are: A—arch dam; E—earthfill dam; E–R earthfill–rockfill, i.e., a rock embankment with earth core; G —gravity dam; G–E gravity dam with earth embankments; MV —multiple arches or vaults; R —rockfill dam.

The ASCE/USCOLD survey shows that relatively few modern dams (built after 1930) have failed. The study concludes that "the greatest decline in this (failure) percentage was made after 1930 and has remained below 0.3 % for each decade since 1930" (ASCE/USCOLD, 1975, p. 77). If one assumes that each batch of dams created during a decade is diminished over subsequent decades by a constant loss rate of 0.3 % per decade, then the loss would follow an exponential decay function. The time over which the population is reduced to half, its *half-life*, is then obtained from $\ln(0.5) = - 0.003\,D$, where D is the half-life measured in decades. Solving for D gives D = 231. This means that if a decade has produced 1,000 dams, after 2,310 years, 500 dams would have failed. To maintain the infrastructure undiminished, maintenance would have to reduce the loss rate of the dam population to zero. Whereas extraordinary longevity has occurred in some cases, it seems unlikely that a zero loss rate can be achieved for a large dam population over the long run. As a population ages, its loss rate tends to increase. The ASCE report card of 2001 reported that from 1998 to 2000, there were some 520 dam incidents including 61 failures. This would amount to some 300 failures per decade. For the total dam population of 78,000 this gives an average loss rate 300/78,000 = 0.0038, or 0.38 % per decade, which is on the order of what the ASCE/USCOLD study considers as low.

1.5 Regulations and Guidelines in Support of Maintenance

1.5.1 Government Regulations

In 1920, the hydropower industry provided 30 % of the total installed capacity and 40 % of the total electric energy. By 1996, the hydro energy share had dropped to 10 %. Additional capacity and energy had been increasingly provided by thermal power plants using coal and nuclear energy. With the decline of the nuclear option, additional power sources have increasingly used the precarious resources of oil and gas. The annual energy output of the hydropower industry is about $310 \cdot 10^6$ MWh which amounts to 96 % of all sustainable or "green" energy production. To replace this contribution by nuclear power would require some 30 nuclear reactors of 1,250 MW each, operating 90 % of the year, or more than 70,000 wind mills at 1 MWh operating 50 % of the year. Thermal power plants convert about 35 % of primary energy into electricity, and discharge the remaining 65 % or more than 1.8 times the useful output as waste heat into the air and/or water environment. In contrast, hydropower converts about 70 % to 90 % of the primary energy with practically no waste energy impacts. This means that even if hydropower gives up 5% to 10 % of its primary energy to other uses, such as multipurpose reservoir management, fish propagation, minimum flow, and others, it still remains a very efficient primary energy converter.

An energy resource as deeply embedded in the environment as hydropower has its share of environmental impacts. Among them are changes of the stream flow quantity and quality regimes, creating barriers to fish migration, change of aquatic habitat, and the creation of mosquito breeding grounds, and exposures to harm to life and property by uncontrolled water release. In order to avoid or at least reduce the exposure of the public to the damaging side effects of hydroprojects it became necessary early on for the governments to impose regulations that owners and/or operators must follow.

In almost all cases government regulations were preceded by failures and disasters. In California, it was the St. Francis Dam collapse of 1928 that triggered the introduction of state oversight in the public interest. The sudden collapse of this less than 2-year old, 62 m high gravity dam, the first time it reached full impoundment, and the high death toll of 450 led the California legislature to a state dam safety law in 1929. The law required that construction and operation of a large dam should be checked by a panel of independent experts. In France, it was the Malpasset Dam collapse; in Italy, it was the Vaiont Dam overtopping; and in the United States, it was a string of dam disasters in the 1970s that preceded legislation (FEMA, 1996).

In 1971, Baldwin Hills Dam failed in Alabama with no loss of life. Then, in February 1972, Buffalo Creek Dam, a mine tailing dam, failed in West Virginia killing 118 people. In June 1972, Canyon Lake Dam in South Dakota failed with 244 fatalities. The disasters were caused by a mix of human error, technical flaws in construction, operation and maintenance, and lack of supervision of dams. The Dam Inspection Act of 1972 (PL 92-367; Miles, 1977, p. 3) authorized the USACE to inspect all dams in the United States except those under the control of other federal agencies. Since no funds were allocated, rather only inventories, no inspections were made until several more disasters had run their course. In June 1976, the Teton Dam failure killed 11 persons and caused $1 billion in damages, and in November 1977, Kelley Barnes Dam failed in Georgia killing 39 people. The latter accident occurred during the tenure of the FCCSET, which had been created to check into federal practices after the BUREC's Teton Dam failure. Federal dams had been exempted from the 1972 dam inspection legislation (see Section 1.4.3). Finally, in 1978, the federal inspection program started with the high-hazard dams identified by the USACE inventory. The USACE also conducted a survey of state and federal agency capabilities, practices, and regulations regarding the design, construction, operation, and maintenance of dams. The work resulted in guidelines for safety inspection and evaluation of dams, and in recommendations for a comprehensive national dam safety program (Jansen, 1980, p. 96). The *Federal Guidelines for Dam Safety* were published in 1979.

The purpose of the dam safety program is to identify and mitigate the risk imposed on the public by dams. This is done by establishing and enforcing acceptable design, construction, operation, and maintenance standards for dams. In March of 1996, a dam failure occurred in New Hampshire (one fatality and $5.5 million in property damage). In October 1996, the National Dam Safety Program was established as part of the Water Resources Development Act of 1996 (Public Law 104-303, Section 215) with the Director of FEMA as its coordinator (http://www.app1.fema.gov/mit/damsafe/ndspact.htm; ASDSO, 1996, p. 11). Its purpose is to "reduce the risks to life and property from dam failure in the United States through the establishment and maintenance of an effective national dam safety program to bring together the expertise and resources of the federal and non-federal communities in achieving national dam safety hazard reduction." In the aftermath of the September 2001 terrorist attack, this act was extended to include additional dam security aspects. FEMA does not own or regulate dams. It coordinates hazard mitigation in the federal and nonfederal (state and local government, and private) sectors.

1.5.2 BUREC's Safety Evaluation of Existing Dams

The BUREC's projects, as all federal projects, were built to serve multiple purposes, such as flood control, hydropower, water supply, and irrigation. A dam's function for these purposes is to impound runoff, to provide a head for hydropower, a navigable water depth, or to divert water out of its natural channel for irrigation and hydropower uses. The dam is just a component of such water projects, but it is the most important and by far the potentially most damaging one. Most water projects stand and fall with their dam. Therefore, dams must receive primary attention in the form of specialized inspection and maintenance programs.

Familiarity with probabilistic concepts is useful in all phases of maintenance for hydroprojects, including data collection, data analysis, maintenance planning, cost estimating, risk assessment, and decision making. As far as the use of probabilistic methods is concerned, it is not relevant whether dams are inspected, diagnosed, analyzed, and subjected to repair decisions by local maintenance personnel, or whether these tasks are delegated to a special office within an organization or to outside consultants. In 1977, the Water and Power Service (BUREC)[2] submitted its program for safety evaluation of dams to the National Resources Council (NRC) for review. Of 13 recommendations, the first was to create an independent dam safety office. It was implemented and placed near the top of the BUREC's administration, under the Assistant Commissioner-Engineering Research (AC-ER) (BUREC, 1980, p. 7). This centralized inspection service has the advantage of pooling the special expertise required for inspecting dams so as to make it available to the inspection of all dams of the organization. It also eliminates possible information barriers between sites and applies equal quality of inspection to all dams. The members of this service should, however, closely cooperate with the local maintenance personnel, as the people on location are most familiar with the special problems of the site. They also are the first responders to emergencies and must be aware of any hazards discovered by inspections. One would expect a full exchange of information between the local maintenance personnel and the special inspection team, including investigation methods used, philosophy of approach, and findings, as this exchange approach will be overall beneficial. The importance of this point is specifically recognized in the program (BUREC, 1980, p. 61; Bock, 1974, p. 51).

The last of the 13 recommendations of the NRC review of the BUREC's dam safety program is of particular interest here, even if it ranks at the bottom of the list: *"Implement a probabilistic or risk-analysis-based program for the purpose of ranking major Water and Power dams in accordance with the hazard potential and the probability of a failure or partial failure of the dam"* (BUREC, 1980, p. 4). This statement is important because it acknowledges probabilistic methods as an engineering tool evidently not yet used at the time by the BUREC. Whereas the recommendation is made in the context of dams, the probabilistic approach lends itself to maintenance planning for the entire water project.

The *Safety Evaluation of Existing Dams* (SEED) of the BUREC requires regular and six-year examinations of dams. The regular examinations are conducted biannually by a team or just one inspector trained in dam safety. The six-year examinations are in-depth examinations performed by a multidisciplinary team that includes civil, mechanical, and electrical engineers, as well as a geologist. This team is supposed to take a fresh look at the safety of the project's dam and appurtenant structures, and examines the dam's state in the light of state-of-the-art standards and

---

[2]The Bureau of Reclamation's name changed to Water and Power Service, then back to Bureau of Reclamation (BUREC). By 2003, the acronym had changed to BOR. Here BUREC is used throughout.

procedures. After a study of project files, the team proceeds with a field examination and describes its findings in a final report. Various field or analytical follow-up activities may be generated by these findings (BUREC, 1980, pp. 12–14). This team, with its mandate of using an open mind when considering methods to be brought to bear on dam safety investigations and analysis, is the logical group for recognizing the ascending importance of probabilistic thinking and approaches in field examinations and safety evaluations of dams and associated facilities. Some examples of what can be addressed by probabilistic methods in the maintenance area are given in Table 1-12.

Table 1-12: Hydraulic Structure Components, Causes of Failure, and Probabilistic Aspects (Columns 1 and 2 are taken from BUREC, 1980, pp. 18–19, with some modifications)

Component	Evolving Hazards	Probabilistic Aspects
Foundations	Deterioration by removal of solid and soluble materials along hidden seepage pathways; undermining	Amount of water flow and pathways; erosion rate; mobility of material; pressurization along joints, faults, and bedding planes; undermining by scouring; rock decomposition; fault reactivation
	Instability; liquefaction; slide; subsidence	Pore water pressure buildup; seismicity; vibration and shock resistance of structures; slide surface identification; foundation subsidence; blowout potential
Spillways	Obstructions; lining breakage; exceedance of capacity; faulty gates and hoists; stilling basin erosion; scouring	Blockage by debris and ice; erosion of gates and lining by abrasion and corrosion; dam overtopping by excessive flood; inaccessibility of gates or gate hoists during flood; binding of gates; overload of hoist mechanism and gates; standby power supply reliability; gate hoist standby; deicing apparatus; stilling basin resistance

Outlets	Obstructions; silt accumulation; faulty gates and hoists; gate position; gate location	Obstruction of deep intakes by rock, silt, and debris; vibration and jamming of gate hoisting mechanism; leakage into outlet pipe or tunnel; pressure leakage from tunnel into embankment; internal erosion; water saturation of embankment; embankment sagging, sliding; liquefaction
Concrete	Alkali-aggregate reaction; concrete swelling; surface and internal cracking; loss of concrete strength; loss of operability	Concrete growth; expansion rate; repetitive outages and repairs; repair cost increase; possibility of starting or ending of growth; separation of dam segments; shear stress buildup; destabilization of dam elements; crack pressurization causing hazard of sliding and overturning; gate, valve, and turbine misalignment; inoperability of mechanical equipment; steel reinforcement failure; structural failure
Concrete dam	Uplift pressure; displacements; settlement; deflections; overstressing	Excessive heel pressure due to foundation interface pressurization; horizontal sliding on pressurized foundation interface; joint or bedding plane pressurization in the underground; foundation blow-out potential; shear failure of toe or heel of dam; ground subsidence by foundation shear failure; induced vibrations by flood evacuation, earthquakes and machinery
Embankment dam	Liquefaction; slope slide; leakage; internal erosion	Lack of embankment drainage; operational embankment liquefaction by fast drawdown; internal erosion under waterside embankment lining; cavity formation in embankments and abutments; exploratory drilling damage in embankments; core filter inadequacy causing piping
Reservoir	Leakage; slope instability; natural barrier slide	Opening up of sinkholes into underground drainage; pressurization of rock bedding planes in valley slopes; pressurization of slide planes in natural abutments and ridges that form part of the reservoir confinement

The third column of Table 1-12 elaborates on the probabilistic nature of events listed in column 2. Turning the keywords in column 3 into questions gives an idea of the probabilistic nature of possible problems. For example, in the category "concrete" one might ask: what is the probability of concrete swelling, given there is a potential for it. What might be the expansion rate of the concrete mass? How many repetitive repairs will be needed and at what cost? How seriously will the various defects interfere with the safety and operability of the structure? Should periodic repairs be continued or should replacement be considered? Will the expansion stop and when? An extended list of possible failure causes is given in (BUREC, 1980, pp. 19–30).

1.5.3 Designer Legacies for Maintenance

The BUREC's in-depth dam inspection every six years includes a review of design, construction, and operation files. During the design and construction of the dam, information is collected on construction materials, the geology of the dam foundation, its physical and chemical characteristics, seepage through abutments and foundations, drilling exploration of the underground, grouting of foundations and abutments, reservoir water-holding capability, embankment dam and reservoir slope slide potential, equipment operation, hydrologic records, design flood determination, seismic records at the dam site or regional area, and so on. The designer prepares the designers' operating criteria (DOC), which should contain all information that has relevance to the survival and safety of the project, as recognized in the design stage, and which becomes part of the project files upon transfer of the project to operations. Among other things, a DOC may specify how appurtenant equipment, such as gates and valves, should be operated and what operational restrictions may apply, such as avoiding certain gate positions (BUREC, 1986, pp. 1, and 35–45). It should list all design and construction liabilities that are being handed over to O & M with the project. The DOC is a supporting document of the *Standing Operating Procedures* (SOPs) discussed in Section 1.5.4.

Design and construction liabilities can haunt maintenance throughout a project's life. An example of such a liability are earthfill dams built by the hydraulic fill method. The potential instability of such dams was recognized when some of them already failed during construction. Others failed during earthquake shaking. Design liability also dates to the principal building period of hydraulic structures, from the 1930s to the 1970s, when ever-larger and bolder structures were built well beyond the experience available at the time. The tools for structural and hydraulic performance predictions were hand calculations or physical scale model tests, at least until the 1960s when computers were used to drastically expand the depth of engineering analysis. Some defects that resulted from design or construction errors made themselves known rather promptly, on first filling before maintenance had a chance to interfere, or in the early years of operation when decisive interference by maintenance could make the difference between a repairable accident and failure.

1.5.4 Standing Operating Procedures

There is a recognized need for providing continuity in safety surveillance, as well as general O & M as project personnel come and go. Each water project has its special characteristics due to location, size, type of structure, purposes of operation, and so on, so that each project requires its own special attention. Structures are designed and built by different companies and different people, and each structure has its legacy from its design and construction stages that carries over into O & M, as has already been mentioned. The BUREC developed an outline for preparing *Standing Operation Procedures* (SOPs) for each project that "should contain all information and instruction necessary for operators to perform their duties" (BUREC, 1986, p. 1; Bock, 1974, p. 57). The SOP should be available at the time the project passes from construction to O&M. The SOPs were initiated for dams and reservoirs, but they also have been prepared for major power plants and pumping stations. Updates are available in the manuals for Facilities Instructions, Standards, and Techniques (FIST manuals) posted on the BUREC web site.

At all projects with power facilities, "a detailed review and analyses of the operation and maintenance procedures" is required periodically (BUREC, 1986, p. 8; BUREC web site, 2002). The updated SOPs are annually approved by the regional director of the BUREC and copies are kept at the project and at the regional office. They serve for reference, training, and emergencies. Their stated purpose is to minimize incidents. These reviews include periodical study of carryover information from design and construction, attendance at the project and access, especially under adverse conditions, standby power, communications, permanent wall-mounted instructions for operators, operating instructions for gates and valves, and identification of valves, pipes, and equipment by nameplates and color coding. The SOPs list 20 supporting documents "that are part of total instructions for operation and maintenance of the dam and reservoir for all offices having any responsibility in the care and operation of the facility." Among these documents are the DOC, the USACE flood control regulations, flood forecasting and operating criteria, power plant operating instructions, a facilities security plan, major maintenance procedures, manufacturers' instructions and drawings, oil or other hazardous fluid spill prevention measures and countermeasures, instrumentation reports and/or results, and others (BUREC, 1986, p. 37). The DOC is prepared by the central office of the BUREC in Denver. It is based on engineering drawings, manufacturers' literature, and regional and project procedures. It contains a description of the structure; O & M instructions for the dam, appurtenant works, and equipment; and selected construction drawings pertinent to O & M (for the latest updated information, go to the BUREC web site).

Also the DOCs are updated periodically to reflect changes in operating equipment and procedures.

The establishment of an SOP for a project also has probabilistic ramifications. The timely organization of all project information with the goal of documenting operation and maintenance procedures cuts down on delays when fast response is critical to prevent major damage or failure and to minimize downtime. Thus, it increases maintainability, the ease of carrying out maintenance. Even preprogrammed routine maintenance activities, such as color coding of pipes and valves, have a probabilistic aspect, as they reduce exposure of equipment and personnel to human error, extend equipment life, and reduce operator stress in emergencies. In critical situations, swift and correct response can reduce the extent of failure and its consequences. The SOPs require description and clear instructions regarding operation and maintenance for outlet works, spillways, and electrical systems and equipment. Specifically cited in the SOP are the following instructions (BUREC, 1986, p. 40):

- Balance discharge from a multi-gate spillway. Unbalanced discharge may produce tailwater eddies that may cause gravel deposition and/or scour in critical areas.
- Properly use a gated drop inlet-outlet structure (also known as glory-hole spillway or shaft spillway). Such a spillway, by pumping action of downflowing water, may produce explosive compressed air release from the shaft causing severe shocks to the entire structure.
- Avoid small gate and valve openings. Small openings may cause excessive outflow jet contraction and spray with high air entrainment, pressure pulsations, vibrations, and cavitation of gates and valves.
- Occasionally exercise emergency gates and valves that are not normally operated. This provides training for operators, and a possibility of detecting and eliminating hitherto undetected design flaws or equipment defects so they can be eliminated before they can interfere with operation under emergency conditions.
- Maintain dam drainage system. A total suppression of all seepage is rarely possible. The next best strategy is to prevent seepage pathways from becoming erosion conduits within the dam or its foundations, and from pressurizing foundation and construction joints.

1.5.5 Review of Operation and Maintenance

In the BUREC jurisdiction, O & M activities are the responsibility of the project's field office staff. In some cases, maintenance results were not satisfactory and required "extensive rehabilitation of structures to restore them to satisfactory condition for continued safe operation" (BUREC, 1991, p. 1). The program was placed under the engineering division in the BUREC's headquarters. The *Review of Operations and Maintenance Program* (RO&M) includes both operation and maintenance procedures "because it is virtually impossible to evaluate maintenance without considering operational procedures and conditions." Similar to the experience with dam inspections, when different user organizations are involved in the operation and maintenance of projects, there is a need for oversight by a centralized, experienced staff that collects experience from all projects.

The O & M reviews were found to be necessary at three-year intervals. Every six years, the staff of the BUREC's engineering division takes the lead and reporting role. The BUREC's Chief of the Division of Water Operations and Maintenance stated that "there is a significant need to provide ... a continuity of investigations over a period of time with adequate documentation and with sufficient support from appropriate experts" (Prichard, 1977, pp. 163–172). The investigation includes the following:

- Evaluate the adequacy of the O&M program at the facility.
- Disclose conditions that might cause disruption or failure of operation.
- Determine the adequacy of the structures and facilities to serve their intended purpose.
- Note the extent of deterioration as the basis for planning maintenance, repair, or rehabilitation.
- Review current operation practices.
- Determine the degree of operational safety at the facility.
- Obtain data for improvement of design, construction, maintenance, and operating practices.

The purpose of O & M reviews is to protect the investment, prolong service life, and reduce breakdown (corrective) maintenance (BUREC, 1991, p. 5). In the particular situation of the BUREC, many different operators are in charge of many small facilities (up to about 2 MW of hydropower). Prichard (1977, p. 163) mentions some 310 dams under the BUREC's administration. In 1980, the BUREC listed 50 hydroplants with a total installed capacity of about 12,000 MW. Checking from time to time on O & M procedures should ensure an acceptable standard of upkeep for all projects in the interest of the clientele that is served by the projects, and of the taxpayer at large.

1.5.6 FERC Project Engineering Guidelines

The FERC was created in 1977 as successor of the Federal Power Commission (FPC). The FERC establishes and administers rates and oversees interstate aspects of the electrical, natural gas, and oil industries. The Office of Hydropower Licensing issues and enforces preliminary permits, exemptions, and licenses for building and operating nonfederal hydroprojects. FERC-regulated projects occupy federal lands, are located on navigable waters, use water or water power at a government dam, and affect the interests of interstate commerce. These requirements subject practically all nonfederal hydroprojects to the jurisdiction of the FERC. A license provides authority to construct and operate a hydroproject. Licenses extend up to 50 years and now require the consideration of other objectives besides hydropower, such as water quality, fish and wildlife habitat, and recreation.

The Division of Dam Safety and Inspections carries out the dam safety program. As part of processing applications for licenses by FERC staff, the Office of Hydropower Licensing issues engineering guidelines. They contain procedures and criteria for review by FERC staff for internal use, but they also are of interest to owners and operators of hydroprojects, as they will be used by the FERC in the review of license applications (FERC, 1987). The topics of the engineering guidelines include:

- Design flood and probable maximum flood
- Arch dams, gravity dams, embankment dams, and other dams
- Geotechnical investigations and studies
- Emergency action plans
- Construction quality control and inspection program
- Instrumentation and monitoring

The engineering guidelines are not in-depth analysis tools but rather a laundry list of things that a licensee must address. Licenses are required for constructed and unconstructed projects both covering about the same topics—project operation; mechanical and electrical equipment; and some hydrologic, geologic, structural, and mechanical details. This information is used by the FERC to evaluate the safety and adequacy of the project. Constructed projects are inspected by FERC representatives. FERC's staff includes specialists to evaluate all phases of constructing, operating, and maintaining nonfederal hydropower projects. Completed projects are inspected annually by a FERC engineer, usually accompanied by licensee staff. Any identified safety hazards must be corrected and remedial work must be carried out by licensees. Spillway gates are test-operated and maintenance work carried out since the last inspection is reviewed. Leakage, settlements, and movements are reviewed and evaluated. Projects with dams higher than 10 m and more than $2.5 \cdot 10^6$ m^3 water storage are required to be inspected by an outside consultant every five years. Some requirements are specified that the consultant must expressly report on, such as the project's spillway adequacy based on the probable maximum flood (PMF). Corrective measures may include increasing spillway capacity, construction of an additional spillway, raising and strengthening of the dam, reduction of uplift, and reduction of operating level (Brown, 1974, pp. 28–31). The consultant's report is reviewed by the FERC.

A major FERC concern is *dam break* analysis (FERC, 1989; www.ferc.fed.us/hydro/docs/engguide/pdf_files/chap2.pdf). The dam break problem is not a maintenance problem but a good example of an analysis that requires a probabilistic approach. It is a mix of probabilistic and deterministic problems. For example, the breach flow is calculated by the broad-crested weir formula, and downstream inundation is calculated by the Saint-Venant equations for unsteady open channel flow. These are prescribed or deterministic relations between independent and dependent variables. The parameters and variables in these equations, however, such as the breach cross section as a function of time, downstream channel roughness and channel cross sections, obstacles and flood plain storage capabilities, which determine the possible flood stages are variables with ranges, in other words, probabilistic variables. Hence, the results will also be variables with ranges. For example, different between breach cross section growth over time leads to different outflow hydrographs which in turn lead to different flood stages. Thus, different problem pathways lead to different problem solutions, some more, others less likely.

The FERC's recommendations on breach width, breach cross section slopes, and time to failure express these uncertainties by giving ranges to these variables. The frequency over a variable range, as a default, can be assumed constant over the range (see Section 2.7.3). The FERC defines the time to failure as the time it takes to fully form the breach. It is short for thin arch dams which most likely fail by sudden collapse, and longer for earth dams where the water must erode the breach. The cause of the breach, overtopping, or low or high level piping makes a big difference in the evolution of the breach and the resulting peak discharge (Rissler, 1988, p. 200). The time to failure does not include the time leading to the start of the breach formation. For example, the time to erode away the dam crest of an earth dam from the downstream edge to the upstream edge is not included. In this situation the time to failure begins with the formation of the breach itself with the lowering of the crest (FERC: Engineering Guidelines, Appendix VI-A: dam break breach parameters, November 1998; see www.ferc.fed.us/hydro/docs/engguide/pdf_files/chap2.pdf). For a well-compacted, or fortified earth dam, the time to the onset of the lowering of the crest may actually outlast the erosive period of the overtopping flow and prevent the dam from failing. The probabilistic approach may show that using worst assumptions for all variables and parameters may lead to highly unlikely and possibly nonsensical results.

1.5.7 Maintenance Activities of the USACE

The USACE is the largest hydroproject operator and hydropower producer
in the United States with 24 % of the total installed hydro capacity. The two runners-
up are the BUREC with 16 %, and the TVA with 6 %. At the beginning of 1988, the
total installed hydrocapacity in the United States was 74,700 MW, with 50,400 MW
in the western states (BUREC, 1991a). An overview of the corps's typical
maintenance work in the three regional divisions (northwestern, mid-Atlantic, and
central, according to the USACE web page) reveals the following activities in support
of hydropower operations (as of March 17, 2000):

- Turbine uprating, rehabilitation, and replacement.
- Turbine runner rebuilding and replacement.
- Exchanging one type of turbine runner for another.
- Refurbishing governors.
- Rewinding and repairing generator stators.
- Cleaning, repairing, and inspecting generator rotors.
- Generator replacement.
- Turbine blade replacement.
- Turbine water passages refurbishment.
- Crane refurbishment, modifications, redesign, and rehabilitation.
- Exciter refurbishment and replacement.
- Thrust bearing rehabilitation.
- Pumps, motors, and gear reducer replacement.
- Wicket gate bushings replacement.
- Generator thrust bearing cooling system repair.
- Powerhouse control systems replacement.
- Transformer replacement.
- Replacement of mechanical and electrical peripheral equipment.
- Rehabilitation of raw water piping system.
- Modifications to the turbine shaft seal.
- Modifications to turbine bearings.
- Transformer replacement.
- Circuit breaker replacement.
- Economic evaluation of best rehabilitation alternative
  based on life cycle costs.
- Replacement of pumps, diesels, and gear reducers.
- Using greaseless bushings.
- Developing survivor curve and hazard functions to estimate
  the probability of failure of exciters, transformers, and circuit breakers.
- Developing cavitation coatings.

- Providing juvenile fish passage.
- Providing barriers and screens for fish protection.

The component of electrical equipment that may require a lengthy and costly repair is the winding insulation system of a generator. The generator consists of a stator providing a magnetic field and a rotor representing the conductor. Electric current is induced in the conductor as it passes through the magnetic field. Both stator and rotor are built by laminae, thin metal sheets insulated from each other, to reduce eddy currents in the metal that would reduce generating efficiency. Each lamina must be connected to its circuit by wiring that either provides magnetic excitation or that drains away induced electric current. The insulation system loses strength and elasticity over time through operational stresses. Eventually, the insulation fails. It is not uncommon for a generator to exceed 200,000 hours of operation before winding failure occurs. Based on 12-hour use per day, this time to failure would represent about 45 years. When the stator winding insulation fails, the high voltage generated in the coils arcs to the surrounding framework, and the generating unit is shut down by protective relaying. The unit must be repaired before it can be restarted. Different types of repairs are possible depending on the age and condition of the generating unit, ranging from a partial winding repair to a complete stator rewind. Replacing deteriorating stator windings on a planned schedule allows appropriate time for repair with the least impact on power generation and reduces the risk of a forced shutdown. A planned rewind can be combined with turbine repairs so that some of the unit disassembly and reassembly time and expenses can be shared by these activities. Reliability analysis and economic analysis of a planned rewind ensure that the job is performed at the time when it is deemed most appropriate. Scheduled rewinds generally provide a cost-effective way to maintain the reliability of old generators (after www.nwp.usace.army.mil/HDC).

1.5.8 Maintenance Activities of the TVA

The TVA operates 47 dams. Thirty have power installations with a total capacity of 4,800 MW, which is winter net dependable capacity. 1,532 MW of this capacity is installed in one pumped storage plant. Ten dams are equipped with navigation locks, and 18 dams have a flood control function. Among the TVA dams are some acquired dams the oldest of which dates to 1911. The dams built by the TVA program date from 1933 to 1979. The youngest power project is the pumped storage plant which dates from 1978, and the youngest dam without power dates from 1979. TVA's dam building period came to a close when one dam, almost finished, had to be removed under environmental pressure.

The TVA has a comprehensive dam safety program that includes analysis, rehabilitation, upgrading, inspection, maintenance, and emergency procedures (Hall,

1993). The TVA also has a program to modernize the hydrosystem that aims at increasing capacity, efficiency, and reliability. This program also considers the need to improve tailwater quality, specifically raising low oxygen concentration in water releases from reservoirs during the summer. From early on, the inspection of dams was delegated to design engineers. They conducted field inspections, analyzed the collected data, prepared reports, and made recommendations for maintenance. Hydroelectric facilities are inspected jointly by design engineers and power production personnel.

The inspection and maintenance functions of each group have been clearly defined and divided into three inspection classes (Lundin, 1974, pp. 41–42):

- <u>Class A</u>: earth and rockfill embankments, concrete dams, spillways, sluiceways, highway bridges that provide access to and over dams, and so on, which affect the safety of the project, but do not require attention for normal plant operation.
- <u>Class B</u>: (1) structural elements and flood control equipment that affect the safety of the dam and require frequent operation and attention of operating personnel, or require power unit shutdowns for inspections of spillway gates, sluice gates and valves, and water conduits.
(2) structural elements and equipment that do not affect the safety of the dam and require frequent operation, such as water power control structures and equipment including intake gates and valves, draft tube gates, gate cranes and hoists, trashracks, and so on. Also included is flood control operating equipment for spillway gates, sluice gates, and valves, which affects the safety of the dam, but requires frequent attention by operating personnel.
- <u>Class C</u>: equipment and facilities that enter into everyday operations and do not affect dam safety, such as lighting, power wiring and controls, turbine generators, and so on.

By this division of surveillance and maintenance responsibilities, the emphasis on public safety decreases from Class A to Class C, whereas emphasis on operational aspects increases from Class A to Class C. Observations by the dam inspection team may lead to recommendations to plant management that can be implemented without further study. If evidence is collected that raises questions about an underlying safety-related problem, instrumentation and field exploration are used to obtain additional information for a diagnostic assessment of the situation. This may mean placing sensors and collecting data on pore pressure, uplift pressure, seepage flow, seepage water quality as represented by temperature, suspended and dissolved matter, drilling additional seepage and pressure observation wells, collection of concrete samples, concrete strength tests, dam alignment surveys, and *in situ* stress measurements.

Major maintenance work is scheduled at least one year in advance to allow time for planning, budgeting, and arranging for special operations (Lundin, 1974, p. 43).

Following the Teton Dam disaster of 1976, federal agencies were directed by President Jimmy Carter to review their safety practices. The TVA participated with other agencies in developing *Federal Guidelines for Dam Safety*. Implementation of these guidelines resulted in a more formalized and structured dam safety program than existed before. The program consists of three major elements:

- Ongoing inspection, operation and maintenance programs.
- Study of dams regarded as possibly deficient with respect to guidelines.
- Preparation of emergency action plans for all TVA dams.

By December 1982, 19 dams were identified as possibly not in compliance with state-of-the-art criteria. Deficiencies were mostly related to spillway capacity for passing the PMF and resistance to the maximum credible earthquake (MCE). All dams were judged by the TVA to be in the nonemergency category. A seven-year in-depth study of these dams was scheduled. The highest priority was given to the study of the Blue Ridge project, a 53 m high dam with a 13 MW power project, built in 1925-30 and acquired by the TVA . The study was completed by 1982, and corrective action was scheduled. Operation and maintenance manuals exist for 36 dams, and emergency action plans were prepared for all dams except seven low-hazard structures (Hall, 1993).

1.6 Probabilistic Maintenance Perceptions of Field Personnel

1.6.1 Survey of Maintenance Personnel

In 1989, the ASCE Task Committee on Probabilistic Approaches to Maintenance of Hydraulic Structures (PAMHS), directed by the author, undertook an investigation to identify areas that could benefit from the use of probabilistic methods. A questionnaire was sent to maintenance practitioners asking them how they perceived the probabilistic aspects of their work. The committee wanted to probe the level of awareness that probabilistic problems existed, and then inform the maintenance community of the availability of methods that could be used to cope with the problems, such as incomplete information, uncertainty, and risk. This activity was the origin of the compilation of material presented in this book. The survey was conducted in the fall of 1990 and about 80 maintenance personnel across the United States participated. The questions addressed the following topics:

1. How do maintenance practitioners in the hydro-sector feel about their present approach to maintenance: are they satisfied, or do they

think a more quantitative and structured approach could be helpful in
various work areas?
2. How do they perceive and deal with the probabilistic aspects
of maintenance?

The experiences of the respondents were quite diverse, as each of them was
dealing with similar but still quite different problems. This is due to the fact that there
are different project types and each project of a given type (e.g., high head and low
head) is again unique. The results of the survey were expected to give a general
picture of attitudes toward the encountered maintenance problems and their
probabilistic aspects. An excerpt of the results of the survey was presented at the
annual ASCE Hydraulic Engineering Conference in 1991 (Wunderlich, 1991).

1.6.2 Selected Survey Questions

The range of judgments used in the questionnaire was from 1 "preferred" to
5 "least preferred." The questions were rather general, but the responses were quite
interesting, even if they probably were off-the-cuff answers. A few of the 27
questions and answers are given in this section[3].

Shares of preventive and corrective maintenance : When asked about their
preference between preventive and corrective action, 69 respondents indicated that
they performed, on average, 55 % preventive and 45 % corrective maintenance (Q1).
In individual instances, the percentages ranged from 10 % to 95 % for preventive
maintenance and from 5 % to 90% for corrective maintenance. The more successful
preventive maintenance becomes, the less corrective maintenance is needed.
Corrective maintenance is clearly motivated by the need to get the system up and
running again. The motivation for preventive maintenance usually is achieving and
maintaining a desirable level of system safety, reliability, and efficiency. Preventive
maintenance puts the maintenance personnel in control, and it is therefore the
preferred mode of operation by respondents who think they are controlled by
excessive corrective maintenance. The ultimate goal of maintenance is the "smart
structure" (Robison, 1992) that is wired to reveal stress excursions, incipient cracks,
and vibrations, as may occur during overloads in floods and earthquakes, or
unfavorable operational conditions, thereby guiding inspection and maintenance to
locations where it is needed. The interpretation of the data collected through thermal,
electrical, fiber-optics, and other sensing devices may be aided by probabilistic
interpretation, if deterministic cause–effect relationships do not exist.

---

[3]The unpublished complete questionnaire is available from the author.

Q1: How much of your maintenance work is
    a. preventive?
    b. corrective (complement to 100 %)?
    c. preventive, but triggered by failure or an experience gained elsewhere
    (assume a. as reference base (100 %))?

The answers were as follows:

Item	Preventive	Corrective	Triggered
Count	69	69	55
Average	55 %	45 %	31 %
Standard Deviation	25 %	24 %	28 %
Range	10–95 %	5–90 %	2–100 %

The count is the number of usable responses. To this question 69 persons responded. An average percentage and a standard deviation were calculated for each category. The ranges of answers are given in the bottom line of the table. The answer to category *preventive* ranged from 10 % to 95 %, to *corrective* from 5 to 90 %, and to *triggered* from 2 to 100 %. The averages are close to the middle of the range. The standard deviations are rather large, about 25 %. The average of *triggered*, in column 3, was estimated from 55 responses as this question was apparently not understood by all respondents.

Q5: Given an observed symptom, e.g., a crack, leak, spalling, overheating, scouring, vibration, etc., what is, on average, the probability of successfully identifying the problem that is causing it? The answers are distributed as follows:

Success	0–10 %	11–30 %	31–60 %	61–90 %	91–99 %	100 %
Response	0 %	5 %	7 %	55 %	32 %	1 %

Of 76 respondents, 0 % claimed a 0 % to 10 % score, and only 1 of 76 (about 1 %) claimed 100 % success. Adding the response percentages for success rates 61 to 90 %, and 91 to 99 % shows that 87 % of the respondents think a successful diagnosis can be made in 61 % to 99 % of cases.

Diagnostic success: Given a damage symptom (crack, leak, spalling, etc.), respondents rated their success with correctly identifying the cause (Q5) as overall better than 50-50. Of 76 respondents, none claimed 0 % success, and only one claimed 100% success; 66 respondents claimed 61 % to 99 % success. A symptom is an indication or manifestation of some underlying cause. Many different symptoms may have the same cause, and many different causes may have the same symptoms. Therefore, the relation between symptom and cause may be probabilistic. In other words, the cause can be predicted with some probability from the symptom or symptoms but cannot be stated with certainty. Mathematical models have been developed, called *expert systems*, which assist in tracing the path from symptoms to the cause, and provide suggestions for addressing the cause of the problem (Maher, 1987). The first such systems were developed as diagnostic aids in medicine. Expert systems are based on *if-then rules* that incorporate the knowledge on a well-defined subject, as elicited from a human expert or group of experts. Typically, such systems may contain some 200 rules. Probabilistic aspects enter expert systems through data uncertainty and uncertainty in the if-then rules. Also, spreadsheets that address uncertainty by using *fuzzy calculus* have appeared in the marketplace (Thorndike, 1994).

Q6: If damage is observed, how sure are you, on average, about the structural integrity? The answers are distributed as follows:

% sure	0	1–10	11–30	31–60	61–90	91–99	100
Response	0 %	1 %	5 %	13 %	39 %	38 %	3 %

Of the 76 respondents, no one claimed to be completely unsure, and only 2 (3 %) claimed to be 100 % sure. The total response of only 99 % is due to rounding error. Adding the percentages of respondents of the 61 % to 90 % "sure" and 91 % to 99 % "sure" class shows that 77 % of respondents claim that they make a correct diagnosis of structural integrity in 61 % to 99 % of cases.

Safety assessment: When asked how sure they were about structural integrity if damage were observed (Q6), of 76 respondents, none claimed being 0 % sure, and 2 claimed being 100% sure. Of the 74 remaining respondents, 15 were from 0 % to 60 % sure, and 59 were 61% to 99% sure. This problem is related to diagnosis and the attempt to find out the true but unknown state of the system. If the relationship between cause and effect (symptom) were precisely known, the structure could be clearly designated as either safe or unsafe, based on the observed symptom(s). But this usually is not the case. Incidents of failures have occurred only hours, days, or weeks after the item was inspected and maintenance was performed. Such incidents usually are damaging to the credibility of the organization and the professionals in charge, as the general public does not appreciate the probabilistic nature of many cause–effect relationships. Usually, the underlying deterioration process can only be influenced, but not completely controlled by maintenance.

Q7: If repair is decided, how sure are you, on average, about the method that should be used? The answers of 76 respondents were as follows:

% sure	0–40 %	41–70 %	71–90 %	91–100 %
% Total Response	1 %	12 %	48 %	39 %

The number of respondents are integers. For example, 1 % of 76 is 0.76, which is interpreted as 1 respondent out of 76. Of the 76 respondents, 39 % were 91 % to 100% sure that they would use the right method, and 87 % were 71 % to 100 % sure of the methods to be used.

Confidence in repair method: When asked about how sure they were, on average, about the method to be used in case a repair becomes necessary (Q7), of 76 respondents, 1 respondent was 0 % to 40 % sure, 46 respondents were 41 % to 90% sure, and 30 respondents were 91 % to 100% sure. Similar to the symptom–cause relationship, probabilistic relations often result between repair method and problem solution. Practically everybody has had the experience of fixing a problem and having the system fail again shortly thereafter. The problem usually is reassessed and added scrutiny is applied to the choice of the repair method so that it is more likely to succeed on second trial. Correct diagnosis, correct repair method, and correct execution of the repair will increase the probability of success. The reliability of maintenance work is, thus, similar to the reliability of serial systems, in which every component must be successful for the total system to be successful.

Q8: How would you generally rank the importance of the following items in your maintenance method selection? Rank from 1 (very important) to 5 (not important):

    a. reliability (1,2,3,4,5)
    b. repair cost (1,2,3,4,5)
    c. high probability of completing the job on time (1,2,3,4,5)
    d. meeting externally imposed requirements (1,2,3,4,5)
    e. other:_____ (1,2,3,4,5)

The answers to a. through d. are summarized as follows:

Ranking	Reliability	Repair Cost	Complete Job on Time	Meet External Requirements
	a.	b.	c.	d.
1	54	15	18	21
2	13	23	34	30
3	2	23	15	16
4	2	10	9	9
5	5	4	0	0
Count	76	75	76	76

The numbers in the columns under a, b, c, and d give the number of respondents who selected a level of importance from 1 (high) through 5 (low) in column 1. For example, reliability received importance ranking 1 from 54 of 76 respondents.

    <u>Ranking of maintenance objectives</u>: Reliability, repair cost, timely completion of job, and externally imposed requirements were to be ranked in terms of importance for selection of maintenance methods (Q8). Sixty-seven of 76 respondents gave reliability above average ranking (1 and 2), repair cost, and timely job completion received average ranking (3). External requirements received below average ranking. Reliability is the probability that a system functions as expected. Reliability came clearly through as the foremost goal of maintenance. Despite the importance of reliability, all factors affecting the selection of maintenance options should be looked at. Reliability was highly ranked by most respondents, but it did not get the highest ranking from all of them.

The following comments refer to Q8e. The numbers in parentheses are the rankings given to the item named by the respondent. Again rankings are from 1 (very important) to 5 (not important):

- Safety (1)
- Maintaining budget (2)
- Dam safety related (1)
- System (demands) load (1)
- Meeting water delivery (no rating)
- Quality of repair (1)
- Monitoring repair & record keeping (1)

- Quality and experience (1)
- Quality of workmanship (1)
- Safety, displacing energy from fossil plants (1)
- Experience (1)
- Getting back in operation (1)
- Maintenance of personnel safety (1).

Q10: When you start a major maintenance job, what is the chance, on average, that you encounter additional problems that may also need correction? The answers are distributed as follows:

Estimated chance of additional problem	Percent						
	0	1–10	11–30	31–60	61–90	91–99	100
% Total response	0	8	30	19	32	8	3

The 77 respondents were divided into two major groups: 38 % think the chances of additional problem are relatively low, 1–30 %; and 40 % think the chances are relatively high, 61–99 %. A smaller group of 19 % thinks they are in the middle range, 31–60 %. One respondent gave the percentage of encountering additional problems as a function of the type of component as follows: civil 30 %, and mechanical 70 %.

The predictability of maintenance work was probed by asking about the probability of opening the proverbial *can of worms*. Once the maintenance work begins, additional damage is discovered that may increase maintenance time and cost (Q10). This uncertainty can be anticipated, at least to some extent, by using probabilistic time and cost estimates.

Q11: Would you feel comfortable making estimates that quantify uncertainty, for example, estimating the probability that a system, a component, or a piece of equipment will reach the next worse state, or possibly even the breakdown state, before the next inspection? Rate yourself from 1 (very comfortable) to 5 (very uncomfortable), or 6 (unable). The results are:

Feeling about making a probability estimate	VC	C	Able	UC	VUC	Unable
Numerical rating of ability	1	2	3	4	5	6
% Total response	9 %	26 %	42 %	11 %	8 %	4 %

Column headings: VC–very comfortable; C–comfortable; UC– uncomfortable; VUC–very comfortable.

Of a total of 74 respondents, 77 % (cumulated ratings 1, 2, and 3) felt from "very comfortable" to "able" in making probability estimates, while 12 % (ratings 5 and 6) felt very uncomfortable or unable to do so. One respondent commented that he is comfortable with predicting or monitoring the decline of a piece of equipment, but not comfortable with predicting the failure of a seemingly sound piece of equipment.

Judgmental quantification of uncertainty: The respondents were asked how comfortable they were with estimating the probability that a system or component will reach the next worse state, possibly the breakdown state, before the next inspection (Q11). Only 9 % of 74 respondents rated themselves very comfortable, but 68 % rated themselves from comfortable to average. Of the remainder, 23 % rated themselves uncomfortable to unable. It usually is hard to make such a prediction. It resembles a weather prediction. Nevertheless, sometimes consequential decisions are made based on such predictions.

Decision analysis. There are diverging opinions about the legitimacy of using subjective quantification of uncertainty, such as the subjective or personal probabilities elicited from experts, such as in Q11, in structured approaches to decision making, called _decision analysis_. On the one hand, such an approach can be construed as a way of making a more complete analysis by including _all_ information;

on the other hand, it can be interpreted as subjectively biasing a decision. A decision that is arrived at without decision analysis, just out of the back of the decision maker's head, is called a *seat-of-the-pants decision*. Such a decision is, of course, very likely to be subjectively biased too, with the additional disadvantage that usually it does not provide an explanation as to how it was reached, and that it cannot be repeated for slightly changed assumptions. In areas where subjective judgment is part of the decision process, subjectivity is unavoidable and not necessarily bad. The advantages of a structured approach to decision making over an unstructured approach include the following (Raiffa, 1970, p. 268):

- The decision problem is analyzed as a whole and quantification of input requires a certain discipline concerning estimates.
- Systematic examination of input information provides feedback for identification, collection, and analysis of data.
- Analysis is an incentive to clearly identify viable alternative actions.
- Analysis provides documentation on how a decision is reached. This may help clarify points of disagreement and rally support for the decision.
- A structured decision analysis can provide a framework for continuous evaluation of recurring problems, such as operational and maintenance problems.

Willingness of taking innovation risks: When asked if they would use a new repair technique (Q21), most respondents said they would if the technique was less costly, promised longer life, was environmentally acceptable, and was less disruptive than conventional techniques. Those who did not want to use new techniques quoted lack of experience, likelihood of premature failure, and lack of data (or information). They did not like the idea of choosing a new technique as an experiment. Many new techniques are basically "experimental," such as the use of new materials. Experimenters may reap benefits from new techniques, but they also must be prepared to pay the price in the form of an increased failure rate or other unexpected problems. In Chapter 5 it is shown that premature demise of a new design does not necessarily mean a financial loss.

Alternatives to present maintenance practice: Q26 asked if alternatives were seen to the present approach to planning and carrying out of maintenance. Of a total of 76 respondents, 47% saw a need for an alternative to their present approach. The desired changes were increased computerization in the areas of data collection, data evaluation, equipment tracking (monitoring), maintenance planning and decision making; installation of maintenance management systems; computer-triggered notices of maintenance needs; and moving away from routine maintenance to dynamic maintenance which is maintenance on an as-needed basis. In some cases, maintenance

Q21: Would you choose a new repair technique over a traditional one?

Of 74 respondents, 69 or 93 % said yes, which indicates a very high level of readiness to use new techniques. Five respondents said no. One respondent said that he would use a new technique if he were reasonably sure of success. The question also asked those who answered yes, to rank the suggested reasons why new repair techniques would be used (from 1 — most likely to 5—not likely):
  a. less costly
  b. has longer expected service life
  c. is environmentally more acceptable
  d. has shorter expected service disruption
  e. is experimental

The responses to Q21a through Q21e are as follows (the numbers are responses that add to the total number of respondents on the bottom line):

Ranking	Less costly	Longer life	Environmentally acceptable	Less disruptive	Experimental
	a.	b.	c.	d.	e.
1	17	42	24	29	1
2	24	21	28	27	1
3	19	6	14	6	7
4	5	0	2	2	19
5	3	0	1	3	39
Count	68	69	69	67	67

The reasons a, b, c, and d for choosing new repair techniques got high approval rankings. The respondents apparently did not like the idea of using a new repair technique that was "experimental." Q21 may have been ill-posed and may not have been understood, as intended. The intention was to find out if a new technique would be used that was still relatively untried and, therefore, somewhat risky, but showed promise.

activity was driven by corrective maintenance, allowing no time to think about a better way. In other cases, the continuing high reliability and availability of equipment did not warrant an effort of instituting better methods, at least not at the time of the survey (1990).

1.6.3 Features Desired by Maintenance Personnel

Based on the responses to the survey, the following desirable developments in the maintenance sector were mentioned by the participants:

- *Increased computerization*. Almost half of the respondents (47 %) thought that there is room for change in their approach to maintenance. One respondent said: "We need a better system. We are still approaching maintenance the same way we did 30 years ago." Frequently mentioned was a desire for increased computerization. The respondents believed that it would enhance data collection, data evaluation, equipment tracking (monitoring), maintenance planning, and decision making. The use of data on labor costs, material costs, and repair times would help planning and cost estimation, work scoping, and work scheduling. Some want entire computerized *maintenance management systems* (MMS). Work is under way in their organizations to establish such MMSs. Also desired are more computer-triggered notices of maintenance needs and a change from routine maintenance to a *dynamic maintenance approach* on an as-needed basis.

- *Increasing preventive maintenance*. Reducing corrective maintenance is a desire expressed by several respondents. The adaptation of preventive maintenance programs that have been developed for other industries (fossil-fired power plants) to the hydro sector were proposed. Maintenance personnel are frustrated by the corrective approach, which puts them in a situation in which they have to respond to random failures instead of being in control. Preventive maintenance is expected to give increased control, even if it will never be full control. Problems seem to have arisen with a low-priority approach to maintenance in some organizations. Some understaffed and underbudgeted maintenance departments are overwhelmed by failures. They have to devote their resources to corrective maintenance. No time is left for planning and carrying out preventive maintenance or for thinking about a better approach.

- *Reliable budgeting and building a competent maintenance team*. Frustration is caused by the unpredictable annual budget allocations in some organizations. This uncertainty foils meaningful long-range maintenance planning and causes the team morale to suffer. In the more fortunate situations, continuing high project availability relegates maintenance to a low-priority problem.

- *If it ain't broke don't fix it*. Slightly more than half of the respondents don't see a need for changing their present approach to planning and implementing maintenance. Many believe a new approach is not needed, just some enhancements. Some see the computer more as a threat than as a helper (in 1990). Apprehension about changing the old ways was expressed by some comments. One respondent mentioned that higher management was imposing a more mathematical approach and computerized data collection. This move was viewed with apprehension instead of as a welcome change in support of future maintenance work. In any case, it is advisable to carefully examine the existing maintenance system before plunging into new approaches. The users themselves must be participants in the development of the new approaches, or they will not think they have a stake in their success.

Resistance to change is a common behavior, a sort of defensive and emotional reaction in the face of the unknown. Combined with some rational motive, it can be a formidable, insurmountable obstacle regardless of its rational merits. The still rather sparse use of probabilistic methods after many decades of promotion in the engineering sector is an example. Some intrinsic difficulties with understanding probabilistic concepts combined with the undeniable computational burden and the usual lack of information are the possible causes that prevented probabilistic methods from becoming a widely accepted engineering tool by now. Resistance to change may be experienced in three basic ways:

- Progressive management proposals may meet resistance from the rank and file.
- Progressive rank and file proposals may meet resistance from management.
- Developers proposing new approaches may meet resistance from one or both of the above when recommending changes in procedures.

The probability of overcoming resistance to change can be markedly increased if there is a clear management policy favoring change, and if this policy can be translated into cooperation, communication, and goodwill among all participants. An alternative to not fixing what is not yet "broke" is "*if it is not perfect make it better*" (anonymous Japanese quality expert).

****

2. Elements of Probabilistic Methods

2.1 Sets, Events, and Probabilities

2.1.1 Sets

A *set* is a group of one or more *elements*. An element is a basic item that makes up a sample or a set. The *universal set*, also called *collective, universe, or population*, Ω, is the total of all sets and elements. If a set has only one element, it is called a *unit set*. If set A consists of a elements, then this is expressed by: $a \in A$ (read: a elements in A). A set is *finite* if it has a countable number of elements. It is *infinite* if it has an uncountable number of elements. If a set has zero elements, it is an empty set, or *null set*, $\phi = 0$. A collective of elements can be created by taking samples from a natural or industrial process. For example, by taking n concrete samples from a concrete production process, forming a test cylinder of each, and measuring their strengths produces a population of n concrete strengths. Another example is the record of repair times for a specific item that reoccurs time and again over a number of years. If several of the same or nearly the same elements are grouped under a label, for example, under the central value of the group, they form a set within a population. For example, the 13 spades or the 4 kings are sets in the population of 52 playing cards.

If all elements of a set A are contained in another set B, then A is a *subset* of B. This is expressed by

$$A \subset B, \tag{2.1-1}$$

which is read: "*A* in *B*." *A* also is a subset of the universal set, Ω, or $A \subset \Omega$. The elements of sets A and B can be joined in a *union*,

$$A \cup B. \tag{2.1-2}$$

This is read "*A or B*." This means that an element can be in *A* or *B*, in other words, the elements of both sets are considered. If some elements belong to set *A* and set *B*, then these elements form an *intersection* of sets *A* and *B*,

$$A \cap B, \tag{2.1-3}$$

which is read "*A and B*." The intersection is a more restricted set than the union.

If the universal set contains one set *A*, then every element **not** in *A* is part of the *complementary set A'*, or the *complement* of *A*. The union of *A* and *A'* makes up the collective or *sample space* on which probabilities will later be defined:

$$A \cup A' = \Omega. \tag{2.1-4}$$

The set *A* and its complement *A'* cannot overlap so that

$$A \cap A' = 0. \tag{2.1-5}$$

Sets with no overlap are mutually exclusive; in other words, an element can belong only to one set or the other.

Calculating with sets is known as Boolean algebra. An easy way to become familiar with such calculations is to use marked elements. A few examples illustrate this method. Let the universal set be a group of numbers,

$$\Omega = \{1, 2, 3, 4, 5, 6, 7, 8, 9\}.$$

The numbers in braces are the elements of the universal set. The null set is $\phi = 0$. Assume the universal set contains two sets, *A* and *B*:

$$A = \{1, 2, 3, 4, 5\} \text{ and } B = \{3, 4, 5, 6, 7\}.$$

Then: $A \cup B = \{1, 2, 3, 4, 5, 6, 7\}$
$A \cap B = \{3, 4, 5\}.$
$(A \cup B)' = \{8, 9\}$
$(A \cap B)' = \{1, 2, 6, 7, 8, 9\}$
$(A \cup B) \cup (A \cup B)' = \{1, 2, 3, 4, 5, 6, 7\} \cup \{8, 9\}$
$\quad = \{1, 2, 3, 4, 5, 6, 7, 8, 9\} = \Omega$
$(A \cup B) \cap (A \cup B)' = \{1, 2, 3, 4, 5, 6, 7\} \cap \{8, 9\} = 0$
$(A \cap B) \cup (A \cap B)' = \{3, 4, 5\} \cup \{1, 2, 6, 7, 8, 9\} = \Omega$
$(A \cap B) \cap (A \cap B)' = \{3, 4, 5\} \cap \{1, 2, 6, 7, 8, 9\} = 0$
$(A \cup B) \cap (A \cap B) = \{1, 2, 3, 4, 5, 6, 7\} \cap \{3, 4, 5\} = \{3, 4, 5\}$

$(A \cup B) \cup (A \cap B) = \{1, 2, 3, 4, 5, 6, 7\} \cup \{3, 4, 5\} = \{1, 2, 3, 4, 5, 6, 7\}$

$\Omega \cap (A \cup B) = \{1, 2, 3, 4, 5, 6, 7, \ 8, \ 9\} \cap \{1, 2, 3, 4, 5, 6, 7\}$
$= \{1, 2, 3, 4, 5, 6, 7\}$

$\Omega \cap (A \cap B) = \{1, 2, 3, 4, 5, 6, 7, 8, 9\} \cap \{3, 4, 5\} = \{3, 4, 5\}$

$A \cup \phi = A = \{1, 2, 3, 4, 5\}$

$A \cap \phi = \phi = 0$

$A' = \{6, 7, 8, 9\}$

$B' = \{1, 2, 8, 9\}$

$A \cup A' = \Omega$

$A \cap A' = 0$

$A' \cup B' = \{1, 2, 6, 7, 8, 9\}$

$A' \cap B' = \{6, 7, 8, 9\} \cap \{1, 2, 8, 9\} = \{8, 9\}.$

Using such number sets, one can easily clarify complex unions and intersections, even when more than two sets are involved. Graphic representations of sets are known as *Venn diagrams*. Figure 2-1 shows a Venn diagram of the union of two mutually exclusive or independent sets, $A \cup B$. The complement is $(A \cup B)'$, which is also $A' \cap B'$. This latter statement can be visualized in the Venn diagram by marking the complements A' and B' with a distinct hatching as in Figure 2-2. Those parts that become cross-hatched are the parts that overlap and thus represent the

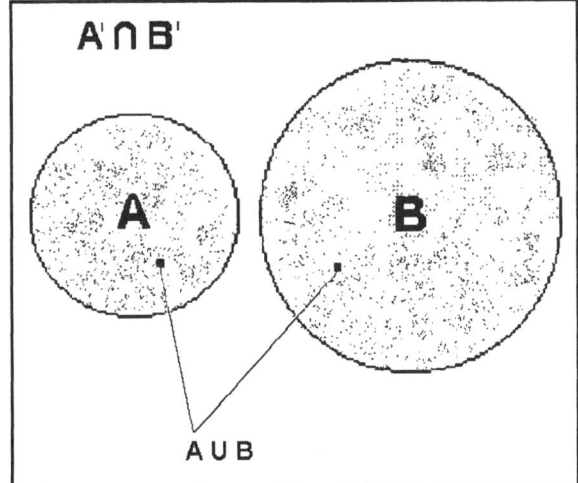

Figure 2-1: Venn diagram of the union of mutually exclusive sets A and B, $A \cup B$. The rectangular box represents the universal set or sample space on which probabilities are defined. The complement of sets A and B to the universal set is $(A \cup B)' = A' \cap B'$.

intersection $A' \cap B'$, that is the complementary area around the two sets. This can be even more simply demonstrated by the previous examples. More details on calculating with sets can be found in statistics texts (Savage, 1972, p. 12; Moan, 1982, p. 4–2; Goldberg, 1986, p. 16; and others).

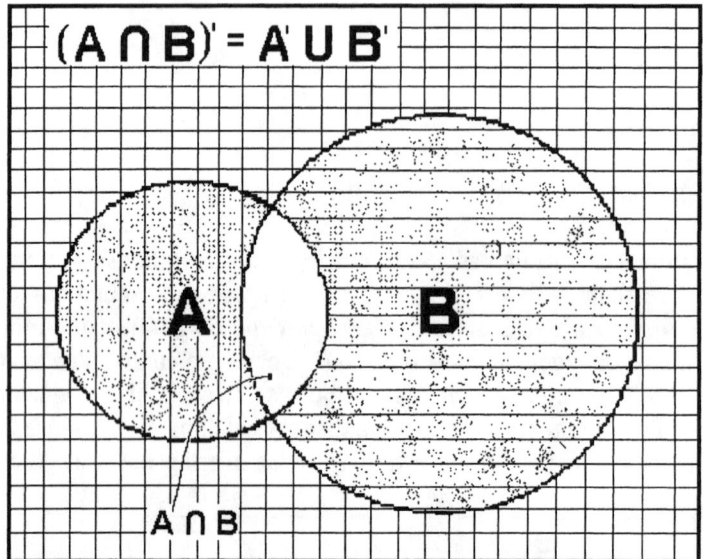

Figure 2-2: The intersection of sets A and B is $A \cap B$, the area without hatching. The complement to the intersection, $(A \cap B)'$, is the area of the rectangle (collective) excluding the lens-shaped intersection of A and B. It is the union of A', the horizontally hatched area, and B', the vertically hatched area, i.e., $A' \cup B'$. The intersection of A' and B', marked by the cross hatched area, is $(A \cup B)'$, the complement of $A \cup B$.

<u>Example</u>: According to the *ASCE 1998 Report Card for the American Infrastructure* (ASCE, 2001), the total number of state-regulated dams is 94,517. Assume this is the universe of dams. Set A consists of 9,133 high-hazard dams (those whose failure can cause loss of life and high property damage). Set B consists of 2,099 unsafe dams. The report card also mentions that there are a number of dams (elements) that belong to both sets. This intersection $A \cap B$, high-hazard *and* unsafe dams, is of greatest concern as a danger to life and property.

2.1.2 Events and Probabilities

An *event* is something that happens, the breakdown of a machine, or the outcome of a coin toss. Often there is uncertainty associated with an event. For a repair event, it is its occurrence in time, its cost, and the length of the repair time. A measure of the uncertainty of an event is its probability. An event must be possible to have a probability of occurring. If it is possible, its probability is greater than zero. If it has a reasonably good probability to occur, it is *probable*. If it has a very small probability, it is *improbable*, or unlikely, but still possible. If it is *impossible*, it cannot occur and its probability is zero. On the other hand, if it is certain, it has probability 1. An event can be a set of possible realizations of values of a variable within its possible range, with a probability of occurrence, or frequency, attached to it. A graphical example of such a set is the bar of a histogram. The collective of all *random realizations* of a variable is the *sample space* of the random variable.

The universal set is also the *sample space*, *S*. The elements of *S* are *simple events*, for example, data items. Sets consist of several of the same or similar elements, and subsets are subdivisions of sets. The unit set consists of just one element, or simple event. A number, or probability, can be assigned to each simple event that must meet the following two criteria (Goldberg, 1986, p. 55):

1. The probability of each element, E_i, of the sample space must be equal to or greater than zero, i.e., it must be a positive number:

$$P(E_i) \geq 0, \quad i = 1, 2, 3, ..., n. \tag{2.1-6}$$

2. The sum of the probabilities of all elements of the sample space must be equal to 1:

$$\sum_{i=1}^{n} P(E_i) = 1. \tag{2.1-7}$$

Probabilities are dimensionless, but they may be interpreted as a number of special occurrences in the possible total of occurrences, the latter being the sample space. This property should always be carefully checked to avoid an erroneous probability assignment.

As already mentioned, probability may be interpreted as a ratio of two related quantities. If 90 light bulbs out of 2,700 fail during a specified time such as a month, then the monthly *absolute frequency* of failure is 90. The 90 failures are special occurrences in a sample space of 2,700. The monthly *relative frequency* of failure or failure probability is 90/2700 = 0.03, or 3 %. The 90 failures out of 2,700 may be visualized by a Venn diagram like Figure 2-1 with a failure set A of 90 elements inside the rectangle defining the sample space of 2,700 elements. The complementary set of non-failure events is 2,700 - 90 = 2,610 and it occupies the remaining sample space with probability 0.97. Both probabilities are defined on the same sample space and add to 1, as required by Equation (2.1-7). A similar diagram can be drawn for each month, and Equation (2.1-7) must be fulfilled for each month. If the failure probabilities of all months were added, they would not meet Equation (2.1-7), because they are not defined on the same sample space. Suppose it is found that annually 1080 bulbs fail out of the installed 2,700 bulbs. Then the annual failure probability is 0.4, and the annual non-failure probability is 0.6. Both probabilities are defined on the same sample space and add to 1.

A quantity that is sometimes mistaken for a probability is the *failure rate* that refers to a quantity other than the one from which the special events for which the probability is defined originate, such as the number of failures per time unit, instead of the number of failures per total population. If the failure rate of elements is defined on some quantity other than the total population of elements, then the failure rate has units, such as occurrences per time, or occurrences per length, and is not a dimensionless quantity or probability. Suppose the failures are pipe breaks per month, which are different numbers in each month. Their distribution over the year is not a probability distribution. Suppose the sample space is repair time for a specific job or maintenance project that has been estimated to range from a lower limit to an upper limit. This repair time distribution can be visualized by a Venn diagram whose rectangle of repair time sample space is divided by vertical lines into a number of time sets that represent repair times of increasing length, each with a probability attached to it. These probabilities must satisfy Equation (2.1-7). In the simplest case one could assign equal probability to all times of the sample space. These probabilities must add to 1, which means each must be $1/n$, where n is the number of time sets that fill the sample space (see also Section 2.7.3).

Assume the probability definition, Equation (2.1-7), is applied to the elements, E_i, of a set which represents event E. Then one obtains the probability of E as the probability of the union of all elements in E. Suppose the number of all elements in the sample space is n, and each is given the probability $P(E_i) = 1/n$, then the probability of the set (event) E is the sum of the probabilities of all elements that belong to set E:

$$P(E) = \sum_{i=1}^{f} P(E_i) = P(E_1) + ... + P(E_f) = \frac{1}{n} + ... + \frac{1}{n} = \frac{f}{n}. \quad (2.1\text{-}8)$$

Since the elements in the set are mutually exclusive, only one can occur at a time, but there are f chances of such an occurrence so that the total probability of the set is f/n. This also means that the occurrence of any of the elements, $E_1, E_2, E_3,$ or ... E_f

makes E occur. Therefore, the probability of E is the probability of the union of f mutually exclusive elements:

$$P(E) = P(E_1 \cup E_2 \cup \ldots \cup E_f) = P(E_1) + P(E_2) + \ldots + P(E_f) \ . \quad (2.1\text{-}9)$$

Equation (2.1-9) holds only as long as none of the elements can occur simultaneously. The property of elements (events) being mutually exclusive is expressed by

$$E_1 \cap E_2 \cap \ldots \cap E_f = \phi = 0 \ . \qquad (2.1\text{-}9a)$$

If there are no intersections, then the set of elements in the intersection is a null set and its probability is zero. In the general case, *ordinary sets* intersect which means they share elements. Figure 2-3 illustrates a multiple intersection of sets. The formula for calculating the probability of the union of intersecting sets is much more complicated than Equation (2.1-9) as it requires elimination of any double counting of shared elements. It will be discussed in Section 2.1.3.

Statement 2 on probability, Equation (2.1-7), has some interesting interpretations:

● The summation over the probabilities of all n elements of the sample space S is

$$P(S) = \sum_{i=1}^{n} P(E_i) = 1. \qquad (2.1\text{-}10)$$

● If all n elements of the sample space are equally likely with probability $1/n$, then their probabilities sum to 1. This property anticipates the fundamental property required of all probability density functions (see Sections 2.6 and 2.7).
● If one element of the sample space has probability 1, then all other elements must have probability 0, that means they do not exist. This is the special case assumed by the deterministic approach, namely only one value of the variable exists.

Sample spaces may be *continuous* or *discrete* by their nature. If elements are obtained by sampling a continuous function, the sample space usually is only a small number of all possible realizations as the total is uncountable. If the elements are obtained by sampling a discrete function, then the sample space consists of a countable number of elements. In both cases the probabilities must be defined so that $P(S) = 1$. Obviously, if the number of elements is uncountable, the probability of each element must approach zero so that the total probability of the sample space does not exceed 1.

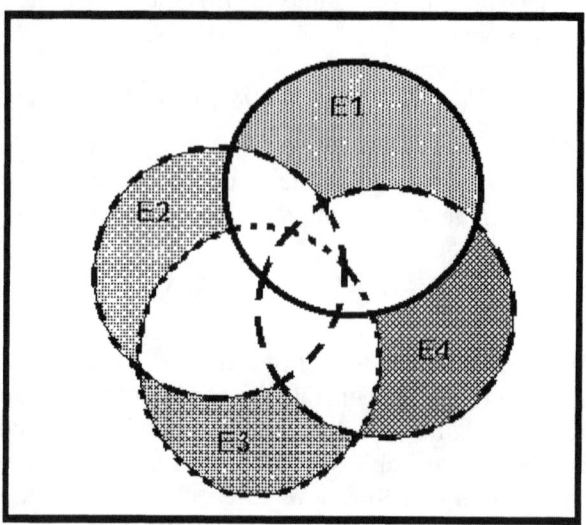

Figure 2-3: Union of four ordinary sets that intersect with each other but are independent of each other like the four units of a power plant. The event "unit on" can occur in four ways. Either E_1 or E_2 or E_3 or E_4 can be on. The probability of these mutually exclusive events is illustrated by the sum of the probabilities of the shaded parts of the four sets. To calculate this probability, $P(E_1 \cup E_2 \cup E_3 \cup E_4)$, the probabilities of all joint events (intersections) must be subtracted from $P(E_1) + P(E_2) + P(E_3) + P(E_4)$ by using Equation (2.1-14).

Sample spaces obtained by tests usually are *discrete*, as tests can only create a countable sample space, one of finite size. The limited number of concrete test data, 100 altogether, in Figure 2-4, represents an actually infinite sample space by a finite number of elements. The sample space has been partitioned into 12 mutually exclusive sets, here classes of concrete strength, with different numbers of elements each. Their relative frequencies computed by Equation (2.1-8) are represented by the bar heights of the histogram. If many more data were obtained, the number of the sets could be increased by reducing the class width until the class width becomes very small and the number of sets becomes very large. With the number of test data or elements approaching infinity, the discrete sample space would converge toward a

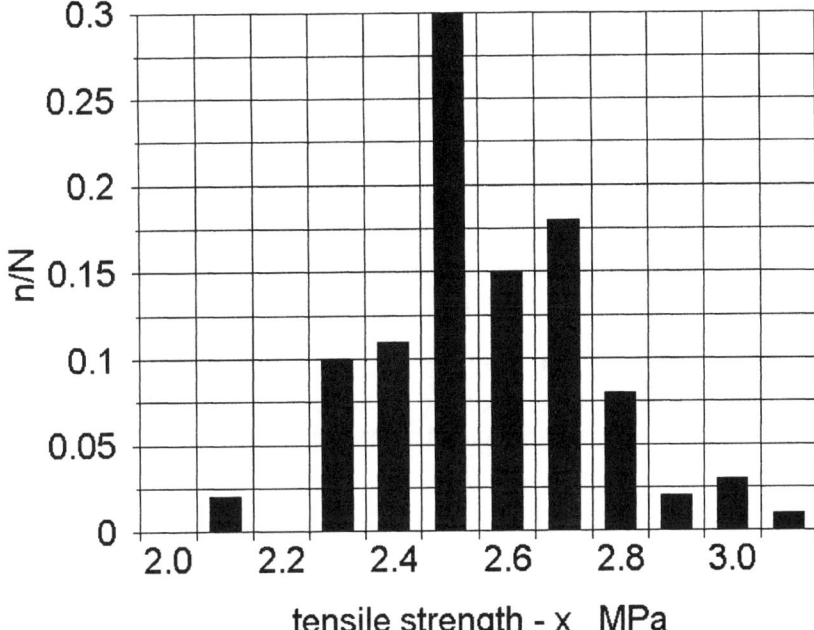

Figure 2-4: Histogram of concrete tensile strengths of test cylinders (data by D. L. Ivey, as referenced by Kreyszig, 1979, p. 840). The class widths represent ranges from 0.05 MPa below to 0.05 MPa above the indicated central value of the class. The classes chosen here produce an empty class (2.2 MPa) and a class with a very large number of elements (2.5 MPa), but the general tendency toward a bell shape is recognizable.

continuous sample space. The sample space shown in Figure 2-4 has two sets with zero elements because of the relatively small size of the sample. If thousand elements had been taken instead of hundred, all sets would probably have elements in them. Statistical sampling theory shows that the characteristics of an *infinite* sample space can be approximated by those of a *finite* sample with sufficient accuracy (see Section 3.4). This statistical property of samples is important because usually only samples of limited size are practically obtainable and economically affordable.

An example of a discrete sample space is created by throwing a coin. The outcomes can be only heads (H) or tails (T), if the unusual outcome of a coin standing on its rim is excluded by the experimental setup. The sample space in this case consists of two partitions H and T, regardless of the number of elements generated. Other experiments can be devised that lead to other partitions of the sample space. For example, three partitions would result according to the count of heads if two coins were thrown simultaneously (0, 1, 2); other partitions would result if one coin were thrown several times in a row. Figure 2-5 illustrates the possible outcomes or partitions of the sample space if a coin is thrown four times in a row. The possible outcomes of each set of sequential throws form a pathway of an *event tree*.

In the coin throw example, each throw is represented by two branches. In a generalized process, if there are more than two branches and more than a few throws, it quickly becomes a chore to calculate the total number of outcomes of the experiment. The *counting* of possible outcomes of an event tree can be calculated by multiplying the number of branches at each branching. The total number of outcomes of an event tree with n_i branches at k branchings is

$$C = n_1 \cdot n_2 \cdot n_3 \cdot \ldots \cdot n_k \qquad\qquad (2.1\text{-}11)$$

where n_1 is the number of branches at the first branching (origin), n_2 is the number of branches at the second branching, and so on. If the number of branches is the same at each branching, as in Figure 2-5, then

$$C = n^k, \qquad\qquad (2.1\text{-}11a)$$

where, in the coin throw example, n is the number of branches (H and T) at each throw and k is the number of throws. For $n = 2$, and $k = 4$, the number of outcomes is $C = 2^4 = 16$. The resulting sample space for the experiment in Figure 2-5 consists of 16 elements.

One sequence of possible events from start to finish of the experiment in Figure 2-5 forms a *trajectory*. As the outcomes are mutually exclusive, only one such trajectory can be realized in an experiment. In the example, after an initial H follows a T, then another H, and another T. Hence, the final outcome has the event sequence HTHT. If the experiment is repeated many times, any of 16 different outcomes of sequences of H and T may be obtained ranging from HHHH to TTTT. These outcomes can be construed as discrete partitions, or mutually exclusive sets of the sample space of this experiment. No matter how many times this particular experiment is repeated, the result will be one of these sets. If the four throws were repeated 16 times, the total number of different outcomes, some trajectories may occur several times, whereas others may not occur at all. Only over the long run will one find that all trajectories have equal probability of occurring. However, the number of occurrences of H and T are found to follow a binomial distribution (see Example 6 and Section 2.7.2).

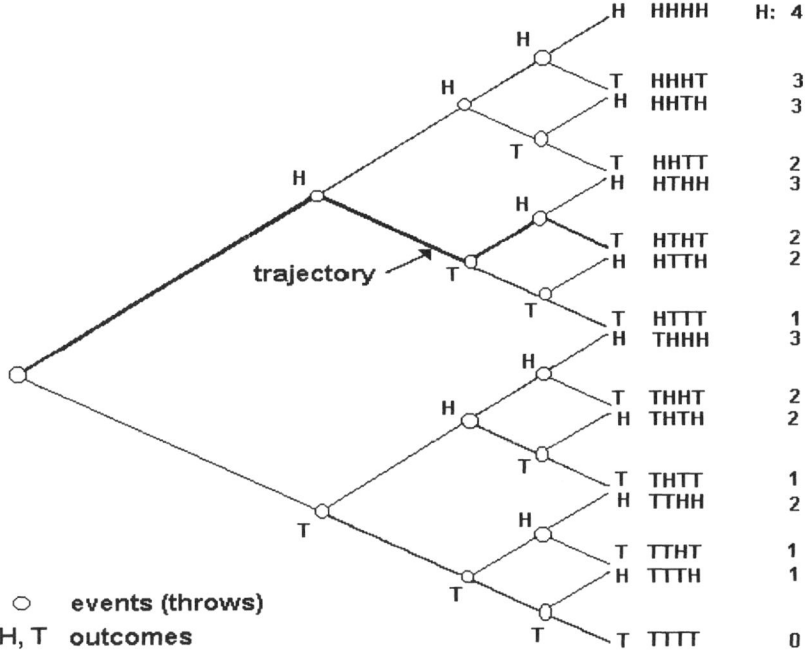

Figure 2-5: Throwing a coin produces two outcomes, H (heads) and T (tails), each with probability 0.5. Throwing a coin repeatedly leads to repeated branchings forming a path in the resulting *event tree*. The tips of the branches are marked with the possible series of events (outcomes) in sequence that each path produces if a coin is thrown four times. The outcomes of each throw are mutually exclusive, as only one or the other can occur at a time. One possible sequence of events is illustrated by the trajectory emphasized by the heavy line that leads to HTHT. Any number of repetitions of the experiment can reproduce only one of the outcomes shown in the column on the right-hand side, all with equal probability. However, the number of occurrences of H and T are found to follow a binomial distribution.

<u>Examples</u>: (1) Records show that a hydroplant operates on the average 340 days during the year. The downtime includes 20 days of scheduled outage, the rest being forced outage. What is the probability of the plant operating when scheduled? <u>Solution</u>: Scheduled outage has probability 1 and cannot be part of the sample space. The sample space consists of the time of the year where operation or outage can occur. These two sets have probabilities of $340/345 = 0.99$ and $5/345 = 0.01$, respectively. The probability of the plant operating when scheduled is 0.99.

(2) Medical diagnostics provides an example of the use of the probabilistic approach to maintenance of a system. Here the system is a person who undergoes observations and testing with the purpose of identifying hidden causes that are not clearly related to symptoms observed and test results obtained. The diagnosis of rheumatoid arthritis (RA) in a person is difficult, and a guide has been developed to diagnose four different levels of likelihood of the presence of RA (Merck, 1982, p. 1178): classic, definite, probable, and possible. Eleven criteria are specified, which, for brevity, are given here only by the designations c_1 through c_{11}, plus two additional criteria, c_{12} and c_{13}. They have to do with joint stiffness, pain, swelling, and persistence of symptoms. They are typical findings a doctor can make by observing, querying, and testing a person. If the person meets a specified set of these criteria, he becomes an element in this set:

1. <u>Classic</u>: must meet 7 of the 11 criteria, plus joint symptoms described in c_1 through c_5 must be persistent for at least 6 weeks.
2. <u>Definite</u>: must meet 5 of the 11 criteria, plus joint symptoms described in c_1 through c_5 must be persistent for at least 6 weeks.
3. <u>Probable</u>: must meet 3 of the 11 criteria, plus in at least one of the criteria c_1 through c_5 the joint symptoms must be continuous for at least 6 weeks.
4. <u>Possible</u>: must meet 2 of the criteria c_1, c_2, c_4, c_6, plus criteria c_{12}, and c_{13}.

The names given to the sets (underlined) are not necessarily in agreement with the definitions used in this text. For example, "possible" seems to mean the lowest level of probability, whereas we use it as having a probability greater zero. "Probable" expresses a higher probability than "possible," and "definite" expresses a still higher probability. Whatever the jargon used, the names define sets to which the patients, the elements of the sample space, are assigned by diagnosis. The four sets are intersecting because they use overlapping subsets of the criteria for their definition. Sometimes, because of the many intersections of the sets, it may not be clear to which set a patient belongs . This creates the possibility of a patient being assigned to the wrong set.

(3) Strength testing of concrete cylinders usually reveals that even after meticulous concrete preparation and accurately following test procedures, the strength measured for each concrete specimen varies over a range (sample space). Once the range of strength is known it can be divided into equal intervals (classes) and the test results (elements of the sample space) assigned to the classes. The resulting groups are non-intersecting sets, or partitions of the sample space that stretches from a lower limit to an upper limit. Counting the elements in each class usually reveals that the classes in the center of the range have more elements than those near the limits, For a sufficiently large number of test results, here 100, the shape of the resulting distribution (histogram) resembles a bell shape, as displayed by Figure 2-4, that reveals a fundamental law of probability (see Section 3.4)

(4) In a concrete strength test, 30 out of 100 test results fell into the class (set) "2.5 MPa." Giving each of the hundred test results (elements) equal probability of occurrence, $P(e_i) = 0.01$, the set or event $E = 2.5$ MPa, as represented by the 30 elements, e_i , has the probability

$$P(2.5) = \sum_{i=1}^{30} P(e_i) = 30 \cdot 0.01 = 0.3$$

(5) A study of some 300 dam failures worldwide (Biswas and Chatterjee, 1971, as cited by Jansen, 1980, p. 100) found the following causes:

Floods in excess of spillway capacity . . . . 35 %
Foundation problems, including seepage, piping, excessive pore pressure, inadequate cutoff, fault movements, settlements, rock slides . . . . . 25 %
Various problems, including design and construction flaws, inferior materials, wave action, acts of war . . . . 40 %.

The sets of failure mechanisms are actually intersecting with sets of dam types which are not shown. Some failure mechanisms are more frequent with some dam types than others. For example, more arch dams may be destroyed by acts of war than rockfill dams, whereas arch dams are not as much, if at all, affected by piping as rockfill dams. If all sets and their intersections were presented, one would get a more complete picture of which dam failure has the highest probability with which type of dam.

(6) Calculate the number of occurrences of H and T that are found in the outcomes of the coin throw experiment illustrated in Figure 2-5. Solution: If the sample space is partitioned into sets labeled with the number of H in the outcomes, one obtains the following result:

Number of H in Outcomes - Sets	0	1	2	3	4	Sum
Absolute Frequency of Sets	1	4	6	4	1	16
Relative Frequency of Sets	1/16	1/4	3/8	1/4	1/16	1

Notes: Row 1 shows the possible number of occurrences of H along a path through the event tree; these numbers designate sets; row 2 gives the absolute frequency of the sets in a total of 16 sets of the sample space; row 3 shows the relative frequency of these sets; the probability of the union of all sets is $P(S) = 1$ (row 3, last column). The numbers in row 2 (except the last) are binomial coefficients (see Section 2.7.2). A similar result is obtained by counting the occurrences of T.

2.1.3 Probabilities of Ordinary Sets

The *Venn diagram* in Figure 2-1 shows two mutually exclusive or non-intersecting sets, whereas Figures 2-2 and 2-3 show the general cases of intersecting or *ordinary sets*. Calculating the total probability of ordinary sets is much more complicated than for mutually exclusive sets. If the sample space contains two or more large sets it is unavoidable that they intersect. The first sign that there is an intersection may be a sum of probabilities greater than 1 obtained from Equation (2.1-8). But also small sets may intersect and there is no such warning signal.

In many cases, the sample space is partitioned in two parts, an event and a complementary event. Examples are an event and a non-event, "breakdown" and "no breakdown," "win" and "lose," and so on. The event and its complementary event are mutually exclusive, or non-intersecting events, and they exhaustively partition the sample space, in other words, they fill it. If A is the event and A' is the complementary event, the probability of their union is

$$P(A \cup A') = P(A) + P(A') = 1 .$$
(2.1-12)

If the probability of the union of two events is 1, one event or the other must happen. For example, if A is the breakdown event and A' is the non-breakdown event, then there is either a breakdown or no breakdown; there cannot be both simultaneously, and there has to be one or the other. If $P(A)$ is the probability of breakdown, and $P(A')$ is the probability of no breakdown, then the sum of their probabilities must fulfill Equation (2.1-12). Usually the "no breakdown" event is much more likely than the breakdown event so that $P(A) << P(A')$. Since the probability of joint occurrence of an event and its complementary event is zero, the intersection of the two sets is a null set: $A \cap A' = \phi$.

Ordinary sets are sets that have elements in common. If one such set occurs also another one may occur. The probability of such *joint events* is $P(A \cap B \cap C$

...). This means that A and B and C, and so on, can occur simultaneously. The probability of the union of two ordinary sets is calculated by the *general addition rule*:

$$P(A \cup B) = P(A) + P(B) - P(A \cap B) \qquad (2.1\text{-}13)$$

Equation (2.1-13) calculates the probability of event A or event B by eliminating the probability of event A and B. $P(A \cap B)$ is the probability of the joint events or joint sets. Joint events can occur independently, or they may depend on each other, or one may depend on the other. If joint events are independent, then the probability of joint events is the product of their probabilities:

$$P(A \cap B) = P(A)\, P(B). \qquad (2.1\text{-}13a)$$

If an earthquake and a flood occur at the same time, just by chance, they would be joint but independent events. An earthquake may trigger the liquefaction of an embankment dam. Then the liquefaction event is a dependent event that may occur if an earthquake occurs. The probability of dependent joint events will be discussed in Section 2.1.4.

If the set of joint events is empty, then $P(A \cap B) = 0$, and Equation (2.1-13) reverts to Equation (2.1-9). Equation (2.1-9) is illustrated by Figure 2-1, and Equation (2.1-13) is illustrated by Figure 2-2. The size of the circles can be construed as representing the probabilities of the sets, and the area of the rectangle including all sets inside represents the probability of the sample space, $P(S) = 1$. The subtraction of $P(A \cap B)$ in Equation (2.1-13) eliminates double counting of the intersecting elements of the sets when strictly the probability of A or B is required.

When the ordinary sets A and B represent only a small part of the sample space, then the sum of their probabilities may be much less than 1. As long as this is the case, the user may erroneously use Equation (2.1-9) instead of Equation (2.1-13) for calculating the probability of the union of sets. If $P(A) = 0.9$, and $P(B) = 0.5$, then it is obvious that the sets must intersect, because Equation (2.1-9) would give a probability exceeding 1; in this case clearly Equation (2.1-13) must be used.

When there are many sets in the sample space and many intersections, it becomes quite complicated to calculate the probability of the union of sets by excluding all double counting of intersections between the sets. Figure 2-3 illustrates how confusing it can become to figure out the probability of intersecting sets by eliminating all double counting of intersecting areas and determining only the area within the enveloping boundaries of sets E_1 through E_4. The general formula for the union of ordinary sets is (Moan, 1982, p. 4-5)

$$P(E_1 \cup E_2 \cup \ldots \cup E_n) = \sum_{i=1}^{n} P(E_i) - \sum_{i=1}^{n-1} \sum_{j=i+1}^{n} P(E_i \cap E_j) + \ldots$$

$$+ \sum_{i=1}^{n-2} \sum_{j=i+1}^{n-1} \sum_{k=j+1}^{n} P(E_i \cap E_j \cap E_k) -$$

$$- \sum_{i=1}^{n-3} \sum_{j=i+1}^{n-2} \sum_{k=j+1}^{n-1} \sum_{l=k+1}^{n} P(E_i \cap E_j \cap E_k \cap E_l) + \ldots$$

$$\ldots + (-1)^{n+1} P(E_1 \cap E_2 \cap \ldots \cap E_n) \tag{2.1-14}$$

where n is the number of sets. For n sets, this formula consists of n terms, with a term meaning one of the expressions under a summation sign, plus the last term. More and more multiple sums have to be included as n increases. The last term represents the intersection of all sets. It stands in lieu of a term under n summation signs, with the first sum from 1 to 1, the second sum from 2 to 2, and so on. Such sums produce only one term with the indices of the sets being the starting counts of the sums, i.e., 1, 2, 3, and so on. This degenerate sum is represented by the last term. Equation (2.1-14), as written, has four sums plus the last term and therefore represents the general formula for the union of five ordinary sets. Not all sets may intersect each other. Some may be disjoint, which means they do not intersect. The intersections in such cases have zero probability and do not contribute to the summation.

The evaluation of Equation (2.1-14) for the union of three ordinary sets produces

$$P(E_1 \cup E_2 \cup E_3) = P(E_1) + P(E_2) + P(E_3)$$

$$- [P(E_1 \cap E_2) + (E_1 \cap E_3) + P(E_2 \cap E_3)] + P(E_1 \cap E_2 \cap E_3). \tag{2.1-15}$$

The first three terms on the right-hand side represent the evaluation of the single sum which is recognized as the probability of three mutually exclusive sets. The probabilities of pairs of set intersections are deducted from this sum, and, finally, the last term adds the intersection of all three sets. This means that the intersection of all three sets is first added in three times with each set, then deducted three times with each intersection, and then added back in one more time. Drawing a Venn diagram for three intersecting sets, similar to Figure 2-3, helps to visualize this calculation. Only single and double sums are evaluated for $n = 3$. For n sets, Equation (2.1-14) produces $(n,1)$ terms by the single sum, $(n, 2)$ terms by the double sum, etc., and $(n,n) = 1$, one last term; $(n,1)$, $(n,2)$, etc., are binomial coefficients. The total number of terms produced by Equation (2.1-14) can be calculated by (Shooman, 1968, p. 24)

$$(n, 1) + (n, 2) + \ldots + (n, n) = 2^n - 1 \qquad (2.1\text{-}15\text{a})$$

where n is the number of sets. For $n = 3$, one obtains $3 + 3 + 1 = 2^3 - 1 = 7$, the number of terms in Equation (2.1-15). This control can avoid errors with the evaluation of Equation (2.1-14).

When one uses complex probability formulas, one should always make checks here and there to avoid errors. The probability of a union requires adding probabilities; the sum must always be $P(A \cup B) = P(A) + P(B) \leq 1$; the probability of an intersection requires multiplying probabilities, $P(A) \cap P(B) = P(A) \ P(B) \leq 1$. If this is not the case, an error has occurred. Another check is to make sure that the probabilities are in some size relation to each other (Goldberg, 1986, p. 72):

$$P(A \cap B) \leq P(A) \leq P(A \cup B) \leq P(A) + P(B) \ . \qquad (2.1\text{-}16)$$

The reader may want to test these conditions on examples using numbers as elements, as shown in Section 2.1.1.

Examples: (1) Proper operation of a floodgate on a dam depends on several factors. The hoist must function, the gate must withstand both the force and possible vibrations induced by the water flow; also the power source (winch) and the power transfer mechanisms (cables) must function. Failure of the entire gate operation mechanism could occur if any of these components fails. Suppose there could be an electrical hoist motor failure, A, or a mechanical failure, B, or both. The sample space of 'failures' can be illustrated by a Venn diagram, such as the one in Figure 2-2. The intersecting sets illustrate that A and B can fail independently or both jointly. The remaining sample space, the complement of the union of A and B, represents the non-failure probability. Suppose maintenance records show that, on average, during an annual period, $P(A) = 0.1$, $P(B) = 0.05$, and $P(A \cap B) = 0.01$. The probability that one or the other failure happens is $P(A) \cup P(B) = P(A) + P(B) - P(A \cap B) = 0.1 + 0.05 - 0.01 = 0.14$. This probability is obviously greater than the probability of each single event. Since the probability of a union is the probability of one or the other event, the joint failure probability, if it exists, must be subtracted. The probability that nothing happens, that the equipment functions as expected, is the probability of the complement of the union, $P\{(A \cup B)'\} = 1 - 0.14 = 0.86$.

(2) Assume the annual flood season consists of 100 consecutive days during which a 100-year flood can occur, whereas earthquakes can occur on any day of the year. Hence, the two event sets intersect on the 100 days of the flood season each year. What is the probability that a 100-year flood and a 10-year earthquake occur on the same day? What is the annual probability of such an event? <u>Solution</u>: The probability of a 10-year earthquake occurring on any day of the year is $P(E) = 1/(10 \cdot 365) = 2.74 \cdot 10^{-4}$. The probability of the flood occurring on any of the 100 days of the flood season of the year is $P(F) = 1/(100 \cdot 100) = 10^{-4}$. Hence, the daily probability of the two independent events occurring on any day of the flood season is $P(F \cap E) = P(F) P(E) = 2.74 \cdot 10^{-8}$. The annual probability could be calculated as the probability of the union of 100 daily events but one would have to eliminate all joint events among 100 sets (occurrences on more than one day of the period, on which the events could occur independently). In such a case, the use of nonoccurrence probability offers a simpler way. The nonoccurrence of the joint event is $1 - P(F \cap E)$; taken to the 100th power and subtracted from 1 gives the probability of occurrence in 100 days as $2.74 \cdot 10^{-6}$. This is the same probability as the probability of the union of 100 mutually exclusive sets, which is $100 \cdot 2.74 \cdot 10^{-8} = 2.74 \cdot 10^{-6}$. This unexpected result is a consequence of the very small probability in this case. One can see this by equating the first calculation to the second: $1 - (1 - p)^n = n p$, with p being the daily occurrence probability. Developing the left-hand side into a series but retaining only the first term, because of the smallness of p, leads to $n p$ on both sides of the equation.

(3) Suppose a sample space has 100 elements: 60 belong to set A, 50 belong to set B, and 30 are shared by both sets. All elements have the same probability. What is the probability of the union $P(A \cup B)$? <u>Solution</u>: The probability of each set is obtained as the sum of the probabilities of the simple events within each set: $P(A) = 60 \cdot (1/100) = 0.6$, $P(B) = 50 \cdot (1/100) = 0.5$, and $P(A \cap B) = 30 \cdot (1/100) = 0.3$. The probability of the union of the two ordinary sets A and B is obtained from Equation (2.1-13): $P(A \cup B) = 0.6 + 0.5 - 0.3 = 0.8$. Since the sets are intersecting, erroneous double counting the probability of the intersecting elements would occur by using the formula for the union of mutually exclusive sets, $P(A \cup B) = P(A) + P(B)$. This formula produces a probability greater than 1, which, of course, is wrong.

(4) A sample space, $S = \{1, 2, 3, 4, 5, 6, 7, 8, 9\}$, consists of two sets with 5 elements each, $A = \{1, 2, 3, 4, 5\}$, and $B = \{3, 4, 5, 6, 7\}$, each element having equal probability. Show that Equation (2.1-16) holds. <u>Solution</u>: Given each element of the sample space has equal probability, $1/9$, one obtains $P(A) = 5/9$, and $P(B) = 5/9$. The intersecting set, $A \cap B = (3, 4, 5)$, has 3 elements so that $P(A \cap B) = 3/9$. The union $A \cup B$ has the elements $(1, 2, 3, 4, 5, 6, 7)$, with elements 3, 4, and 5 being shared by A and B. The probability of the union of two ordinary sets is $P(A \cup$

B) = $P(A)$ + $P(B)$ - $P(A \cap B)$ = 5/9 + 5/9 - 3/9 = 7/9. The same result is obtained by counting the overlapping elements only once: $P(A \cup B)$ = (10 - 3)/9 = 7/9. The inequality, Equation (2.1-16), becomes $P(A \cap B) \leq P(A) \leq P(A \cup B) \leq P(A)$ + $P(B)$, or 3/9 ≤ 5/9 ≤ 7/9 ≤ 10/9.

(5) Given are three ordinary sets A, B, and C, which contain the following subsets: A = {2, 4, 7, 9, 10}, B = {2, 6, 8, 10}, and C = {3, 5, 9, 10}. Subset 1 does not belong to any of the ordinary sets. The subsets and their probabilities (in parentheses) are: 1 (0.08), 2 (0.02), 3 (0.05), 4 (0.13), 5 (0.05), 6 (0.15), 7 (0.10), 8 (0.25), 9 (0.12), and 10 (0.05). Calculate the probability of the union of A, B, and C. Solution: First the probabilities of the sets and their intersections are calculated, each as the union of its elements. Here the elements are construed as subsets with probabilities: $P(A)$ = 0.02 + 0.13 + 0.10 + 0.12 + 0.05 = 0.42, $P(B)$ = 0.47, $P(C)$ = 0.27. The joint set, ($A \cap B$), consists of subsets 2 and 10, hence, $P(A \cap B)$ = 0.07; similarly the probabilities of the other joint sets are obtained as the union of their subsets: $P(A \cap C)$ = 0.12, $P(B \cap C)$ = 0.05, and $P(A \cap B \cap C)$ = 0.05. For n = 3, the probability of the union of A, B, and C is obtained by Equation (2.1-15): $P(A \cup B \cup C)$ = 0.37 + 0.47 + 0.27 - 0.07 - 0.12 - 0.05 + 0.05 = 0.92. As a control, the simple sum of the probabilities of subsets 2 through 10 also gives 0.92, which shows that Equation (2.1-15) indeed eliminates double counting of the probabilities: The first three additive terms of Equation (2.1-15) add the probabilities of sets 2 and 9 twice and the probability of subset 10 three times. Then the three subtraction terms eliminate the probabilities of sets 2 and 9 once and the probability of subset 10 three times. The last term adds the probability of set 10 back in so that the probability of each subset is counted once. All subsets 2 through 10 remain part of the union of A, B, and C. $P(A \cup B \cup C)$ is the probability that at least one element (subset) of the union occurs. Graphically this probability is illustrated by the shaded areas of the four intersecting sets in Figure 2-3.

(6) A power outage could be caused by the failure of a control gate (A), a turbine blade (B), or the generator (C). Over a 20-year period, there have been five control gate failures, three turbine blade failures, and two generator failures. Gate and turbine have failed together twice, gate and generator have failed together once, turbine and generator have failed together once, and all three have failed together once. What is the probability that there is a failure of at least one of the components per year? Solution: The annual average probabilities are $P(A)$ = 5/20 = 0.25, $P(B)$ = 3/20 = 0.15, $P(C)$ = 2/20 = 0.1. $P(A \cap B)$ = 0.1, $P(A \cap C)$ = 0.05, and $P(B \cap C)$ = 0.05; $P(A \cap B \cap C)$ = 0.05. Then, according to Equation (2.1-15), $P(A \cup B \cup C)$ = 0.25 + 0.15 + 0.1 - 0.1 - 0.05 - 0.05 + 0.05 = 0.35. There is a 35 % probability of at least one of the three components A, B, or C failing. Equation (2.1-15) eliminates the probabilities of joint occurrences and leaves only the probabilities

of the mutually exclusive occurrences, as illustrated by the shaded areas in Figure 2-3.

(7) A hydroplant has four units, two newer ones and two older ones. The newer units have an average unexpected outage of 5 days per year; the older ones have average unexpected outages of 10 days per year. What is the probability (a) that all units operate on a given day; and (b) as one of the older units goes into rehab, what is the probability that at least one of the three remaining units will operate? Solution: The sample space is the 365 days of the year. The probabilities of the outage events are $P(E_1) = P(E_2) = 5/365 = 0.014$, and $P(E_3) = P(E_4) = 10/365 = 0.027$. (a) The probability of operating for each unit is the complement of the failure probability: $P(E_1') = P(E_2') = 1 - 0.014 = 0.986$, and $P(E_3') = P(E_4') = 1 - 0.027 = 0.973$, where E' denotes 'unit operating' and E denotes 'unit not operating'. If each unit is visualized as a set in a Venn diagram, the sample space is almost filled by one unit, so that the 4 sets of 'units operating' must overlap. Given the two newer units operate on the average 360 days and the two older units 355 days, then most of the time all units must be operating simultaneously. The probability of all units operating, given they are independent of each other, is according to Equation (2.1-13a):

$$P(E_1' \cap E_2' \cap E_3' \cap E_4') = P(E_1') \, P(E_2') \, P(E_3') \, P(E_4')$$

$$= 0.986 \cdot 0.986 \cdot 0.973 \cdot 0.973 = 0.920.$$

The resulting joint probability represents the intersection of four sets 'units operating' of the sample space.

(b) Suppose E_4 is on scheduled outage. The probability of at least one of three units operating, is the probability of the union of the three ordinary sets E_1', E_2', and E_3':

$$P(E_1' \cup E_2' \cup E_3') = P(E_1') + P(E_2') + P(E_3') -$$

$$P(E_1' \cap E_2') - P(E_1' \cap E_3') - P(E_2' \cap E_3') + P(E_1' \cap E_2' \cap E_3')$$

If all events are independent, then by Equation (2.1-13a) and Equation (2.1-15) the probability of at least one unit operating is

$$P(E_1' \cup E_2' \cup E_3') = P(E_1') + P(E_2') + P(E_3')$$

$$- P(E_1') P(E_2') - P(E_1') P(E_3') - P(E_2') P(E_3') + P(E_1') P(E_2') P(E_3')$$

$$= 0.986 + 0.986 + 0.973 - 0.986 \cdot 0.986 - 0.986 \cdot 0.973 - 0.986 \cdot 0.973$$

$$+ \ 0.986 \cdot 0.986 \cdot 0.973 = 0.99999.$$

As the number of sets increases it may be easier to obtain this result by using the complementary probability. Using Equation (2.1-13a) one can calculate the probability of unit 1 *and* 2 *and* 3 not operating and then take the complementary probability. Thus the probability of at least one unit operating is

$$P(E_1' \cap E_2' \cap E_3') = 1 - P(E_1) \ P(E_2) \ P(E_3)$$

$$= 1 - (1 - 0.986)(1 - 0.986)(1 - 0.973) = 0.99999.$$

This approach avoids evaluating Equation (2.1-14) when the sets are independent. The joint probabilities of dependent sets are more difficult to calculate (see Section 2.1.4).

2.1.4 Conditional Probabilities

If events can occur jointly, they may do so independently of each other, or they may be dependent on each other. If they are independent, the joint probability of two independent events is given by Equation (2.1-13a). If the events depend on each other, then the joint probability of A and B is

$$P(A \cap B) = P(A) \ P(B|A) \quad , \tag{2.1-17}$$

or, the joint probability of B and A is

$$P(B \cap A) = P(B) \ P(A|B) \ . \tag{2.1-17a}$$

Equation (2.1-17) represents the case that event A occurs and event B has some likelihood to occur given A has occurred, or B is conditioned on A. Equation (2.1-17a) represents the case of A being conditioned on B. Sometimes B may be conditioned on A but A may not be conditioned on B. Leaving such subtleties aside for the moment, if the left-hand sides of the two equations are assumed equal, equating both equations leads to

$$P(A) \ P(B|A) = P(B) \ P(A|B) \ . \tag{2.1-18}$$

Solving Equation (2.1-17a) for $P(A|B)$ gives

$$P(A|B) = \frac{P(A \cap B)}{P(B)}, \qquad\qquad\qquad (2.1\text{-}19)$$

where $P(B)$ is a probability greater than zero. $P(A|B)$ is the probability of A, given B has occurred. It is the *conditional probability* of A. This relation is explained as follows: sets A and B of the sample space have elements that are common to both sets. The probability of this intersecting set $(A \cap B)$ is $P(A \cap B)$. The probability of the intersecting set is calculated as the ratio of the number of elements in $(A \cap B)$ to the number of elements in the sample space. Since the probability of event A is conditioned on set B, the reference base of the intersecting set is changed from using the sample space to using set B as base, which is expressed by dividing by $P(B)$. This is seen by using some general numbers. Suppose a sample space with N elements each with probability $1/N$ contains two sets A and B with n_A and n_B number of elements, respectively. The sets intersect and have n_{AB} elements in common. Then the probabilities in Equation (2.1-19) are calculated as follows:

$$P(A) = \frac{n_A}{N}, \quad P(B) = \frac{n_B}{N}, \quad P(A \cap B) = \frac{n_{AB}}{N}. \qquad\qquad (2.1\text{-}20)$$

Substituting Equation (2.1-20) into Equation (2.1-19) gives

$$P(A|B) = \frac{n_{AB}/N}{n_B/N} = \frac{n_{AB}}{n_B}. \qquad\qquad\qquad (2.1\text{-}21)$$

Equation (2.1-21) only exists if $n_B > 0$. The number of intersecting elements, n_{AB}, is a subset of n_A and n_B. Therefore it is maximally equal to, but usually smaller than n_A or n_B. For $P(A|B)$ to exist, $n_{AB} \geq n_B$. Also, $P(A|B) \geq P(A \cap B)$, because $P(B) \leq 1$. Equation (2.1-21) shows that $P(A|B)$ is defined on a reduced sample space, n_B.

An expression similar to Equation (2.1-21) for the conditional probability of B is

$$P(B|A) = \frac{P(B \cap A)}{P(A)} = \frac{n_{AB}}{n_A}. \qquad\qquad\qquad (2.1\text{-}21\text{a})$$

Comparing Equations (2.1-21) and (2.1-21a) shows that $P(A \cap B) = P(B \cap A) = n_{AB}$, an assumption that was made in the derivation of Equation (2.1-18).

Expanding Equation (2.1-17) to three joint events that depend on each other gives

$$P(A \cap B \cap C) = P(A) \ P(B|A) \ P(C|A \cap B) \ . \qquad (2.1\text{-}22)$$

By induction, the joint probability of n events, or the probability of n intersecting sets, A_i, $i = 1, 2, \ldots, n$, is (Moan, 1982, p. 4-6):

$$P(A_1 \cap A_2 \cap A_3 \cap \ldots \cap A_n) = P(A_1) \ P(A_2|A_1) \ P(A_3|A_1 \cap A_2) \ldots$$

$$(2.1\text{-}23)$$

$$\ldots P(A_n|A_1 \cap A_2 \cap A_3 \cap \ldots \cap A_{n-1})$$

In Equation (2.1-23), A_1, the first event has no precursor. All the following events depend on precursor events or joint precursor events. The conditional probability of an event can be quite high, given the precursor event B occurs, whereas the precursor event may be very rare. For example, the probability of a dam failing by liquefaction may be very high, if the earthquake that has the strength believed to make the dam fail occurs. Suppose A is the failure event, and $P(A|B) = 0.99$ is the conditional probability of failure, given earthquake B occurs. Suppose $P(B) = 0.01$, then $P(A \cap B) = P(A|B) \ P(B) = 0.99 \cdot 0.01 \approx 0.01$. In this case, the joint occurrence of A and B practically coincides with the occurrence of the precursor event B, and A becomes almost a deterministic consequence of B. Maintenance should focus on reducing the probabilities it can reduce, for example, the conditional probability $P(A|B)$ by strengthening the dam. If B are events amenable to reduction, for example, human events such as war or terror, then increasing security can also reduce the triggering event B. In the case of more than two events, the analysis has to consider more aspects. If joint events do not occur, or are prevented from occurring, then the set of joint elements is empty, $A \cap B = 0$, and $P(A \cap B) = 0$. This case is illustrated by Figure 2-1.

If conditioning can be removed, then $P(A)$ becomes an independent event but joint events may still occur. In this case, $P(A \cap B) = P(A) \ P(B)$, which is Equation (2.1-13a). If there are numerous events A_i that may occur simultaneously but independently, then there is no conditioning of subsequent events on precursor events, and Equation (2.1-23) simplifies to

$$P(A_1 \cap A_2 \cap A_3 \cap \ldots \cap A_n) = P(A_1) \ P(A_2) \ P(A_3) \ldots P(A_n) \qquad (2.1\text{-}24)$$

Equation (2.1-24) is the *product rule* of probability (Moan, 1982, p. 4-7). It applies wherever independence exists or can be assumed. It will be encountered in the calculation of the reliability of serial systems (Section 4.3.3). Since probabilities by definition are equal to or less than 1, the joint probability of many events usually is small. The probability of joint events is always smaller than the probability of each individual event. This is simply due to the fact that a joint occurrence of events requires one event <u>and</u> another, <u>and</u> so on, to happen at the same time, which is increasingly unlikely.

In a well-maintained system, the probability of a major accident usually is small because such an event requires several rather unusual events to occur simultaneously, or one or more hazardous conditions to exist in the system. The same is true with major natural events that are the result of the simultaneous occurrence of corroborating events. Major floods, for example, require a spatially extensive and high yield storm event to move in a specific direction across the drainage basin; prior rain that has filled depression storage and ground water storage; and reduced flood retention storage of lakes and reservoirs because of previous high flows. The probability of all these conditions occurring simultaneously can be calculated as the product of their probabilities, which renders the probability of the resulting flood event very small.

Human experience with joint probabilistic events is condensed in the adage "*one disaster strikes seldom alone* ." It expresses the probabilistic relationship between a root cause event that may trigger others or make them more likely. This experience is expressed analytically by conditional probability. Once the independent precursor event happens, one or several follow-up events may be triggered by the original event and by joint events, as expressed by Equation (2.1-23), creating a tailspin to disaster. In the Johnstown dam disaster of 1889, the casualties were due to flood <u>and</u> fire. The fire broke out after the flood had ripped wooden houses from their foundations and piled them up against a bridge. Hanging electrical lines then caused short circuits that set off the blaze. Many who had survived the flood were killed by the fire. In the San Francisco earthquake of 1906, as much as 80 % of the damage was caused by fires triggered by the earthquake and only 20 % was due to ground motion (Bolt et al., 1975, p. 47).

Nothing can usually be done about the probability of the precursor events if they are natural, external events, such as earthquakes, hurricanes, or floods. Efforts must be concentrated on reducing the conditional probability of followup or triggered events, such as the probability of cracking, sliding, fires, and so on. In many cases, this amounts to reducing or eliminating hazards that may increase the probability and severity of followup events. Also, the installation of warning networks can reduce the severity of damage and loss if the probability of followup damage cannot be sufficiently reduced.

<u>Examples</u>: (1) Assume a sample space of $N = 1,000$ disasters contains among others a set F (500 fires) and a set E (100 earthquakes). In 80 % of earthquakes there is also a fire. (a) What is the probability of a fire, and (b) what is the probability of a fire given there is an earthquake? <u>Solution</u>: (a) The probability of a fire is $P(F) = 500/1000 = 0.5$. (b) Based on Equation (2.1-21), the conditional probability of a fire, given there also is an earthquake, is $P(F|E) = P(F \cap E)/P(E) = n_{FE}/n_E = 80/100 =$

0.8, where n_{FE} is the number of joint elements, and n_E is the set used as reference for conditioned probability. The formula confirms what one may have concluded from the data anyway. One should note that the probability of a fire is 0.5, whereas the probability of a fire, given there is an earthquake, has increased to 0.8.

(2) A hydraulic fill dam is subject to liquefaction, if a sufficiently strong earthquake occurs. If such an earthquake occurs, there is a 90 % probability of liquefaction failure. Show the difference between the conditional probability of failure and the joint probability of failure by the simultaneous occurrence of a strong earthquake and liquefaction. Solution: The conditional probability of liquefaction failure, according to Equation (2.1-21), is $P(L|E) = n_{LE}/n_E = 0.9$, where n_{LE} is the number of sufficiently strong earthquakes out of a total number of N earthquakes; and n_E is the number of sufficiently strong earthquakes that may cause liquefaction out of a total number of N earthquakes; in other words, 90 % of sufficiently strong earthquakes may cause liquefaction and failure. The total number N cancels from $P(L|E)$. The joint probability of a sufficiently strong earthquake and of liquefaction, according to Equation (2.1-20), is $P(L \cap E) = P(L|E) P(E)$. Suppose $P(E) = n_E/N = 0.01$. Then $P(L \cap E) = (n_{LE}/n_E)(n_E/N) = 0.9 \cdot 0.01 = 0.009$, which is approximately the probability of the earthquake; in other words, liquefaction is a rare event if earthquakes of sufficient size are rare, but once a strong earthquake occurs, liquefaction is very likely.

(3) A process uses two components, A and B. They have their independent failure mechanisms, but if an event E occurs, A may fail and then also B may fail. Calculate the probability of failure of A and B if E occurs. Solution: The probability of the three events occurring jointly is obtained from Equation (2.1-22):

$$P(E \cap A \cap B) = P(E) P(A|E) P(B|E \cap A)$$

where $P(E)$ is the probability of event E; the conditional probability of failure of A given E is

$$P(A|E) = \frac{P(E \cap A)}{P(E)} = \frac{n_{EA}}{n_E}$$

Applying this formula again to the third term on the right-hand side of Equation (2.1-22) gives

$$P(B|E \cap A) = \frac{P(B \cap (E \cap A))}{P(E \cap A)} = \frac{n_{EAB}}{n_{EA}}$$

where n_{EAB} is the number of elements in the triple intersection of the sets E, A, and B, and n_{EA} is the number of elements in the intersection of sets E and A. With $P(E) = n_E/N$, the probability of the joint occurrence of E, A, and B becomes

$$P(E \cap A \cap B) = \frac{n_E}{N} \cdot \frac{n_{EA}}{n_E} \cdot \frac{n_{EAB}}{n_{EA}} = \frac{n_{EAB}}{N}$$

This shows that the probability of joint dependent events is determined by the relative size of the highest (here triple) intersection with respect to the total sample space. The more intersections, the smaller will be the joint set size and the smaller will be the resulting probability.

(4) The joint probability of failure of two items, A and B, that depend on each other, is $P(A \cap B) = P(B \cap A) = 0.03$. The individual failure probabilities are $P(A) = 0.3$ and $P(B) = 0.2$. What are the conditional failures of each item? Solution: The conditional probability that item A fails, given B has failed, is given by Equation (2.1-19):

$$P(A|B) = \frac{P(A \cap B)}{P(B)} = \frac{0.03}{0.2} = 0.15$$

and the conditional probability that item B fails, given A has failed, is

$$P(B|A) = \frac{P(B \cap A)}{P(A)} = \frac{0.03}{0.3} = 0.1 \ .$$

In both cases the conditional probabilities, $P(A|B)$ and $P(B|A)$, are smaller than $P(A)$ and $P(B)$ because of the small joint set $A \cap B$ compared to the sets A and B. A Venn diagram can be used to illustrate this case of two sets with a small intersection.

(5) An air traveler wants to know the probability of being killed in an airplane crash. Solution: According to Equation (2.1-21), the probability of being killed, given that the plane crashes, is $P(A|B) = P(A \cap B)/P(B) = n_{AB}/n_B$, where n_{AB} is the number of people being killed in a crash and n_B is the number of people being involved in a crash. If 90 % of the people involved in a crash are killed, then $P(A|B) = 0.9$. Only the low probability of the crash itself prevents a high probability of loss of life.

(6) Items are taken from a sample of 30 items at random for destructive testing. The sample has, on average, 6 bad items among 30. What is the probability that the first bad item will be found on the third test? Solution: Picking a good item on the first try

is denoted with G_1, and picking a bad one with B_1. The probability of picking a good item is then $P(G_1) = 1 - P(B_1) = 1 - 6/30 = 0.8$. The probability of picking a good item on the second try, given the first item was good and is not replaced in the sample, is $P(G_2|G_1) = 1 - 6/29 = 0.793$. The probability of a good item on first and second try without replacement is

$$P(G_1 \cap G_2) = P(G_1)\, P(G_2|G_1) = 0.8 \cdot 0.793 = 0.634.$$

The probability of picking a bad item, B_3, on the third try, given items 1 and 2 were good and have been removed from the sample, is $P(B_3|G_1 \cap G_2) = 6/28 = 0.214$. The probability of picking a bad item on the third try is then

$$P(G_1 \cap G_2 \cap B_3) = P(G_1)\, P(G_2|G_1)\, P(B_3|G_1 \cap G_2)$$

$$= 0.8 \cdot 0.793 \cdot 0.214 = 0.136.$$

This probability is smaller than the probability of picking a bad item on the first try, $P(B_1) = 6/30 = 0.2$. If G_2 and G_1 are independent of each other, then the conditioning disappears, and $P(G_2|G_1) = P(G_2)$, $P(G_2 \cap G_1) = P(G_2) \cdot P(G_1)$, and $P(B_3|G_1 \cap G_2) = P(B_3)$. The probability of failure on the third try for independent elements is

$$P(G_1 \cap G_2 \cap B_3) = P(G_1)\, P(G_2)\, P(B_3) = 0.8 \cdot 0.8 \cdot 0.2 = 0.128.$$

2.1.5 Bayes' Formula

When events are related to each other, the probability of one event can be significantly sharpened if information on the probability of the other event is obtained. If event A is conditioned on event B, then its conditional probability is obtained from Equation (2.1-18) as

$$P(A|B) = \frac{P(B|A)}{P(B)}\, P(A). \tag{2.1-25}$$

Equation (2.1-25) can be interpreted as a revision of probability $P(A)$, called the *prior probability* of A to a probability $P(A|B)$, called *posterior probability* of A, with the help of a *modifying factor*,

$$f = \frac{P(B|A)}{P(B)} \qquad\qquad\qquad (2.1\text{-}26)$$

where $P(B)$ is the probability of an additional event that is related to A and that can be conditioned on A by $P(B|A)$. The posterior probability of A is a probability conditioned on B.

The updating of probabilities by additional information is the essence of *Bayes' formula*, which is a generalization of Equation (2.1-25). For its derivation we proceed by considering a sample space that is partitioned into sets, $\{A_1, A_2, \ldots, A_n\}$. These sets are disjoint and exhaustive, which means they do not intersect and they fill the sample space completely. Such a sample space is shown in Figure 2-6. The sets have probabilities $P(A_1)$, $P(A_2)$, and so on, which add to 1. The additional event, E, is overlaid on the partitions which means that E occurs jointly with some or all of the events A_i. The resulting intersecting sets $E \cap A_i$ are partitions of the event space E, as illustrated in Figure 2-6. This can be expressed by E being a union of n mutually exclusive sets:

$$E = (E \cap A_1) \cup (E \cap A_2) \cup \ldots \cup (E \cap A_n). \qquad\qquad (2.1\text{-}27)$$

The total probability of E is the sum of the probabilities of all these joint events E and A_i :

$$P(E) = P(E \cap A_1) + P(E \cap A_2) + \ldots + P(E \cap A_n). \qquad\qquad (2.1\text{-}28)$$

By using Equation (2.1-17) one can write Equation (2.1-28) in terms of conditional probabilities as

$$P(E) = P(A_1)\,P(E|A_1) + P(A_2)\,P(E|A_2) + \ldots + P(A_n)\,P(E|A_n), \qquad (2.1\text{-}28a)$$

and in abbreviated form,

$$P(E) = \sum_{i=1}^{n} P(A_i)P(E|A_i). \qquad\qquad\qquad (2.1\text{-}29)$$

Equation (2.1-29) represents the total probability of E. If there are multiple additional events, E_k, then the total probability of each E_k becomes

$$P(E_k) = \sum_{i=1}^{n} [P(A_i)P(E_k|A_i)], \ k = 1, 2, 3 \dots K \tag{2.1-29a}$$

where n is the number of partitions of the sample space and K is the number of the additional events. Substituting Equation (2.1-29a) for $P(B)$ in Equation (2.1-25) and making other notational changes leads to

$$P(A_i|E_k) = \frac{P(E_k|A_i)}{P(E_k)} P(A_i), \tag{2.1-30}$$

where $i = 1, \dots, n$, denotes the partitions of the sample space, A_i, and $k = 1, \dots, K$, denotes an additional event E_k on which the A_i are conditioned. The posterior probabilities, $P(A_i|E_k)$, are different for each event E_k, but add to 1 for each event E_k, as do the prior probabilities, $P(A_i)$. Equation (2.1-30) is known as *Bayes' formula*.

The overlaying of event E on the partitions of a sample space A that has A_i partitions, or mutually exclusive sets, is one explanation of Bayes' formula, but there is yet another method that helps its understanding. The A_i's can be construed as root branches of a *probability tree* emanating from a common origin, as illustrated by Figure 2-7a. As exhaustive sets of the sample space the A_i represent all possible branches issuing from the origin. In practical problems, the null-event, which represents the event of everything staying as it is, is a part of a partitioned sample space, and also may have the highest probability. The next step is to append the $P(E|A_i)$, the probability of event E happening, given root branch A_i is taken, to the root branches. Hence, the product $P(A_i) P(E|A_i)$ is the probability that E occurs via root branch A_i and, thus, represents a path through the probability tree. Equation (2.1-29) represents the total probability of E happening via all of the root branches and their follow-ups, $P(A_i) P(E|A_i)$; in other words, each pathway i through the probability tree leading to E has a probability $P(A_i) P(E|A_i)$ and one will be realized. The sum of the $P(A_i)$ over all i must be 1, as they are mutually exclusive sets that exhaustively fill the sample space. If there are additional events, E_k, then the probability tree is expanded by additional pathways leading to them. For example, if E has a complementary E', then the conditional probability $P(E|A_i)$ has a complement $P(E'|A_i)$ and both add to 1. Sometimes the complementary branches attached to each root branch are not part of an application but the user must not lose sight of their potential existence. Figure 2-7a shows a complete probability tree for a sample space partitioned in two sets A and A', and two additional events E and its complement E'. The total probability of E is

$$P(E) = P(A)P(E|A) + P(A')P(E|A'), \tag{2.1-31}$$

and similarly, for the complementary event E' or another event E_k,

$$P(E') = P(A)P(E'|A) + P(A')P(E'|A') \quad .$$

(2.1-31a)

Equation (2.1-31) is called the *rule of elimination* because it eliminates the conditioning from E. Its general form is given by Equation (2.1-29).

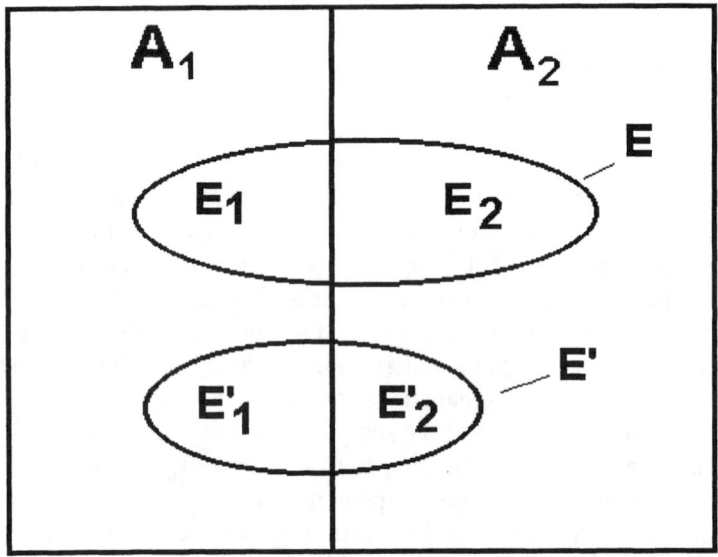

Figure 2-6: The sample space, S, is exhaustively partitioned into mutually exclusive sets, here A_1 and A_2. The additional events (observations) E and E' provide information on A_1 and A_2 which is represented by their intersections with A_1 and A_2 creating the joint sets $E_1 = E \cap A_1$, $E_2 = E \cap A_2$, and so on. Additional information in the form of the probabilities $P(E \cap A_1)$, $P(E \cap A_2)$, $P(E' \cap A_1)$, and so on, is used to calculate $P(E)$ and $P(E')$, and to revise the prior probabilities, $P(A_i)$, into the posterior probabilities $P(A_i|E)$ and $P(A_i|E')$ using the Bayes' formula.

The generalized expression for the modifying factor becomes

$$f_{ki} = \frac{P(E_k|A_i)}{P(E_k)} \, ,$$

(2.1-32)

where $f_{k\,i}$ is the modifying factor for additional events E_k and prior probability A_i; $P(E_k|A_i)$ denotes the probability of E_k occurring, given A_i, which can be large or small; $f_{k\,i}$ can be less than or greater than 1. The posterior probability becomes

$$P(A_i|E_k) = f_{k\,i}\;P(A_i)\,, \tag{2.1-33}$$

where i denotes the i-th partition of the sample space in Figure 2-6, or the i-th root branch in Figure 2-7a. If $f_{k\,i} > 1$, the prior probability $P(A_i)$ is revised upward, otherwise it is revised downward. Hence, the posterior probability $P(A_i|E_k)$ can become larger or smaller that the prior probability.

A major difficulty with applying Bayes' formula to practical problems is to construe a practical problem in terms of the algorithm. The following template can be used as a guide for the steps to take when applying the procedure:

(1) Identify the partitions of the sample space (root branches in the probability tree); here two partitions (root branches) are assumed: A_1 and A_2.

(2) Identify the (*prior*) probabilities of the partitions (the probabilities of the root branches), which must add to 1: $P(A_1) + P(A_2) = 1$.

(3) Identify the additional (intersecting) events E_k, in many cases it is just E and its complement E'. These events must be probabilistically related to the events represented by the partitions. This property is graphically expressed by the intersections, or the conditioning of E_k on A_i; their probabilities $P(E \cap A_i)$, $P(E' \cap A_i)$ etc., can be construed as forming follow-up branches on the root branches leading to E and/or to an alternative or complementary event, E'.

(4) Find the conditional probabilities of the intersections of all events E with the partitions of A (probabilities on the follow-up branches):

For A_1: $P(E|A_1)$ and $P(E'|A_1)$; for A_2: $P(E|A_2)$ and $P(E'|A_2)$, and so on.

Since $P(E|A_1)$ and $P(E'|A_1)$ are the probabilities of follow-up branches that split from the same root branch A_1, their sum must be 1. The same is true for root branch A_2:

$$P(E|A_1) + P(E'|A_1) = 1\,; \text{ and } P(E|A_2) + P(E'|A_2) = 1$$

Sometimes, only event E is used; then only $P(E|A_1)$ and $P(E|A_2)$ are used, and the other branches, for example, those to E', are ignored.

(5) Calculate the joint probabilities of each partition, A_i, and events E_k:

$$P(E \cap A_1) = P(E|A_1)P(A_1)$$
$$P(E' \cap A_1) = P(E'|A_1)P(A_1)$$
$$P(E \cap A_2) = P(E|A_2)P(A_2)$$
$$P(E' \cap A_2) = P(E'|A_2)P(A_2)$$

(6) Calculate the total probabilities of E and E':

$$P(E) = P(A_1 \cap E) + P(A_2 \cap E) = P(A_1)P(E|A_1) + P(A_2)P(E|A_2)$$

$$P(E') = P(A_1 \cap E') + P(A_2 \cap E') = P(A_1)P(E'|A_1) + P(A_2)P(E'|A_2)$$

(7) Calculate the revised (*posterior*) probabilities of the partitions, A_1 and A_2, conditioned on the events E and E':

$$P(A_1|E) = \frac{P(A_1 \cap E)}{P(E)} = \frac{P(A_1)P(E|A_1)}{P(E)} \qquad (2.1\text{-}34)$$

$$P(A_2|E) = \frac{P(A_2 \cap E)}{P(E)} = \frac{P(A_2)P(E|A_2)}{P(E)}$$

$$P(A_1|E') = \frac{P(A_1 \cap E')}{P(E')} = \frac{P(A_1)P(E'|A_1)}{P(E')}$$

$$P(A_2|E') = \frac{P(A_2 \cap E')}{P(E')} = \frac{P(A_2)P(E'|A_2)}{P(E')}$$

(8) The posterior probabilities for event E_k again must add to 1:

$$P(A_1|E) + P(A_2|E) = 1$$
$$P(A_1|E') + P(A_2|E') = 1$$

The posterior probabilities depend on the event on which they are conditioned:

$$P(A_1|E) \neq P(A_1|E')$$

$$P(A_2|E) \neq P(A_2|E')$$

A difficulty with applying the Bayes' formula is the identification of one or several events E that interact with the existing sets A_i and produce probabilities of E conditioned on A_i. The two graphics of the method, Figure 2-6 and Figure 2-7a, combined with some experience, can help overcome this problem. Also, one must not confuse $P(E|A_i)$ and $P(E \cap A_i)$. $P(E|A_i)$ is the probability of a successive branch in the probability tree whereas $P(E \cap A_i)$ is the joint probability of both branches, or the probability of a path through the probability tree from root branch A_i to E. Since $P(E \cap A_i)$ is the product of probabilities it is smaller than $P(E|A_i)$ Once $P(E)$ is known, tracing back the path from E to A_i gives the posterior probability $P(A_i|E) = P(E \cap A_i)/P(E)$.

Examples: (1) For a numerical check of the conditional probabilities used in connection with Bayes' formula assume a sample space with 200 elements; 50 belong to set A, 100 belong to set B, and 20 to the intersection of sets A and B. Verify Equation (2.1-25). Solution: $P(A) = 50/200 = 0.25$; $P(B) = 100/200 = 0.5$; and $P(A \cap B) = P(B \cap A) = 20/200 = 0.1$. Then, $P(A|B) = P(B|A) P(A)/P(B) = P(A \cap B)/P(B) = 0.1/0.5 = 0.2$. Also, $P(B|A) = P(A|B) P(B)/P(A) = P(B \cap A)/P(A) = 0.1/0.25 = 0.4$.

(2) The probability $P(A)$ can be construed as a first guess of a revised probability $P(A|B)$, the latter being obtained by using a modifying factor based on additional information provided by B. Use example (1) to determine this modifying factor and find the revised probability. Solution: According to Equation (2.1-26) and the numbers of example (1), the modifying factor is $P(B|A)/P(B) = 0.4/0.5 = 0.8 < 1$. This means that $P(A)$ is revised downward from $P(A) = 0.25$ to $P(A|B) = 0.2$.

(3) Assume a sample space A has 100 elements and is partitioned into four sets or elements (events): $A_1 = 20$, $A_2 = 10$, $A_3 = 25$, and $A_4 = 45$ elements. An event E occurs on this sample space. It has 5 elements in common with A_1, 6 with A_2, 9 with A_3, and 30 with A_4. This can be construed as if a test is being made to check the distribution of the elements of the sample space over the sets A_i. Calculate the prior probabilities, the posterior probabilities, and the modifying factors that revise prior probabilities into posterior probabilities. Solution: The prior probabilities are the probabilities of the original sets: $P(A_1) = 20/100 = 0.2$, $P(A_2) = 0.1$, $P(A_3) = 0.25$, $P(A_4) = 0.45$. The intersections of E with the sets are $(E \cap A_1) = 5$, $(E \cap A_2) = 6$, $(E \cap A_3) = 9$, and $(E \cap A_4) = 30$ elements. The conditional probabilities of E, given A_i, are $P(E|A_1) = 5/20 = 0.25$, $P(E|A_2) = 6/10 = 0.6$, $P(E|A_3) = 9/25 = 0.36$,

$P(E|A_4) = 30/45 = 0.67$. The total probability of E is $P(E) = 0.2 \cdot 0.25 + 0.1 \cdot 0.6$ + $0.25 \cdot 0.36$ + $0.45 \cdot 0.67 = 0.50$. $P(E) \le 1$ is due to the fact that not all elements of each set A_i also belong to E. Bayes' formula with emphasis on the modifying factors is now used to calculate the revised probabilities for the sets A_i:

$$i = 1 : P(A_1|E) = \frac{P(E|A_1)}{P(E)} P(A_1) = \frac{0.25}{0.50} \cdot 0.20 = 0.50 \cdot 0.20 = 0.10$$

$$i = 2 : P(A_2|E) = \frac{P(E|A_2)}{P(E)} P(A_2) = \frac{0.60}{0.50} \cdot 0.10 = 1.20 \cdot 0.10 = 0.12$$

$$i = 3 : P(A_3|E) = \frac{P(E|A_3)}{P(E)} P(A_3) = \frac{0.36}{0.50} \cdot 0.25 = 0.72 \cdot 0.25 = 0.18$$

$$i = 4 : P(A_4|E) = \frac{P(E|A_4)}{P(E)} P(A_4) = \frac{0.67}{0.50} \cdot 0.45 = 1.34 \cdot 0.45 = 0.60$$

The three number columns to the right of the formulas summarize the results: the column written in the form of quotients gives the modifying factors for set probabilities i and event E. They are either greater than or less than 1. If $f_i < 1$ (index k omitted as there is only one event), the prior probabilities, $P(A_i)$, are revised downward; if $f_i > 1$, they are revised upward. The prior probabilities, $P(A_i)$ as well as the posterior probabilities for E, $P(A_i|E)$ add to 1, as required. The posterior probabilities become the revised probabilities of the sample space partitions. If another event E is observed, the posterior probabilities can be used as the prior probabilities, and a new update is made.

(4) In a maintenance center, server lines L_1, L_2, and L_3 account for 45 %, 30 %, and 25 %, respectively, of the center's throughput. The probabilities of a defective repair job coming from line L_1, L_2, and L_3, respectively, are 0.004, 0.006, and 0.01. What is the probability of a defective repair job coming out of the center, and what is the probability that it comes from L_1? (adapted from an example by Freund, 1993, p. 163).Solution: The sample space consists of partitions represented by the server lines. The percentages of the work performed by the server lines are the probabilities of the partitions and add to 1. The probability of a job being done by line L_1 is $P(L_1)$ = 0.45, and the probability of a defect, given the job was done by line L_1 is $P(D|L_1)$ = 0.004. Similarly, the probabilities can be written for lines L_2 and L_3. Then the total probability of a defect leaving the center is

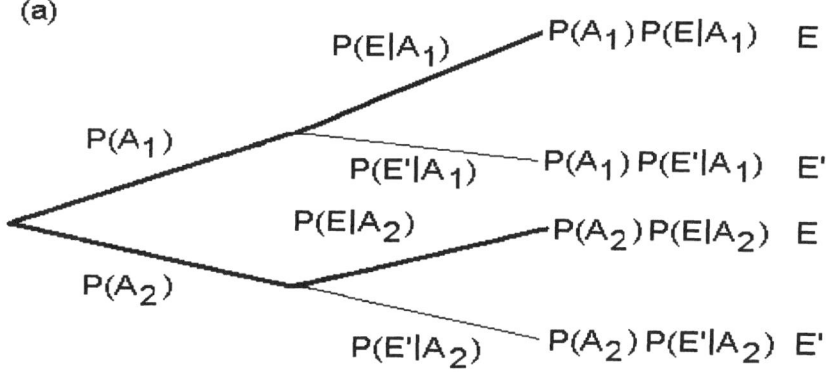

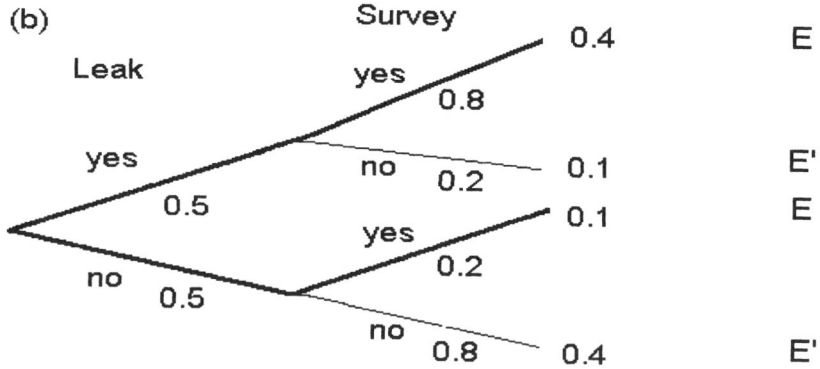

Figure 2-7: Probability tree: (a) general; (b) leak prediction example. The prior probabilities of "leak" or "no leak" are the root branches of the tree. The conditional probabilities on the heavy branches connect to the "survey-says-yes" event, E; the thin branches connect to the "survey-says-no" event, E'. Here, $P(E)$ = $P(E')$ = 0.5. The path probabilities divided by these event probabilities give the posterior probabilities which are, however, different for E and E': $P(A_1|E)$ = 0.8, $P(A_2|E)$ = 0.2, and $P(A_1|E')$ = 0.2, $P(A_2|E')$ = 0.8; each pair adds to 1. The posterior probabilities are not shown in the figure; a reversed tree with root branch probabilities $P(E)$ and $P(E')$, respectively, can be drawn each branching into follow-up branches which have the posterior probabilities.

$$P(D) = P(L_1)P(D|L_1) + P(L_2)P(D|L_2) + P(L_3)P(D|L_3)$$

$$= 0.45 \cdot 0.004 + 0.30 \cdot 0.006 + 0.25 \cdot 0.01 = 0.0061.$$

The probability of a defective repair job coming out of the center is 0.6 %, and the probability of a defect coming out of the center that originated from line L_1 is

$$P(L_1|D) = \frac{P(D|L_1)}{P(D)} P(L_1) = \frac{0.004}{0.0061} 0.45 = 0.66 \cdot 0.45 = 0.295.$$

Similarly, the probabilities of defects coming from the other lines are

$P(L_2|D) = 0.006 \cdot 0.3/0.0061 = 0.295$, and

$P(L_3|D) = 0.01 \cdot 0.25/0.0061 = 0.410.$

The three probabilities, $P(L_i|D)$, $i = 1, 2, 3$, are the updated probabilities of the partitions of the sample space and must add to 1. This example can be interpreted as the sampling of markers from job streams (instead of defects) that serve to update the percentage of work performed by each line.

(5) A dam has a 50 % probability of developing a major leak. The dam owner orders a geologic survey by drilling and permeability tests, but he knows that the results may not be entirely conclusive. If there should be a major leak, the survey will predict it 80 % of the time. If there is no major leak, the survey also will predict this result 80 % of the time. The questions are: (a) Given the survey predicts a major leak, what is the probability that this is true? In this case, a multimillion dollar curtain wall construction program would have to be launched. (b) If the survey predicts no major leak, what is the probability that a major leak will occur after all? (c) What is the probability of no leak, given a leak has been predicted? (This example is adapted from an oil drilling example by Lapin, 1978, p. 144.) Solution: The sample space consists of two partitions *leak* (L) and *no leak* (L'). The (prior) probabilities of the partitions are $P(L) = 0.5$, $P(L') = 0.5$. The event is the geologic survey, T if it predicts a leak, and T' if it does not predict a leak. The probability of the survey predicting a leak, given there is one, is $P(T|L) = 0.8$, and the probability of not predicting a leak, given there is no leak is $P(T'|L') = 0.8$, with the prime designating a complementary event. The complementary probability of predicting a leak, given there is none, is $P(T|L') = 0.2$. The probability tree of this problem is illustrated in Figure 2-7b. The total probability of a correct prediction of a leak is

$$P(T) = P(L)P(T|L) + P(L')P(T|L') \quad = 0.5 \cdot 0.8 + 0.5 \cdot 0.2 = 0.5.$$

The probability of a wrong prediction becomes

$$P(T') = P(L)P(T'|L) + P(L')P(T'|L') \quad = 0.5 \cdot 0.2 + 0.5 \cdot 0.8 = 0.5.$$

(a) the probability of a major leak, given a major leak has been predicted:

$$P(L|T) = \frac{P(T|L)}{P(T)} P(L) = \frac{0.8}{0.5} 0.5 = 0.8 .$$

(b) the probability of a major leak, given no major leak has been predicted:

$$P(L|T') = \frac{P(T'|L)}{P(T')} P(L) = \frac{0.2}{0.5} 0.5 = 0.2 .$$

(c) the probability of no leak, given a major leak has been predicted:

$$P(L'|T) = \frac{P(T|L')}{P(T)} P(L') = \frac{0.2}{0.5} 0.5 = 0.2 .$$

The revised probabilities of the partitioned sample space, L and L', are 0.8 for a major leak, up from 0.5, and 0.2 for the no leak, down from 0.5, given a positive leak prediction in both cases. The geologic survey has improved the odds from 50/50 to 80/20 for investing in a cutoff wall. The decision maker has to decide if the uncertainty is acceptable or if more tests are required.

(6) The planning for a major maintenance/rehabilitation job has led to estimates of repair times that can be grouped in one of the following classes: short duration, T_1 = 15 months; medium duration T_2 = 25 months; and long duration T_3 = 35 months. The planning study also assigns probabilities to these classes: $P(T_1)$ = 0.3, $P(T_2)$ = 0.6, and $P(T_3)$ = 0.1. These classes can be construed as a histogram of probabilistic repair times, similar to Figure 2-4. Planning a maintenance shutdown can be like planning to open the proverbial can of worms. The project manager has ordered the additional examination of scenarios anticipating what may happen when the job gets underway. Any number of such scenarios could be examined but the decision is made to examine three. Each of the three scenarios, Z_1, Z_2, and Z_3, favors one of the three time estimates, T_1, T_2, and T_3, but cannot exactly determine what the repair time will be. All that can be done is to associate the repair times and the scenarios by a

probabilistic relationship. The study produces probabilities of a scenario Z_k coinciding with a repair duration T_i, $P(Z_k|T_i)$, as follows:

Scenarios Z_k Favors Time Estimate T_i		Repair Time Estimates		
		Short Duration 15 Months	Medium Duration 25 Months	Long Duration 35 Months
		$P(T_1) = 0.3$	$P(T_2) = 0.6$	$P(T_3) = 0.1$
		$P(Z_k|T_1)$	$P(Z_k|T_2)$	$P(Z_k|T_3)$
$k = 1$	Z_1 favors T_1	0.7	0.3	0.2
$k = 2$	Z_2 favors T_2	0.3	0.5	0.2
$k = 3$	Z_3 favors T_3	0.0	0.2	0.6
Total		1.0	1.0	1.0

What is the probability of the time T_i, given scenario Z_k prevails. Solution: The sample space is the repair time and these are partitioned into three sets or classes T_1, T_2, and T_3. The probabilities of these partitions add to 1. The possible scenarios Z_k provide new time estimates based on specific assumptions of what may be encountered during the maintenance job. The illustrations in Figures 2-6 and 2-7a can now be used to envision the problem. According to Figure 2-6, the sample space is envisioned as consisting of three partitions, T_i, $i = 1, 2, 3$, over which the three scenarios, Z_k, $k = 1, 2, 3$, stretch from left to right with the exception of Z_3 which does not intersect with T_1. According to Figure 2-7a, each of the root branches, T_i, splits into secondary branches, $P(Z_k|T_i)$, two for T_1 and three each for T_2 and T_3 which creates three paths each to Z_1 and Z_2, and two paths to Z_3. With these visualizations the remaining evaluation of the posterior probabilities should be straightforward.

First, the total probability of scenario Z_k is calculated by using Equation (2.1-29a), where $i = 1, 2, 3$ is the number of partitions, and $k = 1, 2, 3$ is the number of scenarios (additional events). The total probability of scenario Z_1 is (taking row probabilities from the table)

$$P(Z_1) = P(T_1)P(Z_1|T_1) + P(T_2)P(Z_1|T_2) + P(T_3)P(Z_1|T_3)$$

$$= 0.3 \cdot 0.7 + 0.6 \cdot 0.3 + 0.1 \cdot 0.2 = 0.41.$$

The revised probabilities of the partitions T_1, T_2, or T_3, given scenario Z_1, are obtained from Equation (2.1-30),

$$P(T_i|Z_k) = P(T_i)\frac{P(Z_k|T_i) \cdot}{P(Z_k)}$$

For $i = 1, 2, 3$, and $k = 1$,

$P(T_1|Z_1) = 0.3 \cdot 0.7/0.41 = 0.512$
$P(T_2|Z_1) = 0.6 \cdot 0.3/0.41 = 0.439$
$P(T_3|Z_1) = 0.1 \cdot 0.2/0.41 = 0.049.$

For scenario Z_2, one obtains $P(Z_2) = 0.3 \cdot 0.3 + 0.6 \cdot 0.5 + 0.1 \cdot 0.2 = 0.41$, and for $i = 1, 2, 3$, and $k = 2$,

$P(T_1|Z_2) = 0.3 \cdot 0.3/0.41 = 0.219$
$P(T_2|Z_2) = 0.6 \cdot 0.5/0.41 = 0.732$
$P(T_3|Z_2) = 0.1 \cdot 0.2/0.41 = 0.049.$

For scenario Z_3 one obtains $P(Z_3) = 0.3 \cdot 0 + 0.6 \cdot 0.2 + 0.1 \cdot 0.6 = 0.18$, and for $i = 1, 2, 3$, and $k = 3$,

$P(T_1|Z_3) = 0.3 \cdot 0/0.18 \quad = 0$
$P(T_2|Z_3) = 0.6 \cdot 0.2/0.18 = 0.667$
$P(T_3|Z_3) = 0.1 \cdot 0.6/0.18 = 0.333.$

The posterior repair time probabilities are summarized as follows:

Scenario Z_k Favors Time Estimate T_i		Revised Probabilities of Repair Time Estimates			Total
		Short Duration 15 Months	Medium Duration 25 Months	Long Duration 35 Months	
		$P(T_1) = 0.3$	$P(T_2) = 0.6$	$P(T_3) = 0.1$	
		$P(T_1|Z_k)$	$P(T_2|Z_k)$	$P(T_3|Z_k)$	
$k = 1$	Z_1	0.512	0.439	0.049	1.0
$k = 2$	Z_2	0.219	0.732	0.049	1.0
$k = 3$	Z_3	0	0.667	0.333	1.0

The columns of the table allow a comparison of the prior and posterior repair time probabilities, the latter being conditioned on the scenarios encountered. The posterior repair time probabilities must add to 1. These revised repair time probabilities can become prior probabilities of another revision, if additional studies are performed that yield new information in the form of new $P(Z_k|T_i)$.

(7) A sample space of 52 dams that failed is partitioned into four dam types: concrete (C), earth (E), rockfill with core (RC), and rockfill with facing (RF), with 5, 25, 10, and 12 elements, respectively. The prior probabilities of failure for each set are $P(C) = 5/52 = 0.096$, $P(E) = 0.481$, $P(RC) = 0.192$, and $P(RF) = 0.231$. An additional investigation has shown that 11 failures were attributed to the formation of fissures: 6 in concrete dams, 2 in rockfills with core, and 3 in rockfills with protective faces. What are the failure probabilities of the four dam types due to fissures? <u>Solution</u>: The conditional event probabilities for fissures occurring in the four dam types: $P(F|C) = 6/11 = 0.545$, $P(F|E) = 0$, $P(F|RC) = 2/11 = 0.182$, and $P(F|RF) = 3/11 = 0.273$. The total event probability of fissures occurring in the dams is

$$P(F) = P(C)\,P(F|C) + P(E)\,P(F|E) + P(RC)\,P(F|RC) + P(RF)\,P(F|RF)$$

$$= 0.096 \cdot 0.545 + 0.481 \cdot 0 + 0.192 \cdot 0.182 + 0.231 \cdot 0.273 = 0.1503.$$

The posterior probabilities for the events "failure given fissures" are:

$$P(C|F) = 0.096 \cdot 0.545/0.1503 = 0.096 \cdot 3.633 = 0.348,$$

up from the prior probability $P(C) = 0.096$;

$$P(E|F) = 0.000,$$

down from the prior probability $P(E) = 0.481$;

$$P(RC|F) = 0.192 \cdot 0.182/0.1503 = 0.192 \cdot 1.213 = 0.232,$$

up from the prior probability $P(RC) = 0.192$;

$$P(RF|F) = 0.231 \cdot 0.273/0.1503 = 0.231 \cdot 1.820 = 0.420,$$

up from the prior probability $P(RF) = 0.231$.

The sum of both prior and posterior probabilities is 1, as required. The comparison of the prior and posterior probabilities shows that probability of failure, given fissures, has drastically increased for concrete dams, it has been reduced to zero for earth dams, and it has been substantially increased for both rockfill types. Of course, there are caveats to be observed. The updated probabilities are only as good as the information being used.

2.1.6 Murphy's Law

The popular *Murphy's Law*, states: *if something can go wrong, it will*. It dates from U.S. Air Force studies in the 1950s and claims that if some accident is possible, it will happen. Matthews (1997) investigated some of the incidents cited in connection with Murphy's Law and came to the conclusion that some events, such as the turning of the buttered toast on its buttered side by the time it hits the floor, can be explained by physical laws. The table is just not high enough, or it just lacks moment for a complete turn. Other incidents are pure chance events, such as the damage that the mail delivery system inflicted on the very issue with the article on Murphy's Law, that a subscriber reported to the editor of *Scientific American*, after he had received 362 issues before, all undamaged. Again, other incidents are remembered by people's selective memory of failures in critical moments, whereas many other cases in which nothing unusual occurred are forgotten. However, most incidents that seem to validate Murphy's Law are of a probabilistic nature.

A comment to the editor in response to the article by Matthews (1997) claims that Murphy and his team, who were working on aircraft pilot safety, supersonic jets, and Apollo landing craft, "were not content to rely on probabilities for their successes. Because they knew that things left to chance would definitely fail, they went to painstaking efforts to ensure success." Probabilistic processes tend to frustrate such efforts. However painstaking the efforts may be, they can reduce but not eliminate the probability of failures. In the space program, utmost care is taken to reduce the probability of failure of complex systems and processes to the lowest conceivable level. Men were shipped several times to the moon and back, and space shuttles were launched into orbit and returned many times. Despite all efforts accidents have occurred, because a residual probability of failure could not be eliminated, as has been demonstrated by the Challenger and the Columbia failures.

If some particular incident is *possible*, then it has a probability greater than zero. This means that it may happen, but also that it may not happen within a limited time and space. Whereas the probability for a particular incident to happen may remain the same over time (usually it increases with the age of the system, unless maintenance keeps the system young), every trial affords a new chance for it to happen, so that over an infinite time and space it *will* happen, and Murphy's Law is fulfilled. The Scottish poet Robert Burns is quoted by Matthews (1997) to have said in 1786: *"The best laid schemes o' mice an' men gang aft agley."* This statement that implies a possibility of failure is in line with a more general formulation of Murphy's Law: *"if something can go wrong, it may."* Grant and Ireson (1970, p. 44) make a statement to this effect in connection with project cost estimates: "Even the most careful estimates are likely to go wrong."

2.2 Probabilistic Variables and Functions

2.2.1 Probabilistic Variables

A variable is a quantity that can take on different values in contrast to a constant that has a single value. In mathematical formulas, an independent variable, x, is given a value to calculate the value of the dependent variable, y. Mathematical formulas relate dependent variables to independent variables. Examples are $y = a x$, or $y = a x^2 + b$. Letters usually stand for coefficients or parameters. In a deterministic formula, coefficients are single-valued numbers which are determined by theory or by experiment. In a deterministic calculation, the independent variable x is given by a series of values and a corresponding series of y's is obtained. In a probabilistic problem, x may be the output of a random process, which when fed into the deterministic formula produces a random output, y. In many probabilistic problems, formulas are deterministic functions whereas input and output are probabilistic quantities. A function becomes probabilistic when the output given an input is no more known with certainty. An example is a formula whose coefficients are random variables. An illustration of this problem is the scattering of points (x, y) around a straight line. In such a case, the coefficients a and b in the function $y = a x + b$ are different for each pair (x, y). The use of such functions can become burdensome, as many different values of the independent variable and of the coefficients have to be tested for an evaluation of y. Methods have been developed to deal with such problems and some of them will be discussed in this and the following chapters.

A *probabilistic variable* can take on many values in the possible range in which it exists, in which its values have a probability greater than zero. If some values are taken on more often than others the values in the range have varying frequency. A variable, x, that can take different values in an uncontrolled way within a range is called a *random variable*, X. According to definition, something is *random* if it lacks a definite plan, purpose, or pattern (Webster, 1993). This is true for a *random realization* of X, which is an actual value x that X can assume at a given time and/or location. The collective of x is expressed as $x \in X$, which reads " x in X," where X stands for all possible realizations of x in time and space by random selection from the range. The random realizations x may occur at uniform or varying frequencies, and the range may be continuous or discrete, finite or infinite. If the range of X contains 100 (discrete and countable) realizations x, and 10 of them happen to have the same value, then the absolute frequency of this random variable X_1 is 10, and the relative frequency is $10/100 = 0.1$. The relative frequencies scale the sample space to a maximum size of 1. Figure 2-8a shows two single-valued deterministic variables, A and B; Figure 2-8b shows two multivalued probabilistic variables, A and B, with ranges.

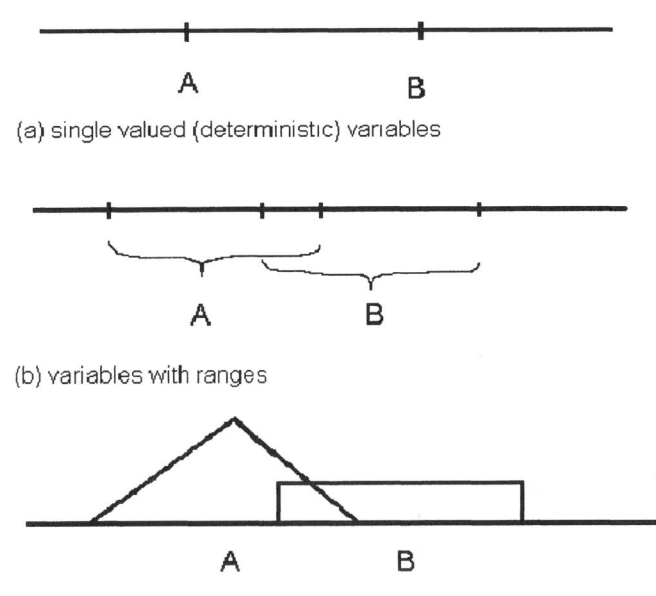

(a) single valued (deterministic) variables

(b) variables with ranges

(c) variables with probability density functions

Figure 2-8: (a) Single-valued deterministic variables, A and B; (b) probabilistic variables A and B with ranges: and (c) probabilistic variables A and B with probability density functions.

Four attributes define a probabilistic variable or random variable, X:

- a population mean of N realizations: μ
- a population standard deviation of N realizations: σ
- a range between a lower and an upper bound of possible realizations x, $x_1 \leq X \leq x_u$.
- a pdf that allocates a frequency to a realization x of X, $f_X(x)$.

The *population mean*, μ, is the mean value of a collective of random realizations. It is the most likely value of all values of the range, assuming it exists. The *population standard deviation*, σ, is a measure of the average departure of realizations from the mean (see Section 2.3). More complex random numbers require additional parameters for their description such as the coefficient of *skewness* (asymmetry) and *kurtosis* (peakedness). Often the collective or entire population is not known or cannot be measured. But if some number, n, of realizations of X is known, say from conducting tests or taking samples, a *sample mean*, m, and a *sample standard deviation*, s, can be calculated. These are estimates of the population mean and the population standard deviation, respectively. Statistical procedures for calculating how closely these sample parameters approach the population parameters are discussed in Section 2.4.5.

The probability of a random variable assuming some discrete value in its range can be stated as follows:

> $P(X = x) = a$: the probability of X being equal to a realization x is equal to a, $0 \le a \le 1$.
>
> $P(X \le x) = b$: the probability of X being less than or equal to a realization x is b, $0 \le b \le 1$, also known as *non-exceedance probability*.
>
> $P(X > x) = c$: the probability of X being greater than a realization x is c, $0 \le c \le 1$, also known as *exceedance probability*.
>
> $P(x_1 < X \le x_2) = d$: the probability of X being in the range $x_1 < x \le x_2$ is d, with $0 \le d \le 1$.

Examples: (1) State the probability of tensile strength 2.5 MPa of the concrete samples in Figure 2-4. Solution: $P(X = 2.5) = 0.3$.

(2) State the non-exceedance probability of 2.5 MPa for the concrete samples in Figure 2-4. Solution: Reading the probabilities of the bars from the graph and adding them (mutually exclusive events) produces $P(X \le 2.5) = 0.02 + 0.10 + 0.11 + 0.30 = 0.53$. The bars of a histogram can be construed as mutually exclusive and exhaustive sets of the total sample space so that the sum of their probabilities does not exceed 1.

(3) State the exceedance probability of 2.5 MPa for the concrete samples in Figure 2-4. Solution: Reading the probabilities of the bars from the graph from 2.6 MPa to 3.1 MPa and adding them (mutually exclusive events) produces $P(X > 2.5) = 0.15 + 0.18 + 0.08 + 0.02 + 0.03 + 0.01 = 0.47$.

(4) State the probability of concrete sample strength ranging from 2.4 MPa to 2.8 MPa in Figure 2-4. Solution: Reading the probabilities of the bars from the graph

from 2.4 MPa to 2.8 MPa, including these boundaries, gives $P(2.4 \leq X \leq 2.8) = 0.11 + 0.30 + 0.15 + 0.18 + 0.08 = 0.82$.

2.2.2 Probabilistic Functions

Each realization of a random variable has a frequency. When plotted over the range of the random variable, the (relative) frequency may display any kind of shape. The function describing this shape is the *probability density function* (pdf). Two examples of such pdf's, a triangular and a uniform pdf, are shown in Figure 2-8c. Some of these functions occur more often than others in the analysis of natural and technical processes. Some of the best-known or easiest to use functions have bell-shape, rectangular, triangular, or exponential shape (see Section 2.7). Functions that describe probabilities are referred to as *probability functions*.

To distinguish a pdf from an algebraic function, $y = f(x)$, the pdf is marked by an index that denotes the random variable to which it refers. For example, the frequency distribution over the range of random variable, X, is denoted as $f_X(x)$, and the frequency distribution over the range of random variable Y is $f_Y(y)$. The capital letters X and Y denote the random variables of which x and y, respectively, are realizations. If the meaning is clear from the context these subscripts are sometimes dropped to simplify the notation. It is important, however, to distinguish a common function from a pdf. The pdf $f_X(x)$ does not represent some other variable y like a common algebraic function but the frequency of realization x.

The pdf may be a discrete or continuous function. If the pdf is discrete, it exists at specific values of x only and not in between. The sum of all discrete relative frequencies over the range of X must add to 1. If the pdf is continuous, its ordinates must be scaled in such a way that the integral over the range of X is 1 (see Section 2.6).

For a discrete pdf, the probability of a realization is

$$P(x) = f_X(x), \tag{2.2-1}$$

and for a continuous pdf,

$$P(x) = f_X(x) \, dx. \tag{2.2-1a}$$

These functions are referred to as *probability functions* and will be discussed in some detail in Section 2.5.

The common function $y = f(x)$ represents a variable y that depends on x by some functional relationship. As x takes on a specific value, y takes on one or more values $f(x)$ prescribed by the function. The relation may be as simple as $y = k\,x$, with k being a constant coefficient. As long as k is a constant, the functional relation is fully determined. Then, $y = f(x)$ is a *deterministic function*, and x and y are *deterministically* related even if they are realizations of random variables. Such a function may serve as a transformation function of one random variable into another. An example is the transformation of random variable x into its logarithms, $y = log(x)$, a sort of scaling that makes some kinds of data easier to analyze. Another example is a transformation prescribed by a physical law, for example, the transformation of a random earthquake acceleration into a random force acting on a structure. If k becomes a random variable, then the function becomes a *probabilistic function*. The general relation between probabilistic variables X and Y is denoted as

$$Y = f(X). \tag{2.2-2}$$

This means that a random realization x of random variable X produces a random realization y of a random variable Y by a deterministic or probabilistic relation.

Sometimes the functions $f_X(x)$ and $f(x)$ can look deceivingly similar. For example, the seasonal distribution of the number of repair jobs over the year may have a bell shape represented by $N_t = f(t)$, where N is a monthly number of repair jobs, or the percent of total annual repair jobs, and t is the time of the year at which N_t occurs. In this case, t is not a random realization but a chronological time progression, and the function represents a distribution but not a pdf. On the other hand, the distribution of repair jobs within a month can be represented by a pdf, $f_N(n)$, where N is the random number of repair jobs in the month and n is a random realization of N. If N can assume any number from 5 to 20 in a particular month, then $f_N(n)$ represents the frequency of n repair jobs that could occur in that month.

2.2.3 Probabilistic Processes

A *random process* is the archetype of a probabilistic process. It is defined as a collective of elements or events with a definite probability of occurrence. Occurrences may be equally spaced or randomly spaced and may be random by size. A process that appears to be random may not really be a *random process*. It may be a philosophical question if there is any process that is random. Often it is the nature of the process that we do not understand and we therefore refer to it as random. A sequence of events may be the manifestation of complex underlying physical processes. Brownian motion is caused by the impact of a resulting force produced by many molecules on a small particle, for example, a small fat particle in milk, that is being pushed around in a random fashion, because the molecules seem to lack aim or guidance. These displacements fit the definition of a random process. In other cases, a large number of well-understood contributions may mimic a random outcome. In such cases, the many contributions may cause such a complex web of interactions that the process defies conceptual modeling in detail, even if the basic process is understood. Therefore, if a mathematical random process satisfactorily mimics the realizations (outcomes) of such a complex process, it may be used as a model for making predictions about future behavior without a complete understanding of why the process behaves the way it does. In some cases, stochastic processes are combined with long-term trends which may actually be stochastic processes on a different time scale. The hydrograph of a river is an example of a mixed process where a predictable depletion function is perturbed by random additions due to rainfall.

A *stochastic process* is another word for a random process and is defined as a collection of random variables (Parzen, 1965, p. 7). A collection of random variables could be the appearance of some quantity of varying size every time the process is observed. Such a process can be presented by

$$\{X_t, t \in T\} = X_1, X_2, X_3, \ldots, X_t, \ldots \tag{2.2-3}$$

where X_t is a collection of random variables and t is an index of an index set, $t \in T$ (read: t in T). No restriction is placed on the nature of T. It can be a time index or a space index. A stochastic process may be discrete or continuous. Only discrete processes are dealt with here. Suppose $T = (1, 2, 3, \ldots, t, \ldots)$ is an index set for a discrete stochastic process over time. At each point in time, the random variable takes on one of its discrete realizations $X = \{x_1, x_2, \ldots, x_t, \ldots\}$. An example would be monthly repair costs caused by independent repair events. According to Parzen (1965, p. 7, footnote) the expressions 'chance process' or 'random process' are used as synonyms for stochastic process. The word *stochastic* is derived from Greek meaning "aim at" or "guess at." It was already used in seventeenth century England and appeared in connection with probability in the 1920s (Webster, 1993). Presently, *stochastic process* is the most popular name for a probabilistic process.

The collective of states a structure may visit during its lifetime, for example, as *as good as new*, *slightly deteriorated*, *seriously deteriorated*, and *shutdown*, can be construed as the variables of a discrete stochastic process, but of a special kind as we will see. Another example would be a set of annual average flows, or a set of annual peak flows for the period of record. Still other examples are the interarrival times of repair jobs in a service department or the strengths of concrete samples, and so on. The tests of concrete samples are conducted according to a meticulous procedure to minimize errors in preparing the sample, curing it, and measuring its strength. Nevertheless, each sample is usually found to have at least a slightly different strength so that each test outcome is a random realization within some range. The statistical processing of the data then leads to the kind of histogram shown in Figure 2-4. The rather orderly arrangement of the sample data hint at some order in apparent chaos which will be further explored in Chapter 3.

If the probability of the process being at some state level of state X_{n+1} depends only on the most recent state, X_n, then the process is a *Markov process*. A Markov process can be discrete or continuous. Only discrete processes are discussed here. A discrete Markov process in which the state of the process depends only on the most recent state is a *Markov chain*. The state probability of a Markov chain is defined as (Parzen, 1965, p. 188):

$$P(X_{n+1} = x_{n+1}|X_1 = x_1, \ldots, X_n = x_n)$$

$$= P(X_{n+1} = x_{n+1}|X_n = x_n) \tag{2.2-4}$$

This statement says that the probability of being in state X at a state level x, at the moment in time $n+1$, given the process has been at other state level at previous times, 0, 1, 2, ... , n, is only conditioned on the most recent state level, which is x at time n. This means that the process has no memory of state levels visited before the last level. Such a process is shown in Figure 2-9. The transition probabilities from the present state level to other levels are conditional probabilities:

$$p(i,j) = P(X_{n+1} = j|X_n = i), \tag{2.2-5}$$

where $p(i, j)$ is a transition probability for stage n, and i, j denote the discrete levels of the state variable X. For example, in Figure 2-9, $i = 2$, and $j = 1, 2, 3, 4$, each connected to i by a transition probability. Since $n + 1 - n = 1$, $p(i, j)$ is called a one-step transition probability. The transitions out of a state are mutually exclusive and exhaustive so that

$$\sum_{j=1}^{J} p(i,j) = 1 \tag{2.2-6}$$

for all i. In Figure 2-9, the process starting in state level $i = 2$ has to move along one or the other of the pathways, each having a transition probability associated with it whose sum must add to 1. Equation (2.2-6) becomes

$$p_{21} + p_{22} + p_{23} + p_{24} = 1. \tag{2.2-7}$$

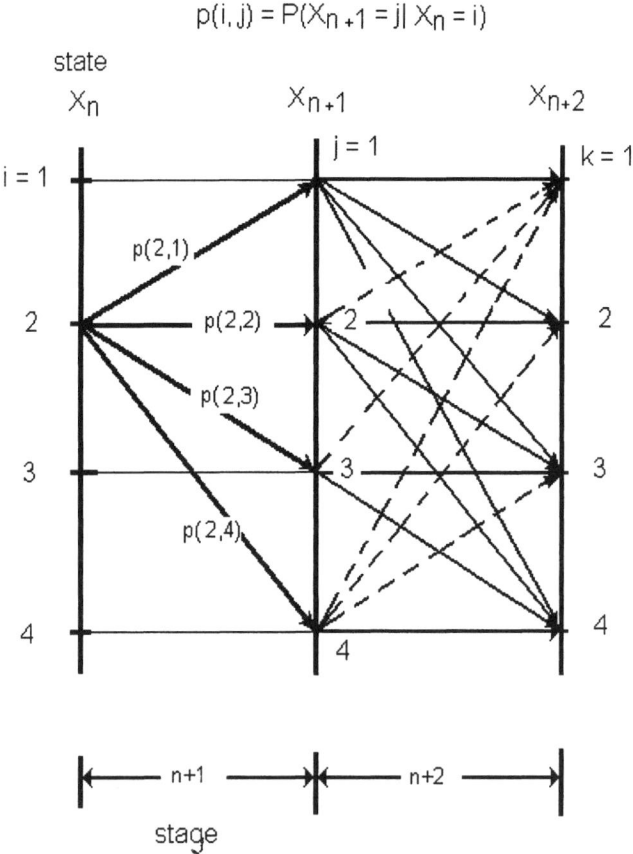

$$p(i, j) = P(X_{n+1} = j | X_n = i)$$

Figure 2-9: A two-stage Markov chain. The state variable
X has four discrete realizations, x_1, x_2, x_3, and x_4. The
transitions from one state realization to another are random
and mutually exclusive. Given the process is in state 2 at
the end of stage n which is also the beginning of stage
$n+1$, designated $X_{n2} = x_2$, transitions can occur only from
this state to one of the state realizations $X_{n+1\,j} = x_j$. If the
states are construed as repair states of a system, only the
as-is state (horizontal pathway) or deteriorated states
(downward pathways) can be accessed, other paths have
transition probabilities zero.

It may be desirable to know the probability with which a multistep Markov chain will end in a state $X_{n+t} = k$ after t transitions. For the process shown in Figure 2-9, given the starting state $X_n = 2$, the probability of ending up in state level $k = 1$ after two transitions is

$$P(X_{n+2} = 1 \mid X_n = 2) = p_{21} p_{11} + p_{22} p_{21} + p_{23} p_{31} + p_{24} p_{41}, \tag{2.2-8}$$

where each probability product represents the probability of going from the beginning state to the ending state by a mutually exclusive path. In each probability product the second number of the index of the first probability is the same as the first number of the index of the second probability, as they connect along the possible path leading from $i = 2$ via $j = 1, 2, 3, 4$ to $k = 1$. For example, $p_{22} p_{21}$ is the joint probability of path 2-2-1. In general notation one can write (Parzen, 1965, p. 194),

$$p_{ik}(s,t) = \sum_{j=1}^{J} p_{ij}(s,u)\, p_{jk}(u,t), \quad \text{for all states } j, \tag{2.2-9}$$

where i and k are the beginning and ending state levels of the chain, respectively, j is an intermediate state level, s marks the beginnig of the first stage, t marks the end of the last stage, and u is the end or beginning of an intermediate stage. Equation (2.2-9) is known as the *Chapman-Kolmogorov equation*. It states that if the entire Markov chain (s, t) is split in two parts at some arbitrary stage u, the two parts (s, u) and (u, t), each evaluated by a two-stage process, can be joined together as if they were just two single stages.

A two-stage Markov chain from an initial state level 2 to an ending level 1 can be written in matrix notation as

$$[p_{21}\ p_{22}\ p_{23}\ p_{24}][p_{11}\ p_{21}\ p_{31}\ p_{41}]^T. \tag{2.2-10}$$

The first set of values in brackets is a row matrix. The second set is a column matrix For convenience of notation, the column matrix is written as the transpose of a row matrix. This is expressed by the superscript T. The row matrix and column matrix multiplication as encountered in Equation (2.2-8) is also known as the dot product. Row matrices and column matrices are also referred to as row vectors and column vectors, where a vector is a quantity that is defined by more than one number in contrast to a scalar, which is a single number. In the column vector of Equation (2.2-10), p_{11} is the top element of the column, and p_{41} is the bottom element. After carrying out the dot product of vector elements with each other, the result shown on the right side of Equation (2.2-8) is obtained. It is a single value, here the total probability of going from level 2 to level 1 via all intermediate levels j in a two-stage (or two-step) process.

If the transition probabilities are written for the paths from all beginning state levels to the same ending state level 1, one obtains four products of the kind of Equation (2.2-10),

$$[p_{11} \quad p_{12} \quad p_{13} \quad p_{14}][p_{11} \quad p_{21} \quad p_{31} \quad p_{41}]^T \qquad (2.2\text{-}11)$$

$$[p_{21} \quad p_{22} \quad p_{23} \quad p_{24}][p_{11} \quad p_{21} \quad p_{31} \quad p_{41}]^T$$

$$[p_{31} \quad p_{32} \quad p_{33} \quad p_{34}][p_{11} \quad p_{21} \quad p_{31} \quad p_{41}]^T$$

$$[p_{41} \quad p_{42} \quad p_{43} \quad p_{44}][p_{11} \quad p_{21} \quad p_{31} \quad p_{41}]^T$$

Each line represents the probabilistic path from one of the four starting levels, $i = 1, \ldots, 4$, to the same ending level $k = 1$, which is also the second number of the index of the transpose elements. The block of probabilities on the left side has the form of a matrix, whereas the block on the right is a repetition of the same vector. In matrix form, the transitions can be written as the product of a 4 by 4 matrix and a 4 by 1 matrix,

$$\begin{vmatrix} p_{11} & p_{12} & p_{13} & p_{14} \\ p_{21} & p_{22} & p_{23} & p_{24} \\ p_{31} & p_{32} & p_{33} & p_{34} \\ p_{41} & p_{42} & p_{43} & p_{44} \end{vmatrix} \begin{vmatrix} p_{11} \\ p_{21} \\ p_{31} \\ p_{41} \end{vmatrix} = \begin{vmatrix} p_{11}p_{11} + p_{12}p_{21} + p_{13}p_{31} + p_{14}p_{41} \\ p_{21}p_{11} + p_{22}p_{21} + p_{23}p_{31} + p_{24}p_{41} \\ p_{31}p_{11} + p_{32}p_{21} + p_{33}p_{31} + p_{34}p_{41} \\ p_{41}p_{11} + p_{42}p_{21} + p_{43}p_{31} + p_{44}p_{41} \end{vmatrix}$$

$$(2.2\text{-}12)$$

On the left is a one-step 4×4 transition probability matrix with 4 rows and 4 columns. The next term on the left-hand side is a 4×1 column vector. The probabilities of each matrix row must add to 1, as they represent all possible exits from one starting state, i, indicated by the first row index number. When multiplying a $m \times n$ matrix with an $n \times p$ matrix, one can figure out the dimension of the product by writing it as $(m \times n) \times (n \times p)$. After cancelling the inner two n's, one obtains a matrix $m \times p$. Equation (2.2-12) gives $(4 \times 4) \times (4 \times 1) = 4 \times 1$, a column matrix. Each row of this column matrix is the total probability of getting from one of the initial state levels, $i = 1, \ldots, 4$, via one of the intermediate levels,

$j = 1, \ldots, 4$, to the ending state level $k = 1$. The top element of this column matrix, according to Equation (2.2-9), for $i = 1$ and $k = 1$, $s = 0$, $u = 1$, and $t = 2$ is

$$p_{11}(0, 2) = \sum_{j=1}^{4} p_{1j}(0, 1) p_{j1}(1, 2) = \qquad\qquad (2.2\text{-}9a)$$

$$p_{11}(0, 1) p_{11}(1, 2) + p_{12}(0, 1) p_{21}(1, 2)$$

$$+ p_{13}(0, 1) p_{31}(1, 2) + p_{14}(0, 1) p_{41}(1, 2)$$

where each transition probability is marked by two pairs of numbers: the indices indicate the pathway and the numbers in parentheses indicate the stage spanned. A similar sum is obtained for each of the other elements of the column vector. The meaning of Equation (2.2-12) and Equation (2.2-9a) becomes clear by looking at the pathways from all beginning state levels, i, to a selected ending state level, k, in Figure 2-9. For clarity only the exits from beginning state levels $i = 2$ are shown in Figure 2-9. The pathways first fan out to all possible state levels j of the intermediate state, at the end of the first stage, and then converge toward $k = 1$ at the end of the second stage.

If a column vector is added for each ending state in Equation (2.2-12), a product of two matrices emerges. For example, in order to get to state $k = 2$, the second index of the vector would have to be 2, and the vector to be added on the left hand side of Equation (2.2-12) is

$$[p_{12} \ p_{22} \ p_{32} \ p_{42}]^T,$$

and so on, one for each ending state. The probabilistic transition from all beginning states to all ending states of a two-stage process is thus represented by the product of two one-step transition probability matrices. If $P(1)$ is such a one-step transition probability matrix of the Markov chain, then an n-step chain is represented by the product of the matrices, one for each stage,

$$P(n) = P(0, 1) \ P(1, 2) \ \ldots \ P(n\text{-}1, n) = P(0, n) \qquad\qquad (2.2\text{-}13)$$

where $P(0, 1), \ldots, P(n\text{-}1, n)$ are one-step transition probability matrices from the start of the process to the nth step (stage); the numbers in parentheses refer to the number of steps the matrix represents. If the process consists of one-step, stationary matrices (not changing from state to state), the indexes can be dropped and the n-step matrix becomes

$$P(n) = P^n .$$ (2.2-14)

A Markov chain that can be represented by the product of n identical matrices is *homogeneous*. If the matrix is different from stage to stage, then the product has to be taken matrix by matrix.

If the beginning state levels are subject to chance, then the n-step matrix must be premultiplied by the probability of the initial state levels to obtain the probability vector of the ending state levels:

$$p(n) = p(0)\, P^n$$ (2.2-15)

where $p(n)$ is the probability row vector of the ending state levels, and $p(0)$ is the probability row vector of the starting state levels.

Examples: (1) Calculate the two-step matrix of a two-state, two-step, homogeneous Markov chain. The one-step matrix rows are p_{11}, p_{12}; and p_{21}, p_{22}. Solution: Multiplying the one-step matrix by itself produces the two-step matrix:

$$P(2) = \begin{vmatrix} p_{11} & p_{12} \\ p_{21} & p_{22} \end{vmatrix} \begin{vmatrix} p_{11} & p_{12} \\ p_{21} & p_{22} \end{vmatrix} = \begin{vmatrix} p_{11}p_{11} + p_{12}p_{21} & p_{11}p_{12} + p_{12}p_{22} \\ p_{21}p_{11} + p_{22}p_{21} & p_{21}p_{12} + p_{22}p_{22} \end{vmatrix}$$ (2.2-16)

The probabilities of each row of a transition probability matrix add to 1, as they represent all possible paths out of the state that is indicated by the first number of the double number index and directed to the state that is indicated by the second number. The sum of the products of probabilities that appear in the elements of the two-step matrix are total probabilities for the complete path from the start of the first stage to the end of the second stage. For example, $p_{11}\, p_{11} + p_{12}\, p_{21}$ is the total probability of two mutually exclusive pathways from the beginning state level 1 to the ending state level 1: the first via the intermediate state level 1, and the second via intermediate state level 2. Altogether 4 pathways are represented by the 4 sums of the resulting 2 $\times$ 2 matrix, two from each beginning state to each ending state. They can be traced in Figure 2-10.

(2) Test the previous example with probabilities $p_{11} = 0.9$, $p_{12} = 0.1$, $p_{21} = 0$, and $p_{22} = 1.0$, and interpret the result. Solution: The probability matrix for two transitions, $P(2)$, is

$$P(2) = \begin{vmatrix} 0.9 & 0.1 \\ 0 & 1 \end{vmatrix} \begin{vmatrix} 0.9 & 0.1 \\ 0 & 1 \end{vmatrix} = \begin{vmatrix} 0.81+0 & 0.09+0.1 \\ 0+0 & 0+1 \end{vmatrix} = \begin{vmatrix} 0.81 & 0.19 \\ 0 & 1 \end{vmatrix}$$

The stochastic process represented by the initial matrix, $P(1)$, has a probability 0.9 that it will stay in its starting state and a probability 0.1 that it will go to state level 2. Once it is in this state, it cannot come back out of it. It is trapped in it and stays there with probability 1. A component that goes into the breakdown state and cannot repair itself, is trapped in this state. $P(2)$ indicates that after two stages (time periods), with a probability of 0.81 the process is still in state 1, and with a probability of 0.19 it has left state 1 for state 2. The probabilities of the resulting matrix rows add to 1 as required. Figure 2-10 provides an illustration of this example.

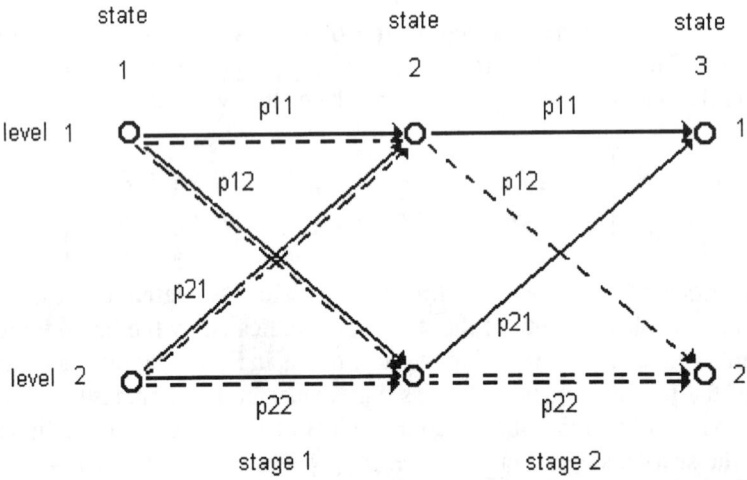

Figure 2-10: Two-step Markov chain with two state levels. Solid paths lead to ending state level 1, and dashed paths lead to ending state level 2.

(3) Given the beginning state levels 1 and 2 in Figure 2-10 have probabilities p_1 and p_2, what are the probabilities of the ending state levels for the two-step process? Use the same numbers as in Example (2). <u>Solution</u>: Begin by multiplying the row vector of the beginning probabilities into the first stage matrix:

$$\begin{vmatrix} p_1 & p_2 \end{vmatrix} \begin{vmatrix} p_{11} & p_{12} \\ p_{21} & p_{22} \end{vmatrix} = \begin{vmatrix} p_1 p_{11} + p_2 p_{21} & p_1 p_{12} + p_2 p_{22} \end{vmatrix} \qquad (2.2\text{-}17)$$

The multiplication of a $1 \times m$ row vector by an $m \times k$ matrix produces a $1 \times k$ row vector, which is the row vector of the probabilities of the ending states after the first transition. The first sum of the new row vector of Equation (2.2-17) is the probability of ending state level 1, and the second term is the probability of ending state level 2, as indicated by the indices. The new row vector also is the new initial probability row vector that is now multiplied by the matrix of the second stage:

$$\left| p_1 p_{11} + p_2 p_{21} \quad p_1 p_{12} + p_2 p_{22} \right| \begin{vmatrix} p_{11} & p_{12} \\ p_{21} & p_{22} \end{vmatrix} \tag{2.2-18}$$

$$= \left| (p_1 p_{11} + p_2 p_{21}) p_{11} + (p_1 p_{12} + p_2 p_{22}) p_{21} \right.$$

$$\left. (p_1 p_{11} + p_2 p_{21}) p_{12} + (p_1 p_{12} + p_2 p_{22}) p_{22} \right|.$$

Equation (2.2-18) produces a 1×2 row vector; the first term represents the total probability of all paths to ending level 1, as indicated by the second index number, from all beginning levels, here 1 and 2; the second term represents all paths to ending level 2 from beginning levels 1 and 2. Now the two components in Equation (2.2-18) are rearranged as follows:

$$\left| p_1 (p_{11}^2 + p_{12} p_{21}) + p_2 (p_{21} p_{11} + p_{22} p_{21}) \right.$$

$$\left. p_1 (p_{11} p_{12} + p_{12} p_{22}) + p_2 (p_{21} p_{12} + p_{22}^2) \right| \tag{2.2-18a}$$

Equation (2.2-18a) still is a row vector, but a comparison of the elements in parentheses with those of $P(2)$ of example (1) shows that it is the product of the row vector of initial probability multiplied by the two-step transition probability matrix, and thus proves that the ending probability vector is $p(2) = p(0) P^2$, as required by Equation (2.2-15).

(4) With the same numbers as in example (2) and the initial probability vector, $p(0)$ = [0.8, 0.2], calculate the ending state probabilities. <u>Solution</u>: According to Equation (2.2-15), using $P(2)$ of example (2), one obtains

$$p(2) = p(0) P^2 = \left| 0.8 \quad 0.2 \right| \begin{vmatrix} 0.81 & 0.19 \\ 0 & 1 \end{vmatrix}$$

$$= \left| 0.8 \cdot 0.81 + 0.2 \cdot 0 \quad 0.8 \cdot 0.19 + 0.2 \cdot 1 \right| = \left| 0.65 \quad 0.35 \right|$$

This means that ending state level 1 is reached after two transitions with probability $p_1(2) = 0.65$, and ending state level 2 is reached with probability $p_2(2) = 0.35$, given the starting state probabilities are $p_1(0) = 0.8$ and $p_2(0) = 0.2$. The beginning and ending probabilities are probabilities of partitions of the state space and thus must add to 1.

2.2.4 Probabilistic Models

A mathematical model consists of one or more mathematical formulas that prescribe a transformation of input into output, called a *system*. A model is used to simulate such a system as it operates on input to produce output, as shown in Figure 2-11. From a cause–effect modeling point of view, there are two basic types of models: *black box models* and *conceptual models*. Black box models are used for both simple as well as complex systems. Both can be modeled by an empirical input-output correlation, such as a regression equation, in cases when the workings of the true process are poorly understood or prohibitively difficult to model. The previously discussed stochastic processes are probabilistic black box models. It would be very difficult to model the underlying process that results in random arrivals of repair jobs of different kinds in a service shop. The number of arrivals and the time between arrivals are both manifestations of a very complex process. Suppose an industrial process produces the same kind of item under stringent quality standards, similar to the concrete samples discussed in connection with Figure 2-4. If samples are subjected to tests, it is very unlikely that they all will be found to have the same strength. It would be very difficult to model the process that leads to these random outcomes.

A conceptual model uses known causal relations including physical laws and empirical formulas that link inputs and outputs of a process and assembles a set of equations that describes the workings of the system. Such a system model usually requires *calibration* by which empirical coefficients are given the most likely value to reproduce the observed data. Once the calibration is satisfactorily completed, *verification* runs are made to simulate observed performance of the process. After the successful conclusion of verification runs, the model is considered to have acquired some degree of reliability. Such a model that is based on deterministic relations is a *deterministic conceptual model*. It can be used on probabilistic input to produce probabilistic output. If the model has internal coefficients that cannot be fixed because of their probabilistic nature, or if the model takes random branches not under the control of the model operator, then the model is a *probabilistic model*.

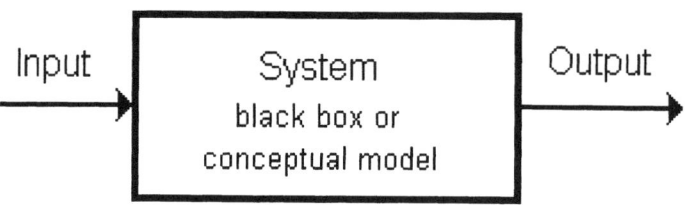

Figure 2-11: A system acts on input to produce output. A mathematical model of a system evaluates one or more mathematical equations and formulas that describe input–output relations and thus numerically mimics the system's transformation of input into output. There are black box models and conceptual models. They can be deterministic or probabilistic. Depending on input and the inner workings of the models, output can be deterministic or probabilistic.

An overview of possible combinations among input types and model types and the resulting output is given in Table 2-1. Three of the four combinations in the table produce probabilistic output. If only one component of the modeling process is probabilistic, the outcome is probabilistic. The variability of the functional relations of a probabilistic model, as well as of the input, increases the computational burden.

Table 2-1: Output of Deterministic and Probabilistic Models

Model Input	Model Type	
	Deterministic	Probabilistic
deterministic	deterministic	**probabilistic**
probabilistic	**probabilistic**	**probabilistic**

Evaluating all possible outcomes can exceed available computer capability. A special sampling method known as *Monte Carlo method* (Hammersley and Hanscomb, 1964 and 1967) has been developed that selects input samples to obtain a representative output sample while limiting the computational burden. Melching and Yoon (1996) suggest that "to reduce the computational burden and to increase the reliability and accuracy of the results of the probabilistic analysis, models describing the system operation should be analyzed to determine the key sources (parameters and variables) of uncertainty. The uncertainty, e.g., the ranges of key parameters and variables, should then be reduced by further study of the system and/or its operation. Detailed uncertainty analysis can reveal that for systems simulated with complex models involving many parameters only a few of these parameters significantly affect the uncertainty of system output and, thus, require further study."

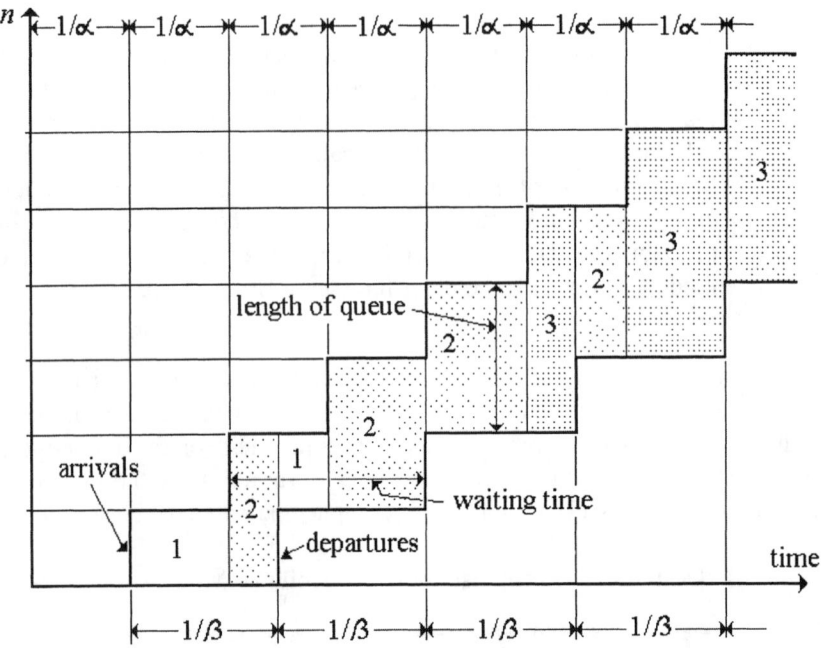

Figure 2-12: A deterministic queuing model with interarrival times that are equal to the inverse of an average arrival rate, α, and service times that are equal to the inverse of an average service rate, β. The vertical distance between the cumulative arrivals and departures (step functions) represents the length of the queue or the number of elements waiting, including the one in service. The horizontal distance between steps represents the waiting time of an element in the queue, including service time. The step height of the arrival curve is the number of arrivals by the end of the arrival interval, and the step height of the departure curve is the number of departures by the end of a service interval. In a probabilistic model, arrivals and departures are random events.

Simple probabilistic models that can be solved analytically do not necessarily increase the computational burden. But even when they do it may be well justified to use the probabilistic approach. The following comparison of a deterministic and a probabilistic waiting line (queuing) model demonstrates *the possibility that deterministic methods produce non-conservative results*. This possibility should be kept in mind as an incentive to use the more general probabilistic methods.

First, consider a deterministic queuing model. An example of such a model is illustrated in Figure 2-12 (after Gross and Harris, 1974, p. 14). Assume a job backlog must be predicted for a maintenance shop to determine manpower requirements. The rate of arriving repair jobs is α, and the rate of servicing jobs is β, both in units of number of jobs per time. The backlog is measured by the number of jobs that are waiting for service at time t. From the start of the period, at time $t = 0$, to time t, n_a jobs arrive and n_s jobs are serviced or completed. Hence, the number of jobs waiting at time t is

$$n(t) = n_a - n_s ,$$

(2.2-19)

where $n(t)$ is the number of jobs in the queue waiting for service, also called the backlog at time t; $n_a = \alpha \cdot t$, and $n_s = \beta \cdot (t - 1/\alpha)$, where $1/\alpha$ is the time that elapses before the first job arrives, as can be seen by setting $n_a = 1$, and solving for t. In terms of arrival rates and service rates, Equation (2.2-19) becomes

$$n(t) = (\alpha \cdot t) - (\beta \cdot t - \frac{\beta}{\alpha}),$$

(2.2-20)

where t is the ending time of time period $(0, \quad t)$ during which jobs arrive. The parentheses around the two terms on the right-hand side of Equation (2.2-20) mean that only integer quantities are taken into account. Service can occur only after the first job has come in, as the number of jobs waiting was assumed to be zero at $t = 0$. Hence, the service time in period $(0, \quad t)$ is reduced by $1/\alpha$, because there are no jobs to be serviced during this time. The delay causes a loss of service β/α. For a more general selection of the time period, one could replace β/α by the backlog at the beginning of a period (t_0, t) by n_{to}, so that

$$n(t) = n_{to} + \alpha t - \beta t .$$

(2.2-21)

If $\alpha > \beta$, the jobs arrive faster than they can be serviced, and the backlog increases. If the jobs are serviced faster than they arrive, $\alpha < \beta$, the backlog— if one exists— is reduced to zero, and a new backlog cannot build up. Suppose $n_{to} = 5$, $\alpha = 1/(6$ h$)$, $\beta = 1/(4$ h$)$, and $t = 8$ h. Then $n(8) = 5 + (8/6) - (8/4) = 5 + 1 - 2 = 4$, with arrivals and departures rounded to integers. For $n(t) = 0 = 5 + (t/6) - (t/4)$, one obtains $t = 60$ h, as the time after which the backlog is reduced to zero.

It is not realistic to assume the arrivals and departures are known with certainty for the estimation of a backlog. It is more likely that the number of arrivals as well as the number of services completed in a defined time period are random numbers. A *probabilistic model* is then required to calculate the probability of a backlog of n items. Assume the simple case of exponentially distributed arrival rates and service rates (see Section 2.7.5):

$$a(t) = \alpha' \, e^{-\alpha' t} \qquad\qquad\qquad\qquad (2.2\text{-}22a)$$

$$s(t) = \beta' \, e^{-\beta' t} \qquad\qquad\qquad\qquad (2.2\text{-}22b)$$

where α' and β' represent an average arrival rate and service rate, respectively; $a(t)$ and $s(t)$ are the probability density functions of interarrival times and service times. Then, the probability of n jobs being in the backlog at time t, given a steady state has developed, is (Gross and Harris, 1974, p. 47)

$$p_n = (\frac{\alpha'}{\beta'})^n \, (1 - \frac{\alpha'}{\beta'}) \, . \qquad\qquad\qquad\qquad (2.2\text{-}23)$$

This queuing model is an analytically derived model and is known as an *M/M/1 model*, where the first "M" stands for an exponential interarrival time distribution and the second "M" stands for an exponential service time distribution. The "1" means that there is one server. The mean arrival rate must be smaller than the service rate, i. e., $\alpha'/\beta' < 1$, for the model to be meaningful. For example, for $\alpha'/\beta' = (1/6)/(1/4) = 2/3$, the probability of having five jobs in the backlog would be $p_5 = (2/3)^5 (1 - 2/3) = 0.044$, or about 4 %.

Example: A spillway rehabilitation usually requires a reliable assessment of the size of the design flood. Flow records are usually too short for reliable extrapolations to very rare and large floods. It has become accepted practice to generate rainfall sequences and calculate the associated floods because it is thought that rainfall events can be more easily and reliably constructed and transformed into floods than just extrapolating flow records. Of course, there are uncertainties related to the transformation of basin rainfall into runoff and streamflow when historically unprecedented quantities of water are produced over and conveyed through the drainage basin. Beard (1974, p. 318) states that "through a great deal of experience or by extensive study of the rainfall variations and the rainfall-runoff variations within a region, it is possible to develop extreme flood magnitudes to which a degree of probability can be assigned with some confidence." Franz et al. (1991) report the use of a stochastic rainfall model to provide input to a deterministic watershed model to produce a synthetic maximum annual flood record for three watersheds, one in California, one in Texas, and one in Georgia. For the Altamaha River at Baxley, Georgia, a synthetic 10,000-year annual flood record was constructed. One may have reservations about such an extreme extension of any record that is based on a relatively short data record whatever these data may be. A comparison of the synthetic flood record with an extrapolation of the actual 66-year streamflow record showed good agreement, whereas the 30-year streamflow record remained below the synthetic record.

2.3 Mean and Variance

2.3.1 Population Mean

The total number of all elements of a population is the sample space. Sometimes the population is a very large and unwieldy entity. It may be inaccessible or unobtainable within practical limits on time and cost. A population may be continuously changing so that it is not ever possible to measure the total population. While it is preferable to work with the total population, it is always possible to work with a sample taken from it. It is important to distinguish between   *population* parameters and *sample* parameters. Here it is assumed that the population is known and consists of a collective of random realizations, x_i, $i = 1, \ldots, N$, the population elements. The *population mean* is

$$\mu = \frac{1}{N} \sum_{i=1}^{N} x_i = E(X),$$

(2.3-1)

where $E(X)$ is the expectation of X, if X is a random variable. $E(.)$ is an operator notation that specifies a certain operation to be performed on X, here taking the expectation of X. From a probabilistic point of view, the factor $1/N$ that is associated with each x_i can be construed as a uniform probability of each element, $p = 1/N$. In the general case, Equation (2.3-1) becomes

$$\mu = \sum_{i=1}^{N} p_i x_i$$

(2.3-2)

where the p_i must satisfy

$$\sum_{i=1}^{N} p_i = 1.$$ (2.3-2a)

The p_i associated with each element of the population are the *unit probabilities*. If each of the N elements of a population has the same probability, then $p_i = 1/N$. If the population can be partitioned into sets of similar elements, then the probability of such a set of f_j elements is

$$p_j = \frac{f_j}{N}$$ (2.3-3)

where f_j could be the number of tests that produce concrete strengths in the subrange from 2.25 MPa to 2.35 MPa of the total range from 2.1 MPa to 3.1 MPa, as illustrated in Figure 2-4. The probabilities of all sets j that fill the sample space must add to 1 and the sum of the set elements must add to N:

$$\sum_{j=1}^{k} p_j = 1, \quad \text{and} \quad \sum_{j=1}^{k} f_j = N.$$ (2.3-3a)

where the second equation is obtained from the first equation by multiplying both sides with N.

When $p_j x_j$ is used in Equation (2.3-2) instead of $p_i x_i$, x_j represents a set of values x_i all with probability p_j. This notation is used in histograms where similar elements x_i are lumped into a class (or set) x_j.

Example: Figure 2-4 represents a histogram of 100 concrete strength test results (elements) distributed over the range from 2.1 MPa to 3.1 MPa. The 100 elements are allocated to 11 sets. The number of elements in the sets are $f_1 = 2, f_2 = 0, f_3 = 10, f_4 = 11, f_5 = 30, f_6 = 15, f_7 = 18, f_8 = 8, f_9 = 2, f_{10} = 3, f_{11} = 1$. Calculate the mean. Solution: The sum of the f_j is equal to $N = 100$, as required by Equation (2.3-3a). The mean is obtained as sum of the products $p_j x_j$, where $p_j = f_j/N$ is the set probability, and x_j is a representative value of set j. Equation (2.3-2) gives $\mu = 0.02 \cdot 2.1 + 0 \cdot 2.2 + 0.10 \cdot 2.3 + 0.11 \cdot 2.4 + 0.30 \cdot 2.5 + 0.15 \cdot 2.6 + 0.18 \cdot 2.7 + 0.08 \cdot 2.8 + 0.02 \cdot 2.9 + 0.03 \cdot 3.0 + 0.01 \cdot 3.1 = 2.56$ MPa. If the set elements are not identical, as is the case here, the representation of each set by an average value x_j introduces some inaccuracies in the average. The arithmetic average of the 100 elements according to Equation (2.3-1) is 2.51 MPa.

2.3.2 Population Variance

The variance is a measure of the dispersion of the data around the mean. The *population variance* is defined as the mean of the squared departures from the population mean. It is obtained by

$$Var(X) = \frac{1}{N}[(x_1 - \mu)^2 + (x_2 - \mu)^2 + \ldots + (x_N - \mu)^2] , \qquad (2.3\text{-}4)$$

where $Var(.)$, like $E(.)$, is an operator that specifies an operation taken on X. The variance also is the square of the *standard deviation*, $Var(X) = \sigma_X^2$, or simply σ^2, and Equation (2.3-4) becomes

$$\sigma^2 = \frac{1}{N}\sum_{i=1}^{N}(x_i - \mu)^2 . \qquad (2.3\text{-}5)$$

Similar to Equation (2.3-2), with a probability assigned to sets of x_i, the variance can be written as

$$\sigma^2 = \sum_{j=1}^{k} p_j(x_j - \mu)^2 \qquad (2.3\text{-}6)$$

where the count j runs over the j sets of the population, and x_j represents all x's in set j. By multiplying out the square term under the sum and taking the sums of all terms one obtains

$$\sigma^2 = \sum_{j} p_j x_j^2 - 2\mu \sum_{j} p_j x_j + \mu^2 \sum_{j} p_j .$$

With $\sum_{j} p_j x_j = \mu$ and $\sum_{j} p_j = 1$, one obtains

$$\sigma^2 = \sum_{j=1}^{k} (p_j x_j^2) - \mu^2 . \qquad (2.3\text{-}7)$$

This is a convenient formula for the variance because the sum of $p_j x_j$ is also needed for calculating μ. The variance also can be written in the form of expectations as

$$Var(X) = E\{[X - E(X)]^2\} .$$ (2.3-8)

The expectation of the simple difference iszero:

$$E[X - E(X)] = 0.$$ (2.3-8a)

But the expectation of the squared difference is a positive number. Expanding the expression in the brackets gives

$$Var(X) = E\{X^2 - 2XE(X) + [E(X)]^2\} .$$

The expectation of a sum or difference is the expectation of the sum or difference of the individual terms (see also Section 2.4.1). The expectation of the first term is $E[X^2]$, the expectation of the second term is

$$E[2 X E(X)] = 2 E[X E(X)] = 2 E(X) E(X) = 2 [E(X)]^2,$$

so that Equation (2.3-8) becomes

$$Var(X) = E(X^2) - [E(X)]^2.$$ (2.3-9)

The *standard deviation* is

$$\sigma = \sqrt{Var(X)} = \sqrt{E(X^2) - [E(X)]^2}   .$$ (2.3-10)

Example: Suppose X has the following random realizations: 1, 4, 6, 5, 8, 2. Calculate $E(X)$, $E[X - E(X)]$, and $E\{[X - E(X)]^2\}$. Solution: The expectation is obtained from Equation (2.3-1), $E(X) = 13/3$. The differences $X - E(X)$ are $-10/3$, $-1/3$, $+5/3$, $+2/3$, $+11/3$, $-7/3$. Their expectation is $E[X - E(X)] = 0$. The expectation of the squares of X is $E(X^2) = 146/6 = 73/3$. The square of the expectation is $[E(X)]^2 = 169/9$. Hence, by Equation (2.3-9), $Var(X) = 219/9 - 169/9 = 50/9$. The same result is obtained by Equation (2.3-8) which requires $E\{[X - E(X)]^2\} = \{[100 + 1 + 25 + 4 + 121 + 49]/9\}/6 = 300/(9 \cdot 6) = 50/9$.

2.3.3 Coefficient of Variation

The standard deviation, having the same dimension as the mean, can be used to form the dimensionless ratio,

$$C_v = \frac{\sigma}{\mu} \tag{2.3-11}$$

where C_v is the *coefficient of variation* . It provides a dispersion measure that expresses σ by multiples or fractions of μ. C_v can be used to characterize dispersion without specifying or knowing the numerical values of μ and σ.

Sometimes it is useful to scale the variable x by dividing it by its mean, μ. This produces a dimensionless variable, x/μ. Dividing Equation (2.3-1) by μ shows that the mean of x/μ is

$$\frac{1}{N}\sum_{i=1}^{N}\frac{x_i}{\mu} = E[\frac{X}{E(X)}] = 1 . \tag{2.3-12}$$

The variance of x/μ is obtained from Equation (2.3-5) by multiplying out, summing terms and dividing by μ^2,

$$Var(\frac{X}{\mu}) = \frac{1}{N}\sum_{i=1}^{N}(\frac{x_i}{\mu}-1)^2 = E[(\frac{X}{\mu})^2]-1 . \tag{2.3-13}$$

Because $E(X/\mu) = 1$, the variance of x/μ is also C_v^2, and the standard deviation of x/μ is C_v.

2.4 Calculating with Expectations and Variances

2.4.1 Individual Data Versus Expectation as Input

As in calculus, some probabilistic problems can be solved elegantly by analytical closed form solutions, e.g., by an integral of a function, whereas others must be solved numerically. Computerization of approaches allows one to deal with problems that are not easily or not at all amenable to a closed form analytical solution. In such cases, the probabilistic method uses random inputs and makes numerous model runs that produce voluminous output which is then statistically processed for its information content. The traditional deterministic approach runs the deterministic model for an estimated "best" or "most likely" input and claims that the output is the best or most likely result. It will be shown here that an average (most likely) input produces an average (most likely) output only if the input–output relationship is linear. Also it is highly unlikely that several variables and coefficients, if they can vary at random, are best or most likely all at the same time. To the surprise of those who have to deal with the result of such erroneous assumptions, what was thought to be the most likely result may actually have a very low probability of occurrence.

(a) *Linear function*: The random variable X is related to random variable Y by a linear function,

$$Y = aX + b,$$

(2.4-1)

where X is the random input, Y is a transformed random output, and a and b are constant coefficients. Equation (2.4-1) is evaluated for a large number of random realizations, $X = \{x_1, x_2, x_3, \dots, x_n\}$. Each realization x_i produces a transformed realization y_i by the deterministic relation Equation (2.4-1):

$$y_1 = a\,x_1 + b$$
$$\vdots$$
$$y_n = a\,x_n + b$$

Summing all outcomes from 1 to n yields

$$\sum_{i=1}^{n} y_i = a \sum_{i=1}^{n} x_i + n\,b.$$

(2.4-2)

Dividing the sums by n leads to

$$E(X) = \frac{1}{n} \sum_{i=1}^{n} x_i$$

(2.4-3)

and

$$E(Y) = \frac{1}{n} \sum_{i=1}^{n} y_i.$$

(2.4-4)

Substituting Equations (2.4-3) and (2.4-4) into Equation (2.4-2) gives

$$E(Y) = a\,E(X) + b.$$

(2.4-5)

Equation (2.4-5) is similar to Equation (2.4-1) with the random variables Y and X replaced by their expectations. This shows that for the linear model, the expected output $E(Y)$ is obtained by making *one* calculation using the expected input, $E(X)$, instead of carrying out many calculations with random inputs and then processing the random outputs to find the expectation of the output.

This result also can be obtained by the *calculus with expectations*. Taking expectations of both sides of Equation (2.4-1), we get

$$E(Y) = E(a X + b) .$$ (2.4-6)

Arithmetic with expectations has already been encountered in Section 2.3.2. Some rules are summarized as follows:

1. The expectation of a sum is equal to the sum of expectations:

$$E(X + Y) = E(X) + E(Y).$$

2. The expectation of a difference is equal to the difference of expectations:

$$E(X - Y) = E(X) - E(Y).$$

3. The expectation of a constant is equal to the constant:

$$E(a) = a.$$

4. The expectation of the product of a   *constant* coefficient and a random variable is equal to the coefficient times the expectation of the random variable:

$$E(a X) = a E(X).$$

5. The expectation of an expectation is equal to the expectation (rule 3):

$$E[E(X)] = E(X).$$

Rules on products of expectations are encountered in Section 2.4.3.

(b) *Nonlinear function*:

$$Y = a X^b$$ (2.4-7)

where a is a constant coefficient and b is a constant exponent. Equation (2.4-7) is evaluated for a large number of random realizations, $X = \{x_1, x_2, x_3, \dots , x_n\}$. Each realization, x_i, produces a realization y_i by Equation (2.4-7):

$$y_1 = a\, x_1^b$$
$$\vdots$$
$$y_n = a\, x_n^b$$

Summing the outcomes of all realizations yields

$$\sum_{i=1}^{n} y_i = a \sum_{i=1}^{n} x_i^b$$

With $E(Y)$ from Equation (2.4-4) and

$$E(X^b) = \frac{1}{n}\sum_{i=1}^{n} x_i^b$$

the expectation of Y becomes

$$E(Y) = a\, E(X^b).\qquad\qquad(2.4\text{-}8)$$

If $E(X)$ were used as input in Equation (2.4-7), the result would be

$$E(Y)' = a[E(X)]^b.\qquad\qquad(2.4\text{-}9)$$

Since $E(X^b) \neq [E(X)]^b$, the expectation of $E(X)$ used as input in Equation (2.4-7) does not produce the same result as using the individual random realizations, x_i.

2.4.2 Variance of Linear and Nonlinear Relations

The expected value alone does not give an indication of the dispersion of possible outcomes. A measure of the dispersion is the variance (Section 2.3.2).

(a) *The variance of a linear relation*: Equation (2.4-1) is substituted into Equation (2.3-9) to obtain the variance of Y:

$$Var(Y) = E[(b + a\, X)^2] - [E(b + a\, X)]^2.$$

Multiplying out the squared terms and making use of the rules of calculus with expectations (Section 2.4.1), one obtains

$$Var(Y) = b^2 + 2\,a\,b\,E(X) + a^2\,E(X^2) - b^2 - 2\,a\,b\,E(X) - a^2\,[E(X)]^2,$$

which simplifies to

$$Var(Y) = Var(a\,X) = a^2\,\{E(X^2) - [E(X)]^2\} = a^2\,Var(X) . \qquad (2.4\text{-}10)$$

Equation (2.4-10) shows that the variance of the linear function, $Y = a\,X + b$, is equal to the square of the coefficient of the random variable, a, times the variance of the random variable. The additive constant, b, makes no contribution to the variance.

(b) *The variance of a nonlinear relation*: Equation (2.4-7) is reduced to a linear form by the substitution $X^b = U$. Then, by Equation (2.4-10), one obtains

$$Var(Y) = a^2\,Var(U) = a^2\,Var(X^b) = a^2\,\{E[(X^b)^2] - [E(X^b)]^2\}. \qquad (2.4\text{-}11)$$

2.4.3 Variance of a Sum or Difference

In engineering calculations, usually each quantity is a variable with some variance. An example is the net benefit of a project. It is the difference between benefit and cost. If both quantities are numbers with a range, it is of interest to know the variance of the difference

$$N_B = B - C. \qquad (2.4\text{-}12)$$

By Equation (2.3-9),

$$Var(N_B) = E[(B - C)^2] - [E(B - C)]^2 \qquad (2.4\text{-}13)$$

The first right-hand side term of Equation (2.4-13), after multiplying out and taking expectations, becomes

$$E[(B - C)^2] = E(B^2) + E(C^2) - 2\,E(B\,C). \qquad (2.4\text{-}13a)$$

The second right-hand side term of Equation (2.4-13), after taking expectations inside the square brackets and then multiplying out, becomes

$$[E(B - C)]^2 = [E(B) - E(C)]^2$$

$$= [E(B)]^2 + [E(C)]^2 - 2\,E(B)\,E(C). \qquad (2.4\text{-}13b)$$

Substituting Equations (2.4-13a) and (2.4-13b) back into Equation (2.4-13) gives

$$Var(N_B) = E(B^2) - [E(B)]^2 + E(C^2) - [E(C)]^2$$

$$- 2\,[E(B\ C) - E(B)\ E(C)].\qquad\qquad(2.4\text{-}13c)$$

With

$$Var(B) = E(B^2) - [E(B)]^2,\ \text{and}\ \ Var(C) = E(C^2) - [E(C)]^2$$

Equation (2.4-13c) becomes

$$Var(N_B) = Var(B) + Var(C) - 2\,Cov(B, C)\qquad\qquad(2.4\text{-}14)$$

with

$$Cov(B, C) = E(B\ C) - E(B)\ E(C)\qquad\qquad(2.4\text{-}15)$$

where $Cov(B, C)$ is the covariance of the variables B and C. If B and C are independent, $Cov(B, C) = 0$, and

$$E(B\ C) = E(B)\ E(C).\qquad\qquad(2.4\text{-}15a)$$

Equation (2.4-15a) is another rule of the calculus with expectations (see Section 2.4.1): the expectation of the product of two independent random variables is the product of their expectations. Thus, if B and C are independent random variables,

$$Var(B - C) = Var(B) + Var(C).\qquad\qquad(2.4\text{-}16)$$

Equation (2.4-16) states that the *variance of the difference* of two independent random variables is the sum of their variances. The same result is also obtained for the sum of two random variables:

$$Var(B + C) = Var(B) + Var(C).\qquad\qquad(2.4\text{-}16a)$$

Equation (2.4-16a) states that the *variance of the sum* of independent random variables is the sum of their variances. Regardless of whether two random variables are added or subtracted, the resulting variance is always the sum of the individual variances. Equations (2.4-16) and (2.4-16a) apply only if the variables are independent.

<u>Example</u>: For the nonlinear function, $Y = a\,X^2$, calculate the difference in expectations between the probabilistic approach that calls for evaluating many realizations x_i to obtain y_i, and their $E(Y)$, and the deterministic approach that uses $E(X)$ to obtain $E(Y)'$. <u>Solution</u>: The expectation of $Y = a\,X^2$ obtained from many pairs (x,y) is $E(Y) = a\,E(X^2)$. The Y obtained from a single value $E(X)$ is $E(Y)' = a\,[E(X)]^2$ (see Section 2.4.1b). The relative difference (or relative error) between the two is

$$\varepsilon = \frac{E(Y) - E(Y)'}{E(Y)'} = \frac{a \cdot \{E(X^2) - [E(X)]^2\}}{a \cdot [E(X)]^2} = \frac{Var(X)}{[E(X)]^2} = C_v^2$$

where C_v is the coefficient of variation, Equation (2.3-11). Suppose $C_v = 0.25$, so the relative error is $\varepsilon \approx 0.06$. This means that the probabilistic evaluation of $E(Y)$ by random pairs (x, y) exceeds the deterministic evaluation $E(Y)'$ that uses the expectation $E(X)$ as input by 6 %.

2.4.4 Sample Mean and Sample Variance

A sample is a random selection of a number of elements taken from a sample space or population. Frequently, the need arises to estimate population parameters from samples. Statistical sampling theory has developed relations between sample parameters and population parameters for this purpose. Because of the importance of understanding these relations, some derivations are given. To be representative, a sample must be unbiased. That means it must not emphasize any specific characteristic of the population, thereby falsifying the conclusions drawn from the sample. Also, if the sampling is done with replacement, then the selected samples are independent of each other.

Assume a population consists of N elements, $x_1, x_2, x_3, \ldots, x_N$. Now assume a sample of n elements is drawn randomly with replacement, which means that the same element can be drawn repeatedly and be present in successive samples. Let the random outcomes of the draws be designated $X_1, X_2, \ldots, X_n$. Then, if the population consists of $S = \{x_1, x_2, x_3, x_4\}$, a random sample of 3 elements drawn from S may consist of $X_1 = x_4$, $X_2 = x_1$, $X_3 = x_2$. A second sample may consist of $X_1 = x_2$, $X_2 = x_4$, $X_3 = x_3$. The X_i denote random numbers and the x_i denote their realizations by the sampling process (see also Lapin, 1978, p. 187).

For each sample, a *sample mean* is obtained by

$$\overline{X} = \frac{1}{n}(X_1 + X_2 + \ldots + X_n) \qquad (2.4\text{-}17)$$

where $\overline{X}$ is a sum of random variables, which makes it again a random variable. Taking expectations on both sides of Equation (2.4-17) and moving n to the left side gives

$$n\,E(\overline{X}) = E(X_1 + X_2 + \ldots + X_n)\ .$$ (2.4-18)

The expectation of the sum of random variables on the right side of Equation (2.4-18) is equal to the sum of expectations so that

$$n\,E(\overline{X}) = [E(X_1) + E(X_2) + \ldots + E(X_n)]\ .$$ (2.4-19)

Equation (2.4-19) means that each random element taken from the same population has the same expectation so that the sum of these expectations is n times the same expectation. Thus, one can equate the expression in square brackets with $n\,\mu$, and after canceling n on both sides of Equation (2.4-19) one obtains

$$E(\overline{X}) = \mu$$ (2.4-20)

where $E(\overline{X})$ is the *expectation of the sample mean*. The expectation of the sample mean is the *unbiased estimator* of the population mean. This means that, on average, the unbiased procedure tends to produce a population parameter, here the population mean (Lapin, 1978, p. 236). If k samples are taken, each with n observations, and k sample means, $\overline{X}_i$, are calculated, then Equation (2.4-19) can be written as

$$k\,E(\overline{X}) = [E(\overline{X}_1) + E(\overline{X}_2) + \ldots + E(\overline{X}_k)]\ .$$ (2.4-19a)

Substituting the k expectations of the sample means, $E(\overline{X}_i)$, by $k\,\mu$ again leads to Equation (2.4-20).

The population variance, which is based on all elements, x_i, of the population, was given in Section 2.3.2 by Equation (2.3-5). It can be proved that by dividing by $N - 1$ instead of by N, the sample variance $Var(X)$ becomes an unbiased estimate of the population variance (Lapin, 1978, p. 237). The sample variance is

$$Var(X) = \frac{1}{n-1} \sum_{i=1}^{n} (X_i - m)^2$$ (2.4-21)

where the X_i are the random elements of the sample, as used in Equation (2.4-17), and $m = \overline{X}$ is an abbreviation for the sample mean for notational convenience.

Multiplying out the square under the sum of Equation (2.4-21) and taking sums of all terms, one can simplify the second term and the third term as follows: since $\sum X_i = n\,m$, the second term becomes $-\sum 2\,X_i\,m = -2\,m\sum X_i = -2\,n\,m^2$; and the third term becomes $\sum m^2 = n\,m^2$, so that Equation (2.4-21) becomes

$$Var(X) = \frac{1}{n-1}\sum_{i=1}^{n} X_i^2 - \frac{n}{n-1}m^2 \; . \tag{2.4-22}$$

Equations (2.4-21) and (2.4-22) are formulas of the sample variance that produce an approximation of the population variance, σ^2.

A sum of random variables can be written as

$$n\,\overline{X} = X_1 + \ldots + X_n$$

Taking the variance on both sides gives

$$Var(n\,\overline{X}) = Var(X_1 + \ldots + X_n) \; . \tag{2.4-23a}$$

According to Equation (2.4-16a), the variance of a sum of independent random variables is the sum of their variances:

$$Var(X_1 + \ldots + X_n) = Var(X_1) + \ldots + Var(X_n) \; . \tag{2.4-23b}$$

According to Equation (2.4-10),

$$Var(n\,\overline{X}) = n^2\,Var(\overline{X}) \; . \tag{2.4-23c}$$

Since the X_i in Equation (2.4-23a) are all sampled from the same population, the variances in Equation (2.4-23b) are all the same, so that $Var(X_1) = \ldots = Var(X_n) = \sigma^2$, and their sum is $n\,\sigma^2$, so that Equations (2.4-23a), (2.4-23b), and (2.4-23c) can be combined to give

$$n^2\,Var(\overline{X}) = n\,\sigma^2 \; . \tag{2.4-24}$$

With $Var(\overline{X}) = \sigma_{\overline{X}}^2$, one obtains the *variance of the sample mean* as

$$\sigma_{\overline{X}}^2 = \frac{\sigma^2}{n} \; , \tag{2.4-25}$$

and the *standard deviation of the sample mean* is

$$\sigma_{\bar{X}} = \frac{\sigma}{\sqrt{n}} \tag{2.4-25a}$$

where σ is the population standard deviation, and n is the sample size, or number of elements used to calculate the sample parameters, such as the sample mean and the sample standard deviation. Equation (2.4-25a) states that the standard deviation of the sample mean (not of the sample elements) decreases with the size of the sample so that the distribution of the sample mean, $\bar{X}$, becomes more and more spiky and tight around the population mean with which the sample mean ultimately coincides. The law of large numbers states that for an infinitely large sample, $n \to \infty$, the probability of the sample mean converging toward the population mean is 1 (see Section 2.4.6). The sample mean is based on a sum of randomly selected elements of the population and, in contrast to the population mean, is itself a random variable. Each sample mean taken from the population is a little different from the other. This is why the standard deviation of the sample mean, $\sigma_{\bar{X}}$, is also called the *standard error of the sample mean*. Equation (2.4-25a) shows that for large sample size it is much smaller than the *population standard deviation, σ*. Not to be confused with these two is the standard deviation of a sample which is calculated from the sample elements and the sample mean, m, by Equation (2.4-21), or similar equation in Section 2.3.2.

The variance of the sample mean, $\sigma_{\bar{X}}^2$, is computed by

$$\sigma_{\bar{X}}^2 = \sum_{i=1}^{k} (p_i \bar{X}_i^2) - [E(\bar{X})]^2 \tag{2.4-26}$$

or

$$\sigma_{\bar{X}}^2 = E(\bar{X}^2) - [E(\bar{X})]^2 \tag{2.4-26a}$$

where $\bar{X}_i$ is the sample mean and p_i is its frequency in a sample that consists of k sets ($\bar{X}_i$, p_i); $E(\bar{X}^2)$ is the expectation of squared sample means that can be calculated by a probabilistic sum, similar to the first term on the right side of Equation (2.4-26). The statistical behavior of the sample mean, or of the sums of random variables, will be discussed in context with the *central limit theorem* in Section 3.4.2.

Example: A general derivation of the standard deviation of the sample mean from sample data is given to further illuminate the differences between population parameters and sample parameters (after a numerical example by Goldberg (1986, p. 223). Suppose the sample space is $S = \{x_1, x_2, x_3\}$, with probabilities p_1, p_2, and p_3, which add to 1. Samples are taken of size $n = 2$. Each random selection, X_1 and X_2, made from S can include any of the x_i of the population. Since replacement is assumed, a repetitive draw of the same element is possible which would produce a sample (x_1, x_1). The p_i's of the elements add to 1, as the elements cover the sample space exhaustively. Calculate (a) the expectation of the sample mean and compare it with the population mean; and (b) calculate the standard deviation of the sample mean and compare it with the population standard deviation. Solution: All possible samples of size $n = 2$ and their probabilities that can be drawn from the population, a total of nine, are included and are as follows:

Sample No.	1	2	3	4	5	6	7	8	9
Elements	x_1,x_1	x_1,x_2	x_1,x_3	x_2,x_1	x_2,x_2	x_2,x_3	x_3,x_1	x_3,x_2	x_3,x_3
Probability	p_1^2	$p_1 p_2$	$p_1 p_3$	$p_2 p_1$	p_2^2	$p_2 p_3$	$p_3 p_1$	$p_3 p_2$	p_3^2

The probabilities of the samples are the joint probabilities of the sample elements. For example, the probability of x_1 being selected twice, as in sample 1, is

$$P(X_1 = x_1) = p_1; \; P(X_2 = x_1) = p_1; \text{ hence, } P[(X_1 = x_1) \cap (X_2 = x_1)] = p_1 p_1.$$

Since the nine samples in the tabulation represent all events of the sample space, their total probability adds to 1. Adding the joint probabilities in the bottom row of the table and separating them into a sum of factors one of which is $p_1 + p_2 + p_3 = 1$ gives

$$p_1(p_1 + p_2 + p_3) + p_2(p_1 + p_2 + p_3) + p_3(p_1 + p_2 + p_3) = (p_1 + p_2 + p_3) = 1.$$

The sample means are calculated from the sample elements. For the first sample, the elements are $X_1 = x_1$ and $X_2 = x_1$. Each sample mean is calculated from

$$\overline{X} = (X_1 + X_2)/2.$$

These means are denoted by m for simplicity of notation. In this manner all nine sample means are calculated:

$$m_1 = (x_1 + x_1)/2, \; m_2 = (x_1 + x_2)/2, \; \dots, \; m_9 = (x_3 + x_3)/2.$$

The expectation of the sample mean, $E(m)$, is the probability-weighted sum of all sample means:

$$E(m) = p_1 p_1 m_1 + p_1 p_2 m_2 + p_1 p_3 m_3 + \dots + p_3 p_3 m_9.$$

Substituting the expression for the means by the elements, x_i, and ordering the resulting sum by elements produces

$$2 E(m) = x_1 (2 p_1 p_1 + p_1 p_2 + p_1 p_3 + p_2 p_1 + p_3 p_1)$$

$$+ x_2 (p_1 p_2 + p_2 p_1 + 2 p_2 p_2 + p_2 p_3 + p_3 p_2)$$

$$+ x_3 (p_1 p_3 + p_2 p_3 + p_3 p_1 + p_3 p_2 + 2 p_3 p_3) \; .$$

With $p_1 + p_2 + p_3 = 1$, one obtains

$$2 E(m) = 2 (p_1 x_1 + p_2 x_2 + p_3 x_3)$$

The sampling process exhaustively samples (draws all possible samples from) the population of the three elements that has a population mean $\mu = p_1 x_1 + p_2 x_2 + p_3 x_3$. Hence, the calculation proves that the expectation of the sample mean is equal to the population mean, $E(m) = \mu$. The number 2 stands for the number of elements n in each sample.

(b) The variance of the sample means is the probability-weighted sum of the squared departures of the sample means from the expectation of the sample mean:

$$Var(m) = p_1 p_1 [m_1 - E(m)]^2 + p_1 p_2 [m_2 - E(m)]^2 + \dots + p_3 p_3 [m_9 - E(m)]^2.$$
Multiplying out each term produces

$$p_1 p_1 m_1^2 - 2 p_1 p_1 m_1 E(m) + p_1 p_1 [E(m)]^2$$

$$+ p_1 p_2 m_2^2 - 2 p_1 p_2 m_2 E(m) + p_1 p_2 [E(m)]^2$$

$$\dots$$

$$+ p_3 p_3 m_9^2 - 2 p_3 p_3 m_9 E(m) + p_3 p_3 [E(m)]^2$$

The sum of the first column represents $E(m^2)$, the sum of the second column represents $2 E(m) E(m) = 2 [E(m)]^2$, and the sum of the third column is $E\{[E(m)]^2\} = [E(m)]^2 = \mu^2$, so that one can summarize

$Var(m) = E(m^2) - \mu^2$.

The next step is to evaluate $E(m^2)$ for the sample elements:

$$E(m^2) = p_1 p_1 (\frac{x_1 + x_1}{2})^2 + p_1 p_2 (\frac{x_1 + x_2}{2})^2 + p_1 p_3 (\frac{x_1 + x_3}{2})^2 + \ldots$$

where the expressions in parentheses are the sample means. Multiplying out the first three terms and summarizing leads to

$$E(m^2) = 0.25 \, [p_1 \, (p_1 + p_2 + p_3) \, x_1^2 + 2 \, p_1 \, x_1 (\, p_1 \, x_1 + p_2 \, x_2 + p_3 \, x_3)$$

$$+ \, p_1 \, (p_1 \, x_1^2 + p_2 \, x_2^2 + p_3 \, x_3^2) + \ldots].$$

Introducing expectations for the terms in rounded brackets leads to

$$E(m^2) = 0.25 \, [p_1 \, x_1^2 + 2 \, p_1 \, x_1 \, E(X) + p_1 \, E(X^2) + \ldots].$$

Adding in the other terms by induction leads to the complete form of $E(m^2)$:

$$E(m^2) = 0.25 \, [p_1 \, x_1^2 + 2 \, p_1 \, x_1 \, E(X) + p_1 \, E(X^2)$$

$$+ \, p_2 \, x_2^2 + 2 \, p_2 \, x_2 \, E(X) + p_2 \, E(X^2) + p_3 \, x_3^2 + 2 \, p_3 \, x_3 \, E(X) + p_3 \, E(X^2)] \ .$$

Collecting terms simplifies the expression to

$$E(m^2) = 0.25 \, \{2 \, E(X^2) + 2 \, [E(X)]^2\} = 0.5 \, [E(X^2) + \mu^2].$$

Substituting into the expression for the sample variance derived before gives

$$Var(m) = E(m^2) - \mu^2 = 0.5 \, [E(X^2) + \mu^2] - \mu^2 = 0.5 \, [E(X^2) - \mu^2].$$

The population variance is obtained as

$$Var(X) = p_1 \, (x_1 - \mu)^2 + p_2 \, (x_2 - \mu)^2 + p_3 \, (x_3 - \mu)^2 = E(X^2) - \mu^2.$$

Hence, the comparison of sample variance and population variance is given by

$$Var(m) = Var(X)/2$$

which is Equation (2.4-25). The 2 in the denominator can be shown to be the sample size $n = 2$. Taking the root of $Var(m)$ leads to Equation (2.4-25a):

$$s(m) = \frac{\sigma}{\sqrt{n}} = \sigma_{\bar{X}} \; .$$

2.4.5 Confidence Bands

After a sample mean has been obtained by sampling a population, it is usually of interest how closely it approximates the population mean. Confidence bands are used to make such estimates. Depending on which parameters are known, different formulations of confidence bands can be used, all of which can be derived from each other. A confidence band for $\bar{X}$ centered on μ is shown in the sketch below. The probability of $\bar{X}$ being within this band or on its boundaries is the confidence level C.

lower bound		upper bound
$\bar{X} = \mu - e$		$\bar{X} = \mu + e$

$$\underline{\;\;\big|\!\rule{3cm}{0.4pt}\!\big|\!\rule{2cm}{0.4pt}\!\big|\rule{1cm}{0.4pt}}$$

$$\bar{X} < \mu \qquad \mu \qquad \bar{X} > \mu$$

(a) If the population mean μ is known, the sample mean $\bar{X}$ can be bracketed by a confidence interval centered on μ:

$$C = P[\mu - e \le \bar{X} \le \mu + e] \; . \tag{2.4-27}$$

Equation (2.4-27) states that $\bar{X}$ is with confidence C within a band of width $\pm\, e$ around μ.

(b) If the population mean is not known, a probability statement can be made of μ being in a band of width $\pm e$ around $\bar{X}$:

$$C = P[\bar{X} - e \le \mu \le \bar{X} + e]. \tag{2.4-28}$$

In this case the band is centered on the random variable $\bar{X}$ and moves back and forth with it while containing μ with probability C.

(c) The sample mean $\overline{X}$ can be assumed to have a normal distribution for reasons to be discussed in Section 3.4.2. If the standard deviation of the sample mean is known, the confidence band width can be expressed by $e = z \sigma_{\overline{X}}$, with z being the normal variable (Section 2.6). Then Equation (2.4-28) becomes

$$C = P[\overline{X} - z \sigma_{\overline{X}} \leq \mu \leq \overline{X} + z \sigma_{\overline{X}}] \; . \tag{2.4-28a}$$

The normal variable z can be selected to determine the band width and the probability of μ being inside it or on its boundaries.

(d) If the population standard deviation σ is known, then the standard deviation of the sample mean can be replaced by the population standard deviation using $\sigma_{\overline{X}} = \dfrac{\sigma}{\sqrt{n}}$:

$$C = P[\overline{X} - z \frac{\sigma}{\sqrt{n}} \leq \mu \leq \overline{X} + z \frac{\sigma}{\sqrt{n}}] \; . \tag{2.4-28b}$$

Equation (2.4-28b) states that with confidence level C the population mean μ is found within or on the boundaries of a band of width $e = z\sigma / \sqrt{n}$ centered on the sample mean. This statement can be used for estimating the population mean:

$$\mu = \overline{X} \pm z \frac{\sigma}{\sqrt{n}} \tag{2.4-28c}$$

where z is a measure of the reliability of this formula (see Section 2.6).

(e) Modifying Equation (2.4-28b) by subtracting $\overline{X}$ from all terms, then multiplying all terms with -1, turning the inequality signs around, and rearranging the terms gives:

$$C = P[-z \frac{\sigma}{\sqrt{n}} \leq \overline{X} - \mu \leq +z \frac{\sigma}{\sqrt{n}}] \tag{2.4-29}$$

which can also be written as

$$C = P[|\overline{X} - \mu| \leq |z| \frac{\sigma}{\sqrt{n}}] \; . \tag{2.4-29a}$$

Equation (2.4-29a) states that the absolute difference between the sample mean and the population mean must not exceed a specified amount. The parameter z can be specified by the analyst and represents a probability measure (see Section 2.6).

Examples: (1) The sample mean drawn from a population is required to be within a specified band around the population mean, $\mu \pm e$, with a confidence level of 0.95. What is the size of e? Solution: The confidence band is centered on the population mean μ. One must know the probability distribution of the sample mean to relate the band width and the confidence level (if this is not the case, the Chebyshev inequality, Section 2.4.6, may be used as an approximation). Since the distribution of the sample mean can be assumed to be normal, for $C = 0.95$, one finds from a table of the normal distribution (Section 2.6.2, Table 2-3, column 5) for an interval centered on the mean of the normal distribution $z = 1.96$. Hence, from Equation (2.4-29),

$e = \pm z\sigma / \sqrt{n} = \pm z\sigma_{\overline{X}}$. If the population standard deviation and the sample size are

known, $e = \pm 1.96\sigma / \sqrt{n}$. If the sample standard deviation is known, $e = \pm 1.96 \sigma_{\overline{X}}$.

(2) What is the confidence level that the population mean μ is within a band of $\pm$ $e = z\ \sigma_{\overline{X}}$ around the sample mean, if $z = 0.5$? Solution: The confidence band is

centered on $\overline{X}$. The central limit theorem (Section 3.4.2) supports the assumption that the distribution of the sample mean is at least approximately normal. Then, for $z = 0.5$, one obtains $C = 0.383$ for the probability of μ being within a band of $\pm$ $z = \pm e/\sigma_{\overline{X}} = 0.5$ around the sample mean (Section 2.6.2, Table 2-3, column 5). Compared to Example (1), the confidence level is much lower because the specified band is narrower and the probability of μ being close to $\overline{X}$ is smaller.

2.4.6 Chebyshev Inequality

If the probability distribution of a random variable is unknown, the *Chebyshev inequality* can be used to bound the variable. One form of expressing it is (Benjamin and Cornell, 1970, p. 141; Lapin, 1978, p. 58):

$$P[\mu - k\sigma \le X \le \mu + k\sigma] \ge 1 - \frac{1}{k^2} . \qquad (2.4\text{-}30)$$

In this form, the Chebyshev inequality states a lower limit on the probability that a data point is within a specified range, $\mu \pm k\sigma$. According to Equation (2.4-30), the probability of X being within a band width of $\pm 1\ \sigma$ centered on the population mean is zero or greater than zero. This means that the Chebyshev inequality can give only rough estimates for wide band widths. Some values of P for various k's are given in Table 2-2, column 3. For comparison, the probability of confidence bands of the same widths using the normal probability distribution (Section 2.6) is given in column 5.

The Chebyshev inequality also can be stated as follows (Goldberg, 1986, p. 193; Bronstein and Semendjajew, 1984, p. 675):

$$P[|X - \mu| \geq e] \leq \frac{\sigma^2}{e^2} \tag{2.4-31}$$

where $X - \mu$ is the departure of X from μ to be tested for its probability of exceeding a specified distance, e. Equation (2.4-31) states that the probability of $|X - \mu|$ exceeding e is σ^2/e^2 or less. The random variable X can be the sample mean, $\overline{X}$. By representing e as multiples of σ, some general probability statements can be made, as shown in Table 2-2, columns 3 and 4. Equation (2.4-30) tests X for being within the band $\pm e$, whereas Equation (2.4-31) tests X for being outside the band.

Table 2-2: Estimating Confidence Bands and Probabilities by the Chebyshev Inequality

k	e	$1 - \sigma^2/e^2$ $= 1 - 1/k^2$ Eq. (2.4-30)	σ^2/e^2 $= 1/k^2$ Eq. (2.4-31)	Normal Non-exceedance Probability
(1)	(2)	(3)	(4)	(5)
1	$1\,\sigma$	0.00	1.00	0.683
2	$2\,\sigma$	0.75	0.25	0.954
3	$3\,\sigma$	0.89	0.11	0.997

Notes: column 1: multiples of the standard deviation used in Equation (2.4-30).
column 2: interval width extending k multiples of σ to each side of the mean.
column 3: lower limit on probability that X is within the range $e = \pm k\,\sigma$ of the population mean.
column 4: upper limit on probability that $|X - \mu|$ exceeds $e = \pm k\,\sigma$.
column 5: probability that X is within the range $e = \pm k\,\sigma$ around the mean, given X is normally distributed (see Section 2.6.2, Table 2-3, column 5).

The normal probabilities are given in column 5 of Table 2-2 for comparison with those of column 3. The differences show that the lack of knowledge about the distribution of the random variable that is assumed by the Chebyshev inequality translates into a considerable loss of prediction accuracy.

The Chebyshev inequality can also be stated for estimating the probability of a discrepancy e between the sample mean and population mean (Goldberg, 1986, p. 226):

$$P[|\overline{X} - E(\overline{X})| \geq e] \leq \frac{\sigma_{\overline{X}}^2}{e^2} \qquad (2.4\text{-}32)$$

where $\overline{X}$ is the individual sample mean; $E(\overline{X})$ is the expectation of the sample mean, which is also μ, the population mean; and $\sigma_{\overline{X}}$ is the standard deviation of the sample mean, which is also $\sigma_{\overline{X}} = \sigma / \sqrt{n}$. Replacing the sample parameters by the population parameters converts Equation (2.4-32) to

$$P[|\overline{X} - \mu| \geq e] \leq \frac{\sigma^2}{e^2 n} \qquad . \qquad (2.4\text{-}33)$$

Equation (2.4-33) states that the probability of the departure of the sample mean from the population mean exceeding a specified e can be made arbitrarily small by increasing the sample size n. With the inequalities turned around, the Chebyshev inequality becomes

$$P[|\overline{X} - \mu| \leq e] > 1 - \frac{\sigma^2}{e^2 n} \qquad . \qquad (2.4\text{-}34)$$

Equation (2.4-34) states that the probability of the departure of the sample mean from the population mean being within a specified range, e, is greater than some specified limiting probability given by the right-hand side of Equation (2.4-34).

Equation (2.4-34) can be used to illustrate the validity of a form of the *law of large numbers* (Goldberg, 1986, p. 226). If $e = k\,\sigma$, and if e is made very small by making k very small, say $k = 0.1$, and if n is made very large, say $n = 10,000$, then

$$P[|\overline{X} - \mu| \leq 0.1\,\sigma] > 1 - \frac{1}{k^2 n} = 0.99 \qquad . \qquad (2.4\text{-}34a)$$

This means that the probability of $\overline{X}$ approaching μ, as n becomes very large, approaches 1. The weak law of large numbers states that as the number of sample elements n goes to infinity the probability of the sample mean approaching the population mean goes toward 1. The *strong law of large numbers* states that as n goes to infinity, $\overline{X}$ becomes μ with probability 1 (Bronstein and Semendjajew, 1984, p. 675).

If the left-hand side of Equation (2.4-34) is abbreviated by R, the reliability, then one can write

$$R = 1 - \frac{\sigma^2}{e^2 n} .$$
(2.4-35)

Solving for n produces

$$n = \frac{\sigma^2}{e^2 (1 - R)} .$$
(2.4-36)

Equation (2.4-36) shows that, all other parameters being equal, if R approaches 1, then n approaches infinity.

Examples: (1) Estimate the change of sample size to achieve a desirable reduction of the probability $P[|\,\overline{X} - \mu\,| \geq e]$. Solution: Reducing the probability expressed by Equation (2.4-33) by a factor $\gamma < 1$ amounts to multiplying the right-hand side by γ. This has the effect of increasing the sample size from n to n/γ. For a 10 % reduction in probability, $\gamma = 0.9$, n increases to $n/\gamma = 1.11\,n$; in other words, as the sample size increases, $P[|\,\overline{X} - \mu\,| \geq e]$ is reduced.

(2) A reliability of $R = 0.95$ or higher is desired for a sample mean, $\overline{X}$, to be within a band width $\pm\,e$ around the population mean, μ. What would the sample size have to be? Solution: According to Equation (2.4-36), for $R = 0.95$, $n = 20\,\sigma^2/e^2$. For $R = 0.99$, $n = 100\,\sigma^2/e^2$. Since n cannot be made unreasonably large, n has some practical upper limit.

2.5 Probability Functions

2.5.1 Probability Density Function

Functions that represent probabilities or probability densities are *probability functions*. A random variable, when sampled, may take on any value in its range. These values may all have the same frequency, or those near the center of the range may have higher frequencies than those near the limits of the range. The distribution of the frequencies over the range of a random variable is the *probability density function*, pdf (see Section 2.2.2). If the range of a random variable is subdivided into a number of nonoverlapping classes, then the relative number of elements in each class represents a discrete frequency. If the range of the variable is continuous, then we speak of a continuous frequency.

Suppose in a sample space of n elements there are n_i elements of the same kind then such a class or set, j, has the relative frequency

$$f_j(x) = \frac{n_i}{n}.$$
(2.5-1)

If the range is subdivided by classes, the resulting sequence of ordinates, $f_j(x)$, represents a *discrete probability density function* of a variable x. A random variable is strictly discrete only if there are no possible values between the countable ordinates. An example of such a discrete variable is one that can take on either 1 or 0, head or tail, on or off, or an integer number set.

The pdf of a discrete random variable X is

$$f(x) = P(X = x)$$
(2.5-2)

where $f(x)$ is the relative frequency associated with a distinct realization x. The discrete pdf, $f(x)$, is a dimensionless number, $0 \le f(x) \le 1$, that is sometimes written as $f_X(x)$ to denote it as a pdf of random variable X; $P(X = x)$ is the probability of the random variable X taking the value x.

The pdf of a continuous variable has ordinates $f(x)$ which do not represent probabilities, but a probability rate per increment,

$$f(x) = \frac{dF(x)}{dx}$$
(2.5-3)

where $f(x)$ is the probability rate per increment of the range; $F(x)$ is the integral of $f(x)$; $dF(x) = f(x)\,dx$ can be interpreted as an infinitely narrow bar of a histogram with bar height $f(x)$ and bar width dx. This bar represents a partition of the sample space and $dF(x)$ is its probability. An illustration of such a partitioned sample space is given by the histogram of Figure 2-4. If an infinite number of samples were taken, an infinite number of partitions could be created for every increment dx of the range. The dx would then become infinitesimally small, and the histogram would converge toward a continuous pdf.

2.5.2 Cumulative Distribution Function

The summing of a discrete pdf and the integration of a continuous pdf produce the *cumulative distribution function* or cdf. Summation and integration are the same in computerized calculations so that one or the other word can be used. The summation over the entire range gives 1; however, the summation from the lower boundary of the range to some specified x of the range, the upper integration limit amounts to adding the probabilities of all partitions to the left of the limit, which gives the probability of the union of all partitions up to this limit. Hence, this sum of probabilities is the probability of any element that belongs to any of the partitions or sets up to this limit. The value of the cdf is its value at its upper bound (summation limit).

(a) The discrete cdf is the sum of all pdf ordinates, $f(x)$, from a lower limit, x_l, up to and including an upper limit, x_u:

$$F(x_u) = P(X \leq x_u) = \sum_{x_l}^{x_u} f(x)$$

(2.5-4)

where $P(X \leq x_u)$ represents the total probability of all mutually exclusive events, X, from and including a lower limit, x_l, to and including an upper limit x_u (Moan, 1982, p. 4-9). Often the function $F(x_u)$ is stated simply as $F(x)$ where x represents the upper summation limit. The sum $F(x)$ is said to be a function of its upper bound. A characteristic of the discrete $f(x)$ is that it represents the *probability* of the discrete random variable realization x. A characteristic of $F(x)$ is that it can be represented as a *step function* that increases at a discrete value x by the step $f(x)$, then stays flat until the next x. This means that it does not grow between discrete values of x but in contrast to $f(x)$ it has a value between two discrete x, namely the value $F(x)$ it has accumulated at the previous x.

A finite difference, $\Delta F(x)$, represents the probability of X being within a specified range Δx: For a *discrete random variable*, X, this probability is

$$P(x_1 < X \leq x_2) = F(x_2) - F(x_1) = \sum_{x_1}^{x_2} f(x) \; ,$$

(2.5-5)

where the sum is taken over the range $x_1 < X \leq x_2$, with the first function value $f(x)$ being the one following x_1 and the last being $f(x_2)$. A discrete pdf as well as its integral are functions represented by discrete function spikes as function values only exist at the discrete values of x. The spikes of the integral $F(x)$ increase by finite steps $f(x_i)$, such as $F(x_1) = f(x_1)$, $F(x_2) = f(x_1) + f(x_2)$, and so on.

(b) The *continuous* cdf is the integral of the continuous pdf, $f(x)$, from some lower limit up to some upper limit. If the integral includes the entire sample space, then the sum of all probabilities must be 1:

$$\int\limits_{-\infty}^{+\infty} f(x)dx = 1 \quad . \tag{2.5-6}$$

This condition must be met by any $f(x)$ in order to qualify as a pdf. The integration limits from $-\infty$ to $+\infty$ are symbolic to indicate that the integration must cover the entire range of the random variable. If the range has finite limits, i.e., x_l and x_u, then the integration is carried out for them.

The probability of X being within a range from x_1 to x_2, is

$$P(x_1 < X \le x_2) = F(x_2) - F(x_1) = \int_{x_1}^{x_2} f(x)\, dx \quad . \tag{2.5-7}$$

In Equation (2.5-7), the probability statements on the left are equated to the integral $\int f(x)\, dx$. Since x usually has a dimension and dx has the same dimension, $f(x)$ cannot be dimensionless, hence it is not a probability. In connection with Equation (2.5-3), $f(x)$ was defined as a probability rate per increment of range, so that $f(x)\, dx$ represents a probability. Hence, the continuous pdf, $f(x)$, has the inverse dimension of x. For example, if x and dx have the units of a length, then $f(x)$ must have the unit of 1/length. Keeping track of units is important for calculations with pdf's because it can avoid serious errors.

Similar to Equation (2.5-4), the integral of the continuous pdf from the lower bound of the range to a specified upper limit is

$$F(x_u) = P(X \le x_u) = \int_{-\infty}^{x_u} f(x)\, dx \tag{2.5-8}$$

where $F(x_u)$ is the *non-exeedance probability* of x_u. The complement of $F(x_u)$,

$$P[X > x_u] = 1 - F(x_u) \tag{2.5-9}$$

is the *exceedance probability* of x_u.

If X represents a variable such as *life length*, or time between breakdowns, then it is of interest to know the probability of such a variable to be exceeded. The exceedance probability is

$$R(x) = 1 - F(x) \tag{2.5-10}$$

where x is a quantity to be exceeded, and $R(x)$ is the *survival probability*, or *reliability*.

Examples: (1) The discrete probability density function only has function values at discrete x. Suppose the discrete pdf has values at x_1, x_2, and x_3 which are $f(x_1)$, $f(x_2)$, and $f(x_3)$. What is the probability $F(x_2) = P(X \leq x_2)$? Solution: According to Equation (2.5-4), $F(x_2) = f(x_1) + f(x_2)$.

(2) Given the same distribution as for Example (1), what is the probability $P(x_1 \leq X \leq x_3)$? Solution: According to Equation (2.5-5), $P(x_1 \leq X \leq x_3) = F(x_3) - F(x_1) = f(x_1) + f(x_2) + f(x_3) - f(x_1) = f(x_2) + f(x_3)$.

2.5.3 Moments of the Random Variable

Moments of the type used in statistical calculations are the first and higher moments of areas around an axis. Since $f(x)$ dx is a strip of area under the pdf, the product of this area and its distance x from the y-axis represents a probability-weighted point of the range of x with respect to the origin of x. Taking the integral of this expression over the range amounts to summing the probability- weighted x-values, which produces the mean. Similar moments are calculated by using the area $f(x)$ dx and higher powers of x or some other distance. Some examples are given.

(a) *Mean of a continuous random variable*: The product of the total area A under a pdf and its average distance from the y-axis is set equal to the sum of moments of area slices and their distances from the y-axis. For a continuous function, this balance is expressed by

$$\mu A = \int_a^b x \, dA \qquad (2.5\text{-}11)$$

where μ is an *a priori* unknown distance of the center of gravity of the range from the y-axis; A is the total area under the pdf which by definition of a pdf is $A = 1$. With $dA = f(x) \, dx$ the integration produces

$$E(X) = \mu = \int_a^b x \, f(x) \, dx \qquad (2.5\text{-}12)$$

where $E(X)$ is the expectation of the random variable X, which is also μ; the integration limits a and b include the entire range of the random variable, which could be from $-\infty$ to $+\infty$. Whereas $f(x)$ has only positive function values, x may have positive or negative values. Equation (2.5-12) is the *first moment* of the random variable because the moment arm x is of first order.

(b) *Variance of a continuous random variable*: The moment of inertia of an area is computed by taking the area times the square of the distance from the moment axis. If this axis is through the gravity center, μ, instead of the y-axis, then we speak of a *centered second moment*. It produces the variance of the random variable:

$$E[(X - \mu)^2] = \sigma^2 = \int_{-\infty}^{+\infty} (x - \mu)^2 f(x)dx \qquad (2.5\text{-}13)$$

where σ^2 is the variance or the expectation of the squared distance of all points of the range from the mean of the range, and σ is the standard deviation. The integration limits express the inclusion of the entire range of the random variable.

(c) *Mean of a discrete random variable*: With x_i being the discrete values of the random variable and $f(x_i)$ being the associated frequencies, one obtains the expectation of the random variable as the probability-weighted sum of all x_i:

$$E(X) = \mu = \sum_{i=1}^{n} x_i f(x_i) \qquad (2.5\text{-}14).$$

(d) *Variance of a discrete random variable*: Using the squared distance from the mean, the moment becomes

$$E[(X - \mu)^2] = \sigma^2 = \sum_{i=1}^{n} (x_i - \mu)^2 f(x_i) \qquad (2.5\text{-}15)$$

where σ^2 is the variance or centered second moment of X.

(e) *The expectation of a function of a random variable*: Suppose the random variable is given as some algebraic function, $g(X)$. Then the expectation of $g(X)$ is the sum of all probability-weighted elements of $g(X)$. For a discrete pdf, $f(X)$, one obtains

$$E[g(X)] = \sum_{i=1}^{n} g(x_i) f(x_i). \qquad (2.5\text{-}16)$$

Equation (2.5-16) is a general form of moment formulas already discussed. The function $g(X)$ can be linear, quadratic, or any other function of the random variable. If $g(X) = X^k$, with X being the quantity displayed on the x-axis, and $k = 1$, then $E(X)$ obtained from Equation (2.5-16) is the first moment of X. If $k = 2$ and X is the (centered) distance from $E(X) = \mu$, then $g(X) = (X - \mu)^2$ is the squared centered distance from the mean, and its probability-weighted sum or integral is the *second central moment,* or variance. Up to four central moments of random variables are used in statistics (Kreyszig, 1979, p. 871; Moan, 1982, pp. 4-10 to 4-12).

2.6 Gaussian Probability Density Function

2.6.1 Derivation of the Probability Density Function

The normal probability density function, also called Gauss-Laplace pdf, or Gaussian, with its characteristic bell shape is the most widely known pdf. It also is the most frequently used and probably also the most important of all pdf's. These statements will be fully appreciated only in Section 3.4.2. The Gaussian is based on the quadratic exponential function

$$f(x) = e^{-\frac{x^2}{2}} .$$

(2.6-1)

The pdf values, $f(x)$, can be easily calculated from Equation (2.6-1), the integral, $F(x)$, however, is analytically obtainable only in special cases. Function values $F(x)$ are usually taken from tables or calculated by numerical summation formulas (numerical integration). Equation (2.6-1) is distinguished from the common exponential decay function, which is linear in x, by having a horizontal tangent at $x = 0$, as is shown by its derivative

$$\frac{df(x)}{dx} = - x\, e^{-\frac{x^2}{2}} .$$

(2.6-2)

For $x = 0$, $df(x)/dx = 0$; for $x \to \infty$, $df(x)/dx = \infty \cdot 0 \to 0$. This result follows from the Bernoulli–L'Hopital rule, which calls for taking the derivative of numerator and denominator separately. This gives

$$\frac{d(-x)/dx}{d(e^{x^2/2})/dx} = \frac{-1}{x\, e^{x^2/2}} .$$

For $x \to \infty$, one obtains $-1/(\infty \cdot \infty) = 0$. This means that the quadratic exponential function has a double curvature. It is concave (at or below its tangent) near the origin, and convex[1] (at or above its tangent) for large x. Its tangent is zero at $x = 0$, then becomes negative (for the positive branch of x) and again approaches zero with $x \to \infty$, there must be a maximum negative slope in between. Taking the derivative of Equation (2.6-2) and setting it to zero gives $x = 1$ as the location of the maximum slope for the positive branch. The positive branch of Equation (2.6-1) is shown in Figure 2-13. The discussed properties give Equation (2.6-1) the well-known bell shape with the y-axis as symmetry axis.

Equation (2.6-1) is a simple algebraic function in x. The exponent x is a pure number. A more general function is obtained by reducing the general variable x to

$$z = \frac{x - \mu}{\sigma} \tag{2.6-3}$$

where z is the normalized variable, μ is the mean, and σ is the standard deviation of the random variable X. Subtracting μ from x moves the symmetry axis to $x = \mu$, and dividing $x - \mu$ by σ makes z a dimensionless number. The function $f(z)$ becomes

$$f(z) = e^{-\frac{z^2}{2}} = e^{-\frac{1}{2}\frac{(x-\mu)^2}{\sigma^2}}. \tag{2.6-4}$$

The first term on the right side of Equation (2.6-4) looks exactly like Equation (2.6-1), with x replaced by z. For $\mu = 0$, and $\sigma = 1$, $x = z$. This is a special case for which Equations (2.6-1) and (2.6-4) produce identical curves centered on $x = z = 0$. For other sets of (μ, σ), for example, $\mu = 1$, and $\sigma = 0.3$, Equation (2.6-4) produces a bell shape that is centered on $x = \mu$, where $z = 0$. Also, the bell shape is reduced in width, as shown by Figure 2-13. The corresponding function value for $f(z)$ at $z = 0$ is $f(0) = 1$, which is the maximum ordinate of $f(z)$. In contrast to Equation (2.6-1), the bell shape produced by $f(z)$ does not depend on negative x. Equation (2.6-4) can shift the bell shape far enough to the right of $x = 0$ so that the entire function, $f(z)$, or at least the part that is of practical importance, is on the positive x-axis. This is important for the use of the Gaussian as a distribution function of X when the negative domain of X is meaningless.

---

[1]A function is convex from below, or briefly *convex* (bowl shaped), in the interval (a, b), if its function values, $f(x_2)$, are at or *above* its tangent values $f(x_1) + f'(x_1)(x_2 - x_1)$, for a tangent in x_1; $f'(x_i) = df(x_i)/dx$ can be taken anywhere in the interval (a, b); x_2 is a location of testing the criterion for the convex shape: $f(x_2) \geq f(x_1) + f'(x_1)(x_2 - x_1)$, for all x_1, x_2 in the interval (a, b). A curve is convex from above, or *concave* (cave shaped), if its function values are at or *below* its tangent, or $f(x_2) \leq f(x_1) + f'(x_1)(x_2 - x_1)$, for all x_1, x_2, in (a, b).

The maximum slope of the bell shape occurs where $f''(z) = 0$. Taking the second derivative of Equation (2.6-4) and setting it zero gives $z = \pm 1$, or $x = \mu \pm \sigma$, in other words, at the distance $\pm \sigma$ from the symmetry axis through μ.

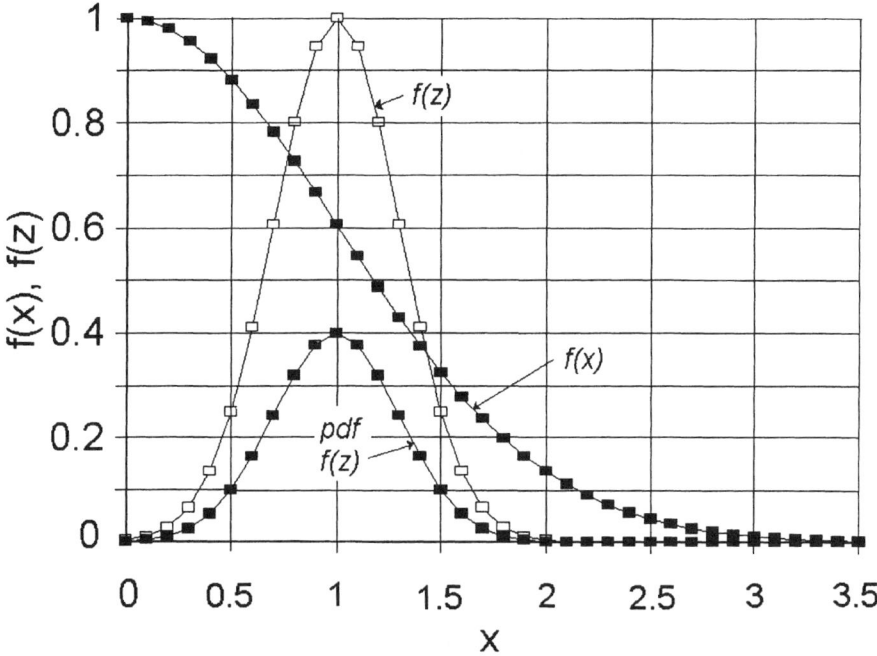

Figure 2-13: The exponential function $f(x) = e^{-x^2/2}$ is transformed into a two-parameter function $f(z)$ by replacing x by the normalized variable $z = (x - \mu)/\sigma$. Here, with $\mu = 1$ and $\sigma = 0.3$, the transformed function, $f(z) = \exp\{-0.5\,[(x - \mu)/\sigma]^2\}$, has its symmetry axis at $z = 0$, or $x = \mu = 1$, where it reaches its maximum value $f(z) = f(0) = 1$; its maximum slope is at $x = \mu \pm \sigma = 1.3$ and 0.7. The transformed variable z produces positive and negative values of z even for only positive x, which allows a symmetric bell shape to form over the positive x-axis. In order to become a pdf, the ordinates $f(z)$ must be scaled by the factor 0.398942 so that the area under the curve integrates to 1. The maximum ordinate of $f(z)$ is thereby reduced from 1 to 0.398942 and becomes the maximum ordinate of the normal pdf. The pdf is usually referred to as $f(z)$. It is shown as the significantly scaled down curve under the unscaled function $f(z)$.

The area under the bell-shaped function $f(z)$ must be 1 for $f(z)$ to be a pdf. This is achieved by scaling the ordinates of $f(z)$. To do this, an auxiliary exponential function, $g(z) = \exp(-z^2/2)$, is integrated over the entire range of z, $-\infty \leq z \leq +\infty$. With the substitution $z/\sqrt{2} = t$ and $dz = \sqrt{2}\,dt$, the integral becomes

$$\int_{-\infty}^{+\infty} g(z)\,dz = \int_{-\infty}^{+\infty} e^{-\frac{z^2}{2}}\,dz = \sqrt{2}\int_{-\infty}^{0} e^{-t^2}\,dt + \sqrt{2}\int_{0}^{+\infty} e^{-t^2}\,dt \quad . \tag{2.6-5}$$

The two integrals on the right-hand side are solved with the help of the gamma function, $\Gamma(x)$ (Abramowitz and Stegun, 1970, p. 255):

$$\Gamma(\frac{1}{2}) = 2\int_{0}^{+\infty} e^{-t^2}\,dt = \sqrt{\pi} = 1.772454 \ . \tag{2.6-6}$$

Making use of the symmetry of $g(z)$ with respect to $z = 0$, the result of Equation (2.6-6) is substituted into Equation (2.6-5) to produce

$$\int_{-\infty}^{+\infty} g(z)dz = \sqrt{\frac{\pi}{2}} + \sqrt{\frac{\pi}{2}} = \sqrt{2\pi} \ . \tag{2.6-7}$$

To meet the requirement that the total area under $f(z)$ equals 1, all ordinates of $g(z)$ are divided by the factor $\sqrt{2\pi}$, which produces the scaled $f(z)$

$$f_Z(z) = \frac{1}{\sqrt{2\pi}} e^{-\frac{z^2}{2}} = 0.398942\,e^{-\frac{z^2}{2}} \tag{2.6-8}$$

where $f_Z(z)$ is the normal pdf of the random variable Z, dimensionless function.

By Equation (2.6-3), z is also a function of x. Differentiation of z versus x gives

$$dz = \frac{1}{\sigma} dx \ . \tag{2.6-9}$$

The dimensionless product $f_Z(z)\,dz$ is the normal probability. The transformation from x to z does not change the probability of x and one can write

$$f_X(x)\,dx = f_Z(z)\,dz \ . \tag{2.6-10}$$

where the indices X and Z indicate that these functions are pdf's. From Equation (2.6-10) one obtains $f_X(x) = f_Z(z)\, dz/dx$, so that with Equation (2.6-8) one obtains

$$f_X(x) = \frac{0.39894}{\sigma} e^{-\frac{(x-\mu)^2}{2\sigma^2}} \tag{2.6-11}$$

The normal probabilities are

$$f_Z(z)\, dz = 0.39894\, e^{-\frac{z^2}{2}}\, dz \tag{2.6-12}$$

and

$$f_X(x)\, dx = \frac{0.39894}{\sigma} e^{-\frac{(x-\mu)^2}{2\sigma^2}}\, dx \; . \tag{2.6-12a}$$

Given Equation (2.6-9), it is obvious that the right-hand sides of Equations (2.6-12) and (2.6-12a) are identical. A check of the units shows that $f_X(x)$, in contrast to $f_Z(z)$, is not a dimensionless quantity. From Equation (2.6-9) it follows that dx has the dimension of σ and from Equation (2.6-11) it is seen that $f_X(x)$ has the reciprocal dimension of σ, whereas $f_Z(z)$ as well as dz are dimensionless. The products $f_X(x)$ and $f(z)\, dz$ are both dimensionless, as probabilities must be.

The maximum function value of $f_Z(z) = 0.39894$ occurs at $z = 0$ and is a constant. In contrast, the maximum function value of $f_X(x)$ occurs at $x = \mu$, and is

$$f_X(x)_{max} = \frac{0.39894}{\sigma} . \tag{2.6-13}$$

Equation (2.6-13) shows that $f_x(x)_{max} \to \infty$ as $\sigma \to 0$. For all other ordinates, Equation (2.6-11) shows that two things happen when $\sigma \to 0$. The factor of the exponential function increases and so does the exponent. Since the exponent is negative, it reduces the ordinates faster than the factor can increase them. This narrows the normal distribution to a spike around the mean. Since $\int f_x dx$ over the entire distribution must still be 1, the probability of $x = \mu$ approaches 1, as $\sigma \to 0$. This property of a normal random variable suggests the interpretation of a deterministic variable as a special case of a random variable whose standard deviation is close or equal to zero.

The normal pdf, $f_Z(z)$ given by Equation (2.6-8) depends only on z and has an invariable shape. It starts out with $f_Z(z)_{max} = 0.3989$ for $z = 0$ and decays with increasing z. In textbooks on statistics or probability theory, $f_Z(z)$ is tabulated for $z \geq 0$ up to about 4 or 5. At $z = 3$, $f_Z(z) = 0.004432 = 4,432 \cdot 10^{-6}$, a value small enough to be frequently used as a practical limit to define a finite range of the distribution. At $z = 4$, $f_Z(z) = 1,338 \cdot 10^{-7}$, and at $z = 4.99$, $f_Z(z) = 1,563 \cdot 10^{-9}$.

For $\sigma = 1$, $f_X(x)$ is formally the same as $f_Z(z)$. But this is strictly not true as σ usually has a unit, and therefore $f_X(x) \neq f_Z(z)$. Sometimes one finds the statement $\mu = 0$ and $\sigma = 1$, so that $z = x$ according to Equation (2.6-3). This statement glosses over the fact that x, μ, and σ usually have a dimension, whereas z is always dimensionless. Suppose a population x has the unit of a length, and $\sigma = 1$ m, then $f_X(x)$ has the units $[m^{-1}]$, whereas $f_Z(z)$ has the unit $[1]$. Keeping track of the units of all quantities may improve understanding and provide insight. It also can avoid serious errors.

Example: Figure 2-13 shows three functions: $f(x)$ by Equation (2.6-1), $f(z)$ by Equation (2.6-4), and $f_Z(z)$ by Equation (2.6-8). Only the last equation represents a pdf and is given the index Z that refers to random variable Z. The function values in Figure 2-13 are calculated from $x = 0$ to $x = 3.5$ at intervals $dx = 0.1$. With assumed values of $\mu = 1$ and $\sigma = 0.3$, verify that Equation (2.6-8) and Equation (2.6-11) are pdf's. Solution: (a) In a spreadsheet, calculate z and $f_Z(z)$ for $0 \leq x \leq 3.5$ at increments $dx = 0.1$ using Equation (2.6-3) and Equation(2.6-8), respectively. The sum of the 36 values of $f_Z(z)$ is 2.9993. If this sum or each individual value $f_Z(z)$ is converted to a probability by multiplying with $dz = dx/\sigma = 0.1/0.3$, the sum becomes $0.9998 \approx 1$. (b) If the ordinates of $f_Z(z)$ are divided by σ one obtains $f_X(x)$ according to Equation (2.6-11). The sum of the 36 ordinates $f_X(x)$ is 9.9978. Multiplying this sum or the individual ordinates by $dx = 0.1$ converts them to probabilities which add up to $0.9998 \approx 1$. Hence, as required by Equation (2.6-10), $f_X(x) dx = f_Z(z) dz$. The departure of the sums of probabilities from 1 is caused by numerical inaccuracy and by the fact that the left tail of the pdf is truncated at $z = -3.3333$, or $x = 0$.

2.6.2 Cumulative Distribution Function and Reliability

The cumulative distribution functions, $F(x)$ or $F(z)$, in order to represent a probability, must be dimensionless and must integrate to 1. In the previous section it was shown that $f(x) dx$ and $f(z) dz$ are interchangeable. For simplicity of notation, the random variable indices are dropped. The differential of $F(x)$ is

$$dF(x) = f(x) dx \approx F(x + \Delta x) - F(x) \tag{2.6-14}$$

Equation (2.6-14) gives the probability of any of the realizations x in an arbitrarily narrow class of width, $dx \approx \Delta x$. The increment $dF(x)$ can be construed as the probability of a set of mutually exclusive realizations x of a sample space partition dx, or its numerical equivalent Δx. Switching to the normalized variable, z, the probability of a realization z to be less than or equal to a limiting value, z_u, is

$$F(z_u) = P(Z \le z_u) = \int_{-\infty}^{z_u} f(z)dz \qquad (2.6\text{-}15)$$

where $P(Z \le z_u)$ is the *non-exceedance probability of z_u*. For the numerical evaluation of such an integral, the meaning of the integral, namely, the sum of narrow strips of areas between the z-axis and $f(z)$ with width dz is used, similar to subdividing the area under a curve over a range x by bars in the form of a histogram. Since the original variable is x, the relation between dx and dz must be observed. For example, suppose $\Delta x = 0.1$ MPa, and $\sigma = 0.19$ MPa, then $\Delta z = \Delta x / \sigma = 0.525$.

The non-exceedance probability of a quantity, z_u, with Gaussian distribution is obtained as

$$F(z_u) = \int_{-\infty}^{z_u} f(z)dz = 0.39894 \int_{-\infty}^{0} e^{-\frac{z^2}{2}} dz + 0.39894 \int_{0}^{z_u} e^{-\frac{z^2}{2}} dz \; . \qquad (2.6\text{-}16)$$

This is a general expression that pivots on $z = 0$ regardless of whether z_u is greater than or less than zero. The problem of the boundaries $-\infty$ or $+\infty$ is resolved by the fact that the integral of the normal distribution from $z = -\infty$ to $z = 0$, as well as from $z = 0$ to $z = +\infty$ is 0.5 by symmetry. Thus, the actual integration range can always be carried out on the positive z-axis, from $z = 0$ to $z = z_u$, regardless of whether z is positive or negative. An integral on the negative z-axis, from $z = 0$ to $z = -z_l$, can be replaced by an integral on the positive axis from $z = 0$ to $z = z_l$ with the integration result receiving a negative sign. Hence,

for $z_u > 0$:
$$F(z_u) = \int_{-\infty}^{z_u} f(z)dz = 0.5 + 0.39894 \int_{0}^{z_u} e^{-\frac{z^2}{2}} dz \qquad (2.6\text{-}17)$$

for $z_l < 0$:
$$F(z_l) = \int_{-\infty}^{-z_l} f(z)dz = 0.5 - 0.39894 \int_{0}^{z_l} e^{-\frac{z^2}{2}} dz \qquad (2.6\text{-}18)$$

Both Equations (2.6-17) and (2.6-18) require finding an integration from $z = 0$ to some upper positive limit

$$F_0(z_u) = 0.39894 \int_0^{z_u} e^{-\frac{z^2}{2}} dz \tag{2.6-19}$$

The integral $F_0(z)$ covers the right half of the bell shape and is tabulated in textbooks. It converges to 0.5, as z goes to infinity. Sometimes also the full integral, $F(z)$, of Equation (2.6-16) is tabulated for positive z which amounts to adding 0.5 to the table values of $F_0(z)$ and thus converges to 1, as z goes to infinity. Illustrations of four integrations using tabular values of $F_0(z)$ are given in Figure 2-14. $F_0(z)$ is always an area under $f(z)$ adjacent to $z = 0$, either to the left or to the right of it.

The complement of the cumulative distribution function, $F(z_u)$, is

$$R(z_u) = 1 - F(z_u) \tag{2.6-20}$$

where $R(z_u)$ is the *exceedance probability* of z_u. It is also the probability that $z \leq z_u$ does not occur. Therefore it also called   *probability of nonoccurrence, survival probability*, and *reliability*. If z is the normalized variable of time to failure, then $R(z_u)$ is the probability that the time to failure will exceed z_u. Selected values of Equations (2.6-19), (2.6-16), and (2.6-20) for the normal distribution are given in Table 2-3, in columns 3, 4, and 6, respectively.

For practical calculations, the most convenient way to evaluate the probability integral is as follows:

1. Calculate the ordinates $f(z)$ from the exponential function, Equation (2.6-8).

2. Make a convenient choice for dz, such as $dz \approx \Delta z = 0.1$.

3. Calculate the integral as the sum of rectangular areas that have the average of two adjacent values $f(z)$ as height and Δz as width. This calculation is also referred to as *Euler's numerical integration formula* (or trapezoidal formula). For $z_u > 0$, the numerical integral is

$$F(z_u) = 0.5 + \Delta z \left(0.5 f_0 + \sum_{i=1}^{n-1} f_i + 0.5 f_n \right). \tag{2.6-21}$$

For $z_u < 0$, the term following 0.5 in Equation (2.6-21) receives a negative sign; $f_0 = f(0) = 0.39894$ at $z = 0$, and $f_n = f(z_u)$ is the ordinate at $z = z_u$; these two ordinates represent the lower and upper boundary of the area $F_0(z_u)$. The other ordinates, f_i, for z-values with indices $i = 1$ to $n - 1$ are in the interior of $F_0(z_u)$. The number of integration steps is $n = z_u/\Delta z$, where z_u is measured on the z-axis from $z = 0$ to the right, and Δz is a finite increment. As long as Δz is kept reasonably small, the numerical integration will be sufficiently accurate. Whatever the sign of z, the integration can always proceed along $z \geq 0$. For $z = 5$, $F_0(5) = 0.499\ 999\ 7$, which is sufficiently close to 0.5 to break off the integration.

Four examples of integrations using table values or calculated $F_0(z)$, all for $z \geq 0$, are illustrated in Figure 2-14. The true function values, $F(z)$, are found as follows:

for $z < 0$:

non-exceedance probability: $F(-z) = 0.5 - F_0(z)$ (2.6-22)

exceedance probability: $R(-z) = 1 - F(-z) = 0.5 + F_0(z)$ (2.6-23)

for $z > 0$:

non-exceedance probability: $F(z) = 0.5 + F_0(z)$ (2.6-24)

exceedance probability: $R(z) = 1 - F(z) = 0.5 - F_0(z)$ (2.6-25)

where 0.5 represents the integral over one half of the bell-shaped $f(z)$, either from $z = -\infty$ to $z = 0$, or from $z = 0$ to $z = +\infty$. For $z = 0$, $F_0(0) = 0$, and for $z \to \pm\infty$, $F_0(z) \to 0.5$. The values $F_0(z)$, usually taken from a table, are always positive, regardless of whether the area is integrated in the $+z$ or $-z$ direction. A sign has to be given to the result according to the direction of the integration. For example, case (b) in Figure 2-14 is represented by Equation (2.6-23) and the probability is $R(-z) = 1 - F(-z) = 1 - [0.5 - F_0(z)] = 0.5 + F_0(z)$; case (d) is represented by Equation (2.6-25) and the probability is $R(z) = 1 - F(z) = 1 - [0.5 + F_0(z)] = 0.5 - F_0(z)$. In both cases, $F_0(z)$ originally has the sign that corresponds to the integration direction, but the sign is reversed when the complementary probability is taken.

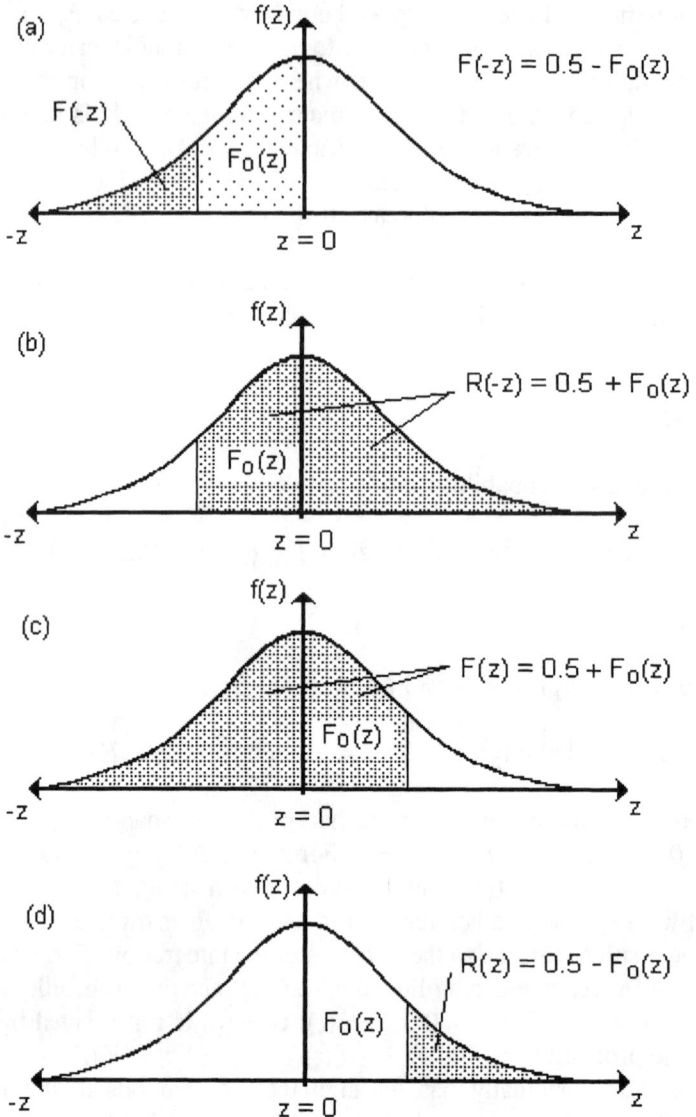

Figure 2-14: Illustration of Gaussian nonexceedance and exceedance probabilities, $F(z)$ and $R(z)$, respectively, as areas under the pdf, $f(z)$, for positive and negative z. $F_0(z)$ is the portion tabulated in Gaussian function tables. The area 0.5 is the half space either to the left or to the right of $z = 0$.

Table 2-3: Selected Values of the Gaussian Function: $f(z)$, $F_0(z)$, $F(z)$, and $R(z)$

z	$f(z)$	$F_0(z)$	$F(z)$	$F(z) - F(-z)$	$R(z)$
(1)	(2)	(3)	(4)	(5)	(6)
0	0.3989	0	0.5	0	0.5
0.5	0.3521	0.1915	0.6915	0.3830	0.3085
0.6745	0.3178	0.25	0.75	0.5	0.25
1.0	0.24197	0.34134	0.84134	0.68268	0.1587
1.64	0.1040	0.4495	0.9495	0.8990	0.0505
1.96	0.05844	0.47500	0.97500	0.95	0.0250
2.0	0.05399	0.47725	0.97725	0.9545	0.0228
2.58	0.01431	0.49506	0.99506	0.99012	0.00494
3.0	0.0044318	0.49865	0.99865	0.9973	0.00135
4.0	0.0001338	0.49997	0.99997	0.99994	0.00003
5.0	$1.4867 \cdot 10^{-6}$	0.499999	0.999999	0.999998	0.000001

Notes: column 1: selected z values.

column 2: $f(z) = 0.398942 \exp(-z^2/2)$.

column 3: integral over $f(z)$ from $z = 0$ to an upper limit z. Table values $F_0(z)$ are found in Abramowitz and Stegun (1970, p. 966).

column 4: integral over $f(z)$ from $z = -\infty$ to the upper limit z_u. The integral $F(z_u)$ can be visualized as the area under the curve $f(z)$ to the left of the ordinate $f(z_u)$. $F(z)$ is also $0.5 + F_0(z)$.

column 5: probability of z being within a band width $\pm z$ around $z = 0$; it is computed by $F(z) - F(-z) = F(z) - [1 - F(z)] = 2 F(z) - 1$. The difference between columns 4 and 5 is the integral from $z = -\infty$ to $-z$, or the left tail area, $F(-z) = 1 - F(z)$, which is the same as the right tail area.

column 6: $R(z) = 1 - F(z)$, the exceedance probability of z, which is the right tail area.

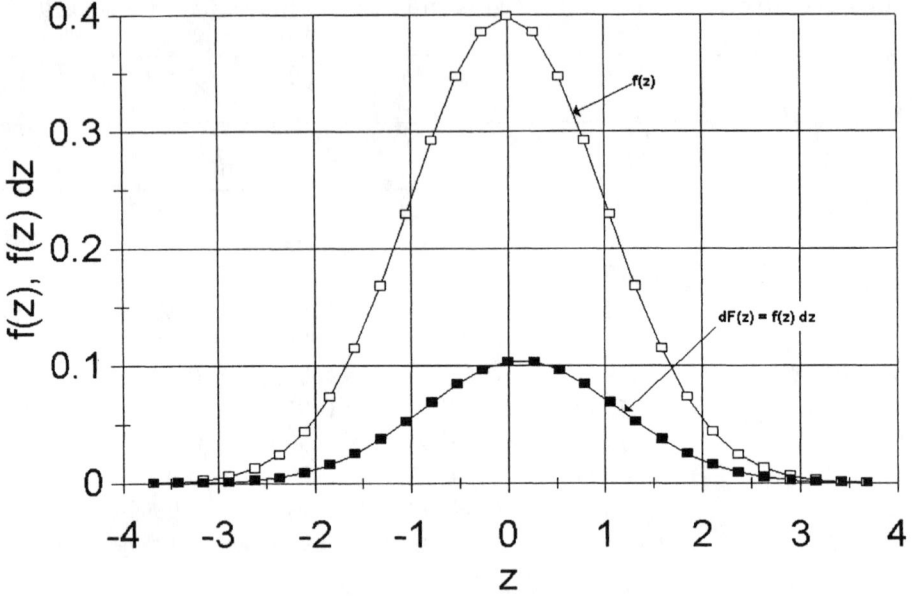

Figure 2-15: The Gaussian pdf $f(z)$ is shown for parameters $\mu = +2.5$ MPa and $\sigma = 0.19$ MPa, and the normalized variable $z = (x - \mu)/\sigma$. The numerical values are given in Table 2-4. Each increment $dF(z) = f(z)dz$ represents the probability of a contiguous data set or partition, similar to a histogram (Table 2-4, column 4). The probability of the union of all partitions (or the integral over the range of $f(z)dz$) is the probability of the sample space, $P(S) = 1$; here, the numerical integration gives 0.9993 (bottom row of Table 4, column 5).

The Gaussian distribution is a two-parameter function. Usually only a limited number of data are available for calculating the parameters, μ and σ. Once the parameters are known, the pdf, $f(z)$, and all other associated functions can be calculated. An example using concrete tensile strength data is given in Table 2-4. The resulting $f(z)$ is shown in Figure 2-15. This calculated shape does not necessarily fit the empirical distribution that is shown as a data histogram in Figure 2-4 because either the data may be insufficient, the parameters may be inaccurate, or the assumed distribution does not apply. Statistical tests exist to reject or not reject an assumed distribution (see Kolmogorov–Smi rnov test in Section 3.4.4 and regression test in Section 3.4.5).

The *numerical integration of a continuous pdf* is similar to summing a discrete pdf. The integral of a discrete pdf is the sum of its ordinates up to a specified limit. The numerical integral of a continuous pdf is the sum of area strips up to and including a specified limiting ordinate. The integration step is

$$\Delta F(z) = f(z_m)\Delta z = \frac{1}{2}[f(z_i) + f(z_{i+1})]\Delta z \tag{2.6-26}$$

where $f(z_m)\,\Delta z$ is the bar area of a histogram, whereas $f(z_m)$ is the average function value at the center of interval, Δz, here the arithmetic average of the values on the interval boundaries, $f(z_i)$ and $f(z_{i+1})$. The integration of a constant pdf, illustrated by a strip $\Delta F(z)$, produces a straight line ascending across the interval (see also Section 2.7.3). For a series of contiguous $\Delta F(z)$ strips this amounts to constructing an integral curve that consists of a contiguous series of straight line segments between the interval boundaries, $f(z_i)$ and $f(z_{i+1})$. The smaller the intervals, the more closely this straight line sequence will follow the true integral curve. The integration amounts to stacking the increment, $\Delta F(z) = f(z_m)\,\Delta z$, on top of the previous value of the integral, $F(z_i)$, which produces the new value $F(z_{i+1})$ at end of the interval $i + 1$.

The numerical integration procedure is demonstrated by integrating two contiguous area strips bounded on the left by $f(z_1)$ and on the right by $f(z_3)$, with $f(z_2)$ in between:

$$\int_{z_1}^{z_3} f(z)dz = \Delta F(z)_1 + \Delta F(z)_2$$

$$= 0.5\,[f(z_1) + f(z_2)]\,\Delta z + 0.5\,[f(z_2) + f(z_3)]\,\Delta z$$

$$= 0.5\,f(z_1)\,\Delta z + f(z_2)\,\Delta z + 0.5\,f(z_3)\,\Delta z$$

$$= [0.5\,f(z_1) + f(z_2) + 0.5\,f(z_3)]\,\Delta z. \tag{2.6-27}$$

If the summation of strips is extended from two to n, Euler's numerical integration formula, i.e., the second term of Equation (2.6-21), is obtained. The numerical integration of a normal pdf is carried out in Table 2-4 by cumulating the $\Delta F(z)$ in column 5. The distribution of the increments $\Delta F(z)$ is shown in Figure 2-15, and the integral curve $F(z)$ and its complement $R(z)$ are shown in Figure 2-16.

Table 2-4: Gaussian Probability Functions $f(z)$, $dF(z)$, $F(z)$, and $R(z)$
for Parameters $\mu = 2.5$ MPa, $\sigma = 0.19$ MPa

x	z	$f(z)$	$dF(z)$	$F(z)$	$R(z)$
(1)	(2)	(3)	(4)	(5)	(6)
1.80	−3.6842	0.0005	0.0000	0.0001	0.9999
1.85	−3.4211	0.0011	0.0002	0.0003	0.9997
1.90	−3.1579	0.0027	0.0005	0.0008	0.9992
1.95	−2.8947	0.0060	0.0012	0.0020	0.9980
2.00	−2.6316	0.0125	0.0024	0.0044	0.9956
2.05	−2.3684	0.0241	0.0048	0.0093	0.9907
2.10	−2.1053	0.0435	0.0089	0.0181	0.9819
2.15	−1.8421	0.0731	0.0153	0.0335	0.9665
2.20	−1.5789	0.1147	0.0247	0.0582	0.9418
2.25	−1.3158	0.1679	0.0372	0.0953	0.9047
2.30	−1.0526	0.2292	0.0522	0.1476	0.8524
2.35	−0.7895	0.2921	0.0686	0.2161	0.7839
2.40	−0.5263	0.3473	0.0841	0.3002	0.6998
2.45	−0.2632	0.3854	0.0964	0.3966	0.6034
2.50	−0.0000	0.3989	0.1031	0.4997	0.5003
2.55	0.2632	0.3854	0.1031	0.6028	0.3972
2.60	0.5263	0.3473	0.0964	0.6992	0.3008
2.65	0.7895	0.2921	0.0841	0.7833	0.2167
2.70	1.0526	0.2292	0.0686	0.8518	0.1482
2.75	1.3158	0.1679	0.0522	0.9040	0.0960
2.80	1.5789	0.1147	0.0372	0.9412	0.0588
2.85	1.8421	0.0731	0.0247	0.9659	0.0341
2.90	2.1053	0.0435	0.0153	0.9812	0.0188
2.95	2.3684	0.0241	0.0089	0.9901	0.0099
3.00	2.6316	0.0125	0.0048	0.9950	0.0050
3.05	2.8947	0.0060	0.0024	0.9974	0.0026
3.10	3.1579	0.0027	0.0012	0.9985	0.0015
3.15	3.4211	0.0011	0.0005	0.9991	0.0009
3.20	3.6842	0.0005	0.0002	0.9993	0.0007

Notes: The two parameters, μ and σ, of this data sample are based on 100 values of splitting tensile strength of concrete cylinder tests by D. L. Ivey, Texas Transportation Institute, College Station Texas, 1965, as reported by Kreyszig, 1979,

p. 840. The calculations are based on the assumption that the Gaussian pdf fits the data. If this should turn out not to be the case, another pdf has to be tried.

column 1: class values of strength, x; class width $dx = 0.05$ MPa; all data of class $1.8 \leq x \leq 1.8499$ are assumed to be represented by $x = 1.8$ MPa; similar assumptions are made for all classes, i.e., rows of the table.

column 2: $z = (x - 2.5)/0.19$; $dz = 0.05/\sigma = 0.263$.

column 3: $f(z) = 0.3989 \exp(-z^2/2)$

column 4: $dF(z) = 0.5 \cdot [f(z_1) + f(z_2)]\, dz$, with $dz = 0.263$; the values of column 4 sum to 0.9992; the values of the classes (area strips) are plotted in Figure 2-15 at the upper limit of their class. This produces a shift of the curve to the right, as can be seen from the asymmetry at the curve maximum. It would be more appropriate to plot $dF(z)$ at the class center which would require another x-scale. In computer applications one can reduce the class width until such inconsistencies become negligible.

column 5: cumulation of column 4 from $z = -3.6842$ to $z = +3.6842$. The sum falls slightly short of 1 because the cumulation starts at $x = 1.8$ which corresponds to $z = (1.8 - 2.5)/0.19 = -3.684$. $F(-3.6842) = 0.00012$. A similar amount is missing beyond $z = 3.6842$. The $F(z)$ are plotted in Figure 2-16 at the upper limit of each class where $dF(z)$ has fully accumulated.

column 6: $R(z) = 1 - F(z)$; the $R(z)$ are plotted in Figure 2-16.

Examples: (1) Evaluate $F(z_u)$ for $z_u = 1$ by numerical integration and compare the result with a table lookup of $F(z_u)$. Solution: If the numerical integration step is chosen as $\Delta z = 0.1$, then the number of integration steps is $n = 1/0.1 = 10$. The $f(z_i)$ for $i = 0$ to 10 are computed from

$$f(z_i) = 0.39894\, e^{-0.5 z_i^2}$$

The numerical integration by the Euler's *numerical integration* formula (there are so many *Euler* formulas that this distinction is necessary), Equation (2.6-21) yields (the function values are abbreviated as f_i)

$$F(1) = 0.5 + 0.1 \cdot \{0.5 f_0 + f_1 + f_2 + f_3 + f_4 + f_5 + f_6 + f_7 + f_8 + f_9 + 0.5 f_{10}\}$$

$$= 0.5 + 0.1 \cdot 10^{-4} \cdot \{3989/2 + 3970 + 3910 + 3814 + 3683$$

$$+ 3521 + 3332 + 3123 + 2897 + 2661 + 2420/2\}$$

$$= 0.5 + 0.3412 = 0.8412.$$

A table lookup of $F_0(z)$ for $z = 1$ gives $F(1) = 0.5 + 0.3413 = 0.8413$.

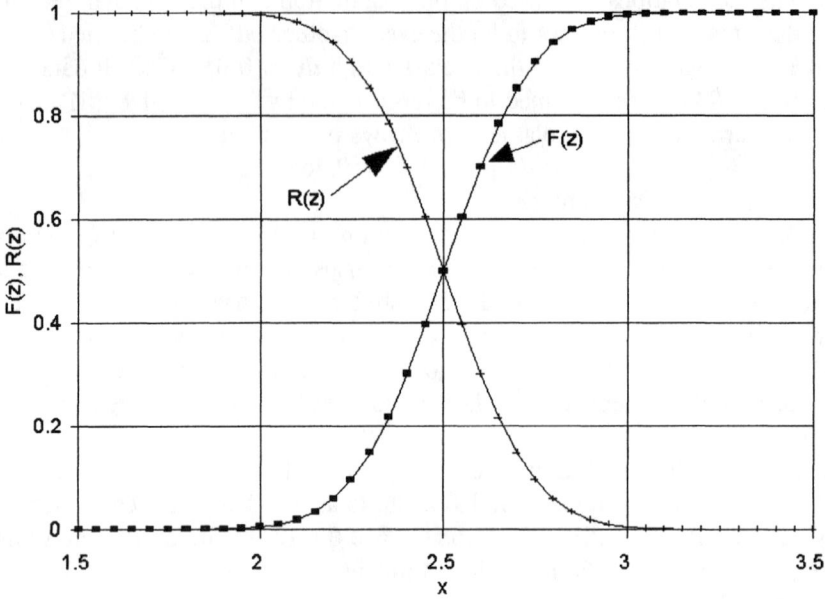

Figure 2-16: Cumulative distribution function, $F(z)$, and reliability function, $R(z) = 1 - F(z)$, based on the Gaussian pdf for $\mu = 2.5$ MPa, and $\sigma = 0.19$ MPa. Numerical data are from Table 2-4, columns 5 and 6, respectively.

(2) What is the probability of z being in the band $-1 \leq z \leq +1$? <u>Solution</u>: From Table 2-3, column 5, one obtains $P(-1 \leq z \leq +1) = F(+z) - F(-z) = F(z) - [1 - F(z)] = 2\ F(z) - 1 = 2 \cdot 0.8413 - 1 = 0.6826$. Using $F_0(z)$, the answer is also $2\ F_0(z) = 2 \cdot 0.3413 = 0.6826$. This means that normally distributed elements are with 68.26 % probability within a band $-1 \leq z \leq +1$ around the mean.

(3) What is the band width that contains half of the sample? <u>Solution</u>: Column 5 of Table 2-3 shows that 50 % of all data fall into the range $-0.6745 \leq z \leq +0.6745$, or within a band $x = \mu \pm 0.6745\ \sigma$, which is a total of $1.349\ \sigma$ wide and centered on the mean. The other 50 % of the data are outside this range. Because of symmetry, 25 % of the data are to the left of $z = -0.6745$, and 25 % are to the right of $z = +0.6745$.

(4) A confidence level used in quality control is 99 %, meaning 1 % may fall outside a confidence band around the mean. What is the width of this confidence band? Assume a table with $F_0(z)$ is available. <u>Solution</u>: For a symmetric band to contain 99 % of all data, $F(z) - F(-z) = 0.99$. With $F(z) = 0.5 + F_0(z)$, and $F(-z) = 0.5 - F_0(z)$, one must solve $F_0(z) = 0.99/2 = 0.495$ for z. From the lookup table one finds for $F_0(z) = 0.495$, $z = 2.58$. Hence the interval extends ± 2.58 σ around the mean.

(5) A practical limit often used for the range of the Gaussian distribution is $z = \pm 3$. What is the meaning of this number, and what is the probability of an element being within these limits? <u>Solution</u>: The standard normal variable is $z = (x - \mu)/\sigma$. If $z = 3$, $x = \mu + 3$ σ. The most distant element x is ± 3 σ from μ. For $z = 3$, a lookup table gives $F_0(z) = 0.49865$. Hence the probability of an element being within the band of ± 3 σ centered on μ is $2 F_0(z) = 0.9973$, or 99.73 %.

2.6.3 Dispersion of the Probability Density Function

The normalized pdf, $f(z)$, has the same shape for all standard deviations. The normalized variable, $z = (x - \mu)/\sigma$, can be interpreted as the distance of x from the mean in multiples or fractions of σ. For example, if $x - \mu = \sigma$, then $z = 1$. At this distance the tangent on the bell shape, after starting out with zero slope at $z = 0$, reaches its maximum negative slope and begins to rise toward zero slope as $z \to \infty$. Distances $\pm x$ from μ, define bands that are centered on μ, and which include specified percentages of the total population or sample. According to Table 2-3, column 5, band widths in multiples of σ extending to both sides of the mean contain the following percentages of the total sample space:

$$P(\mu - \sigma < X \le \mu + \sigma) = 0.6826, \text{ or } 68.26 \% \qquad (2.6\text{-}28a)$$

$$P(\mu - 2\sigma < X \le \mu + 2\sigma) = 0.9545, \text{ or } 95.45 \% \qquad (2.6\text{-}28b)$$

$$P(\mu - 3\sigma < X \le \mu + 3\sigma) = 0.9973, \text{ or } 99.73 \% \qquad (2.6\text{-}28c)$$

where X represents the normal random variable with all its possible realizations. If the two parameters μ and σ are known, then the preceding predictions can be made about its distribution.

The dispersion of population elements x is reflected by the shapes of $f(x)$ according to Equation (2.6-11). In Figure 2-17, three pdf's are shown for the same μ and three different σ. Table 2-5 provides the computational values for Figure 2-17. For the same z, $f(z)$ remains unchanged, regardless of σ, but for the same x, $f(x)$ changes in a complex way in response to σ. The factor $1/\sigma$ stretches the ordinates for small σ and compresses them for large σ. At the same time, a small σ increases the

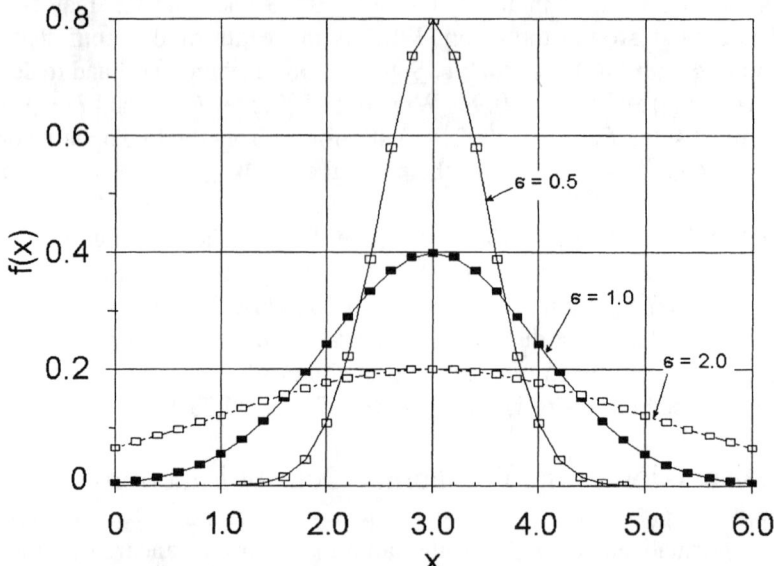

Figure 2-17: Normal pdf's $f(x)$ for three standard deviations $\sigma = 0.5$, 1.0, and 2.0, and $\mu = 3$.

exponent and causes a steep, slender bell shape, whereas a large σ reduces the slope and causes a flat and wide bell shape.

Table 2-5: Dispersion for Three Normal Distributions with Mean $\mu = 3$, and Standard Deviations $\sigma = 0.5$, 1, and 2

x	σ	μ	$z = (x - \mu)/\sigma$	$f(x)$	$f(z)$	$F(z)$	$F(z) - F(-z)$
(1)	(2)	(3)	(4)	(5)	(6)	(7)	(8)
0.0	0.5	3.0	-6.00	0.0000	0.0000	0.0000	–
0.5	0.5	3.0	-5.00	0.0000	0.0000	0.0000	–
1.0	0.5	3.0	-4.00	0.0003	0.0001	0.0000	–
2.0	0.5	3.0	-2.00	0.1080	0.0540	0.0228	–
3.0	0.5	3.0	0.00	0.7979	0.3989	0.5000	0.0000
4.0	0.5	3.0	2.00	0.1080	0.0540	0.9772	0.9544
5.0	0.5	3.0	4.00	0.0003	0.0001	0.9999	0.9999
6.0	0.5	3.0	6.00	0.0000	0.0000	0.9999	0.9999
0.0	1.0	3.0	-3.00	0.0044	0.0044	0.0014	–
0.5	1.0	3.0	-2.50	0.0175	0.0175	0.0062	–
1.0	1.0	3.0	-2.00	0.0540	0.0540	0.0228	–
2.0	1.0	3.0	-1.00	0.2420	0.2420	0.1587	–
3.0	1.0	3.0	0.00	0.3989	0.3989	0.5000	0.0000
4.0	1.0	3.0	1.00	0.2420	0.2420	0.8413	0.6826
5.0	1.0	3.0	2.00	0.0540	0.0540	0.9772	0.9544
6.0	1.0	3.0	3.00	0.0044	0.0044	0.9987	0.9973
0.0	2.0	3.0	-1.50	0.0648	0.1295	0.0668	–
0.5	2.0	3.0	-1.25	0.0913	0.1826	0.1056	–
1.0	2.0	3.0	-1.00	0.1210	0.2420	0.1587	–
2.0	2.0	3.0	-0.50	0.1760	0.3521	0.3085	–
3.0	2.0	3.0	0.00	0.1995	0.3989	0.5000	0.0000
4.0	2.0	3.0	0.50	0.1760	0.3521	0.6915	0.3830
5.0	2.0	3.0	1.00	0.1210	0.2420	0.8413	0.6826
6.0	2.0	3.0	1.50	0.0648	0.1295	0.9332	0.8664

Notes: column 1: selected x-values
column 2: selected σ: 0.5, 1.0, and 2.0.
column 3: $\mu = 3$ for all cases, which fixes the symmetry axis of the pdf at $x = 3$.
column 4: $z = (x - \mu)/\sigma$.
column 5: $f(x) = (0.3989/\sigma)\exp(-z^2/2)$; has the reciprocal dimension of σ.
column 6: $f(z) = 0.3989 \exp(-z^2/2)$; is dimensionless; $f(z)/f(x) = \sigma$.
column 7: by table lookup: for $z < 0$: $F(z) = 0.5 - F_0(z)$; for $z > 0$: $F(z) = 0.5 + F_0(z)$. column 8: for $z > 0$: $F(z) - F(-z) = 2 F(z) - 1 = 2 F_0(z)$.

The band $z = \pm 1$ around the center, $z = 0$, or $x = \mu$, of the distributions in Figure 2-17 contains 68.26 % of all population elements. However, when measured in terms of x, the band width varies according to $x = \mu \pm z \sigma$. As σ increases, with μ and z remaining constant, the same percentage of data occupies an increasingly wider band. If the right boundary is $x_2 = \mu + z \sigma$, and the left boundary $x_1 = \mu - z \sigma$, the band width is

$$x_2 - x_1 = 2 z \sigma \qquad (2.6\text{-}28d)$$

A summary of data taken from Table 2-5 shows the increasing band width x for the same percentage of elements:

σ	z	Population Inside Band %	Band Boundaries x_2, x_1	Band Width $2 z \sigma$
0.5	1	68.26	$2.5 \leq x \leq 3.5$	1
1.0	1	68.26	$2 \leq x \leq 4$	2
2.0	1	68.26	$1 \leq x \leq 5$	4

Similar data can be extracted from Table 2-5 for other data percentages and the resulting band widths, e.g., for $z = 2$ or 3. However, if the band width is kept the same, with increasing σ, fewer and fewer data are found inside it. This is shown in Table 2-5 by the values in columns 1, 4, and 8. Reading along the second to last row in each subdivision, one can see that the fraction of total data within the range $1 \leq x \leq 5$ shrinks from 99.99 % ($z = 4$) for $\sigma = 0.5$, to 95.44 % for $\sigma = 1.0$ ($z = 2$), and to 68.26 % for $\sigma = 2$ ($z = 1$). This is illustrated by the spread of the distributions in Figure 2-17. Whereas the population is practically entirely within the range $1 \leq x \leq 5$ for $\sigma = 0.5$, only 68 % are within it for $\sigma = 2$.

The practical use of the Gaussian distribution is sometimes hindered by the fact that it extends from $-\infty$ to $+\infty$. If the random variable X cannot take on negative values, then the normal distribution can be used only if the mean is sufficiently far to the right of zero so that there is no significant portion of the probability distribution reaching over to negative x-values. Visualize μ as a positive x-value to the right of $x = 0$. Then, if the sample points are positive quantities only and are to be fitted with a normal distribution, it is desirable that $\mu - 3 \sigma > 0$. This requirement can be transformed into an upper bound for the coefficient of variation:

$$C_v = \frac{\sigma}{\mu} < \frac{1}{3} . \qquad (2.6\text{-}28e)$$

For example, concrete strength data are all positive. The average strength, μ, is a positive value to the right of zero. For $\mu = 2.5$ MPa and $\sigma = 0.19$ MPa, $C_v = 0.076$ < 1/3. In this case, μ is 13 σ to the right of $x = 0$, so that the probability of a data point falling onto the negative x-axis is practically zero. In cases when μ is close to zero, X may naturally assume positive and negative values. For example, a net benefit may be positive (gain) or negative (loss), even if the mean net benefit is positive.

2.6.4 Polynomial Approximation and Inversion of the cdf

Since the Gaussian pdf, $f(z)$, is not amenable to integration, it is not possible to derive an explicit cdf, $F(z)$. The cdf can always be obtained by stepwise integration using a numerical integration method (Sections 2.5.2 and 2.6.2). In some cases, it is necessary to calculate function values analytically, for example, when one wants to invert given values $F(z)$ to find z. Polynomial approximations have been constructed for $F(z)$ and for the inverse, $z_F = I[F(z)]$. These polynomials are described next.

(1) Polynomial approximation of $F(z)$ (Abramowitz and Stegun, 1970, p. 932, No. 26.2.16):

(a) for $-\infty < z \leq 0$ (negative z): calculate the auxiliary variable

$$t = \frac{1}{1 + 0.33267|z|} \qquad (2.6\text{-}29)$$

where $|z|$ is the absolute value of z, the normalized variable. Then $F(z)$ is obtained by

$$F(z) = 0.3989423 e^{-0.5z^2} (0.436184t - 0.120168t^2 + 0.937298t^3) \qquad . \qquad (2.6\text{-}30)$$

(b) for $0 \leq z \leq \infty$ (positive z):

$$t = \frac{1}{1 + 0.33267\, z} \qquad (2.6\text{-}31)$$

and

$$F(z) = 1 - 0.3989423 e^{-0.5z^2} (0.436184t - 0.120168t^2 + 0.937298t^3) \qquad .(2.6\text{-}32)$$

The residual error of $F(z)$ for any z is $e < 0.00001$. More accurate formulas are available. A comparison of the computed $F(z)$ with numerical integration and table lookup is given in Table 2-6.

Table 2-6: Comparison of $F(z)$ Obtained by Polynomial, Numerical Integration, and Table Lookup

z	$F(z)$ by polynomial	$F(z)$ by integral	$F(z)$ from table
-3	0.001355	0.001360*	0.001350
0	0.500000	0.499999	0.500000
3	0.998645	0.998638	0.998650

*The numerical integration is started with $z = 4.9$ and proceeds with increments $\Delta z = 0.1$ to $z = 3.0$, similar to Example (1) of Section 2.6.2. Twenty values $f(z)$ are used.

(2) Inversion of $F(z)$. This procedure is used to find the argument z, given a function value $F(z)$:

$$z_F = I\,[F(z)] \tag{2.6-33}$$

where $I\,[F(z)]$ is the inversion operator that finds z from a known $F(z)$; the notation z_F is used to identify the z calculated from $F(z)$ in contrast to $z = (x - \mu)/\sigma$ calculated from data. Abramowitz and Stegun (1970, p. 952; p. 933, No. 26.2.23) give the following procedure: For the Gaussian non-exceedance probability,

$$F(z) = u = \frac{1}{\sqrt{2\pi}} \int_{-\infty}^{z} e^{-0.5t^2}\, dt \quad, \tag{2.6-34}$$

(a) for $0 < u \le 0.5$, calculate the auxiliary variable

$$t = \sqrt{\ln\frac{1}{u^2}}\ . \tag{2.6-35}$$

Then obtain the inverse by

$$z_F = -t + \frac{c_0 + c_1 t + c_2 t^2}{1 + d_1 t + d_2 t^2 + d_3 t^3} + e \; . \tag{2.6-36}$$

(b) for $u > 0.5$, calculate the auxiliary variable

$$t = \sqrt{\ln \frac{1}{(1-u)^2}} \; . \tag{2.6-37}$$

Then obtain the inverse by

$$z_F = t - \frac{c_0 + c_1 t + c_2 t^2}{1 + d_1 t + d_2 t^2 + d_3 t^3} + e \; , \tag{2.6-38}$$

Table 2-7: Inversion $z_F = I\,[(F(z)]$ and Comparison with Table Lookup

u	t	z_F	z (table)
(1)	(2)	(3)	(4)
0.001	3.717	-3.091	-3.090
0.010	3.035	-2.327	-2.326
0.100	2.146	-1.282	-1.282
0.200	1.794	-0.841	-0.842
0.300	1.552	-0.524	-0.524
0.400	1.354	-0.253	-0.253
0.499	1.179	-0.002	-0.002
0.501	1.179	0.002	0.002
0.600	1.354	0.253	0.253
0.700	1.552	0.524	0.524
0.800	1.794	0.841	0.842
0.900	2.146	1.282	1.282
0.990	3.035	2.327	2.326
0.999	3.717	3.091	3.090

Notes: column 1: assumed values $u = F(z)$.
column 2: auxiliary variable for the inversion formula.
column 3: z_F calculated by the inversion formula.
column 4: table lookup of z_F from Abramowitz and Stegun, 1970, p. 976.

The residual error of the approximation is $e = 4.5 \cdot 10^{-4}$. The coefficients for Equations (2.6-36) and (2.6-38) are:

$$c_0 = 2.515517 \qquad c_1 = 0.802853 \qquad c_2 = 0.010328$$
$$d_1 = 1.432788 \qquad d_2 = 0.189269 \qquad d_3 = 0.001308.$$

A calculation sample is given in Table 2-7. Column 1 gives the selected function values $u = F(z)$ to be inverted. Column 2 gives the auxiliary variable, and column 3 gives the inversion z_F. The formulas for t and z_F change as u changes from $u \le 0.5$ to $u > 0.5$. Column 4 gives inversion values z from a table lookup (Abramowitz and Stegun, 1970, p. 976). The comparison of columns 3 and 4 shows that z_F and z are very close.

Example: A number of concrete samples, 15 cm × 30 cm (6" by 12 "), have been tested for compressive strength and the test data are to be checked for normal distribution. The data are from the Bureau of Reclamation's Concrete Manual (BUREC 1975, p. 173, Figure 70, 288-D-2639) and are given in Table 2-8. Solution: The parameters of the concrete strength data are:

mean: $m = 28.764$ MPa   class width: $dx = 0.6826$ MPa
coefficient of variation: $C_v = 0.1165$ normalized class width: $dz = 0.2037$
sample standard deviation: number of classes: $k = 26$
$s = 3.351$ MPa   number of elements: $n = 210$

Columns 1 and 2 of Table 2-8 define the 26 classes of the data histogram. Column 3 gives the center of each class, x_j. Column 4 gives the number, n_j, of items x_i in each class, j. The class probability $p_j = n_j/n$ is listed in column 5. The sample mean is calculated from the class probabilities:

$$m = E(X) = \sum_{j=1}^{k} p_j x_j .$$

The sum is taken over all k classes. This mean is not as accurate as it would be had the sum been taken over all 210 elements x_i instead of using the 26 representative class means, x_j. The sample variance is calculated by

$$s^2 = \sum_{j=1}^{k} p_j (x_j - m)^2$$

where $j = 1, \ldots k$ is the number of classes, as before. The normalized variable for each class representative x_j is calculated using the sample parameters, m and s:

$$z_j = \frac{x_j - m}{s} \; .$$

The experimental non-exceedance probability, $F(z)$, is obtained by cumulating the p_j as if they were a discrete pdf. This is done in column 6 of Table 2-8. The cumulation proceeds from the lower to the upper limit of the range. The experimental $F(z)$ contains information on the shape of the experimental data distribution. Since the p_j are the probabilities of the 26 partitions of the sample space, their probabilities add to 1. By inverting this experimental $F(z)$, one obtains an experimental z_F in column 10. The comparison of z obtained from the sample parameters, and z_F from $F(z)$ reveals how well the sample fits the assumed normal distribution.

Table 2-8: Concrete Strength Data (from Figure 70, BUREC, 1975, p. 173)

Compressive Strength class from	to	Central Value of Class x_j	No. of Items n_j	Class Frequ p_j	Cumul. Frequ. $F(z)$	z from x_i, m, and s	$1 - F(z)$	Auxil. variable t	$z_F = I[F(z)]$
(1)	(2)	(3)	(4)	(5)	(6)	(7)	(8)	(9)	(10)
19.7	20.3	20.0	0	0.0000	0.000	-2.6181	1.000		
20.3	21.0	20.7	2	0.0095	0.010	-2.4123	0.990	3.051	-2.3459
21.0	21.7	21.4	2	0.0095	0.019	-2.2066	0.981	2.815	-2.0746
21.7	22.4	22.1	4	0.0190	0.038	-2.0008	0.962	2.556	-1.7732
22.4	23.1	22.7	3	0.0143	0.052	-1.7950	0.948	2.429	-1.6218
23.1	23.8	23.4	3	0.0143	0.067	-1.5893	0.933	2.327	-1.5003
23.8	24.5	24.1	7	0.0333	0.100	-1.3835	0.900	2.146	-1.2801
24.5	25.2	24.8	12	0.0571	0.157	-1.1778	0.843	1.924	-1.0041
25.2	25.8	25.5	10	0.0476	0.205	-0.9720	0.795	1.781	-0.8222
25.9	26.5	26.2	11	0.0524	0.257	-0.7662	0.743	1.648	-0.6494
26.5	27.2	26.9	12	0.0571	0.314	-0.5605	0.686	1.521	-0.4810
27.2	27.9	27.6	17	0.0810	0.395	-0.3547	0.605	1.363	-0.2636
27.9	28.6	28.3	15	0.0714	0.467	-0.1489	0.533	1.235	-0.0828
28.6	29.3	29.0	16	0.0762	0.543	0.0568	0.457	1.251	0.1065
29.3	30.0	29.6	22	0.1048	0.648	0.2626	0.352	1.444	0.3764
30.0	30.7	30.3	10	0.0476	0.695	0.4684	0.305	1.542	0.5080
30.7	31.4	31.0	21	0.1000	0.795	0.6741	0.205	1.781	0.8222

31.4	32.1	31.7	12	0.0571	0.852	0.8799	0.148	1.956	1.0446
32.1	32.7	32.4	6	0.0286	0.881	1.0857	0.119	2.063	1.1780
32.8	33.4	33.1	9	0.0429	0.924	1.2914	0.076	2.269	1.4302
33.4	34.1	33.8	5	0.0238	0.948	1.4972	0.052	2.429	1.6218
34.1	34.8	34.5	2	0.0095	0.957	1.7029	0.043	2.510	1.7183
34.8	35.5	35.2	3	0.0143	0.971	1.9087	0.029	2.667	1.9026
35.5	36.2	35.8	3	0.0143	0.986	2.1145	0.014	2.915	2.1904
36.2	36.9	36.5	2	0.0095	0.995	2.3202	0.005	3.270	2.5944
36.9	37.6	37.2	1	0.0048	1.000	2.5260	0.001	3.717	3.0926

Notes: columns 1 and 2: Compressive strength, MPa, lower and upper class limits.
column 3: central value x_j of class j, MPa.
column 4: number of elements x_i per class j.
column 5: class frequency, $p_j = n_j/n$, $n = 210$.
column 6: cumulative frequency, $F(z)$; cumulation of the p_j of column 5 starting at the lower boundary of the range 19.7 MPa $\leq x \leq$ 37.6 MPa.
column 7: normalized variable $z_j = (x_j - m)/s$, with $m = 28.76$ MPa and $s = 3.35$ MPa.
column 8: complementary probability, $1 - F(z)$.
column 9: auxiliary variable t for calculating the inverse. For $F(z) < 0.5$, $F(z)$ in column 6 is used, and for $F(z) > 0.5$, $1 - F(z)$ of column 8 is used in Equations (2.6-35) and (2.6-37), respectively.
column 10: $z_F = I[F(z)]$ is calculated by Equations (2.6-36) and (2.6-37), respectively, according to the criteria used for column 9.

The z_F-values of column 10 are plotted against the z-values of column 7 in Figure 2-18. The z-values of column 7 are plotted against themselves to produce a diagonal with which z_F can be compared. The z-values and the z_F-values are plotted at the centers of the classes for which they are calculated. The plot in Figure 2-18 is related to the traditional method of presenting test data in *probability paper*. Using the inverse obviates the need for such paper.

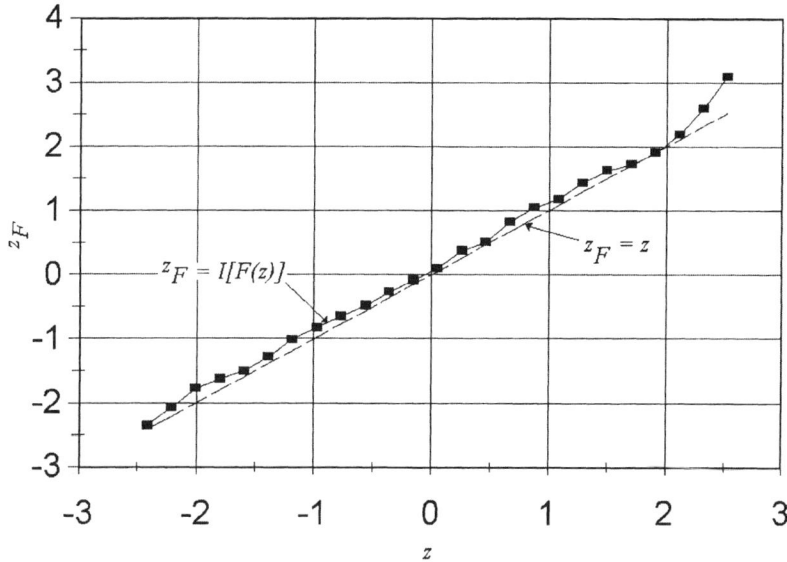

Figure 2-18: Comparison of $z = (x - m)/s$, based on concrete strength, x, sample mean, m, and sample standard deviation, s, and the inverse obtained from the empirical cdf, $F(z)$, $z_F = I[F(z)]$, for a normal distribution. Refer to Table 2-8 for the data.

2.6.5 Judgmental Establishment of the pdf

The bell shape of the Gaussian pdf extends to infinity in both directions of the z-axis. Apart from these tails, the main body of the pdf approximates many data distributions found in practice. In many cases these outer tails are just ignored because of the very small probabilities associated with these parts of the range. There are two properties of the Gaussian pdf that help to establish an analytical expression of the pdf from just a few data points: symmetry and two parameters, μ and σ.

(1) Estimation of the Mean and Standard Deviation from a Practical Data Range

From Table 2-3 it can be seen that 99.73 % of all data lie within ± 3 σ, or a band width of 6 σ. Hence, for practical purposes, the indefinite boundaries, $-\infty$ and $+\infty$, can be replaced by $z = \pm 3$, or $z = \pm 4$ (includes 99.99 %). If the range limits

(x_1, x_2) of a random variable X are known, or can be specified by data or by the nature of the problem, a Gaussian distribution can be constructed for this range by calculating the mean and the standard deviation from these limits:

$$\mu = \frac{x_1 + x_2}{2},$$
(2.6-39)

and, if $z = \pm 3$ is used as range,

$$\sigma = \frac{x_2 - x_1}{6}.$$
(2.6-40)

Alternatively, the 6 in Equation (2.6-40) would be replaced by 8 if a range of $z = \pm 4$ were used. Since μ and σ are the only distribution parameters, the normal pdf, $f(z)$, is defined and frequencies can be calculated for all realizations of the data range. Also the cdf can be established by integration of $f(z)$. Such an estimate of probability functions may be of importance for exploratory studies.

(2) Estimation of the Standard Deviation from the 50-Percentile Range

The middle 50 % range in the center of the normal distribution is also called the interquartile range, because it lies between the outer quartiles of 25 % of the total sample space. Table 2-3, column 3 shows that $F_0(z) = 0.25$ for $z = 0.6745$. This means that 50 % of the total sample lies in the interval $-0.6745 < z \le +0.6745$. From $z = (x - \mu)/\sigma$, for $z = 0.67$, one finds (Lapin, 1978, p. 771):

$$\sigma = \frac{x - \mu}{0.67}.$$
(2.6-41)

If an estimate can be made for the 50 % band width centered on the mean, $2(x - \mu)$, the standard deviation σ is obtained from Equation (2.6-41). The knowledge of μ and σ is all that is needed to construct a normal distribution.

(3) Quartiles and Fractiles that Describe the Shape of the cdf

The shape of the cumulative distribution function, $F(z)$, has the typical S-shape, as shown in Figure 2-16. It represents the non-exceedance probability that increases from 0 to 1 over a practical abscissa range of $-3 \le z \le +3$. Dividing the ordinate in half, then moving to the right over the abscissa value $z = 0$ defines the midpoint of the S-shape, $x = \mu$, or $z = 0$. The half range $-3 \le z \le 0$ corresponds to the 0.5 *percentile*, if percent is used as unit, or to the 0.5 *fractile*, if a fraction of 1 is used as unit. Dividing the lower half of the ordinate in half again and moving from the 0.25 percentile, or *first quartile*, or *0.25 fractile*, to the right over the abscissa value $z = -0.6745$ defines the first quartile point on $F(z)$. Similarly the third quartile point defines $z = +0.6745$. The quartiles, or 0.25 fractiles, and their corresponding ranges are given in Table 2-9. The second and third quartile occupy the central portion of the range. Lying between the outer quartiles, this subrange is also called the *interquartile range*. It has a width of $2 \cdot 0.674 \ \sigma \approx 1.35 \ \sigma$ centered on μ, or 22 % of the total practical range of 6 σ, and 50 % of all elements lie within it. Column 3 of Table 2-9 represents the cumulation of successive quartiles, or the value of $F(z)$ at the end of the quartile ranges given in column 2.

Table 2-9: Quartiles of the Normal Distribution

Quartile	Range of z	$F(z)$
1.	$-\infty < z \leq -0.6745$	0.25
2.	$-0.6745 < z \leq 0$	0.50
3.	$0 < z \leq +0.6745$	0.75
4.	$+0.6745 < z \leq +\infty$	1.00

An empirical pdf must always integrate to 1. A normal pdf integrates to 0.5 at the midpoint of the range. When constructing an empirical pdf, and $F(z) = 0.5$ is allocated to a value that is not at the midpoint of the range, then the pdf is skewed and may not be normal. However, as will be seen in Section 3.2.2, some asymmetrical distributions can be transformed into a normal pdf by a variable transformation. The construction of a general $F(z)$ may proceed by deciding how much of the range is represented by fractiles of $F(z)$, including the 0.5 fractile, the 0.25 fractile, the 0.125 fractile, and the 0.0625 fractile, as well as multiples of them (Lapin, 1978, p. 773). The smaller subdivisions of $F(z)$ are used to sharpen the curvature of the S-shape at its outer fringes, but smaller than 0.0625 fractiles are hard to visualize. A summary of fractiles and the fraction they represent, given the distribution is normal is given in Table 2-10.

Table 2-10: Fractiles of the Normal Distribution

Fraction of $F(z)$	Fractile	z
0	0	$-\infty$
1/16	0.0625	-1.53414
1/8	0.125	-1.15035
1/4	0.25	-0.6745
1/2	0.5	0.000
3/4	0.75	0.6745
7/8	0.875	1.15035
15/16	0.9375	1.53414
1	1.0	∞

Notes: column 2: read "0.0625 fractile;" the fractiles represent non-exceedance probabilities $F(z)$.
column 3: z is the inverse of $F(z)$.

.

2.6.6 Quality Control Based on the pdf

Production of concrete for hydraulic structures must meet uniform strength requirements that also reflect on other concrete properties, such as aggregate quality, durability, watertightness, and resistance against abrasion and erosion. Even if the concrete production process is well controlled, it usually is not possible to produce concrete of exactly the same strength. Samples taken from concrete batches usually show small variations in concrete strength. A small range of strength variation in the test results indicates good control of concrete production, whereas large variations indicate poor control. Many random factors come into play causing deviations from the mean. It is not surprising, therefore, to find the sample strengths data distributed around the mean by a Gaussian pdf. The central limit theorem provides an explanation of this phenomenon (Section 3.4.2). A measure for the dispersion of the test results is the coefficient of variation, $C_v = \sigma/\mu$. A low C_v means good control and a high C_v means poor control. The BUREC requires that a specified percentage of the test specimens must exceed the design strength of the structural concrete (BUREC, 1975, p. 173).

A quality control requirement may be stated by requiring a percentage of the sample to exceed the mean strength. Suppose the required exceedance percentages are

75 %, 80 %, and 85 %, and the data have been found to be normally distributed. Then this requirement is satisfied if the required mean strength of x lies to the left of the sample mean by a distance $-z \, \sigma$, such that the exceedance probability of sample concrete strength is one of these percentages. The random concrete strength of a sample can be expressed by $x = \mu - \sigma z$. This case can be visualized by Figure 2-14 (a), with x being located at $-z$ and μ at $z = 0$. Dividing by μ leads to $x/\mu = 1 - z \, C_v$, and solving for μ gives

$$\mu = \frac{x}{1 - z \, C_v} \; . \tag{2.6-42}$$

For an exceedance probability $1 - F(z) = 0.75 > 0.5$, $F(-z) = 0.25$, which is the area under the pdf to the left of $-z$. From Figure 2-14 (a) one obtains $F_0(z) = 0.5 - F(-z) = 0.25$, and $z = 0.675$. The actual negative sign of z is taken care of by the formulation $x = \mu - \sigma z$. Hence the sample mean required for 75 % exceedance of x is

$$\mu = \frac{x}{1 - 0.675 \, C_v} \; . \tag{2.6-42a}$$

Suppose $C_v = 0.3$, then $\mu = 1.25 \, x$. Thus, the sample mean, μ, must exceed the required mean strength by 25 %. The smaller the coefficient of variation, the closer together the sample mean and x will be. The higher the specified exceedance probability, the greater the distance between x and the sample mean. The factor z in Equation (2.6-42) varies with the specified exceedance probability:

$1 - F(z)$	0.75	0.80	0.85
z	0.675	0.842	1.036

Examples of design strength, x, and sample mean, μ, are given in Table 2-11. For 10 MPa, the sample mean must be 10.4 MPa for the smallest C_v and 12.0 MPa for the largest C_v to meet a 75 % exceedance requirement.

Table 2-11: Quality Control for Concrete Sample (after BUREC, 1975, p. 174)

Design Strength	Exceedance Probability	Normal Variable z	Required Sample Means for Various Coefficients of Variation, C_v		
MPa	%	1	0.05	0.15	0.25
(1)	(2)	(3)	(4)	(5)	(6)
10	75	0.675	10.3	11.1	12.0
10	80	0.842	10.4	11.4	12.7
10	85	1.036	10.5	11.8	13.5
20	75	0.675	20.7	22.3	24.1
20	80	0.842	20.9	22.9	25.3
20	85	1.036	21.1	23.7	27.0
40	75	0.675	41.4	44.5	48.1
40	80	0.842	41.8	45.8	50.7
40	85	1.036	42.2	47.4	54.0

2.7 Other Probability Distributions

2.7.1 Selected Distributions

There are numerous analytical functions that can serve as pdf's. These pdf's summarize the information content of data sets in the form of a mathematical statement. Only a few such functions are discussed here and their treatment is not exhaustive. The intention is to give examples of the characteristics of such functions and an indication of what they can be used for. A comprehensive summary of pdf's and related functions can be found in Moan (1982, p. 4-19 to 4-30). Eight distributions that are frequently encountered in practice are discussed:

Binomial	Beta
Uniform	Poisson
Triangular	Weibull
Exponential	Student's t

One additional distribution, the lognormal distribution, is discussed in connection with variable transformation in Section 3.2.2. Most of the aforementioned pdf's can be relatively easily constructed, even if large amounts of data are not available. In probabilistic approaches, these distributions can be used without any data, if reasonable estimates of the parameters can be made. The binomial distribution is the prototype of a discrete pdf. The uniform pdf is used when no better information than equal frequency over the range is known. The triangular distribution in its symmetric form is used when it is obvious that values at a boundary have lower probability than in the interior. In its asymmetric form, it can express a very low probability at one boundary of the range and very high probability at the other. The exponential distribution is used if one can assume that small variable values, such as life lengths, are more likely than big ones. The beta distribution is adaptable to different shapes of the pdf. The Poisson distribution, a discrete distribution, is known to simulate random arrivals to service queues. The Weibull distribution can model different shapes of pdf's, including exponential function shapes and asymmetric bell shapes that rise at zero and only extend along the positive variable axis; it has great versatility because it is a three-parameter distribution. The Student's t distribution applies in lieu of the normal distribution when the samples are small. Some selected distributions can be used if the normal distribution is inappropriate because of definite range limits and lack of symmetry. Before using any of the functions, the user is advised to test the integral over the range of the random variable to see if it meets the criterion $\int f(x)dx = 1$. If this is not the case, the function must be scaled, as shown in Section 2.6.1. Other pdfs can be found in handbooks (Abramowitz and Stegun, 1970, p. 925; Moan, 1982, pp. 4-19 - 4-49).

2.7.2 Binomial Distribution

There are many random processes that have only two possible outcomes: a coin toss produces a head or a tail, a device either works or does not work, an action succeeds or fails, an accident occurs or does not occur. A random process with two complementary outcomes is a *Bernoulli process*. The sequential actions that produce new outcomes are also called *trials*. The Bernoulli process has three characteristics (Lapin, 1978, pp.162 -163): (1) the probabilities of the outcomes are complementary; (2) the process probability stays the same for all trials, unlike processes where the failure rate increases with the number of trials, due to aging, fatigue, etc.; (3) the trial outcomes are independent of what has happened before. A failure does not become more frequent because several failures have happened before.

Graphically, the Bernoulli process can be visualized as a branching process with a *process probability*, or *branching probability*, p, and a complementary probability $1 - p$. In an unbiased coin throw, the branching probability is $p = 0.5$. If a coin is tossed repeatedly, the outcome of every toss is either heads or tails each time with probability p. If heads is defined a success and there are k heads in n trials, then the *success rate* is $p_k = k/n$. The success rate in four trials is the number of heads in four outcomes. Since the process can take one of two branches at each trial, the total number of possible outcomes is n^2. The 16 possible outcomes of four trials are shown on the right-hand side of Figure 2-5. They range from HHHH to TTTT, meaning four heads in four trials to no heads in four trials. If $k = n$, the success rate is $p_k = n/n = 1$; if $k = 0$, the success rate is $p_k = 0$. The Bernoulli process produces a discrete distribution of p_k, as only a countable number of p_k's can occur. In the four trials of Figure 2-5, there can be at best four successes, or at worst, four failures. Including the last outcome as a zero-success rate, one counts five success rates: 0/4, 1/4, 2/4, 3/4, and 4/4 (see also Example (6) of Section 2.1.1). The number of occurrences of a p_k. is given by the binomial number, $B(n,k)$. Dividing $B(n,k)$ by the total number of process outcomes, here 16, gives the probabilities of the p_k's. For the 4-trial Bernoulli process one obtains

Possible p_k's: k/n	0/4	1/4	2/4	3/4	4/4	Sum
Number of p_k's in n outcomes: $B(n, k)$	1	4	6	4	1	16
Frequency of the p_k's: $B(n, k)/2^n$	1/16	1/4	3/8	1/4	1/16	1

The probability that $k = 0, 1, 2, ..., n$ special events occur in n trials is given by the *binomial distribution* or *binomial formula* (Kreyszig, 1979, p. 874; Lapin, 1978, p. 164)

$$f(k, n, p) = B(n, k)\ p^k q^{n-k} \tag{2.7-1}$$

where $f(k,n,p)$ is the probability of k events in n trials, given the Bernoulli process probability is p, the same in every trial, $q = 1 - p$, is the complement of p; $B(n,k)$ is the binomial coefficient, or Bernoulli number,

$$B(n,k) = \frac{n!}{k!(n-k)!} \tag{2.7-2}$$

where $n! = 1 \cdot 2 \cdot ... \cdot n$. Binomial coefficients with large n and k are conveniently evaluated by cancelling parts of the products that appear in the numerator and denominator of Equation (2.7-2). This leads to

$$B(n,k) = \frac{n(n-1)(n-2)....(n-k+1)}{k!}. \tag{2.7-2a}$$

For error control, the number of factors in the numerator and denominator of Equation (2.7-2a) must be the same.

A sample of sequences of binomial coefficients is given next. They all start and end with 1, and the numbers of the following rows are sums of two numbers of the preceding row. This simple relation is helpful for remembering binomial coefficients. The row sum of the binomial coefficients is the number of outcomes of an n-step two branch sequence as the one shown in Figure 2-5.

n	$B(n, k)$						Sum 2^n
	$k = 0$	$k = 1$	$k = 2$	$k = 3$	$k = 4$	$k = 5$	
1	1	1					2
2	1	2	1				4
3	1	3	3	1			8
4	1	4	6	4	1		16
5	1	5	10	10	5	1	32

The left-hand side of Equation (2.7-1) can also be written as

$$f(k,n,p) = P(p_k = \frac{k}{n}) \tag{2.7-3}$$

where $P(p_k = k/n)$ is the probability of the success rate p_k, which is defined as

$$p_k = \frac{k}{n}. \tag{2.7-3a}$$

The success rate, p_k, is the probability of k outcomes in n trials. It also is a random outcome among the 2^n possible outcomes of an n-trial Bernoulli process, as shown in the last column giving the sum of outcomes in the preceding $B(n,k)$ table. As such, p_k has the probability distribution given by Equation (2.7-1).

The *expectation of successes* of the Bernoulli process is

$$E(k) = n\,p. \tag{2.7-4}$$

According to Equation (2.7-4), if a coin is thrown 1,000 times, given $p = 0.5$, the number of heads will tend toward 500.

The *expectation of the success rate* p_k is

$$E(\frac{k}{n}) = \frac{1}{n} E(k) = E(p_k) = p .$$ (2.7-5)

Equation (2.7-5) can be used to calculate the p of the Bernoulli process. If observations have shown that the expected success rate is $E(k) = 500$, then dividing by the number of trials by which it was obtained, $n = 1,000$, gives $p = 0.5$.

The *variance of k* is

$$Var(k) = n p q,$$ (2.7-6)

and the *variance of* p_k is

$$Var(\frac{k}{n}) = \frac{1}{n^2} Var(k) = \frac{pq}{n}$$ (2.7-7)

where $q = 1 - p$. These characteristics of the binomial distribution are illustrated by a binomial distribution with parameters $p = 0.6$ and $n = 10$, calculated in Table 2-12. An illustration of two distributions, $P(p_k)$, for $p = 0.6$, and $p = 0.5$, and $n = 10$, is shown in Figure 2-19. It shows a shape change from symmetric to asymmetric as p changes from 0.5 to 0.6.

The probability that up to k successes occur in n trials is the cumulative probability of the frequencies, $f(k,n,p)$, given by Equation (2.7-1), from 0 up to and including k successes:

$$F(k) = \sum_{i=0}^{k} B(n,i) p^i q^{n-i}$$ (2.7-8)

where $F(k)$ is the non-exceedance probability of k and represents the sum of all probabilities for $i = 0, 1, ..., k$ successes in n trials. If k is defined as the number of failures, then p must be the failure probability of the process.

Special cases of the binomial distribution:

1. Zero failures in n trials: It is assumed that p is the average *failure* probability at each trial. For no failure, $k = 0$, $B(n,0) = 1$, and $p^0 = 1$. Then, the probability of *no failure* in n trials is

$$f(0, n, p) = f(0) = q^n$$ (2.7-9)

where $q = 1 - p$ is the complementary non-failure or success probability. It is seen that Equation (2.7-9) produces the joint probability of n independent non-failure events, E_1, E_2, ..., E_n, each with the probability $P(E_i) = q = 1 - p$:

$$P(E_1 \cap E_2 \cap E_3 \ldots \cap E_n) = P(E_1)\,P(E_2)\ldots P(E_n)$$

that was derived as Equation (2.1-24) in Section 2.1.4.

The meaning of p must be clearly defined and carefully tracked through the computation to avoid mistakes. Suppose the probability of an event to occur is p. Then the probability that it will not occur, $k = 0$, in n trials is

$$f(0,n,p) = B(n,0)p^0(1-p)^n \ . \tag{2.7-9a}$$

With $f(0, n, p) = P(p_0)$ one obtains the probability of zero successes in n trials

$$P(p_0) = (1 - p)^n. \tag{2.7-9b}$$

The complementary probability that an event of probability p will occur at least once in n trials is

$$P = 1 - (1 - p)^n \ . \tag{2.7-10}$$

Equation (2.7-10) is used to estimate the probability of a rare event in the lifetime of a project. For example, a 100-year flood with $p = 0.01$, in the 100-year lifetime of a project, $n = 100$, has the probability of occurrence $P = 0.63$.

2. One failure in n trials: For one failure, $k = 1$, an average failure probability p, $B(n,1) = n$, and $p^1 = p$, the probability of one failure in n trials is

$$f(1, n, p) = P(p_1) = n\,p\,q^{n-1}. \tag{2.7-11}$$

3. Approximation by the normal distribution: When n becomes very large, the value of $f(k,n,p)$ can be approximated by the normal distribution. This is convenient as the normal distribution is easier to evaluate than the binomial distribution for large k and n (Bronstein-Semendjajew, 1984, p. 676):

$$\lim_{n \to \infty} [f(k,n,p)] \approx \frac{1}{\sqrt{2\pi npq}} \cdot e^{-0.5z^2} \ . \tag{2.7-12}$$

The normalized variable is

$$z = \frac{k - np}{\sqrt{npq}}$$
  (2.7-12a)

with $np = E(k)$, and $npq = Var(k)$, as given by Equations (2.7-4) and (2.7-6), respectively.

Examples: (1) A plant lighting system has an hourly failure rate of 0.06. What is the probability that 5 bulbs fail per day if maximally one bulb can fail per hour? Solution: Visualize the failure process similar to the branching process in Figure 2-5. In the first hour, a bulb can fail or not. The same can happen in the second hour, and so forth, over the next 24 hours. With $k = 5$, $n = 24$; and the average failure probability $p = 0.06$, the probability of five failures per day follows from Equation (2.7-1) as $P(p_5) = B(24, 5) \cdot 0.06^5 \cdot (1 - 0.06)^{19} = 42,504 \cdot 0.778 \cdot 10^{-6} \cdot 0.3086 = 0.010$. Binomial coefficients for medium sized n and k are found in Abramowitz and Stegun (1970, p. 828).

(2) For four trials with an average process probability of $p = 0.5$ find the expectation of k and p_k. Solution: For $n = 4$, $k = 0, 1, ... , n$. The expectation of k is the probability-weighted sum of all possible k with $k = 0$ making no contribution: $E(k) = f(1, 4, 0.5) \cdot 1 + f(2, 4, 0.5) \cdot 2 + f(3, 4, 0.5) \cdot 3 + f(4, 4, 0.5) \cdot 4 = B(4, 1) \cdot 0.5 \cdot 0.5^3 \cdot 1 + B(4, 2) \cdot 0.5^2 \cdot 0.5^2 \cdot 2 + B(4, 3) \cdot 0.5^3 \cdot 0.5^1 \cdot 3 + B(4, 4) \cdot 0.5^4 \cdot 0.5^0 \cdot 4 = 4 \cdot 0.5^4 \cdot 1 + 6 \cdot 0.5^4 \cdot 2 + 4 \cdot 0.5^4 \cdot 3 + 1 \cdot 0.5^4 \cdot 4 = 0.5^4 (4 \cdot 1 + 6 \cdot 2 + 4 \cdot 3 + 1 \cdot 4) = 32/16 = 2$. This is also equal to Equation (2.7-4), $E(k) = 4 \cdot 0.5 = 2$. Equation (2.7-5) is verified by the same calculation but replacing k by k/n, which leads to $E(k)/n = E(p_k) = 0.5 = p$.

(3) For $n = 10$, and $p = 0.6$, calculate (a) the expectation of p_k, and (b) the variance of p_k. Solution: The calculations are carried out in Table 2-12. (a) The sum of probabilities of p_k in column 6 is equal to 1 (last row), which serves as a routine check for the correctness of the probability calculation. The probability-weighted sum of p_k's in column 7 is equal to $p = 0.6$, according to $E(p_k) = p$.

(b) The variance of k is according to Equation (2.7-6)

$$Var(k) = \sum_{k=0}^{n} [P(k)(k - np)^2]$$
  (2.7-6a)

where $P(k)$ is the same as $P(p_k)$, as dividing by n does not change the probability of k; $P(k)$ is calculated in column 6 using Equation (2.7-3) in the binomial formula, Equation (2.7-1). The variance of p_k is obtained by using Equation (2.7-7):

$$Var(\frac{k}{n}) = \frac{1}{n^2} Var(k) = \sum_{k=0}^{n} [P(p_k)(p_k - p)^2].$$ (2.7-7a)

The sums for Equation (2.7-7a) are calculated in column 8 and for Equation (2.7-6a) in column 9 of Table 2-12. The results are given in the last row of Table 2-12 as $Var(k/n) = 0.024$ and $Var(k) = 2.4$. The variance of k by Equation (2.7-6) is

$$Var(k) = n\,p\,(1 - p) = 10 \cdot 0.6 \cdot 0.4 = 2.4,$$

and the variance of k/n by Equation (2.7-7) is

$$Var(\frac{k}{n}) = \frac{1}{n^2} Var(k) = \frac{2.4}{100} = 0.024.$$

These results verify Equations (2.7-6) and (2.7-6a) for $Var(k)$, and Equations (2.7-7) and (2.7-7a) for $Var(k/n) = Var(p_k)$.

Table 2-12: Characteristics of a Binomial Distribution for $p = 0.6$ and $n = 10$

k	$n - k$	k/n	$B(10,k)$	$p^k (1 - p)^{n-k}$	$P(p_k = k/n)$	$P(p_k)(k/n)$	$Var(k/n)$	$Var(k)$
(1)	(2)	(3)	(4)	(5)	(6)	(7)	(8)	(9)
0	10	0	1	0.00010	0.00010	0.00000	0.00004	0.00377
1	9	0.1	10	0.00016	0.00157	0.00016	0.00039	0.03932
2	8	0.2	45	0.00024	0.01062	0.00212	0.00170	0.16987
3	7	0.3	120	0.00035	0.04247	0.01274	0.00382	0.38221
4	6	0.4	210	0.00053	0.11148	0.04459	0.00446	0.44591
5	5	0.5	252	0.00080	0.20066	0.10033	0.00201	0.20066
6	4	0.6	210	0.00119	0.25082	0.15049	0.00000	0.00000
7	3	0.7	120	0.00179	0.21499	0.15049	0.00215	0.21499
8	2	0.8	45	0.00269	0.12093	0.09675	0.00484	0.48373
9	1	0.9	10	0.00403	0.04031	0.03628	0.00363	0.36280
10	0	1	1	0.00605	0.00605	0.00605	0.00097	0.09675
Sum					1	0.6	0.024	2.4

Notes: column 1: $k = 0, \ldots, 10$ are the discrete subdivisions of the sample space.
column 3: success rate k per total number of trials; $k/n = p_k$ for $n = 10$.
column 4: binomial coefficient by Equation (2.7-2).

column 5: product of powers of process probability p, as used in Equation (2.7-1).
column 6: probability of p_k by Equation (2.7-1).
column 7: probability-weighted p_k's used for calculating $E(p_k)$; the sum of this column is p.
column 8: the variance of p_k: $P(p_k) (p_k - p)^2$, computed according to Equation (2.7-7a). The sum of this column is the variance of p_k, by Equation (2.7-7a).
column 9: the variance of k: $P(p_k) [k - E(k)]^2$; with $E(k) = n p$, the column items are $n^2 \cdot P(p_k) \cdot (k/n - p)^2 = n^2 Var(k/n) = Var(k)$; for $n = 10$, column 9 is equal to $100 \cdot$ column 8.

(4) What is the probability that no more than six special events occur in 10 trials?
<u>Solution</u>: Using Equation (2.7-8) for the cumulative probability, for $k = 6$, $n = 10$, $p = 0.6$, one obtains

$$F(6) = \sum_{k=0}^{6} B(10,k) \, 0.6^k \, 0.4^{10-k}$$

$$= B(10,0) \cdot 0.6^0 \cdot 0.4^{10} + B(10,1) \cdot 0.6^1 \cdot 0.4^9$$

$$+ B(10,2) \cdot 0.6^2 \cdot 0.4^8 + B(10,3) \cdot 0.6^3 \cdot 0.4^7 + B(10,4) \cdot 0.6^4 \cdot 0.4^6$$

$$+ B(10,5) \cdot 0.6^5 \cdot 0.4^5 + B(10,6) \cdot 0.6^6 \cdot 0.4^4$$

$$= 1 \cdot 1 \cdot 0.00010 + 10 \cdot 0.6 \cdot 0.00026 + 45 \cdot 0.36 \cdot 0.000655$$

$$+ 120 \cdot 0.216 \cdot 0.001638 + 210 \cdot 0.1296 \cdot 0.004096$$

$$+ 252 \cdot 0.07776 \cdot 0.01024 + 210 \cdot 0.04666 \cdot 0.0256$$

$$= 0.00010 + 0.00156 + 0.01062 + 0.04246 + 0.11148$$

$$+ 0.20066 + 0.25084 = 0.61772.$$

The numbers in the last row before the final result show the contributions of the individual terms to the final result. Table 2-12, column 6, contains the same mutually exclusive probabilities of events $k = 0, 1, 2, 3, 4, 5,$ and 6. Their sum, including $k = 6$, is the non-exceedance probability that 6 or less events occur, $F(6) = P[k \leq 6] = 0.6177 \approx 0.62$.

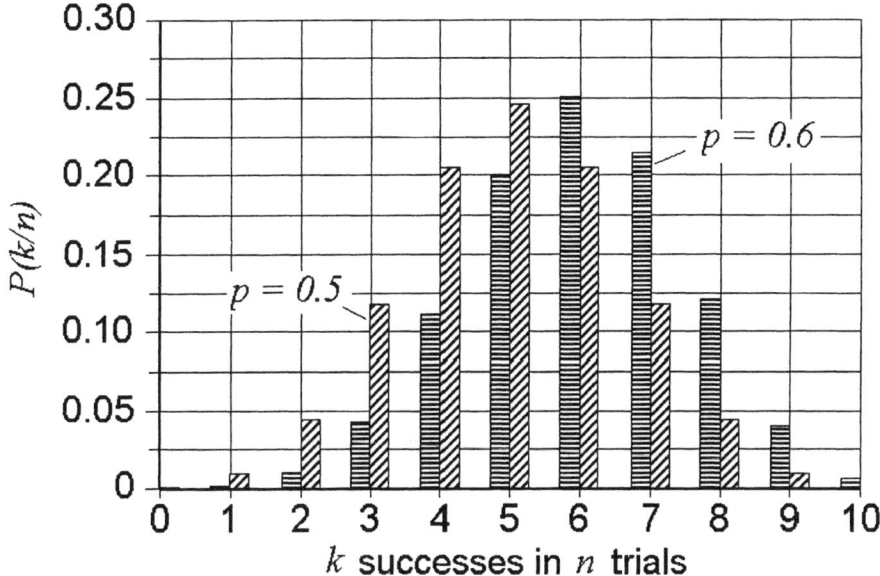

Figure 2-19: Binomial distribution of k successes in $n = 10$ trials for Bernoulli process probabilities $p = 0.5$ (inclined stripes) and $p = 0.6$ (horizontal stripes). The band of width $2\ \sigma$ centered on the mean $E(k)$ = 5 for $p = 0.5$ includes $k = 4$, 5, and 6, and accounts for 65.6 % of all occurrences; the band centered on the mean $E(k) = 6$ for $p = 0.6$ includes $k = 5$, 6, and 7, and accounts for 66.6 % of all occurrences. The p_k-distribution tops out over the k that corresponds to the p for which the frequencies of $p_k = k/n$ are calculated. For $p > 0.5$, the maximum moves to the right of the center of the range, and for $p < 0.5$, it moves to the left. This is illustrated by the shift to the right of the p_k-distribution for $p = 0.6$, in contrast to the symmetric distribution for $p = 0.5$. For $p = 0.1$, the maximum would be located over $k = 1$.

(5) Given is a set of 9 pennies. After the pennies are thrown simultaneously, each shows either a head (H) or a tail (T), if the case of pennies standing on their rim is ruled out. This experiment is repeated 10 times and each time the count k of heads is taken. Whether one penny is thrown 9 times and the total count k of the sequence is recorded, or whether the 9 pennies are thrown simultaneously and k is recorded, does not make a difference as k is a count of independent random events. The recorded k for the 10 experiments may look like the numbers in row 1 of Table 2-13

(a). Calculate the experimental probability distribution $P(p_k)$ and compare it with the one calculated by the binomial formula, Equation (2.7-1). Also calculate the expectation of p_k. Solution: The process probability is $p = 0.5$. This means each penny can produce either an H or a T with equal probability, assuming the pennies are unbiased. The numbers, 4, 7, 7, ... , etc., in row 1 of Table 2-13 (a) are realizations of the discrete range, $0 \leq k \leq 9$. The possible realizations are listed in row 1 of Table 2-13 (b). The k's counted in the 10 experiments are listed in row 2. Remember that each experiment is just one random path through the nine-step event tree, similar to the one marked in Figure 2-5. For a sequence of 9 branchings there is of a total of $2^9 = 512$ such paths leading to 512 outcomes with numbers of k ranging from 0 to 9. In the experiments, $k = 5$ was produced 4 times. Therefore, 4 is listed under $k = 5$. Dividing the experimental k in row 2 by 10, the number of experiments, gives the average experimental probability of a particular k, or of the success (or failure) rate $p_k = k/n$. This experimental probability $P(p_k)$ is shown in row 3 of Table 2-13 (b). For example, for $k = 5$, which appeared 4 times in 10 experiments, $P(p_k = 5/9) = 4/10 = 0.4$.

The average of k for all 10 experiments is the sum of all k in Table 2-13 (a), row 1, divided by the number of experiments, one obtains $E(k) = 49/10 = 4.9$. According to Equation (2.7-4), $E(k) = 9 \cdot 0.5 = 4.5$. The experimental process probability is $p = E(k)/n = 4.9/9 = 0.544$, whereas according to Equation (2.7-5), $E(k)/n = E(p_k) = p = 4.5/9 = 0.5$. Since $p = 0.5$ is the process probability, one would expect the most frequent k to be 4 or 5, and $p_k = 0.5$. This is indeed the case. Table 2-13 (b), row 3 shows how the experimental p_k's group around $p_k = 0.5$, ranging from $3/9 = 0.333$ to $7/9 = 0.777$. Because of the relatively small number of 10 trials, several p_k's did not occur at all, such as p_0, p_1, p_2, p_6, p_8, and p_9. The theoretical probabilities of Table 2-13 (b), row 4, are calculated by Equation (2.7-1):

$$P(p_k = \frac{k}{n}) = B(9,k)\, 0.5^k\, 0.5^{9-k} = B(9,k)\, 0.5^9 \;.$$

The formula represents the distribution if all possible 9-step paths are evaluated leading to 512 outcomes. This complete set of outcomes includes among others p_0 and p_9 which did not occur in the experiment. For example, the case of heads occurring nine times in a row (nine successes in nine decisions each representing success or failure with probability 0.5) has the probability

$$P(p_9 = \frac{9}{9}) = B(9,9)\, 0.5^9 = 0.5^9 \;.$$

This probability of n successes (or failures) in a row also is the probability of the joint occurrence of n successes each with probability $p = 0.5$. In the 10 experiments, the probability of this case is zero, as shown by Table 2-13(a), row 1.

Table 2-13: Experimental p_k Obtained from 10 Experiments for a Process Probability $p = 0.5$ and Comparison with Binomial Distribution

(a) Heads k by experiment:

Row	Item	Experiment										Σ
		1	2	3	4	5	6	7	8	9	10	–
1	k	4	7	7	5	4	5	5	5	4	3	49

Notes: title row: number of experiment, 1 to 10.
row 1: for each experiment, the number of k is recorded. For example, experiment 1 produced $k = 4$ heads among $n = 9$ elements; experiment 2 produced 7 heads among 9 elements, and so on. The 10 experiments produced 90 outcomes of which 49 were heads. Hence, the experimental process probability is $p_{exp} = \Sigma k_i/n = 49/90 = 0.54$, where $i = 1, ... , 10$.

(b) Experimental distribution of p_k and comparison with binomial distribution:

1	Possible k	0	1	2	3	4	5	6	7	8	9	Σ
2	Exp. k	0	0	0	1	3	4	0	2	0	0	10
3	Exp. $P(p_k)$	0	0	0	0.1	0.3	0.4	0	0.2	0	0	1
4	*Theoret.* $P(p_k)$.	0.002	0.018	0.070	0.164	0.246	0.246	0.164	0.070	0.018	0.002	1

Notes: row 1: the possible number of k in each experiment.
row 2: the experimental k picked from Table 2-13 (a), row 1; $k = 0, 1, 2, 8, 9$ did not occur; $k = 3$ occurred once; $k = 4$ occurred 3 times; $k = 5$ occurred 4 times, and so on; the row sum is the total number of experiments, 10.
row 3: occurrences in row 2 divided by the total number of experiments, 10; row sum is 1.
row 4: theoretical binomial probability of p_k for the k in row 1, as computed by Equation (2.7-1) for $k = 0, ... , n$; $n = 9$, $p = 0.5$; the row sum is 1.

(6) Calculate the probability of the special case of *no* item of 50 being defective for a process probability of the item being defective of $p = 0.01$. <u>Solution</u>: The probability of being nondefective is $q = 1 - p$. The probability of 50 items being nondefective is obtained from Equation (2.7-9) for the special case of $k = 0$:

$$f(0, 50, 0.01) = (1 - p)^{50} = 0.99^{50} = 0.605.$$

This is the joint probability of 50 joint events being "nondefective." Note that $k = 0$ refers to probability p. The probability for 50 nondefective items would be $f(50, 50, 0.99)$ with the same result.

(7) Calculate the probability of *more* than two defective items of 50 items if the average probability of being defective is $p = 0.01$. <u>Solution</u>: The probability of more than 2 defective items is the cumulation of all probabilities for $k > 2$. This probability can be conveniently obtained by calculating the complementary probability of up to two defective items. The probability of up to and including two defective items is

$$F(2, 50, p) = F(2)$$

$$= B(50, 0)\, p^0\, (1 - p)^{50} + B(50, 1)\, p^1\, (1 - p)^{49} + B(50, 2)\, p^2\, (1 - p)^{48}$$

$$= (1 - p)^{50} + 50\, p^1\, (1 - p)^{49} + 1225\, p^2\, (1 - p)^{48}$$

$$= 0.6050 + 0.3056 + 0.0756 = 0.9862.$$

The probability of more than two defective items is $P(k > 2) = 1 - F(2) = 1 - 0.9862 = 0.0138$.

(8) Calculate the probability of 20 failures in 5,000 trials if the binomial process failure probability is $p = 0.005$. Use (a) the binomial formula, and (b) the normal approximation. <u>Solution</u>: (a) The binomial formula, Equation (2.7-1), requires the evaluation of

$$f(20, 5000, 0.005) = B(5000, 20) \cdot 0.005^{20} \cdot 0.995^{4980}.$$

The binomial coefficient is evaluated using Equation (2.7-2a):

$$B(5000, 20) = (5000/1) \cdot (4999/2) \cdot (4998/3) \cdot \ldots \cdot 4981/20 = 3.77355 \cdot 10^{55},$$

so that $f(20,5000,0.005) = 3.77355\ 10^{55} \cdot 9.537 \cdot 10^{-47} \cdot 1.442\ 10^{-11} = 0.0519$.

(b) For $n = 5,000$, $k = 20$, $p = 0.005$, one finds $E(k) = n p = 5000 \cdot 0.005 = 25$, and $Var(k) = n p q = 5000 \cdot 0.005 \cdot 0.995 = 24.875$. Then,

$$z = \frac{k - E(k)}{\sqrt{Var(k)}} = \frac{20 - 25}{\sqrt{24.875}} = -1.0025 .$$

The normal approximation, Equation (2.7-12), produces

$$f(k,n,p) \approx \frac{1}{\sqrt{2\pi Var(k)}} e^{-0.5z^2} = \frac{0.39894}{\sqrt{24.875}} e^{-0.5(-1.0025)^2} = 0.0484 .$$

2.7.3 Uniform Distribution

For some processes, all random realizations of the range have equal probability. In this case, the pdf takes the form of a straight line parallel to the x-axis, from the lower bound a to the upper bound b of the range, as shown in Figure 2-20. This pdf is known as the *uniform* pdf. The integral of the pdf over the range must equal 1. This amounts to setting the area of the rectangle of length $b - a$, and height $f(x)$, equal to 1. The condition that the area under the pdf must be 1 gives

$$(b - a) f_X(x) = 1$$

where the index X identifies $f(x)$ as the pdf of the random variable X, a reminder of the nature of $f(x)$ used in connection with pdf's. The frequency function follows as

$$f_X(x) = \frac{1}{b - a} \tag{2.7-13}$$

where $f_X(x)$ is not a frequency and not a probability because it has the inverse dimension of $b - a$. Only $f_X(x)dx$ is a relative frequency or probability of a continuous function.

The non-exceedance probability $F(x) = P(X \le x)$ of x in the range $a \le x \le b$ is

$$F(x) = \int_a^x f_X(x)\,dx = \frac{1}{b-a} \int_a^x dx = \frac{1}{b-a} x\big|_a^x = \frac{x - a}{b - a} . \tag{2.7-14}$$

The cdf, $F(x)$, is a linear function of x over the range $a \le x \le b$, with $F(a) = 0$ and $F(b) = 1$. In order to simplify the notation, the index X is left off because the meaning of $F(x)$ as cdf of X is obvious.

The mean of x is found as the probability-weighted sum of all x. With $f(x)dx$ being the probability of x, the product, $x\,f(x)dx$, is one incremental term of this sum, but it can also be interpreted as the first *moment* of the area $f(x)\,dx$ with respect to the origin of x (see Section 2.5.3). Hence the mean is

$$\mu = \int_a^b x f(x)dx = \int_a^b \frac{x}{b-a}dx = \frac{1}{b-a}\frac{x^2}{2}\Big|_a^b = \frac{1}{2}\frac{b^2-a^2}{b-a} = \frac{a+b}{2}. \quad (2.7\text{-}15)$$

In this simple case, one may have guessed without integration that μ is the average of the upper and lower range boundaries.

The variance is found as the second moment of $f(x)\,dx$:

$$\sigma^2 = \int_a^b (x-\mu)^2 f(x)dx = \int_a^b \frac{(x-\mu)^2}{b-a}dx . \quad (2.7\text{-}16)$$

The second moment leads to a more complicated integral which is solved by making the substitution, $u = x - \mu$. With the differential $du = dx$, the integral produces

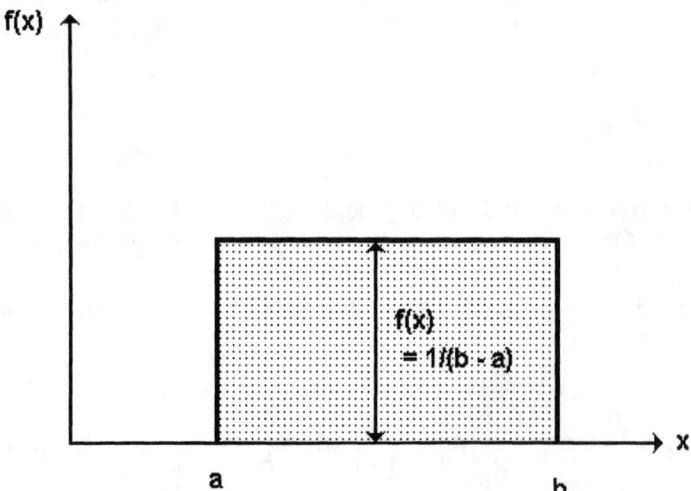

Figure 2-20: Uniform probability density function

$$\sigma^2 = \frac{(x-\mu)^3}{3(b-a)}\Big|_a^b$$

Substituting Equation (2.7-15) for μ and evaluating the integral leads to

$$\sigma^2 = \frac{(b-a)^2}{12} . \tag{2.7-17}$$

The moment integrals for most pdf's are quite difficult to solve. The formulas for the parameters of the most important pdf's can be found in textbooks (Moan, 1982, pp. 4-13 to 4-57).

Example: (1) A sample space contains n elements. They are all equally likely. Calculate the mean. Solution: The sample space with a countable number of elements is discrete. The mean is obtained as the first moment of a discrete pdf. For a discrete distribution, the moment is calculated as the sum of elementary moments. With $f_X(x_i)$ = $1/n$ being the constant probability of each element i one obtains

$$\mu = \sum_{i=1}^{n} x_i f_X(x_i) = \frac{1}{n}\sum_{i=1}^{n} x_i .$$

This expression is the arithmetic mean.

(2) Suppose a random variable has a range $5 \le X \le 15$, in which it can assume any value. This is an infinite sample space. Due to lack of better knowledge, it is assumed that all realizations in this range have the same probability. Calculate (a) the frequency of the distribution; (b) the non-exceedance probability of $x = 10$; and (c) the parameters μ and σ of the distribution. Solution: (a) The frequency is a dimensionless quantity. For a continuous function, the relative frequency is $f(x)dx$. From Equation (2.7-13) one obtains

$$f(x)dx = \frac{dx}{15-5} = 0.1\,dx .$$

(b) For $x = 10$, one obtains the non-exceedance probability from Equation (2.7-14)

$$F(x) = \frac{10-5}{15-5} = 0.5 .$$

This result means that 50 % of all X are smaller than or maximally equal to 10, they don't exceed 10. (c) Equation (2.7-15) gives for the mean $\mu = (5 + 15)/2 = 10$; the standard deviation is obtained from Equation (2.7-17) as $\sigma = (15 - 5) / \sqrt{12} = 2.89$.

2.7.4 Triangular Distribution

The *triangular* pdf allows a maximum frequency to occur somewhere in the range or on the boundaries of the range, (a, b). If the maximum frequency occurs on one boundary, then the linearly rising or falling curve mimics a rising or falling exponential distribution. If the midpoint of the range has the maximum frequency the distribution is symmetric. But also asymmetric triangular distributions occur.

(a) Symmetric triangular distribution. The pdf is again constructed by requiring the area enclosed by the pdf to have the value 1:

$$\frac{1}{2} f_{max} (b - a) = 1 .$$

$\qquad\qquad\qquad\qquad\qquad\qquad\qquad\qquad\qquad\qquad\qquad$ (2.7-18)

Solving for f_{max} gives

$$f_{max} = \frac{2}{(b - a)} .$$

$\qquad\qquad\qquad\qquad\qquad\qquad\qquad\qquad\qquad\qquad\qquad$ (2.7-19)

The maximum ordinate f_{max} marks the value x with the highest frequency of the range, called the *mode*. In Figure 2-21, the range limits have probability zero. If this requirement is too restrictive, it can be satisfied by extending the range somewhat beyond the possible boundaries a and b, and giving the extended limits $a' < a$ and $b' > b$ probability zero. For the case of zero probability for a and b, the pdf is defined as:

$$f_X(x) = 0 \qquad\qquad\qquad \text{for } x < a$$

$$f_X(x) = 4 \frac{x - a}{(b - a)^2} \qquad\qquad \text{for } a \le x \le \frac{a + b}{2} \qquad\qquad (2.7\text{-}20a)$$

$$f_X(x) = 4 \frac{b - x}{(b - a)^2} \qquad\qquad \text{for } \frac{a + b}{2} < x \le b \qquad\qquad (2.7\text{-}20b)$$

$$f_X(x) = 0 \qquad\qquad\qquad \text{for } x > b.$$

Applying the moment integrals to this piecewise continuous pdf, the mean and variance are obtained as

$$\mu = \frac{a+b}{2} \qquad\qquad (2.7\text{-}21)$$

and

$$\sigma^2 = \frac{(b-a)^2}{24} . \qquad\qquad (2.7\text{-}22)$$

The mean is the same as for the uniform pdf, and the variance is one half of the variance of the uniform pdf because the triangular pdf is more concentrated around the center. This result also demonstrates that only the variance, not the mean, is sensitive to the spread of the random variable realizations over the possible range.

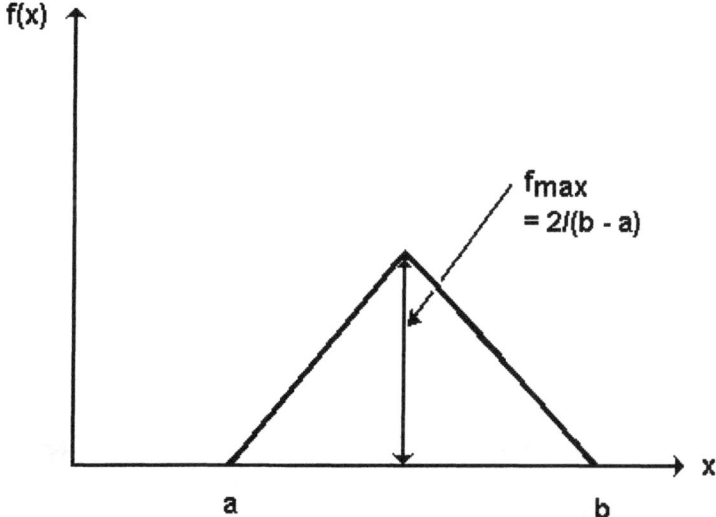

Figure 2-21: Triangular probability density function

(b) Asymmetric triangular distribution. An additional parameter c is needed to mark the point of highest frequency or *mode* within the range. The maximum frequency ordinate is again $f_{max} = 2/(b-a)$. The mean is

$$\mu = \frac{1}{3}(a+b+c),$$ (2.7-23)

and the variance is

$$\sigma^2 = \frac{1}{18}(a^2+b^2+c^2-ab-ac-bc).$$ (2.7-24)

For the special case of a symmetric distribution with the mode in the center of the range, $c = (a + b)/2$, and Equations (2.7-23) and (2.7-24) reduce to Equations (2.7-21) and (2.7-22), respectively.

2.7.5 Exponential Distribution

The *exponential distribution* has many practical applications, especially in business and maintenance, and also is easy to handle, both by calculus and by numerical methods. Quantities that in small amounts have a high probability and in large amounts have a low probability may have an exponential distribution. An example is the life span of mechanical equipment that has a high probability of surviving the early years and a small probability to last very long. The exponential distribution has the form of an exponential decay function:

$$f(t) = ce^{-\lambda t}$$ (2.7-25)

where λ is the average failure rate, which may be the number of failures per time unit, and t is time measured from some point t_0, which is assumed zero in Equation (2.7-25). The coefficient c can be interpreted as the value of $f(t)$ at time $t = 0$, $f(0) = c$. The probability of t, $dF(t) = f(t)\, dt$ must be dimensionless; hence $f(t)$ has the inverse dimension of t. The choice of t as a time random variable is arbitrary. Usually t denotes a time, e.g., the life length of a component, but it also can be a distance or a length, x, and so on. For this function to be a pdf, the area under $f(t)$, or the integral over its range, must be 1. This requirement provides the equation to calculate c. The integral over the range, here $0 \le t \le +\infty$, is set equal to 1:

$$F(\infty) - F(0) = 1 = c \int_0^{+\infty} e^{-\lambda t} dt$$ (2.7-25a)

For $t = 0$, the lower range limit, $F(0) = 0$, and for $t = +\infty$, the upper range limit, $F(\infty) = 1$. For convenience, an auxiliary variable $u = -\lambda t$ is introduced, which gives $du = -\lambda\, dt$. Hence, the integral gives

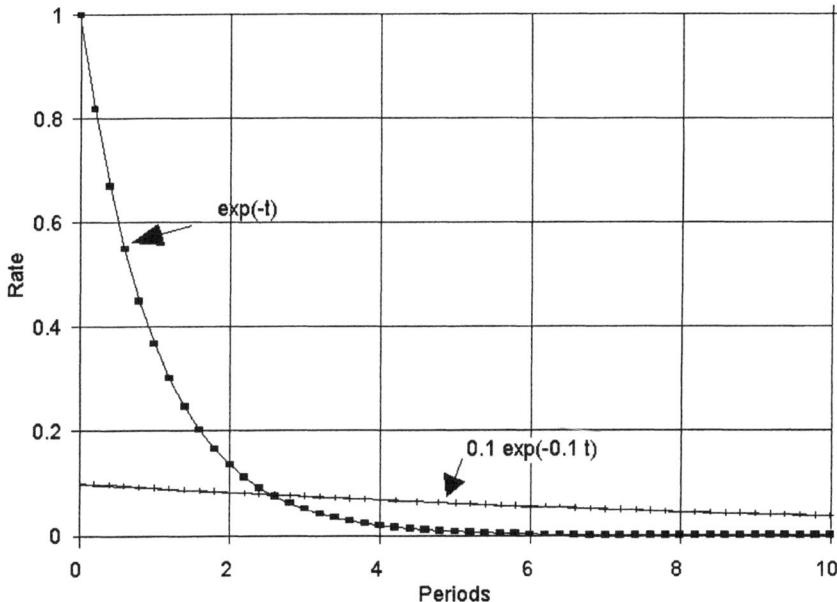

Figure 2-22: Exponential probability density function for parameters $\lambda = 1$ and 0.1 as function of period length t.

$$1 = -\frac{c}{\lambda} \int_0^{+\infty} e^u du = -\frac{c}{\lambda} e^{-\lambda t} \big|_0^{+\infty} = \frac{c}{\lambda} \qquad (2.7\text{-}25b)$$

and $c = \lambda$. The exponential pdf becomes

$$f(t) = \lambda e^{-\lambda t}. \qquad (2.7\text{-}26)$$

The exponential pdf, Equation (2.7-26), has only one parameter, λ, and $f(t)$ has the dimension of λ. Equation (2.7-26), for two choices of the rate, $\lambda = 0.1$ and $\lambda = 1$, is shown in Figure 2-22. The function starts with the value $f(0) = \lambda$. For small λt, $f(t)$ is nearly linear. This is due to the fact that for small exponents, $e^{-\lambda t} \approx 1 - \lambda t$. This linearization is valid as long as approximately $\lambda t \leq 0.2$.

The mean of the exponential distribution is obtained by

$$\mu = \int\limits_0^\infty t f(t) d t = \int\limits_0^\infty t \lambda e^{-\lambda t} d t \,.$$

(2.7-26a)

With the variable transformation $u = -\lambda t$ and with $du = -\lambda dt$, Equation (2.7-26a) becomes

$$\mu = \frac{1}{\lambda} \int\limits_0^\infty u \ e^u d u = \frac{1}{\lambda} \ [e^{-\lambda t}(-\lambda t - 1)]|_0^\infty = \frac{1}{\lambda} \,.$$

(2.7-26b)

If λ is the average number of failures per time, then $1/\lambda$ is the average time per failure or time between failures. For example, if $\lambda = 1$ failure /100 days, then $1/\lambda$ = 100 days /1 failure. The rate λ must always have the inverse dimension of t.

The standard deviation can be found by a similar moment procedure applied to the variance:

$$\sigma^2 = \int\limits_0^\infty (t - \frac{1}{\lambda})^2 \lambda e^{-\lambda t} dt$$

(2.7-26c)

By multiplying out the square, the integral can be split into three integrals which can be relatively easily solved and then evaluated at their boundaries. This leads to products $0 \cdot \infty$, which must be evaluated by the Bernoulli-L'Hopital rule. The products produce zero and the integration gives $\sigma^2 = 1/\lambda^2$, and the standard deviation becomes

$$\sigma = \frac{1}{\lambda} \,.$$

(2.7-26d)

Introducing an additional parameter, a, as starting time, and using $\lambda = 1/b$, the exponential pdf takes on the form

$$f(t) = \frac{1}{b} e^{-\frac{t-a}{b}}$$

(2.7-27)

with the parameters $\mu = a + b$, and $\sigma = b$ (Abramowitz and Stegun, 1970, p. 930). The exponent, $(t - a)/b$, has the form of a reduced variable, similar to z for the normal distribution.

The non-exceedance probability of the exponential distribution is obtained by the integral of Equation (2.7-27):

$$F(t_f) = \frac{1}{b} \int_a^{t_f} e^{-\frac{t-a}{b}} dt \qquad (2.7\text{-}28a)$$

where a is the lower limit and t_f is the upper limit of the integral. Using an auxiliary variable $u = -(t - a)/b$, and $du = -dt/b$, the integration gives

$$F(t_f) = 1 - e^{-\frac{t_f - a}{b}}. \qquad (2.7\text{-}28b)$$

where $F(t_f)$ is the non-exceedance probability of t_f.

Example: Determine (a) the probability of equipment failure by time t_f that is counted from a start-up time; and (b) the probability of survival beyond t_f. Solution: (a) According to Equation (2.7-28b), with $a = 0$, the probability of equipment failure during period $0 \le t \le t_f$ is

$$P(T \le t_f) = F(t_f) = 1 - e^{-\frac{t_f}{b}}.$$

(b) The complement to $F(t_f)$, or the probability of survival of period t_f, is

$$R(t_f) = P(T > t_f) = 1 - F(t_f) = e^{-\frac{t_f}{b}}$$

where $R(t_f)$ is the reliability, or the probability, that the equipment does not fail during period t_f.

2.7.6 Beta Distribution

The beta pdf is a two-parameter distribution. It can assume very different shapes depending on the values of its two parameters α and β:

$$f(t) = \frac{(\alpha + \beta + 1)!}{\alpha! \beta!} t^\alpha (1 - t)^\beta \qquad (2.7\text{-}29)$$

where t is the random variable; for $0 \le t \le 1$, $\alpha > -1$, and $\beta > -1$. The first of the three factors of Equation (2.7-29) is the normalizing factor; in other words, it makes the integral over the range of t equal to 1. It is valid as long as α and β are integers. If this restriction is dropped, Equation (2.7-29) becomes

$$f(t) = \frac{\Gamma(\alpha + \beta + 2)}{\Gamma(1 + \alpha)\Gamma(1 + \beta)} t^{\alpha}(1 - t)^{\beta} \qquad (2.7\text{-}29a)$$

where $\Gamma(...)$ are gamma functions (Benjamin and Cornell, 1970, p. 287; Abramowitz and Stegun, 1970, p. 255).

The following parameter choices provide different shapes of the beta distribution:

(a) $\alpha = \beta = 0$: Equation (2.7-29) gives $f(t) = 1$, a uniform pdf over the range $0 < t < 1$.

(b) $\alpha = \beta = -1/2$: Equation (2.7-29a) requires the evaluation of gamma functions:

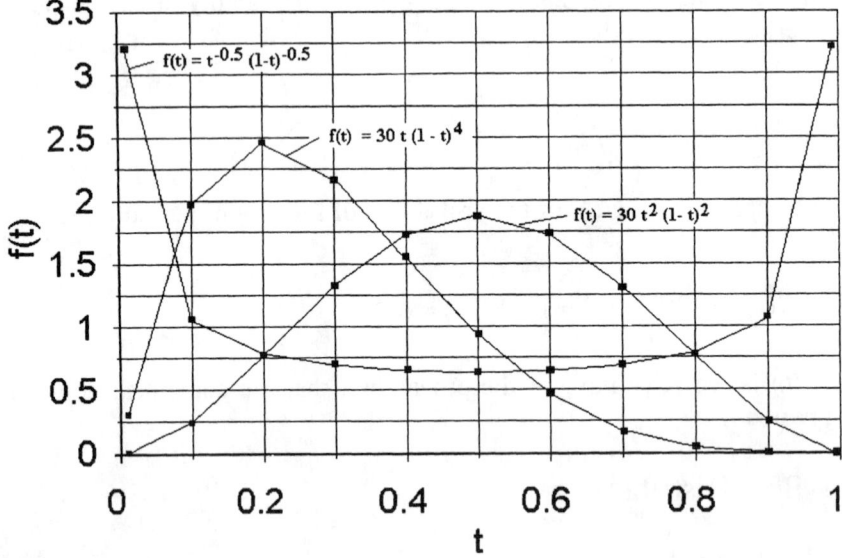

Figure 2-23: Various forms of probability density functions based on the beta distribution. The special cases shown are bathtub, asymmetric bell shape, and symmetric bell shape.

$\Gamma(n + 1) = n!$; hence, for $n = 0$, $\Gamma(1) = 0! = 1$. $\Gamma(1/2) = \pi^{1/2}$. Substituting these expressions into Equation (2.7-29a) gives

$$f(t) = \frac{1}{\pi} t^{-\frac{1}{2}} (1-t)^{-\frac{1}{2}}.$$ (2.7-30)

Equation (2.7-30) has the shape of a symmetric bathtub curve, as shown in Figure 2-23. The minimum is located by solving $df(t)/dt = 0$ for t. The result is $t = \alpha/(\alpha + \beta) = 0.5$, which means the extremum being a minimum in this case is located in the center of the range.

(c) $\alpha = 0$, $\beta = 1$: Equation (2.7-29) gives

$$f(t) = 2 (1 - t).$$ (2.7-31)

Equation (2.7-31) is a linearly decreasing function with maximum ordinate at the start of the range, $f(0) = 2$, and a minimum ordinate at the end of the range, $f(1) = 0$. The pdf forms a one-sided triangular distribution. The area under the pdf is 1, as required.

(d) $\alpha = 1$, $\beta = 4$: Equation (2.7-29) gives

$$f(t) = 30\, t\, (1 - t)^4.$$ (2.7-32)

Equation (2.7-32) is an asymmetric bell shape in the range $0 \le t \le 1$ that looks like a pulse response function, with $f(0) = 0$ and $f(1) = 0$, as shown in Figure 2-23. The maximum is at $t = \alpha/(\alpha + \beta) = 1/5$. Note that this point does not coincide with the mean which is $\mu = 2/7$ (see Equation (2.7-35)).

(e) $\alpha = 4$, $\beta = 1$: The factor in Equation (2.7-29) is again 30 and the pdf becomes

$$f(t) = 30\, t^4\, (1 - t).$$ (2.7-33)

Equation (2.7-33) is an asymmetric bell shape with maximum at $t = \alpha/(\alpha + \beta) = 4/5$, and the mean $\mu = 5/7$.

(f) $\alpha = 2$, $\beta = 2$: The factor of Equation (2.7-29) is $5!/[\,2!\,2!\,] = 30$; the pdf becomes

$$f(t) = 30 t^2 (1-t)^2 .$$ (2.7-34)

Equation (2.7-34) is a symmetric bell shape confined within the interval boundaries $0 \le t \le 1$. The maximum is located at $t = \alpha/(\alpha + \beta) = 0.5$. By changing α and β, the bell shape can be flattened or steepened. Figure 2-23 demonstrates that the beta distribution may be suitable for random variables with bell shape distributions within finite boundaries. Examples are cost distributions, repair time distributions, and distributions of fluctuating reservoir levels (Yevjevich, 1972, p. 149). The pdf's given by Equations (2.7-30) through (2.7-34) integrated over the range $0 \leq t \leq 1$ produce 1, as required.

Some of the *moments* of the beta distribution can be rather easily obtained by integration. For example, for Equation (2.7-34), the first moment is obtained by

$$\mu = \int_0^1 tf(t)dt = 30\int_0^1 t^3(1-t)^2 dt = 30[\frac{t^4}{4} - 2\frac{t^5}{5} + \frac{t^6}{6}]|_0^1 = \frac{30}{4} - \frac{60}{5} + \frac{30}{6} = \frac{1}{2} .$$

The mean is at the center of the range for this symmetric pdf.

Formulas for moments of the beta distribution can be found in handbooks (Moan, 1982, pp. 4-40 to 4-41). For example, the k-th moment of the beta function is

$$V_k = \frac{(\alpha + \beta + 1)!(\alpha + k)!}{(\alpha + \beta + k + 1)!(\alpha)!} \qquad (2.7\text{-}35)$$

For the mean, or first moment, $k = 1$, and with $\alpha = 2$, and $\beta = 2$,

$$V_1 = \mu = \frac{5!\,3!}{6!\,2!} = \frac{1}{2} .$$

This is the same result as obtained by the integration of Equation (2.7-34).

Examples: (1) Show that Equation (2.7-30) is a pdf.   Solution: The integral of Equation (2.7-30) over the range of the random variable $0 \leq t \leq 1$ is

$$\int_0^1 f(t)dt = \frac{1}{\pi}\int_0^1 \frac{1}{\sqrt{t}}\frac{dt}{\sqrt{1-t}}$$

We use the substitution

$u = \sqrt{1-t}$, which also is $t = 1 - u^2$. Introducing $\quad -2\,du = \dfrac{dt}{\sqrt{1-t}}$

and the substitution of $\sqrt{t}$ into the integral gives

$$-\frac{2}{\pi}\int\frac{du}{\sqrt{1-u^2}} = -\frac{2}{\pi}\arcsin(u) = -\frac{2}{\pi}(\arcsin\sqrt{1-t})\big|_0^1 = -\frac{2}{\pi}(0-\frac{\pi}{2}) = 1 \quad .$$

This proves that Equation (2.7-30) is a pdf.

(2) For Equation (2.7-30), calculate the mean by the moment method and also by Equation (2.7-35). Solution: The moment method proceeds similar to the integration in Example (1). The first moment is

$$\mu = \int_0^1 t\, f(t)dt = \frac{1}{\pi}\int_0^1\frac{t}{\sqrt{t}}\frac{dt}{\sqrt{1-t}} \quad .$$

With the same substitutions as in Example (1) the integral is transformed and solved for the transformed integration limits: for $t = 0$, $u = 1$; and for $t = 1$, $u = 0$. The result is

$$\mu = -\frac{2}{\pi}\int\sqrt{1-u^2}\,du = -\frac{2}{\pi}\left[\frac{u}{2}\sqrt{1-u^2} + \frac{1}{2}\arcsin(u)\right]_{u=1}^{u=0} = \frac{1}{2} \quad .$$

The mean, computed by Equation (2.7-35), for $k = 1$, $\alpha = -1/2$, and $\beta = -1/2$, requires the evaluation of several factorials: $0! = 1$, $(1/2)! = (1/2)(\pi/2)$, $1! = 1$; and $(-1/2)! = \pi/2$ (see the factorial function in Abramowitz and Stegun, 1970, p. 255). Substituting these results into Equation (2.7-35) one obtains $\mu = 1/2$, the same as by the moment method. Equation (2.7-30) is a symmetric pdf and the mean is located in the center of the range.

2.7.7 Poisson Distribution

The Poisson distribution is used to model arrival rates and service rates in queuing models. There is a connection between the *exponential* and the *Poisson* distributions. Probabilistic (or stochastic) queuing models use exponential distributions for interarrival time and service time modeling or, equivalently, the Poisson distribution for arrival rate and service rate modeling (Gross and Harris, 1974, p. 23). The number of arrivals during a period extending from time zero to t is a random variable. The probability of n arrivals during a period t, with $n \geq 0$ being positive integers, is derived by setting up so-called stochastic *differential-difference equations*. This calculus sheds light on probabilistic mathematics, but the derivation is somewhat involved and cannot be repeated here. Suffice it to say that the results of the derivation are the *stochastic differential equations* of arrival probabilities (Gross and Harris, 1974, p. 25). For zero arrivals,

$$\frac{dp_0(t)}{dt} = -\lambda p_0(t),$$ (2.7-36a)

where the index relates to the number of arrivals in time t. For arrivals $n \geq 1$, the stochastic differential equation is of the form

$$\frac{dp_n(t)}{dt} = -\lambda p_n(t) + \lambda p_{n-1}(t).$$ (2.7-36b)

Equation (2.7-36b) is a first order linear differential equation in which the dependent variable and its derivative are of the first degree. A solution method consists of finding an *integrating factor* (Nielsen, 1962, p. 48). First Equation (2.7-36b) is rewritten in the form

$$\frac{dp_n(t)}{dt} + \lambda p_n(t) = \lambda p_{n-1}(t).$$ (2.7-36c)

Suppose an integrating factor, $g(t)$, has been found. Then, multiplying the equation with it gives

$$g(t)[\frac{dp_n(t)}{dt} + \lambda p_n(t)] = g(t)\lambda p_{n-1}(t).$$ (2.7-36d)

The integrating factor is of such a nature that it complements the left-hand side of the equation into an exact differential. The integral of an exact differential is readily obtained whereupon only the right-hand side needs to be integrated. The probability $p_{n-1}(t)$ is a constant that is known from a previous calculation so that the right-hand side can also be easily integrated. A hint of an integrating factor is obtained by the solution of the homogeneous equation, Equation (2.7-36a),

$$\frac{dp_n(t)}{dt} + \lambda p_n(t) = 0.$$

The solution is obtained by separation of variables and integration,

$$p_n(t) = ce^{-\lambda t}$$

where c is an integration constant. Moving the e-function to the left side gives

$$e^{\lambda t} p_n(t) = c .$$

Taking the total differential of this equation gives

$$d[e^{\lambda t} p_n(t)] = e^{\lambda t}[\frac{dp_n(t)}{dt} + \lambda p_n(t)] = 0$$

This shows that $e^{\lambda t}$ is an integrating factor that complements the left side of Equation (2.7-36d). With $g(t) = e^{\lambda t}$, the left-hand side of Equation (2.7-36d) can be written as an exact differential $d[e^{\lambda t} p_n(t)]$ and the integral of the equation becomes

$$\int_0^t d[e^{\lambda t} p_n(t)] = \int_0^t e^{\lambda t} \lambda \, p_{n-1}(t) dt \quad .$$

(2.7-36e)

An equation explicit in the dependent variable $p_n(t)$ is obtained by carrying out the integration of the left-hand side. The product of functions on the left-hand side exists only for $t > 0$. Freeing $p_n(t)$ of its factor gives

$$p_n(t) = e^{-\lambda t} \int_0^t e^{\lambda t} \lambda \, p_{n-1}(t) dt$$

(2.7-36f)

where the factor in front of the integral is the result of the integration of the left-hand side. Equation (2.7-36f) is valid for $n \geq 1$. It is a recursive formula which can be used to find the probabilities for $n > 0$ once $p_0(t)$ is found.

For $n = 0$, which means zero arrivals during period t, Equation (2.7-36a) applies. By separation of variables one obtains

$$p_0(t) = e^{-\lambda t}$$

(2.7-37a)

where λ is the mean arrival rate of items per time unit, and λt is the mean number of arrivals in period t.

For $n = 1$, one obtains from Equation (2.7-36f) by integrating from 0 to t,

$$p_1(t) = e^{-\lambda t} \int_0^t e^{\lambda t} \lambda \, p_0(t) dt = e^{-\lambda t} \int_0^t e^{\lambda t} \lambda \, e^{-\lambda t} dt = \lambda t \, e^{-\lambda t} \quad .$$

(2.7-37b)

For $n = 2$: $p_2(t) = \dfrac{(\lambda t)^2}{2} e^{-\lambda t}$.   (2.7-37c)

For $n = 3$: $p_3(t) = \dfrac{(\lambda t)^3}{3!} e^{-\lambda t}$.   (2.7-37d)

The probability of $n = k$ arrivals in period t is

$$p_k(t) = \frac{(\lambda t)^k}{k!} e^{-\lambda t} \ .$$ (2.7-38)

Equation (2.7-38) is the pdf of the Poisson arrival process. It is a discrete probability distribution, where $p_k(t)$ is the discrete probability of $k = 0, 1, 2, 3, ... , k$ arrivals in period t. The Poisson distribution approaches the binomial distribution for large n and small p, with $\lambda t = n p$. The random variable does not have to be time. If λ is the number of accidents per kilometer of highway, then the average number of accidents over 10 km is $a = 10 \lambda$.

The parameter $a = \lambda t$ in Equation (2.7-38) is the only parameter of the Poisson distribution. It is also the mean as well as the standard deviation of the distribution: $\mu = \sigma = a$. The mean can be interpreted as expected number of jobs arriving at a repair shop during a specified period t, whereas the pdf gives the frequencies of random realizations that may actually occur causing job queues. Two examples of Equation (2.7-38) for $a = 5$ and $a = 10$ are shown in Figure 2-24.

The non-exceedance probability of the Poisson distribution is

$$F(k) = e^{-\lambda t} \sum_{i=0}^{k} \frac{(\lambda t)^i}{i!} \ .$$ (2.7-39)

where $F(k)$ is the discrete cdf of the Poisson distribution whose ordinates only exist for integer numbers.

A stochastic process with practical applications is the   *compound Poisson process*. A practical example is the sum of costs that accrue by the random occurrence of repair jobs during some time period, such as a month (Parzen, 1965, p. 128),

$$X(t) = \sum_{n=1}^{N(t)} C_n$$

where $X(t)$ is the stochastic variable, i.e., the total of $N(t)$ random costs, C_n, that occur during a month t by a Poisson arrival process of repair jobs.

Examples: (1) Calculate the probability of $n = 4$ arrivals in period t with $p_3(t)$ given by Equation (2.7-37d). Solution: Using Equation (2.7-36f), one obtains

$$p_4(t) = e^{-\lambda t} \int_0^t e^{\lambda t} \lambda \, p_3(t) \, dt = e^{-\lambda t} \int_0^t e^{\lambda t} \lambda \frac{(\lambda t)^3}{3!} e^{-\lambda t} dt = \frac{(\lambda t)^4}{4!} e^{-\lambda t}.$$

(2) What is the probability that up to and including six arrivals occur in period $t = 10$ h, given the average arrival rate is $\lambda = 0.6$ h^{-1}. Solution: The average number of arrivals is $\lambda t = 0.6 \cdot 10 = 6$. The non-exceedance probability of six arrivals is according to Equation (2.7-39)

$$F(6) = p_0(10) + p_1(10) + p_2(10) + p_3(10) + p_4(10) + p_5(10) + p_6(10) =$$

$$= (6^0/0!) \, e^{-6} + (6^1/1!) \, e^{-6} + (6^2/2!) \, e^{-6} + (6^3/3!) \, e^{-6}$$

$$+ (6^4/4!) \, e^{-6} + (6^5/5!) \, e^{-6} + (6^6/6!) \, e^{-6}$$

$$= 0.00248 \, (1 + 6 + 18 + 36 + 54 + 64.8 + 64.8) = 0.6066.$$

(3) The binomial cdf, Equation (2.7-8), is adapted to Example (2) to find the probability of 6 successes in 10 trials, or the probability of 6 arrivals (successes) in 10 hours ($n = 10$ trials). For $n = 10$, $k = 6$, $p = 6/10 = 0.6$, Equation (2.7-8) becomes

$$F(6) = \sum_{i=0}^{6} [B(10,i) \cdot 0.6^i \cdot 0.4^{10-i}]$$

$$= B(10,0) \cdot 0.6^0 \cdot 0.4^{10} + B(10,1) \cdot 0.6^1 \cdot 0.4^9 + B(10,2) \cdot 0.6^2 \cdot 0.4^8$$

$$+ B(10,3) \cdot 0.6^3 \cdot 0.4^7 + B(10,4) \cdot 0.6^4 \cdot 0.4^6 + B(10,5) \cdot 0.6^5 \cdot 0.4^5$$

$$+ B(10,6) \cdot 0.6^6 \cdot 0.4^4$$

$$= 1 \cdot 1 \cdot 0.000\ 105 + 10 \cdot 0.6 \cdot 0.000\ 262 + 45 \cdot 0.36 \cdot 0.000\ 655$$

$$+ 120 \cdot 0.216 \cdot 0.001\ 64 + 210 \cdot 0.129\ 6 \cdot 0.004\ 096$$

$$+ 252 \cdot 0.077\,76 \cdot 0.01024 + 210 \cdot 0.046\,66 \cdot 0.025\,6$$

$$= 0.618 \approx 0.62$$

This result was also obtained by Example (4) of Section 2.7.2.

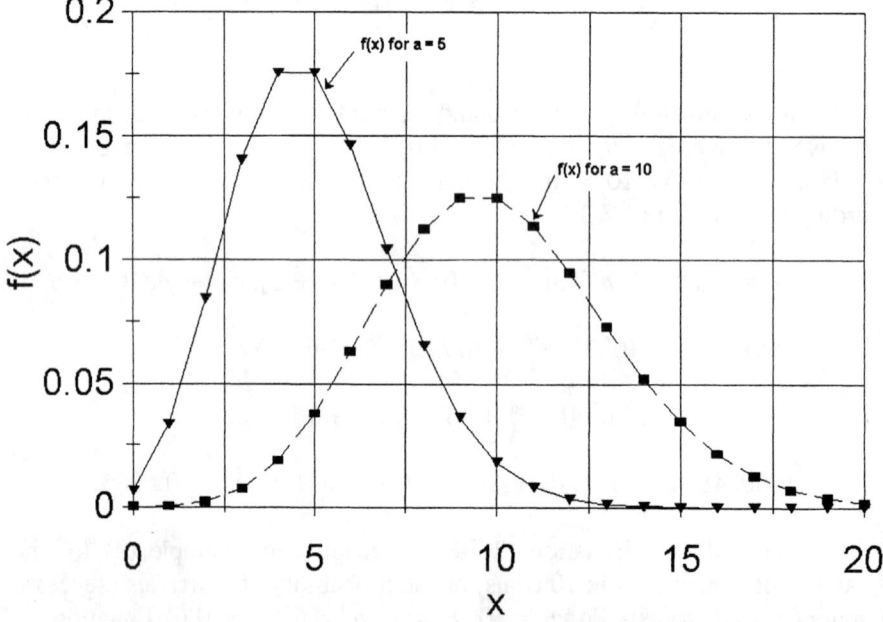

Figure 2-24: Poisson pdf's for random arrivals k in the form of $f(x) = a^x e^{-a}/x!$ where $a = \lambda t$, and $x = k$ for average arrivals $\mu = \lambda t = 5$ and 10. The arrival probabilities are highest around the mean. The random variable $x = k$ is a discrete variable which exists only for integers $x \geq 0$. The $f(x)$ are discrete probabilities, which should be shown as bars over the x's for which they exist. The line connections between the $f(x)$ represent only visual aids.

(4) Calculate the probability of up to two defective items arriving in a production stream over a 50 h period if the average defect rate is $\lambda = 0.01$ h^{-1}. Compare the result with the binomial distribution. <u>Solution</u>: The probability of up to two defective items is given by $F(2)$. For $\lambda t = 0.01 \cdot 50 = 0.5$,

$$F(2) = p_0(50) + p_1(50) + p_2(50)$$

$$= (0.5^0/0!) \, e^{-0.5} + (0.5^1/1!) \, e^{-0.5} + (0.5^2/2!) \, e^{-0.5}$$

$$= 0.6065 + 0.3033 + 0.0758 = 0.9856.$$

This result is very close to the first part of Example (7) of Section 2.7.2, which produced $F(2) = 0.9862$. It demonstrates that the Poisson distribution approaches the binomial distribution for large n and small p. The Poisson probabilities are easier to calculate than the binomial probabilities.

2.7.8 Weibull Distribution

The Weibull distribution was first proposed for "a statistical representation of fatigue failures in solids" (Moan, 1982, p. 4-47), but it also is used in hydrology because it is applicable for positive only random variables (Stedinger et al., 1993, p. 18.13). It is a three-parameter distribution and therefore is quite flexible in the shapes it can simulate. Various formulas are given in the numerous references. Here, the pdf is given as referred to by Parzen (1965, p. 169):

$$f(x) = \frac{k}{v-c} \left(\frac{x-c}{v-c} \right)^{k-1} e^{-\left(\frac{x-c}{v-c} \right)^k} \tag{2.7-40}$$

where $f(x)$ exists only for $x \geq c$, and is zero for $x < c$; c is the location parameter, k is the shape parameter, and v is the scale parameter. Equation (2.7-40) looks somewhat complex, but it is actually quite simple and can be easily integrated to obtain the non-exceedance probability. For this purpose, we introduce the substitution

$$u = \left(\frac{x-c}{v-c} \right)^k .$$

The differential of u is

$$du = k \left(\frac{x-c}{v-c} \right)^{k-1} \frac{dx}{v-c} .$$

Substituting u and du into the integral of the cdf, one obtains

$$F(x) = \int_c^x f(x) dx = -\int_c^x e^{-u} d(-u) = -e^{-u} \Big|_c^x .$$

Evaluating the boundaries gives

$$F(x) = 1 - e^{-(\frac{x-c}{v-c})^k} . \tag{2.7-41}$$

The simple relation between $f(x)$ and $F(x)$ allows easy use of the Weibull distribution in the calculation of the hazard function (Section 3.6).

A graphical determination of parameters is based on taking the double logarithm of a form derived from Equation (2.7-41):

$$\frac{1}{1 - F(x)} = e^{(\frac{x-c}{v-c})^k} .$$

This leads to

$$\ln(\ln \frac{1}{1 - F(x)}) = k \ln(x - c) - k \ln(v - c) . \tag{2.7-42}$$

Equation (2.7-42) is a straight line in a double-log diagram:

$$Y = k X + C \tag{2.7-43}$$

where

$$Y = \ln(\ln \frac{1}{1 - F(x)}) , \tag{2.7-44}$$

$$X = \ln(x - c) , \tag{2.7-45}$$

and

$$C = -k \ln(v - c) . \tag{2.7-46}$$

where Y and X are the coordinates of the double log diagram, C is the empirical constant or intercept, and k is the slope of the line (Moan, 1982, p. 4-49).

Special Cases of the Weibull distribution: The general form of the three-parameter Weibull distribution, Equation (2.7-40), can be simplified for special cases:

(1) If the distribution starts at $x = 0$, then $c = 0$, and Equation (2.7-40) takes the form

$$f(x) = \frac{k}{v}(\frac{x}{v})^{k-1} e^{-(\frac{x}{v})^k} , \qquad (2.7\text{-}47)$$

and

$$F(x) = 1 - e^{-(\frac{x}{v})^k} \qquad (2.7\text{-}48)$$

where x, v, and $k > 0$. This form is frequently used. Its mean and variance are

$$\mu = v\,\Gamma(1 + \frac{1}{k}) \qquad (2.7\text{-}48a)$$

and

$$\sigma^2 = v^2 \{\Gamma(1 + \frac{2}{k}) - [\Gamma(1 + \frac{1}{k})]^2\} \qquad (2.7\text{-}48b)$$

where $\Gamma(.)$ is the gamma function (Stedinger et al., 1993, p. 18.13; Benjamin and Cornell, 1970, p. 284). For $k = 2$, $\Gamma(3/2) = 0.5\ \pi^{0.5} = 0.88623$, and for integers, $\Gamma(n+1) = n!$ For $n = 1$, $\Gamma(2) = 1$. Hence, for $k = 2$, $\mu = 0.8862\ v$, and $\sigma^2 = 0.2146\ v^2$. Both μ and σ^2 are proportional to v and v^2, respectively, as is shown by Equations (2.7-48a) and (2.7-48b), with the gamma functions providing the proportionality coefficients. The parameters k and v can be found from a data fit using Equations (2.7-44) to (2.7-46). According to Equation (2.7-43), k is the slope of the line fitting the data. More on gamma functions is found in Abramowitz and Stegun (1970, p. 255).

(2) For $k = 1$, and $c = 0$, $\lambda = 1/v$, Equation (2.7-40) becomes

$$f(x) = \lambda e^{-\lambda x} \qquad (2.7\text{-}49a)$$

and

$$F(x) = 1 - e^{-\lambda x} . \qquad (2.7\text{-}49b)$$

This is the exponential distribution discussed in Section 2.7.5. The formulas for the mean and the standard deviation, Equation (2.7-48a) and (2.7-48b), respectively, can be used to calculate its parameters. For $k = 1$, and $v = 1/\lambda$ one obtains $\mu = 1/\lambda$, and $\sigma^2 = 1/\lambda^2$, as obtained for the exponential function (Section 2.7.5).

(3) The similarity of the first derivative of the normal distribution and the Weibull distribution is too close to be overlooked. The first derivative, i.e., the tangent of the positive branch of the normal pdf, starts with zero at $x = \mu$. Then, its minus sign disregarded, the tangent reaches a maximum at $x = \mu + \sigma$ and decreases exponentially toward zero. This gives the absolute value of the first derivative a shape that resembles the Weibull distribution. Of course, in such a new distribution, μ and σ of the normal pdf are no more valid. The first derivative of the Gaussian pdf, Equation (2.6-11), is

$$f'(x) = -\frac{0.3989}{\sigma^2}\frac{x - \mu}{\sigma}e^{-\frac{1}{2}(\frac{x-\mu}{\sigma})^2} . \tag{2.7-50}$$

Taking the absolute value of $f'(x)$, setting $\mu = 0$, $v = \sigma\sqrt{2}$, and $k = 2 \cdot 0.3989/\sigma$, the new distribution $f^*(x) = |f'(x)|$ becomes

$$f^*(x) = \frac{k}{v}\frac{x}{v}e^{-(\frac{x}{v})^2} , \tag{2.7-51}$$

which is the Weibull distribution, Equation (2.7-47), for $k = 2$.

Example: Analyze two histograms using the Weibull distribution and the plotting method. Solution: The histograms consisting of groups of n_i elements in classes of width 10 covering a range (0, 200) are given in columns (3) and (8) of Table 2-14. They are converted to relative frequency distributions by dividing each absolute class frequency, n_i, by the total number of elements n of each sample in columns (4) and (9), respectively, $f(x_i) = n_i/n$. The two sets of $f(x_i)$ are plotted at the center of their classes in Figure 2-25. The set "weeks 14–26" is more concentrated than "weeks 1–13." The cdf is calculated by cumulating columns (4) and (9) resulting in $F(x)$ in columns (5) and (10), respectively. Both $F(x)$ sum to 1. The two cdf's are converted to the variable Y and plotted at the upper class limits in Figure 2-26. The abscissa X is calculated in column (6) from the class limits by using Equation (2.7-45) with $c = 0$. Y is calculated from $F(x)$ of the two samples by Equation (2.7- 44) in columns (7) and (11). With the slope k taken from the graph, the intercept C is found by Equation (2.7-46).

Table 2-14: Testing Two Sets of Data Against the Weibull Distribution

		Weeks 14–26					Weeks 1–13			
Lower	Upper	n_i	n_i/n	$F(x)$	X	Y	n_i	n_i/n	$F(x)$	Y
Bounds										
(1)	(2)	(3)	(4)	(5)	(6)	(7)	(8)	(9)	(10)	(11)
0	10	14	0.014	0.014	2.303	−4.234	9	0.009	0.009	−4.669
10	20	197	0.202	0.217	2.996	−1.409	59	0.061	0.071	−2.615
20	30	263	0.270	0.487	3.401	−0.404	113	0.117	0.188	−1.570
30	40	181	0.186	0.673	3.689	0.112	166	0.172	0.360	−0.807
40	50	124	0.127	0.801	3.912	0.478	158	0.164	0.524	−0.298
50	60	74	0.076	0.877	4.094	0.739	107	0.111	0.635	0.007
60	70	48	0.049	0.926	4.248	0.957	86	0.089	0.724	0.253
70	80	26	0.027	0.953	4.382	1.116	74	0.077	0.801	0.478
80	90	16	0.016	0.969	4.500	1.247	44	0.046	0.846	0.628
90	100	6	0.006	0.975	4.605	1.309	38	0.039	0.886	0.775
100	110	4	0.004	0.979	4.700	1.357	27	0.028	0.914	0.897
110	120	8	0.008	0.988	4.787	1.481	20	0.021	0.935	1.004
120	130	4	0.004	0.992	4.868	1.569	16	0.017	0.951	1.106
130	140	4	0.004	0.996	4.942	1.704	13	0.013	0.965	1.207
140	150	1	0.001	0.997	5.011	1.755	13	0.013	0.978	1.342
150	160	2	0.002	0.999	5.075	1.929	6	0.006	0.984	1.426
160	170	0	0.000	0.999	5.136	1.929	4	0.004	0.989	1.498
170	180	0	0.000	0.999	5.193	1.929	5	0.005	0.994	1.625
180	190	0	0.000	0.999	5.247	1.929	3	0.003	0.997	1.753
190	200	1	0.001	1.000	5.298	—	3	0.003	1.000	—

Notes: Explanations of the table columns are given in the text. The sample of Weeks 14–26 has 973 elements (column 3) and the sample of Weeks 1–13 has 964 elements (column 8).

By graphically fitting a straight line to the Weeks 1–13 data in Figure 2-26 one obtains

$$Y = 2.78 \, X - 11.125.$$

According to Equation (2.7-43) this means that $k = 2.78$ and $C = -11.125$. By setting $c = 0$ in Equation (2.7-46) and solving for v gives

$$v = e^{-\frac{C}{k}} = e^{\frac{11.125}{2.75}} = 54.69 \ .$$

The pdf, Equation (2.7-47), becomes

$$f(x) = 0.051(\frac{x}{54.69})^{1.78} e^{-(\frac{x}{54.69})^{2.78}} \ . \tag{2.7-47a}$$

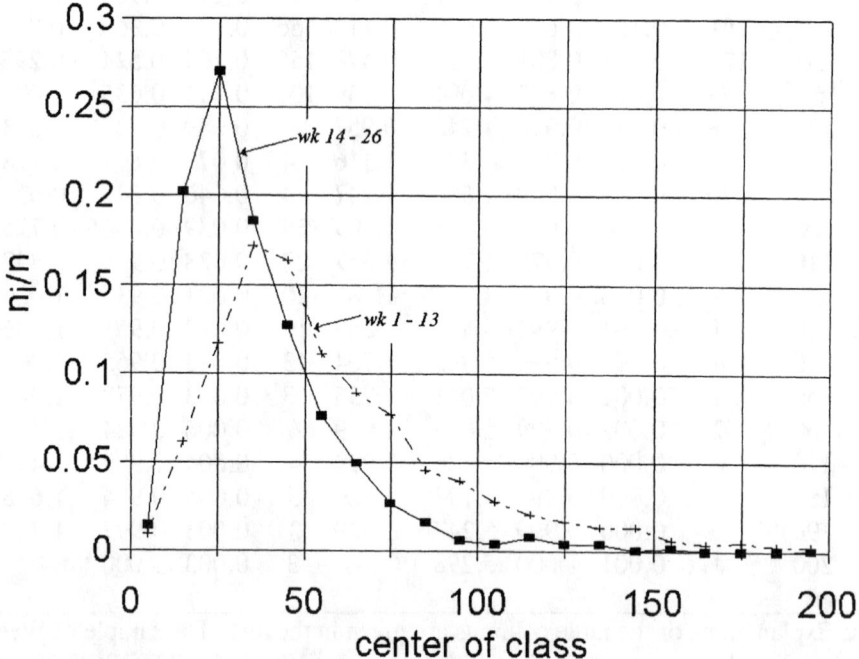

Figure 2-25: Histograms of weekly flow probabilities. The observed probabilities, $n_i/n = f(x) \ dx$, are plotted at the center of classes of width 10 (flow units) for "wk 1 – 13" and "wk 14 – 26;" n is the total number of elements per wk-group, e.g., 13 times the number of years of record, and n_i is the number of elements per class; $f(x)$ is the ordinate of a continuous pdf and $dx = 10$ is the class width. The sum of the n_i/n of each wk-group is 1.

The relation between the experimental discrete pdf developed in Table 2-14 and the continuous Weibull pdf deserves some comments. The experimental values, $f(x) = n_i/n$, in Table 2-14, are the heights of histogram bars. They are plotted in Figure 2-25 at the class center as dots; they represent discrete $f(x_i)$. In contrast, the $f(x)$ computed by Equation (2.7-47a) are continuous function values. The discrete function ordinates, $f(x_i)$, are probabilities, whereas the continuous function values $f(x)$ must be multiplied by dx to become probabilities. For example, with $dx = 10$, for $x = 45$, Equation (2.7-47a) gives $f(x) \, dx = 0.051 \cdot 0.707 \cdot 0.559 \cdot 10 = 0.20$. Table 2-14, for Weeks 1–13 and class 40–50, gives $f(x) = 158/964 = 0.164$. For $x = 25$, Equation (2.7-47a) gives $f(x) = 0.051 \cdot 0.248 \cdot 0.893 = 0.011$, and $f(x) \, dx = 0.11$, whereas Table 2-14 gives $f(x) = 113/964 = 0.117$ for class 20–30.

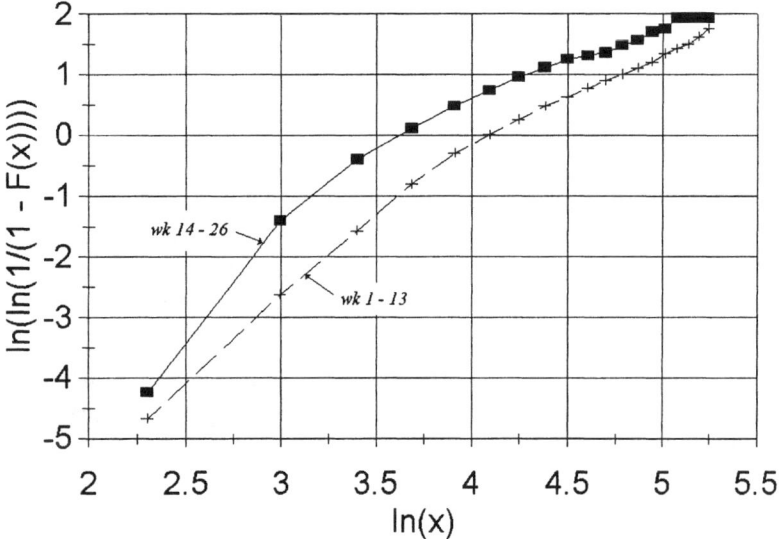

Figure 2-26: Graphical presentation of the Weibull cdf by the linear form $Y = k \, X + C$, where $Y = \ln(\ln(1/(1 - F(x))))$ and $X = \ln(x)$. Only the left lower portion of curve "wk 1 – 13" comes close to meeting this requirement. Other distributions must be tested to find a better fit.

The cdf based on the parameters derived from Figure 2-26 is

$$F(x) = 1 - e^{-(\frac{x}{54.69})^{2.78}}. \qquad\qquad (2.7\text{-}48c)$$

For $x = 50$ (upper class limit of class 40–50), and Weeks 1–13, Equation (2.7-48c) gives $F(x) = 0.541$; from Equation (2.7-44), one obtains $Y = -0.249$, and from Equation (2.7-45), $X = \ln(x) = 3.91$. From Figure 2-26, one reads $Y = -0.27$, and Table 2-14 gives $Y = -0.298$. $F(x)$ is plotted at the upper class limit. Figure 2-26 shows reasonable agreement between the fitted functions and the data, but only in the range $2.25 \le X \le 4$, where the data line up along a straight line. Beyond this range, another straight line would have to be fitted or another distribution has to be found.

Some discrepancies notwithstanding, the reduction of the complicated shape of the Weibull pdf, as shown in Figure 2-25, to a straight line in the form of Equation (2.7-43) offers the possibility of creating an entire data distribution from a hunch, or of correcting and augmenting existing data. Once the straight line is established, an analytical form can be derived, such as Equation (2.7-47) and (2.7-48), for further use in probabilistic calculations (see Section 3.6).

2.7.9 Student's t-Distribution

The Student's t-distribution is used with normal random variables. When inferences must be drawn from small samples, which are samples with numbers of elements $n < 30$, the normal distribution becomes inaccurate and the *Student's t-distribution* is used. The Student's t-distribution is flatter than the normal distribution having a lower center and thicker tails because it deals with small samples, which are less representative of a population than large samples. These differences will be discussed.

The sample mean, $\overline{X}$, is the sum of n random realizations x_i divided by n and is itself a random variable (see Section 3.4). If the population mean, μ, and the population standard deviation, σ, are known, the normalized variable of the sample mean is

$$z = \frac{\overline{X} - \mu}{\sigma / \sqrt{n}} \ .$$

If the population standard deviation is not known, the sample standard deviation s is used instead and z is replaced by the variable

$$t = \frac{\overline{X} - \mu}{s / \sqrt{n}} \tag{2.7-52}$$

where t is Student's t, or the Student t test statistic. Both $\overline{X}$ and s are approximations of the often unknown population parameters, μ and σ.

The t-distribution has its origin in sampling statistics. It is of interest here for its use in determining confidence intervals for the estimation of population means based on small sample means. The Student's t-distribution is given by (Moan, 1982, p. 4-55)

$$f(t) = \frac{1}{\sqrt{\pi n}} \frac{\Gamma(\frac{n+1}{2})}{\Gamma(\frac{n}{2})} (1 + \frac{t^2}{n})^{-\frac{n+1}{2}}$$

(2.7-53)

where $\Gamma(...)$ is the gamma function; n is the number of sample elements. In the limit, for $n \to \infty$, $f(t)$ approaches the normal pdf. For practical purposes, the normal distribution can be used for sample sizes $n \geq 30$. A few examples are given in Figure 2-27, which demonstrate this asymptotic behavior.

Equation (2.7-53) requires evaluations of the gamma function (Abramowitz and Stegun, 1970, pp. 255–256):

$$\Gamma(n+1) = n\Gamma(n) = n!.$$

(2.7-54)

This formula can be recursively applied. If the argument $z = n - 1$ is substituted for n in Equation (2.7-54), one obtains

$$\Gamma(n) = (n-1)\Gamma(n-1) = (n-1)!$$

and so on. Other function values needed for evaluating Equation (2.7-53) are:

$$\Gamma(n + \frac{1}{2}) = \frac{1 \cdot 3 \cdot 5 \cdot 7 \cdot ... \cdot (2n-1)}{2^n} \Gamma(\frac{1}{2}) \ ,$$

(2.7-55)

and $\quad \Gamma(\frac{1}{2}) = \sqrt{\pi}$

(2.7-56)

Some evaluations of Equation (2.7-53) are given that demonstrate the effect of n on the function shape:

(1) For $n = 1$, one obtains for the ratio of gamma functions in Equation (2.7-53) by using Equations (2.7-54) and (2.7-56):

$$\Gamma(\frac{1+1}{2}) \Big/ \Gamma(\frac{1}{2}) = 1/\sqrt{\pi} \ .$$

Substitution into Equation (2.7-53) gives

$$f(t) = \frac{0.31831}{1+t^2} \cdot$$

(2.7-57)

Equation (2.7-57) is shown in Figure 2-27. It is a symmetric function similar to the normal distribution. At the center of the symmetric distribution, $t = 0$, and $f(0) = 1/\pi = 0.31831$. The maximum ordinate of the *normal* distribution at $z = 0$ is $f(0) = 0.39894$. The Student's t distribution is lower in the center and higher in the tails than the normal distribution. Equation (2.7-57) represents the largest deviation of the Student's t from the normal distribution.

(2) For $n = 5$, the ratio of gamma functions is

$$\Gamma(\frac{5+1}{2}) / \Gamma(\frac{5}{2}) = \Gamma(3) / \Gamma(2 + \frac{1}{2}) = 2 / \left(\frac{3}{4}\sqrt{\pi}\right)$$

Substitution into Equation (2.7-53) gives

$$f(t) = \frac{0.37961}{(1+t^2/5)^3} \cdot$$

(2.7-58)

(3) For $n = 30$, one obtains from Equation (2.7-55)

$$\Gamma(\frac{31}{2}) = \Gamma(15 + \frac{1}{2}) = \frac{1 \cdot 3 \cdot 5 \cdot 7 \cdots 29}{2^{15}} \Gamma(\frac{1}{2}) = 1.88912 \cdot 10^{11} \sqrt{\pi} \quad.$$

From Equation (2.7-54), one obtains

$$\Gamma(\frac{30}{2}) = \Gamma(15) = 14! = 8.71783 \cdot 10^{10}.$$

The ratio of the gamma functions becomes 2.16696 $\sqrt{\pi}$, and Equation (2.7-53) becomes

$$f(t) = \frac{1}{\sqrt{30\pi}} 2.16696\sqrt{\pi}(1+\frac{t^2}{30})^{-15.5} = \frac{0.39563}{(1+t^2/30)^{15.5}} . \qquad (2.7\text{-}59)$$

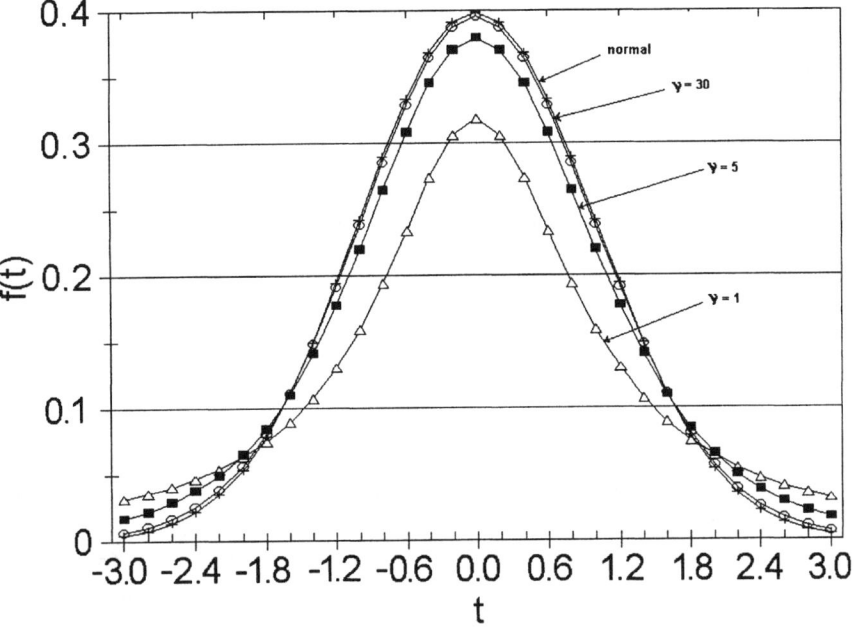

Figure 2-27: Student's t-distributions for parameters $\nu = 1, 5,$ and 30 (degrees of freedom). For $\nu \to \infty$, the Student's t-distribution approaches the normal pdf which is also shown for $\mu = 0$, $\sigma = 1$, and $z = t$.

Equations (2.7-57), (2.7-58), and (2.7-59) for $n = 1, 5,$ and 30, respectively, are shown in Figure 2-27. With increasing n the Student's t-distribution becomes higher in the center and lower in the tails and converges to the normal distribution for $n \to \infty$. The differences in the tail areas are especially significant. For selected one-sided tail areas α and double-sided tail areas $\alpha/2$ the corresponding Student's t statistics t_α and $t_{\alpha/2}$ are given as functions of n in Table 2-15. It is seen that the thick tails for small n move the Student's t statistics away from the center. For $n \to \infty$, the Student's t statistic corresponds to z of the normal distribution. Therefore, the t values for $n \to \infty$ in the last column of Table 2-15 provide a comparison with the normal distribution. For example, for a confidence band centered on the mean that contains 95 % of all elements, the normal distribution requires $z = \pm 1.96$. A one-sided 5 % exceedance probability (elements falling into the right tail beyond t_α) corresponds to $z = +1.645$. The Student's t values for $n < 30$ increase over the normal distribution z's which indicates that predictions based on small samples for the same exceedance probabilities have a greater spread.

Table 2-15: Student's t Values for Exceedence Probability α and Selected Sample Sizes n (after Bronstein-Semendjajev, 1984, p. 22)

Row	t_α and $t_{\alpha/2}$	n			
		1	5	30	∞
1	$t_{\alpha/2} = t_{0.025}$	12.706	2.571	2.042	1.960
2	$t_\alpha = t_{0.05}$	6.314	2.015	1.697	1.645
3	$t_{\alpha/2} = t_{0.0005}$	636.6	6.869	3.646	3.291
4	$t_\alpha = t_{0.001}$	318.3	5.983	3.386	3.090

Notes: row 1: The half band width expressed by $t_{0.025}$, as defined by Equation (2.7-52). The probability of elements falling inside the band is 0.95; $t_{\alpha/2} = t_{0.025}$ means 2.5 % of the total area under the pdf is in the two tails. The value of the normal distribution, for which t is replaced by z, is shown in the last column for $n \rightarrow \infty$.
row 2: One-sided band $t_\alpha = t_{0.05}$; 5 % of the total area under the pdf is under the right tail. Since the one-sided band for 5 % exceedance probability requires twice the tail area of 2×2.5 %, $t_\alpha < t_{\alpha/2}$.
rows 3 and 4: The same as rows 1 and 2, respectively, but for smaller α, hence larger t's.

In confidence estimates, the confidence level γ, also called confidence coefficient (Moan, 1982, p. 4-60), is the probability that a random variable, here the sample mean, $\overline{X}$, falls into a desirable band that is bounded by an upper and a lower confidence limit. The area outside this band is the complementary probability, $\alpha = 1 - \gamma$, that $\overline{X}$ falls outside the confidence interval. A *centered* or *double-sided* confidence interval is bounded by an upper bound and a lower bound which exclude a right (or upper) tail area and a left (or lower) tail area under the distribution. A *one-sided* confidence interval is bounded by an upper bound   or a lower bound and excludes only the one tail area. The probability statements can be formulated as follows:

(a) double-sided band:

$$P(L < t < U) = \gamma \tag{2.7-60}$$

(b) one-sided band:

$$P(t > L) = \gamma \tag{2.7-60a}$$

or

$$P(t < U) = \gamma \tag{2.7-60b}$$

where t is the Student's t statistic of Equation (2.7-52), and U and L are the upper bound and lower bound, respectively. Based on Equation (2.7-52), one can formulate the following confidence statement:

$$P(-t_{\alpha/2} \leq \frac{\overline{X} - \mu}{s/\sqrt{n}} \leq +t_{\alpha/2}) = \gamma \tag{2.7-61}$$

If one wants to bound the location of the unknown population mean, then the confidence interval is formulated as

$$P(\overline{X} - t_{\alpha/2} \frac{s}{\sqrt{n}} \leq \mu \leq \overline{X} + t_{\alpha/2} \frac{s}{\sqrt{n}}) = \gamma \tag{2.7-61a}$$

where α is the total exceedance probability that $\overline{X}$ falls into one of the tail areas; $t_{\alpha/2}$ is a function of n, as shown in Table 2-15. If the population standard deviation is not known, an estimate s based on the sample is used to calculate the standard deviation of the sample mean, $\sigma_{\overline{X}} = s/\sqrt{n}$, for use in Equations (2.7-61) and (2.7-61a) (see Section 2.4.5).

The exceedance probabilities $\alpha/2$ used to set the parameters t in Equation (2.7-61) and (2.7-61a) can be calculated by integrating the Student's t-distribution, Equation (2.7-53), for a given sample size n and confidence level γ. A different curve has to be integrated for any combination of these two parameters. Lookup tables have been published in textbooks for generally occurring parameter combinations. Here the integrations are carried out for the functions represented by Equations (2.7-57), (2.7-58), and (2.7-59). The symmetry of the t-distribution, as shown in Figure 2-27, allows one to calculate the upper tail area by (see Figure 2-14d)

$$P(t \geq t_\alpha) = \alpha = 0.5 - \int_0^{t_\alpha} f(t)\,dt \qquad\qquad (2.7\text{-}62)$$

where α is the area under the upper tail of Student's t to the right of t_α. In the numerical integration, $f(t)\,dt$ is replaced by $\Delta F(t)$, for which ordinates $f(t)$ are calculated at intervals $\Delta t = 0.1$ for the range $0 \leq t \leq 3.0$:

$$\Delta F(t) = f(t)\,\Delta t = \frac{1}{2}(f_1 + f_2)\cdot 0.1 \qquad\qquad (2.7\text{-}63)$$

where $\Delta F(t)$ is the numerical increment of the integral, and f_1 and f_2 are the function values $f(t)$ at the left and right sides of the increment $\Delta t = 0.1$. The number of numerical integration steps from $t = 0$ to t_α is $K = t_\alpha / \Delta t$:

$$\alpha_t = 0.5 - \sum_{i=1}^{K} \Delta F(t) \qquad\qquad (2.7\text{-}64)$$

where α_t is the area under the t-distribution between t and infinity; the numerical integration is performed for the area between $t = 0$ and t; the total number of steps leading from $t = 0$ to $t = 3.0$ ($z = 3$ was chosen here) is $K = 30$. The integration may have to be continued beyond $t = 3.0$ if n is small and if small α are used. Some examples for the required t's for assumed α's are given in Table 2-15.

The results of the integrations for $n = 1, 5$, and 30 are shown in Figure 2-28. The exceedance probability $P(t \geq t_\alpha) = \alpha$ is given as function of t with n as a parameter. As t increases, the area under the right tail decreases and the exceedance probability gets smaller. The curves of Figure 2-28 do not allow precise readings. Statistical tables can be used when accurate numerical values are needed. The curves with their downward slope resemble the upper tail of $R(z)$ shown in Figure 2-16. If Figure 2-28 is given a z-abscissa, $R(z)$ passes through $z = 0$ at $x = \mu$, where $R(z) = 0.5$. Reading the curves at $t = 3$, one obtains the results shown in Table 2-16. The α-values for sample size $n = 5$ show that the exceedance probability of the normal distribution differs from the exceedance probability of the Student's t-distribution by a factor of $(0.017/0.0014 =)\ 12$. At sample size $n = 30$, this factor has shrunk to less than 2.

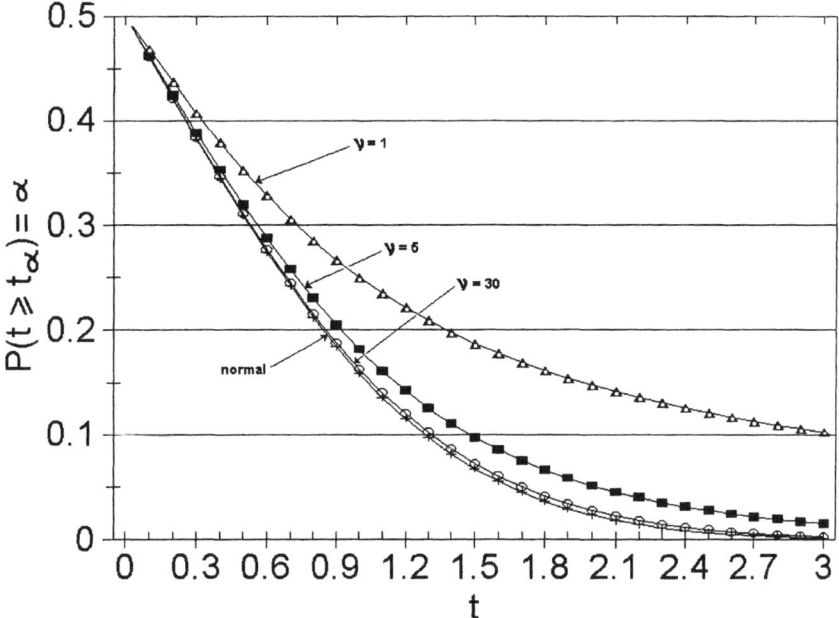

Figure 2-28: Exceedance probabilities $P(t \geq t_\alpha) = \alpha$ by Student's t-distributions for $n = 1, 5$, and 30. The exceedance probability of the normal distribution forms the lower envelope of Student's t-based exceedance probabilities as $n \to \infty$.

Table 2-16: Exceedance Probabilities $P(t \geq t_\alpha) = \alpha$ of the Student's t-Distribution for One-Sided Tail

n	1	5	30	∞
t	3.0	3.0	3.0	3.0
α	0.1	0.017	0.0025	0.0014

Notes: n is number of elements in the sample.
For the normal distribution, $P(z \geq 3) = 0.0013499$.

Examples: (1) Assume a decision must be based on a small sample with $n = 5$ elements to find the location of the population mean. Find the interval size around the *sample* mean such that the *population* mean can be expected to be within it with confidence level 100 γ. Solution: For a double-sided interval, Equation (2.7-61a) applies. The confidence level is $0 \leq \gamma \leq 1$ and the significance level is $\alpha = 1 - \gamma$.

Exceedance of the interval may occur at the lower bound and at the upper bound. If the total exceedance probability is α, then the interval limits are set at $\pm t\alpha/2$. Suppose 100 γ = 90 %, then γ = 0.9, and $\alpha/2$ = 0.05. For a sample of size n = 5 (n degrees of freedom), a statistical table gives $t_{0.05}$ = 2.015. The confidence statement becomes

$$P(\overline{X} - 2.015\frac{s}{\sqrt{n}} \leq \mu \leq \overline{X} + 2.015\frac{s}{\sqrt{n}}) = 0.9$$

The statement can be expressed in words as follows: given a sample mean, $\overline{X}$, and the sample standard deviation, s, one can be 90 % confident that μ is found within $\overline{X}$ $\pm$ 2.015 $s/\sqrt{n}$ (Moan, 1982, p. 4-61, suggests this wording instead of "with probability 90 %"). In comparison, a sample with n = 30 requires an interval $\overline{X}$ $\pm$ 1.697 $s/\sqrt{n}$; for the larger sample, $s/\sqrt{n}$ would be smaller. The value for $t_{0.05}$ can be checked by using the t-distribution for n = 5 and n = 30 in Figure 2-28. One enters the diagram on the left at α = 0.05 and moves over horizontally. First the intersection with the t-distribution for n = 30 is reached for which one reads on the abscissa $t \approx 1.67$; then one moves on to the t-distribution for n = 5 with $t \approx 2.0$. The left most curve is the normal distribution for which one would read $t \approx 1.65$.

(2) Suppose only the upper bound is critical. In this case, the total exceedance probability is represented by the upper tail. The confidence level is again γ = 0.9, and α = 0.1. Determine the upper bound for μ for a sample size n = 5. Solution: For α = 0.1, and n = 5, a statistical table (Lapin, 1978, p. A-25) gives $t_{0.1}$ = 1.476. Equation (2.7-60b) becomes

$$P(\frac{\overline{X} - \mu}{s/\sqrt{n}} \leq 1.476) = 0.9$$

The statement says that for the sample mean, $\overline{X}$, and the sample standard deviation, s, one can be 90 % confident that μ does not exceed $\overline{X}$ + 1.476 $s/\sqrt{n}$. The value for $t_{0.1}$ can be checked by using the t-distribution for n = 5 in Figure 2-28. One enters the diagram on the left at α = 0.1 and moves horizontally over to the intersection with the t-distribution for n =5; on the abscissa one reads $t \approx 1.48$; for an infinitely large sample one would read $t \approx 1.28$.

****

3. Calculating with Random Variables

3.1 Prediction

3.1.1 Prediction of a Random Variable

A random variable, X, consists of a collective of realizations, x, that are produced by a chance process. An example is pipe failure incidences in a water distribution system over time. The possible values of x, or the number of incidences over a specified period, may be confined to a range of size, and each x may have associated with it a frequency of occurrence. The random variable may be discrete or continuous, and the number of realizations may be finite or infinite. The random realizations may occur at some point in time, over some period of time, or over some defined space. Their occurrence and size may not be predictable individually, but some statements can be made about the characteristics of the collective. Such characteristics include the mean and the dispersion around the mean, and the probability of a particular realization relative to others of the collective. Such characteristics can be used to make some limited predictions on the behavior of the random process. A summary follows.

A. Statistical parameters:

1. Mean: $\mu = E(X) = \dfrac{1}{n} \sum\limits_{i=1}^{n} x_i$ 　　　　　　　　　　　　　　(3.1-1)

where x_i is the ith of n elements of the sample space or population.

2. Variance: $Var(X) = \sigma^2 = E(X^2) - [E(X)]^2$ (3.1-2)

where $E(X^2) = \dfrac{1}{n}\displaystyle\sum_{i=1}^{n} x_i^2$,   (3.1-3)

and $[E(X)]^2$ is the square of the expectation.

3. Standard deviation: $\sigma = \sqrt{E(X^2) - [E(X)]^2}$.   (3.1-4)

4. Coefficient of variation: $C_v = \dfrac{\sigma}{\mu}$.   (3.1-5)

5. Mean deviation, or absolute error: $e = E(|X - \mu|)$.   (3.1-6)

6. Upper limit of range: x_u.

7. Lower limit of range: x_o.

8. Range: $r = x_u - x_o$.   (3.1-7)

B. <u>Probabilities</u>:

9. The probability of a realization x of a *continuous* random variable with range $x_o \le X \le x_u$ is defined for an incremental range, dx:

$$P(x \le X \le x + dx) = F(x + dx) - F(x) = dF(x) = f(x)dx \qquad (3.1\text{-}8)$$

where $F(x)$ is the integral from the lower limit of the range, x_o, up to and including the realization x, and $f(x)$ is the probability density function (pdf). For a point, $dx \to 0$, and $f(x)\, dx \to 0$. This means that for a continuous variable, the probability of a particular point or element out of an infinite number of points being realized is infinitely small, approaching zero. This is a very important characteristic of a continuous random variable. A related property of the continuous distribution is that the infinite number of realizations x, share the total probability of the collective, which cannot exceed 1. This means that the cumulative probability over the sample space in the form of an integral or by the sum of an infinite number of probabilities must add to 1. Hence, each of these individual probabilities must be infinitely small. Therefore, a finite probability of a continuous variable can be stated only as an incremental probability $dF(x) = f(x)\, dx$, which is the subtotal of the probabilities of all elements in an increment of space or time, dx or dt.

10. The non-exceedance probability of a realization x_1 of a continuous random variable X is

$$P(X \le x_1) = F(x_1) = \int_{x_o}^{x_1} f(x)dx \qquad (3.1\text{-}9)$$

where $F(x_1)$ is the integral of $f(x)$ from the lower limit of the variable range, x_o, up to and including x_1. $P(X \le x_1)$ is the probability that a realization x of X is found somewhere in the range $x_o \le X \le x_1$, including the boundaries.

11. The *median* of a continuous random variable, X, is the value, x_{med} that satisfies

$$F(x_{\text{med}}) = 0.50. \qquad (3.1\text{-}9a)$$

12. The exceedance probability of realization x_1 of a continuous random variable X is

$$P(X > x_1) = 1 - F(x_1) = \int_{x_1}^{x_u} f(x)dx \qquad (3.1\text{-}10)$$

where x_u is the upper limit of the distribution, which can be finite or infinite. $P(X > x_1)$ is also the complementary probability of $F(x_1)$.

13. The probability of a realization x of a *discrete* random variable with a range $x_o \le X \le x_u$ is

$$P(X = x) = f(x) \qquad (3.1\text{-}11)$$

where $f(x)$ is the probability density of realization x of a discrete random variable X.

14. The non-exceedance probability of a discrete random variable, x_k, can be stated as

$$P(X \le x_k) = F(x_k) = \sum_{i=1}^{k} f(x_i) \qquad (3.1\text{-}12)$$

where $f(x_i)$ are function values that are frequencies, or probabilities.

15. The *median* of a discrete sample space is the value x_{med} that satisfies

$$F(x_{med}) = 0.5. \tag{3.1-12a}$$

If this value does not exist in the discrete sample space, the nearest value is used as an approximation.

16. The exceedance probability of a discrete random variable realization, x_k, is the sum of probabilities of elements ranked above x_k, including the function value of the highest ranking element, $f(x_n)$:

$$P(X > x_k) = 1 - F(x_k) = \sum_{i=k+1}^{n} f(x_i) \tag{3.1-13}$$

where $f(x_n)$ is the frequency of the last element of the discrete sample space that has been sorted in ascending order.

When information on pdf values is unavailable, an experimental function can always be assigned by assuming a plausible distribution, e. g., a uniform pdf, if it is meaningful to give the same frequency to all elements of the sample space. If a particular element type occurs more frequently than others, then this particular element assumes a higher frequency than others, and a variable frequency distribution can be assigned (see Section 2.7).

Examples: (1) Twelve years of monthly pipe repair incidences for urban water mains were reported by a study of the Washington Suburban Sanitary Commission (Habibian, 1988). The data show seasonal variation with high frequency of repairs in fall and winter and low frequency in spring and summer. Each set of monthly repair incidences for 12 years represents a set of random variables. There may be a multiannual carryover effect in the monthly data, as indicated by the winter repair incidences of 1977/78, 1978/79, and 1979/80, which were unusually low. It has been hypothesized by the investigators that these low repair incidences were caused by the unusually cold winters of 1975/76 and 1976/77, which caused pipes to fail that otherwise would have failed in the following years. Table 3-1(a) shows seasonal trends along the rows and random monthly failures along the columns. Table 3-1(b) shows parameters of the random variables for each column of Table 3-1(a). Table 3-1(c) shows repair (or failure) incidences in descending order in the form of cdfs, and Table 3-1(d) gives the annual total repair incidences.

Some characteristics of the data are presented in Figure 3-1. The graph on the top left gives the seasonal distribution of monthly means which shows high numbers of incidences during the winter and low numbers of incidences in the summer (not a probability distribution). The graph on the top right gives the cdf of annual total numbers of incidences. The four graphs below give cdfs of monthly repair incidences for two winter and two summer months (January, April, August, and December). Figure 3-1 could also be consolidated into a three-dimensional plot in which the seasonal distribution of the upper left corner forms the x–y plane, with x being time and y being the number of incidents. The third dimension is provided by the cdfs of the monthly random numbers of repair or failure incidents in vertical planes at right angles to the time axis, one for each month. Because of the paucity of data, cdfs were used instead of pdfs, as they are easier to draw. The cdfs cross the curve of seasonal means in the x–y plane at $F(x) = 0.5$, where x is the monthly mean, more precisely, the median. The difference between mean and median is illustrated by row 1 of Table 3-1(b), which gives the means, and the number that would correspond to the interval between rows 6 and 7 in Table 3-1(c). The steep cdfs for the warmer months of the year indicate small dispersion, small standard deviations, and small ranges, whereas the more sloped cdfs for the cold months indicate larger dispersion, large standard deviation, and large ranges. The range, r, puts brackets on the most likely minimum and maximum number of repair jobs occurring in each month. The upper limit, x_u, of the range is the number of repair incidences that most likely will not be exceeded in each month, whereas the lower limit, x_o, is the number of repair incidences that most likely will be exceeded in each month.

Table 3-1: Number of Monthly Pipe Repairs for the Central Zone of the Sanitary District of the Washington Suburban Sanitary Commission (Habibian, 1988)

(a) Monthly pipe repair incidences

Year	1	2	3	4	5	Month 6	7	8	9	10	11	12
1975	3	0	5	27	21	11	14	8	2	9	30	55
1976	104	31	23	26	30	24	3	25	18	17	40	45
1977	80	44	15	27	21	23	35	22	33	27	9	31
1978	42	37	9	7	13	16	11	5	8	18	24	36
1979	33	44	22	12	17	10	16	13	12	8	15	26
1980	24	21	8	8	0	1	14	9	19	23	26	45
1981	61	30	11	15	9	7	9	8	14	32	24	53
1982	65	18	6	8	16	8	12	14	12	24	26	21
1983	57	19	6	6	8	9	10	11	15	23	35	64
1984	68	4	17	15	14	18	11	8	18	16	38	10
1985	69	30	7	12	11	6	11	17	11	40	15	47
1986	46	24	11	8	17	9	14	17	11	32	43	46

(b) Statistical parameters computed from the columns of Table 3-1(a)

						Month						
	1	2	3	4	5	6	7	8	9	10	11	12
μ	54	25	12	14	15	12	13	13	14	22	27	40
σ^2	660.4	176.6	35.6	59.4	53.0	44.8	52.7	34.7	51.9	82.9	104.2	221.6
σ	26	13	6	8	7	7	7	6	7	9	10	15
C_v	0.47	0.53	0.51	0.54	0.49	0.57	0.54	0.45	0.50	0.41	0.38	0.37
e	21	11	5	6	6	6	4	5	5	7	8	13
x_u	104	44	23	27	30	24	35	25	33	40	43	64
x_o	3	0	5	6	0	1	3	5	2	8	9	10
r	101	44	18	21	30	23	32	20	31	32	34	54

Notes: row 1: monthly mean: $\mu = E(X)$; shown as integer value.
row 2: variance: $\sigma^2 = E(X^2) - [E(X)]^2$.
row 3: standard deviation: $\sigma = \{E(X^2) - [E(X)]^2\}^{1/2}$; shown as integer value.
row 4: coefficient of variation, $C_v = \sigma/\mu$.
row 5: mean deviation: $e = E(|X - \mu|)$; shown as integer value.
row 6: upper limit of range, x_u.
row 7: lower limit of range, x_o.
row 8: range: $r = x_u - x_o$.

(c) Monthly pipe repair incidences ranked in descending order

							Month						
Rank	$F(x)$	1	2	3	4	5	6	7	8	9	10	11	12
(1)	(2)	(3)	(4)	(5)	(6)	(7)	(8)	(9)	(10)	(11)	(12)	(13)	(14)
1	1.00	104	44	23	27	30	24	35	25	33	40	43	64
2	0.92	80	44	22	27	21	23	16	22	19	32	40	55
3	0.83	69	37	17	26	21	18	14	17	18	32	38	53
4	0.75	68	31	15	15	17	16	14	17	18	27	35	47
5	0.67	65	30	11	15	17	11	14	14	15	24	30	46
6	0.58	61	30	11	12	16	10	12	13	14	23	26	45
7	0.50	57	24	9	12	14	9	11	11	12	23	26	45
8	0.42	46	21	8	8	13	9	11	9	12	18	24	36
9	0.33	42	19	7	8	11	8	11	8	11	17	24	31
10	0.25	33	18	6	8	9	7	10	8	11	16	15	26
11	0.17	24	4	6	7	8	6	9	8	8	9	15	21
12	0.08	3	0	5	6	0	1	3	5	2	8	9	10

Notes: column 1: rank number $m = 1$ given to the highest monthly repair number, and $m = 12$ given to the smallest number (see Section 3.7-1).

column 2: discrete non-exceedance probabilities starting with $F(x) = 1.00$ for the highest number of repair incidences. After giving each element of a monthly column the same probability, $1/12$, the non-exceedance probability for each column value is obtained by subtracting $1/12$ from the next higher probability. This calculation is equivalent to using the formula $F(x) = (n - m + 1)/n$, where $n = 12$, and m is the rank in column 1. Since the top value is not exceeded by any value, it has non-exceedance probability 1. The computed non-exceedance probabilities are only approximate because of the small data set.

columns 3 - 14: ranked monthly pipe repair incidences for 12 years. Some numbers appear more than once. For example, in column 4, 44 and 30 appear twice. This means that the monthly population instead of having twelve subsets with one element each has two subsets with two elements and with correspondingly higher subset probability $2 \cdot 1/12 = 1/6$. Each monthly population has its probability density function (pdf) and cumulative distribution function (cdf) of repair incidences.

(d) Annual repair incidences

Year	75	76	77	78	79	80	81	82	83	84	85	86
Total	185	386	367	226	228	198	273	230	263	237	276	278
Mean	15	32	31	19	19	16	23	19	22	20	23	23

Notes: row 1: years 1975 - 1986.
row 2: annual number of repair incidences.
row 3: total annual repair incidences as a monthly average.

(2) Calculate the non-exceedance probability of the top value of column 4, Table 3-1(c). Solution: The value $x = 44$ appears twice at the top of column 4. Since it is not exceeded, its probability of non-exceedance includes all values $x = 44$. Using Equation (3.1-12) and integrating to include all $x = 44$ gives

$$P(X \le 44) = F(44) = \sum_{i=1}^{12} f(x_i) = 12 \cdot (1/12) = 1.0.$$

(3) Calculate the median of the random variable of column 3, Table 3-1(c). Solution: Find the value x_{med} that satisfies $F(x_{med}) = 0.5$. Solution: The condition is satisfied by $x = 57$:

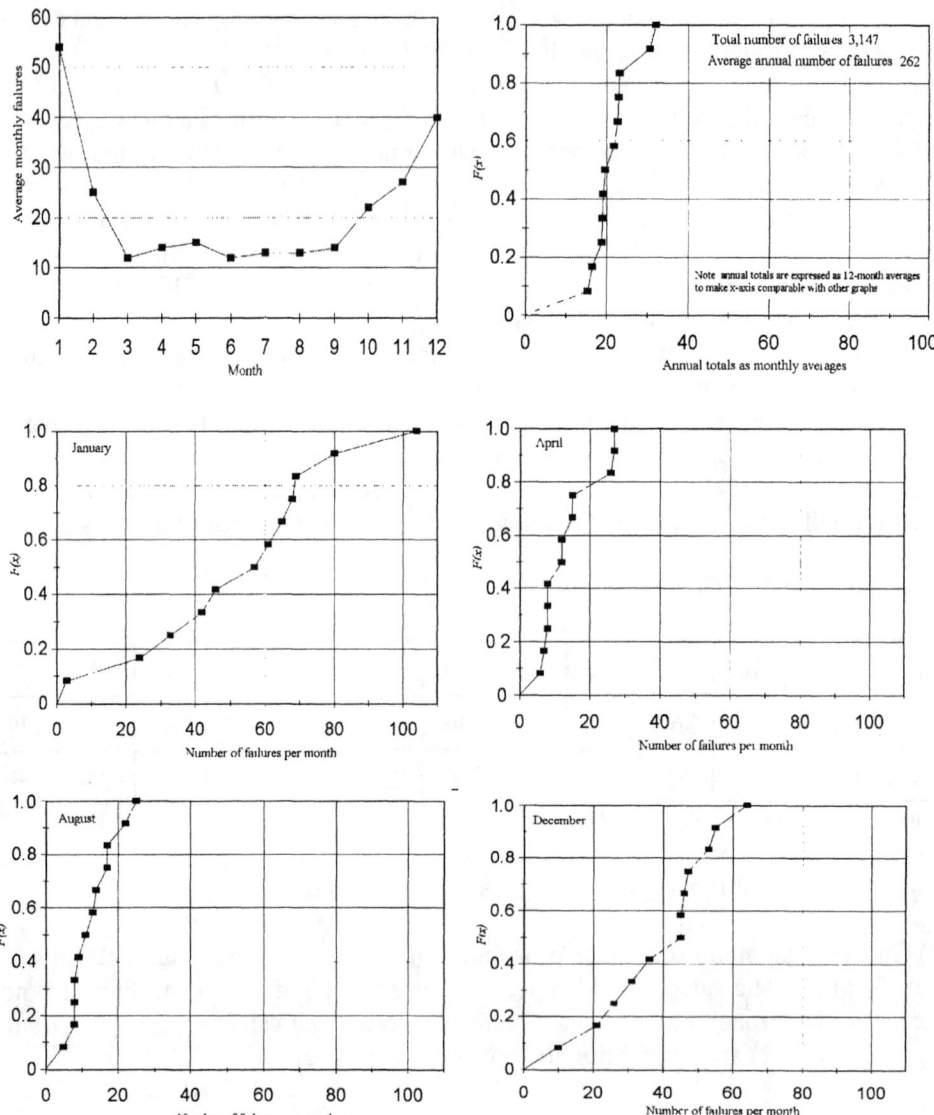

Figure 3-1: The graph on the top left shows the seasonal distribution of average monthly pipe repair incidences from Table 3-1(b). The graph next to it shows the cdf of the annual pipe repair incidences of Table 3-1(d) expressed as average monthly rates; the remaining graphs show cdfs of monthly repair incidences for January, April, August, and December taken from Table 3-1(a) (data after Habibian, 1988).

$$P(X \leq 57) = F(57) = \sum_{i=1}^{6} f(x_i) = 6 \cdot (1/12) = 0.5$$

The horizontal line in Table 3-1(c) divides the elements of column 3 into two parts, those less than or equal to $x = 57$ and those greater than $x = 57$. Similar properties are exhibited by the other columns.

3.1.2 Prediction Using Least Squares

Assume that a number of observations have been made and numerous pairs (x_i, y_i) have been obtained. If these data pairs are plotted in a graph, they usually do not line up in a perfect straight line or curve, but result in a *point cloud* or *scatter diagram*. Sometimes it is possible to draw a straight line or curve through such a cloud that gives it some approximate analytical representation. A method for establishing an analytical function, $y = f(x)$, is to search for a function that best accommodates all points of the cloud. Such a curve that is an average representation of all points is also called a best fit or *regression* curve. Unless a known physical law or systematic variation of the data suggests the use of a specific function that should be fitted to the data, the simplest function, i. e., a linear function should be tried first. The *method of least squares* specifies the objective "minimize the square of all departures from the function to be fitted" and goes about finding that function.

It is often convenient to use the means of x and y to reduce the variables from their absolute values to their departures from their respective means, (m_X, m_Y), which amounts to centering the variables on their means. In statistical texts, capital letters are used for the absolute variables, and lower case letters are used for the centered variables. For example, X stands for the absolute variable and $x = X - m_X$ stands for the centered variable. Here, X and Y have been used and will continue to be used as indicators of random variables, whereas x and y refer to specific realizations of these random variables. Whenever centered variables are used, this will be shown explicitly in the formulas.

A straight line is defined by two points, here the mean of the point cloud, (m_X, m_Y), and another point, (x_1, y_1). These two points define the straight line that must also pass through a general point, (x, y). Equating the slopes through two points at a time gives

$$\frac{y - m_Y}{x - m_X} = \frac{y_1 - m_Y}{x_1 - m_X}. \tag{3.1-14}$$

Solving for y produces

$$y - m_Y = b (x - m_X) \tag{3.1-14a}$$

the centered form of the straight line that passes through a new origin (m_X, m_Y) with slope b. Solving for y gives

$$y = a + b(x - m_X) \tag{3.1-15}$$

where

$$a = m_Y, \tag{3.1-15a}$$

and

$$b = \frac{y_1 - m_Y}{x_1 - m_X} \tag{3.1-15b}$$

where b is the slope of the line through an arbitrary point (x_1, y_1) and the mean of all points, (m_X, m_Y). Equation (3.1-15b) is the justification of fitting a line through a point cloud by inspection. Since it has to go through (m_X, m_Y), one rotates an arbitrarily drawn line through this point until it best fits all points.

Data points with coordinates, (x, y), if plotted in an x-y diagram rarely line up along a straight line, but usually are scattered around it in a *point cloud*. An analytical approach to find the line that best represents *all* points is to formulate an optimization problem that minimizes the sum of the squares of all y-ordinate sections between each point and the line. This approach is known as the *method of least squares*. The function to be minimized is the sum of the squares of all departures from the yet unknown line:

$$S = \sum_{i=1}^{n} \{y_i - [a + b(x_i - m_X)]\}^2 \tag{3.1-16}$$

where S is taken over the squared differences of all point ordinates, y_i, minus the straight line ordinates, $a + b(x_i - m_X)$, of the data set of $i = 1,..., n$ points. The unknowns to be determined are the coefficients a and b. The coordinates of the mean are

$$m_X = \frac{1}{n} \sum_{i=1}^{n} x_i \tag{3.1-17}$$

and

$$m_Y = \frac{1}{n}\sum_{i=1}^{n} y_i \, . \tag{3.1-18}$$

Equation (3.1-16) is solved for a and b such that S is minimized. The *necessary* conditions for an extreme value of S are the first derivatives versus a and b being zero. This produces

$$\frac{\delta S}{\delta a} = 0 \tag{3.1-19a}$$

and

$$\frac{\delta S}{\delta b} = 0 \, . \tag{3.1-19b}$$

The necessary and sufficient conditions for a minimum also require that the second derivative be greater than zero. This check is omitted here. When carrying out a differentiation under a sum sign, it is helpful to write out a few terms of the sum, differentiate them, and re-form the sum. The result is

$$\frac{\delta S}{\delta a} = \sum 2[y_i - a - b(x_i - m_X)](-1) = 0 \tag{3.1-20a}$$

$$\frac{\delta S}{\delta b} = \sum 2[y_i - a - b(x_i - m_X)][-(x_i - m_X)] = 0 \tag{3.1-20b}$$

where the sums are taken over all elements, $i = 1,..., n$, of the data set. In both equations, the terms under the sum can be simplified by dividing all terms by common factors, such as 2 and -1. Making use of $\sum y_i = n\, m_Y$, $\sum a = n\, a$, $b\sum x_i = n\, b\, m_X$, and $\sum b\, m_X = n\, b\, m_X$, Equation (3.1-20a) becomes

$$a = m_Y \tag{3.1-21}$$

an expression already obtained by Equation (3.1-15a).

In Equation (3.1-20b), the factor ($x_i - m_X$) is part of the summation. Substituting $a = m_Y$, multiplying through with $x_i - m_X$, and solving for b, designated as b_1, produces

$$b_1 = \frac{\sum\limits_{i=1}^{n}[(x_i - m_X)(y_i - m_Y)]}{\sum\limits_{i=1}^{n}(x_i - m_X)^2} \tag{3.1-22}$$

Inserting $a = m_Y$, and b_1 into Equation (3.1-14) produces the straight line of "best fit"

$$\hat{y} = m_Y + b_1(x - m_X) \tag{3.1-23}$$

where $\hat{y}$ is the predicted y for a given x. Equation (3.1-23) is called the *estimated regression equation*, and b_1 is the *estimated regression coefficient*. But b_1 is also the slope of the straight line of best fit. The least squares method can be applied to any function that is chosen as the *best fit function* or regression function. One must specify as many conditions as there are unknown coefficients (Draper and Smith, 1966, pp. 34 and 263).

Using centered variables and normalizing them by dividing them by the standard deviation transforms Equation (3.1-23) into

$$\frac{\hat{y} - m_Y}{\sigma_Y} = b_1 \frac{\sigma_X}{\sigma_Y} \frac{x - m_X}{\sigma_X}. \tag{3.1-24}$$

The proportionality coefficient, or the ratio of the normalized variables, is

$$\rho = b_1 \frac{\sigma_X}{\sigma_Y} \tag{3.1-25}$$

where ρ is the *sample correlation coefficient* between the normalized variables. Equation (3.1-23) can be rewritten as

$$\hat{y} = m_Y + \rho \frac{\sigma_Y}{\sigma_X}(x - m_X). \tag{3.1-26}$$

The standard deviations σ_X and σ_Y are obtained from (see also Section 2.3.2)

$$\sigma_X = \sqrt{\frac{\sum\limits_{i=1}^{n}[x_i - E(X)]^2}{n-1}} \tag{3.1-27a}$$

and

$$\sigma_Y = \sqrt{\frac{\sum_{i=1}^{n}[y_i - E(Y)]^2}{n-1}} \qquad (3.1\text{-}27\text{b})$$

where $E(X)$ and $E(Y)$ are simply other notations for m_X and m_Y.

The sample correlation coefficient, ρ, is obtained from Equation (3.1-25) by substituting b_1 from Equation (3.1-22), and σ_X and σ_Y from Equations (3.1-27a) and (3.1-27b)

$$\rho(X,Y) = \frac{\sum\{[x_i - E(X)][y_i - E(Y)]\}}{\sqrt{\sum[x_i - E(X)]^2}\sqrt{\sum[y_i - E(Y)]^2}} \qquad (3.1\text{-}28)$$

where the sums are taken over all n elements of the sample or population. Multiplying numerator and denominator under the root by $(n-1)$ and re-substituting the standard deviations gives

$$\rho(X,Y) = \frac{Cov(X,Y)}{\sigma_X \sigma_Y} \qquad (3.1\text{-}29)$$

with

$$Cov(X,Y) = \frac{\sum\{[x_i - E(X)][y_i - E(Y)]\}}{n-1} \qquad (3.1\text{-}30)$$

where the sums are again taken over all n elements of the sample. $Cov(X,Y)$ is the covariance of X and Y. From Equation (3.1-23), one can see that the slope of the regression line, b_1, is also

$$b_1 = \frac{Cov(X,Y)}{\sigma_X^2}. \qquad (3.1\text{-}31)$$

If Y represents a random process of monthly failures, and y is one of its realizations, and $x = 1,...,10$ is the month of January for a 10-year observation period, then Equation (3.1-23) is the *trendline* of monthly failures over the 10-year period, with b_1 being the average rate of change from year to year, or from one January to the next.

The data points may scatter more or less about the trendline, as there may be increases, decreases or no change from one January to the next. It is desirable to formulate a criterion that expresses the goodness of fit, or the reliability of using the line for prediction purposes. If y represents the observed value, and $\hat{y}$ is the predicted value, then the departure of the observed value from the predicted value can be expressed by (Draper and Smith, 1966, p. 13)

$$y_i - \hat{y}_i \equiv [y_i - E(Y)] - [\hat{y}_i - E(Y)].$$ (3.1-32)

The right-hand side of Equation (3.1-32) is simply an expansion of the left-hand side by introducing the mean $E(Y)$. Taking the squares of both sides for each value i and summing over all n terms gives

$$\sum (y_i - \hat{y}_i)^2 = \sum [y_i - E(Y)]^2 + \sum [\hat{y}_i - E(Y)]^2$$

$$- 2\sum [y_i - E(Y)][\hat{y}_i - E(Y)]$$ (3.1-32a)

The third term on the right-hand side of Equation (3.1-32a) that is a cross product of the other terms is transformed by first substituting Equation (3.1-23) for the second factor under the sum, $\hat{y}_i - E(Y) = b_1[x_i - E(X)]$. By using Equation (3.1-22), the product under the sum becomes

$$- 2b\sum [(x_i - E(X)][y_i - E(Y)] = -2b_1^2 \sum [x_i - E(X)]^2 .$$

By using Equation (3.1-23) again one obtains

$$- 2b_1^2 \sum [x_i - E(X)]^2 = -2\sum [\hat{y} - E(Y)]^2 .$$

Thus Equation (3.1-32a) becomes

$$\sum [y_i - E(Y)]^2 = \sum (y_i - \hat{y}_i)^2 + \sum [\hat{y}_i - E(Y)]^2 .$$ (3.1-32b)

where the sums are taken over all i (Draper and Smith, 1966, p. 14). Equation (3.1-32b) states that the sum of squares about the mean ($SSAM$) is equal to the sum of squares about the regression ($SSAR$) plus the sum of squares due to the regression ($SSDR$), or

$$SSAM \equiv SSAR + SSDR.$$ (3.1-33)

With *SSAM* being known, the *SSAR* should be as small as possible, so that *SSDR*/*SSAM* approaches 1. Solving Equation (3.1-33) for this ratio gives

$$R^2 = \frac{SSDR}{SSAM} = 1 - \frac{\sum (y_i - \hat{y}_i)^2}{\sum [y_i - E(Y)]^2}$$

(3.1-33a)

where R^2 is the *sample coefficient of determination*. It measures the fraction of the variation about the mean explained by the regression. R is the *sample regression coefficient*. Preference is usually given to R^2 because it is a better indicator of the strength of association (Lapin, 1978, p. 352 and p.357).

Example: A study of breach formations in earth dams found that the amount of earth removed from the breach, V_e, is correlated to the product of the volume of water flowing out, V_w, and the depth of the water above the breach base, h (MacDonald and Langridge-Monopolis, 1984, p. 577).The product $V_w \cdot h$, in m^4, multiplied by the unit weight, ρg, in N/m^3, can be construed as the potential energy of the outflowing water, $E_p = \rho g V_w \cdot h$, in Nm. The volume removed is plotted in Figure 3-2 against the potential energy indicator $V_w \cdot h$. The data points and the trendline indicate that some correlation between the chosen variables exists. The trendline obtained by regression is

$$\hat{y} = 0.0984 x^{0.6991}$$

where $\hat{y} = V_e$ is given by the trendline; $x = V_w \cdot h$; $R^2 = 0.9055$; $E(X) = 2.58 \cdot 10^9$ m^4, and $E(Y) = 2.83 \cdot 10^5$ m^3. The ratio of the sums of the squared differences in Equation (3.1-33a) is zero if the trendline fits all points. In this case, $R^2 = 1$. If the trendline is not better than the average, $E(Y)$, than $R^2 = 0$. Here, the regression is said to explain about 90 % of the scatter around the mean.

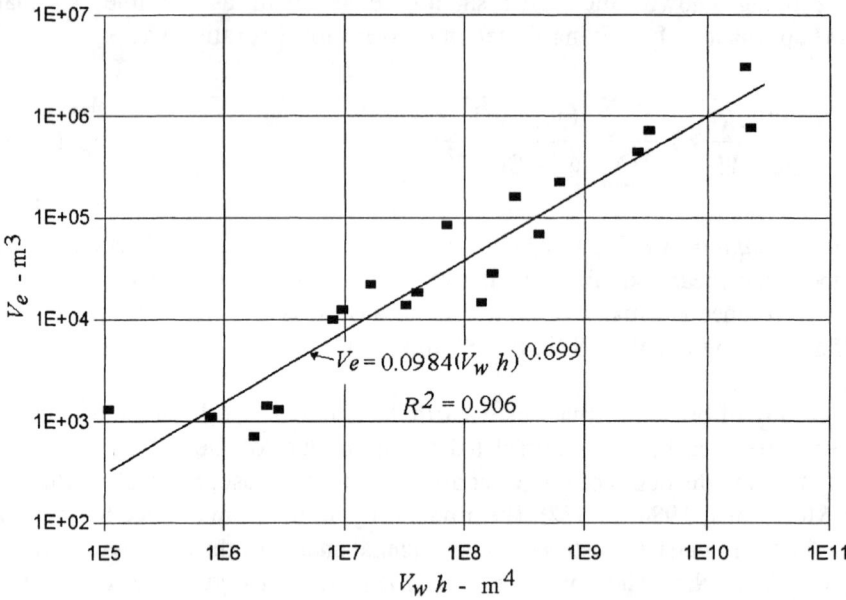

Figure 3-2: Earth volume, V_e, m³, removed from a dam breach of earthfill dams versus a total potential energy indicator, $V_w \cdot h$, m⁴, the product of water volume and initial head above breach base (data after MacDonald and Langridge-Monopolis, 1984).

3.1.3 Prediction Using a Stochastic Model

The Markov process of Section 2.2.3 had the characteristic that the probability of a realization of Y_j was only a function of the previous Y_i, with the transition from Y_i to Y_j being random. In other words, the transition from state i to state j is subject to a transition probability, p_{ij}, where i is the index denoting the given state and j is the state to be visited next. Such a process is illustrated by the flow transition probabilities of Table 3-2.

Table 3-2: Flow Transition Probabilities for Weekly Flows During Weeks 1 to 13 for the Tennessee River at Chickamauga Dam (Based on Data by the Tennessee Valley Authority, Knoxville, TN)

Present State Entry i	Class	Next State j - Exit										
		1 0-10	2 10-20	3 20-30	4 30-40	5 40-50	6 50-60	7 60-70	8 70-80	9 80-90	10 90-100	11 >100
(1)	(2)	(3)	(4)	(5)	(6)	(7)	(8)	(9)	(10)	(11)	(12)	(13)
1	0-10	0.45	0.44	0.00	0.00	0.00	0.00	0.00	0.00	0.11	0.00	0.00
2	10-20	0.03	0.46	0.22	0.10	0.10	0.00	0.02	0.00	0.02	0.02	0.03
3	20-30	0.00	0.14	0.27	0.23	0.13	0.04	0.06	0.05	0.02	0.01	0.05
4	30-40	0.00	0.01	0.26	0.22	0.19	0.10	0.06	0.04	0.04	0.02	0.06
5	40-50	0.00	0.02	0.08	0.26	0.20	0.13	0.09	0.04	0.06	0.03	0.09
6	50-60	0.00	0.00	0.06	0.30	0.19	0.18	0.09	0.07	0.01	0.05	0.05
7	60-70	0.00	0.00	0.02	0.14	0.24	0.15	0.16	0.08	0.07	0.05	0.09
8	70-80	0.00	0.00	0.01	0.07	0.16	0.15	0.12	0.20	0.08	0.05	0.16
9	80-90	0.00	0.00	0.02	0.09	0.16	0.14	0.11	0.05	0.05	0.05	0.33
10	90-100	0.00	0.00	0.00	0.08	0.26	0.16	0.05	0.21	0.03	0.05	0.16
11	>100	0.00	0.00	0.00	0.01	0.07	0.12	0.17	0.14	0.07	0.07	0.35
Entry	975	6	52	108	167	162	111	92	76	44	35	122
Prob	1.0	0.006	0.053	0.111	0.171	0.166	0.114	0.094	0.078	0.045	0.036	0.126
Exit	975	9	59	113	166	158	107	86	74	44	38	121
Prob	1.0	0.009	0.061	0.116	0.170	0.162	0.110	0 088	0.076	0.045	0.039	0.124

Notes: Only an abbreviated table is shown; all rows and columns beyond 100 are lumped into one column and row " >100". The table contains 975 elements.

column 1: present state number; corresponding numbers are given in the table heading for the next state.

column 2: class or range of present state; corresponding ranges are given in the table heading for the next state.

columns 3 through 12: transition probabilities from the present state (present week's average flow) to the next state (following week's average flow) averaged for the first quarter of the year of a 75-year record. The transition probabilities are obtained by dividing the number of flow transitions from the present state i to the next state j by the total number of transitions emanating from the present state i. For example, the first row for present state $i = 1$, or class 0-10, has a total of 9 transitions emanating from it (see explanations for bottom rows). Four times the flow transits from 0-10 to 0-10 (stays the same), four times it transits to 10-20, and once it transits to 80-90. Thus, the transition probabilities are $4/9 = 0.45$, $4/9 = 0.44$, and $1/9 = 0.11$, respectively, after distributing the rounding error.

column 13: to reduce table size, all states beyond 100 (100,000 ft³/s) are lumped into one state " > 100"; the transition probability from the lumped state, $i = 11$, to the lumped state, $j = 11$, is the sum of the transition probabilities included in the lumped state.

Bottom rows: First double row (Entry/Prob): The first number of the first row is the total number of transitions from a state i to a state j, which is 975; the other numbers give the total number of transitions from all states i to a state j: 6 go to state 1, 52 to state 2, and so on. The second row gives the total probability of transition to state 1, 2, and so on, from any of the initial states; it is obtained by dividing the numbers of the previous row by 975. These are the probabilities of being in ending state j regardless of starting state i; they are the marginal ending state probabilities. The row of probabilities sums to 1.

Second double row (Exit/Prob): The first number of the first row is again the total number of transitions, 975, in this case exits from each starting state, 9 from state 1, 59 from state 2, and so on. The second row gives the total probability of coming from a starting state, regardless of where the transition is directed. It is obtained by dividing the previous row by 975. The row of starting state probabilities also adds to 1. The starting state probabilities should be shown as the last column of the table as this column represents the row sum of exits from state i transitions divided by 975; for example, for the first row, 9/975 = 0.009, and for the second row, 59/975 = 0.0605. The limited display of numbers was chosen because of space limitations.

The transition probabilities of Table 3-2 are conditional probabilities except for the marginal probabilities of the bottom rows. If x_i is the present state, and the y_j are the possible next states then $f(y_j|x_i)$ is the probability of the next state being y_j, given the present state is x_i. Using the state designation $x_1 = (0,10)$ instead of "0–10", the sum of the possible transition probabilities for the first row of Table 3-2 is

$$\sum_j f(y_j|x_1) = f(y_1|x_1) + f(y_2|x_1) + ... + f(y_9|x_1) + ... =$$

$$= f[(0,10)|(0,10)] + f[(10,20)|(0,10)] + f[(80,90)|(0,10)] =$$

$$= 4/9 + 4/9 + 1/9 = 1$$

which means that the transition probabilities, p_{ij}, emanating from state i to all states j add to 1. Table 3-2 shows that the row sums of probabilities are 1. This expresses the fact that the process must move somewhere, given it starts in state i, during the defined period (here one week). This includes moving to the same state, $j = i$. The transition probabilities to the same state, p_{ii}, are located on the diagonal of Table 3-2 that runs from the upper left to the lower right. The entire field of transition probabilities of Table 3-2 is the *transition probability matrix*.

In the previous example, states $j = 1$, 2, and 9, or classes 0–10, 10–20, and 80–90, are visited from state $i = 1$. The fact that some states are not visited is explained either by the short record or by physical impossibility. In other words, there is no transition path between some states i and j, and the corresponding p_{ij} are zero.

The probabilities of joint and conditional events were discussed in Section 2.1.4. The pdf of two joint random variables can be written as (Hoel, 1971, p. 35)

$$f(x,y) = f(x)f(y|x) \tag{3.1-34}$$

where x is a given value on which possible values of y are conditioned. For example, in Table 3-2, for a given "present state" x_1 there is a number of "next states" y_j that can be visited. If sums are taken of both sides one can write

$$\sum_y f(x,y) = \sum_y f(x)f(y|x) = f(x)\sum_y f(y|x) = f(x) \tag{3.1-35}$$

where $f(x,y)$ are products of probabilities of x and y, $f(x)$ is the *marginal distribution* of x which is the same for all y of a row. The sum of $f(y|x)$ is the row sum of Table 3-2 which was shown to be 1 in the previous example so that Equation (3.1-35) is fulfilled. Thus, the row sum of joint probabilities, $f(x,y)$, is the marginal probability of the present state x. This is best illustrated by an example.

The first row of Table 3-2 shows that the "present state" x_1 is the starting state 9 times out of 975, and, thus, has the probability $f(x) = 9/975 = 0.009$. Four transitions occur from state x_1 to state y_1, four transitions occur from state x_1 to state y_2, and one transition occurs from x_1 to y_9. Summing the joint probabilities according to Equation (3.1-35) gives

$$\sum_{all\ y} f(x,y) = \sum_{all\ y} f(x)f(y|x)$$

$$= (9/975)\,(4/9) + (9/975)\,(4/9) + (9/975)\,(1/9) = 9/975 = 0.009.$$

Hence, the sum of transitions in a row of the matrix divided by the total number of elements produces a column value of marginal probabilities $f(x)$ that should be shown at the right-hand margin of the table (here shown as fourth bottom row of Table 3-2). These values $f(x)$ are the average starting probabilities for each state, and the second bottom row are the average ending probabilities for each state.

Examples: (1) What is the probability of next week's flow state being 10–20, given the present flow state is 10–20; and what is the unconditional probability of this same flow state? Solution: From Table 3-2 one finds the probability of the next state being 10–20, given the present state is 10-20 as 0.46. The unconditional probability of the next flow state being 10–20 is given in the second bottom row as 0.053.

(2) What is the probability of the flow remaining about the same from one week to the next during the first quarter of the year? Solution: The diagonal of the matrix in Table 3-2 from the upper left to the lower right shows these persistence probabilities. By following this diagonal along the numbers 0.44, 0.46, 0.27, etc., one recognizes that the probability of the flow being sustained in the same class declines with increasing flow rate. An exception is the lumped state of flows greater than 100. The probability of a flow greater than 100 to persist for the week is 0.37.

(3) Given the flow is in the 0–10 state. What is the probability that the flow will not exceed the 10–20 state? Solution: Table 3-2 allows estimates of exceedance and non-exceedance probabilities by adding parts of a row. Given the flow is in the 0–10 state, the sum $0.44 + 0.44 = 0.88$ gives the non-exceedance probability of the 10–20 state.

(4) Columns 3 through 13 of Table 3-2 represent an 11×11 one-stage transition probability matrix. In Section 2.2.3 it was shown by Equation (2.2-14) that the n-stage transition probability matrix is obtained as the nth power or the one-stage transition probability matrix, or $P(13) = P^{13}$, with $P(13)$ being the thirteen-stage transition probability matrix and P being the 1-stage transition probability matrix. $P(13)$ gives the probabilities of going from any of the initial states to any of the ending states over thirteen transitions. Using a program such as Matlab® one obtains for $P(13)$:

0.00	0.04	0.11	0.17	0.17	0.12	0.10	0.08	0.05	0.04	0.12
0.00	0.04	0.11	0.17	0.17	0.12	0.10	0.08	0.05	0.04	0.12
0.00	0.04	0.11	0.17	0.17	0.12	0.10	0.08	0.05	0.04	0.13
0.00	0.04	0.11	0.17	0.17	0.12	0.10	0.08	0.05	0.04	0.13
0.00	0.04	0.11	0.17	0.17	0.12	0.10	0.08	0.05	0.04	0.13
0.00	0.04	0.11	0.17	0.17	0.12	0.10	0.08	0.05	0.04	0.13
0.00	0.04	0.11	0.17	0.17	0.12	0.10	0.08	0.05	0.04	0.13
0.00	0.04	0.11	0.17	0.17	0.12	0.10	0.08	0.05	0.04	0.13
0.00	0.04	0.11	0.17	0.17	0.12	0.10	0.08	0.05	0.04	0.13
0.00	0.04	0.11	0.17	0.17	0.12	0.10	0.08	0.05	0.04	0.13
0.00	0.04	0.11	0.17	0.17	0.12	0.10	0.08	0.05	0.04	0.13

The rows represent the probabilities of the ending states for each starting state. The ending state probabilities are seen to be the same regardless of starting state. The probabilities of each row add to 1 because they represent the mutually exclusive and exhaustive pathways from an initial state to any of the ending states. The probabilities of the rows of $P(13)$ are about the same as those of the last bottom row of Table 3-2 from which the effect of the initial state was eliminated. The matrix approach provides transition probability matrices for any desired number of stages. Here $n = 13$ is the maximum number of stages because it is assumed that the matrix changes for the next thirteen transitions due to seasonal changes.

3.1.4 Prediction Using an Autoregressive Model

A statistical data analysis may reveal that a correlation exists between two system states, e.g., between the flow in the present period and the flow in the previous period. A mathematical model for the correlation between two variables x and y is Equation (3.1-26). With some modification, it can be written as

$$y = m_Y + \rho \frac{\sigma_Y}{\sigma_X}(x - m_X) + \varepsilon_y \qquad (3.1\text{-}36)$$

where y, m_Y, and σ_Y refer to the the present period i, and x, m_X, and σ_X refer to the previous period i -1. The term ϵ_y is some random addition or subtraction that is related to the present period.

Fiering and Jackson (1971, p. 51) suggested such a model for period by period data generation. If the model is applied to subperiods of one and the same season for which the population parameters are the same, then $\sigma_Y/\sigma_X = 1$, $m_Y = m_X = \mu$, $y = y_i$, and $x = y_{i-1}$. Then Equation (3.1-36) becomes an *autoregressive* model,

$$y_i = \mu + \rho_1(y_{i-1} - \mu) + \varepsilon_i' \qquad (3.1\text{-}37)$$

where ρ_1 is the *lag-one autocorrelation coefficient*, and ϵ_i' is a random component that occurs in period i. The lag-one autocorrelation coefficient is calculated as

$$\rho_1 = \frac{E[(y_i - \mu)(y_{i-1} - \mu)]}{\sigma^2} \qquad (3.1\text{-}38)$$

where μ and σ are the mean and standard deviation of the population to which both y_i and y_{i-1} belong (Fiering and Jackson, 1971, p. 28). Equation (3.1-38) is similar to Equation (3.1-31). The lag-0 correlation is the correlation of a value with itself, which makes $\rho_0 = 1$. The lag-1 correlation coefficient, ρ_1, usually is less than 1. For a collective of data with the same parameters, say annual average flow, the autocorrelation coefficient is an average for the data set. If there is persistence in the series, the predicted $y_i - \mu$ tends to have the same sign as $y_{i-1} - \mu$, and ρ_1 is either positive or negative, but stays in the range $-1 \le \rho \le +1$. If there is no persistence, the sign of the terms may be positive as often as negative; in other words, successive flows will be either above or below the average, and ρ is zero; then no autocorrelation exists, and the sequence (y_i, y_{i-1}) is a random process.

Equation (3.1-37) is called a Markov model (Fiering and Jackson, 1971, p. 51; O'Donovan, 1983, p. 53). It is, however, not a Markov process but simply a stochastic process that consists of adding a random realization ϵ_i' to a deterministic term $\mu + \rho_1 (y_{i-1} - \mu)$. If $\rho_1 < 1$, then each successive term $y_i = \mu + \rho_1 (y_{i-1} - \mu)$ is less than the previous one unless the random component lifts y_i above the previous value y_{i-1}. Equation (3.1-37) is also known as an AR(1) model, or a *first-order autoregressive model*, as the prediction of the next value is based only on the known immediately preceding value plus a random perturbation which is a *white noise* component taken from a normal distribution. For $\rho_1 = 1$, the model is known as *random walk*.

The model has been quite successful in practical applications (O'Donovan, 1983, p. 59). In hydrology it has been applied to the simulation of annual, monthly and even weekly flows. The random component simulates input by rainfall whereas the autoregressive component simulates the depletion of the drainage basin.

A trace based on Equation (3.1-37) is shown in Figure 3-3. Starting with a given value y_{i-1}, the new value y_i is the net result of a random component ϵ_i' followed by an autoregressive component using $\rho_1 = 0.9$. The random perturbation consists of a term that is modeled as a random fraction t of the general form of the standard deviation of the random residuals of regressions (Draper and Smith, 1966, p. 94). Fiering and Jackson (1971, p. 50) suggested

$$\varepsilon_i' = t_i \sigma \sqrt{1 - \rho^2} \qquad (3.1-39)$$

where t_i is a random variate. The normal random variate would have a range $-\infty \leq t \leq +\infty$; a uniform random variate could have a range $0 \leq t \leq 1$. The random variable used in Figure 3-3 has a range of $-1 \leq t \leq +1$. Samples of uniform and normal random numbers are given in Fiering and Jackson (1971, pp. 20–21). They can also be found in statistical text books, or they can be generated by computer. In Figure 3-3, from $n = 4$ on, a random perturbation is added. It is based on a generated random number with a range $(-1, +1)$. Depending on the size of the random term used, the generated $y_i - \mu$ may keep or change its sign. According to Equation (3.1-39), the random disturbance is controlled by the data characteristics embodied in σ and ρ. Every spreadsheet run produces a new random number or a new sequence of them, so that the trace shown in Figure 3-3 is unique.

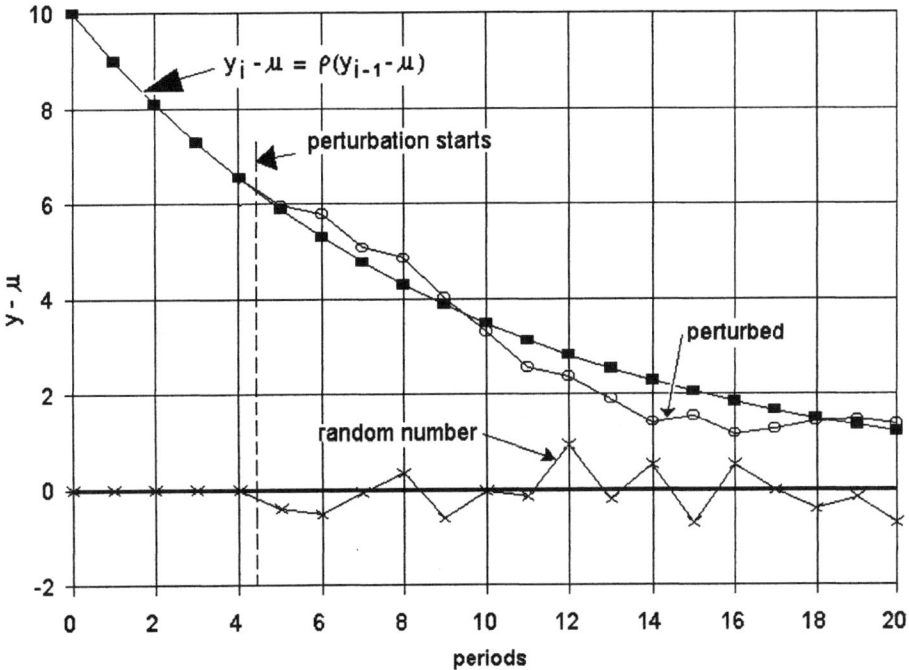

Figure 3-3: Autoregressive prediction model consisting of a deterministic decay component and a random component.

3.1.5 Joint Distribution of Two Normal Random Variables

Joint and marginal distributions for a discrete sample space were discussed in Section 3.1.3. In this section, joint and marginal distributions for two continuous variables are discussed (Hoel, 1971, p. 157; Parzen, 1965, p. 54). The joint distribution of two random variables is

$$f(x,y) = f(y)f(x|y) = f(x)f(y|x) \quad .$$

(3.1-40)

If x and y are independent of each other, then the joint distribution is just the product of the distributions:

$$f(x,y) = f(x)f(y) \quad .$$

(3.1-41)

More precisely, the pdfs $f(x)$ and $f(y)$ should be written as $f_X(x)$, $f_Y(y)$ and $f_{XY}(x,y)$ but this notation is only used when necessary to avoid confusion.

The joint distribution of two independent normal random variables has the normalized variables

$$u = \frac{x - \mu_X}{\sigma_X} \tag{3.1-42a}$$

and

$$v = \frac{y - \mu_Y}{\sigma_Y}. \tag{3.1-42b}$$

Then, according to Equation (3.1-41),

$$f_{XY}(x,y) = \frac{1}{2\pi\sigma_X\sigma_Y} e^{-0.5 \cdot (u^2 + v^2)}. \tag{3.1-43}$$

In the more general case in which x and y are not independent, a correlation coefficient enters the equation:

$$f_{XY}(x,y) = \frac{1}{2\pi\sigma_X\sigma_Y\sqrt{1-\rho^2}} e^{-\frac{u^2 + v^2 - 2\rho uv}{2(1-\rho^2)}} \tag{3.1-44}$$

with

$$\rho = \frac{Cov(X,Y)}{\sigma_X\sigma_Y}. \tag{3.1-45}$$

Similar to the summation (numerical integration) of the discrete distribution $f(x,y)$ by Equation (3.1-35), in Section 3.1.3, the *marginal distribution*, $f(x)$, of the continuous distribution is found by integrating $f_{XY}(x,y)$ over its range. For the normal distribution, the range is from $-\infty$ to $+\infty$, and integration leads to (Hoel, 1971, p. 152)

$$f_X(x) = \frac{1}{\sqrt{2\pi}\sigma_X} e^{-0.5u^2} \tag{3.1-46}$$

where $f_X(x)$ is the pdf of the normal distribution of the continuous normalized variable $u = (x - \mu_X)/\sigma_X$. It should be remembered that the continuous $f_X(x)$ is not a probability, but only the ordinate of the pdf. For the normal pdf, it has the dimension of the reciprocal of σ_X.

3.1.6 Conditional Distribution of a Random Variable

In a probabilistic (x,y)-relation, a given x may have associated with it an entire pdf of y's. Then $f(y)$ is a pdf of y conditioned on x (Hoel, 1971, p. 146; Parzen, 1965, p. 54):

$$f_Y(y|x) = \frac{f_{XY}(x,y)}{f_X(x)}. \tag{3.1-47}$$

All function indices are shown here to make sure the functions are properly interpreted as pdf's of one or the other random variable, or both: $f_{XY}(x,y)$ is the joint pdf of X and Y, $f_X(x)$ is the pdf of X, or marginal distribution of X, and $f_Y(y|x)$ is the conditional pdf of Y, given a realization x. Assume X and Y are normally distributed. With Equations (3.1-44) and (3.1-46) of Section 3.1.5, the conditional probability $f_Y(y|x)$ is obtained as the quotient of the two functions. The resulting new function's factors are the ratios of the factors of $f_{XY}(x,y)$ and $f_X(x)$, and the exponent is the difference of the exponents of $f_{XY}(x,y)$ and $f_X(x)$:

$$f_Y(y|x) = \frac{1}{\sqrt{2\pi}\sigma_Y\sqrt{1-\rho^2}} e^{-\frac{u^2+v^2-2\rho u v}{2(1-\rho^2)} - \frac{u^2}{2}}. \tag{3.1-48}$$

Simplifying the exponent leads to

$$f_Y(y|x) = \frac{1}{\sqrt{2\pi}\sigma_Y\sqrt{1-\rho^2}} e^{-\frac{(v-\rho u)^2}{2(1-\rho^2)}}. \tag{3.1-48a}$$

After substituting for u and v using Equations (3.1-42a) and (3.1-42b), the exponent becomes

$$-\frac{y-\mu_Y-\rho\dfrac{\sigma_Y}{\sigma_X}(x-\mu_X)}{2\sigma_Y^2(1-\rho^2)}. \tag{3.1-49}$$

From the general form of the normal variate of the Gaussian function, $z = (x - \mu)/\sigma$, one can deduce that the mean of the conditioned variable, Y, is

$$\mu_Y | x = \mu_Y + \rho \frac{\sigma_Y}{\sigma_X}(x - \mu_X)$$
(3.1-50)

with standard deviation

$$\sigma_Y | x = \sigma_Y \sqrt{1 - \rho^2} \ .$$
(3.1-51)

Equation (3.1-50) means that a regression can predict for some random realization $X = x$, a μ_Y that is the mean of a distribution of the y-values. The least squares fit, Equation (3.1-26) of Section 3.1.2, produced a $\hat{y}$, but no attempt was made at predicting the distribution of the Y that are scattered around μ_Y (Hoel, 1971, p. 154). Since the parameters μ_Y, μ_X, ρ, σ_Y, and σ_X are calculated from all points of the sample space, the normal distributions, $f_Y(y | x)$, are identical in shape for all x. Only the means of the distributions, $\mu_Y | x$, are a function of x and are predicted by Equation (3.1-50). For a given x, there exists a random population Y with a frequency distribution $f_Y(y | x)$. The regression function is the locus of the conditional means of Y for a particular x, $\mu_Y | x$. An illustration of the distributions $f_Y(y | x)$ is similar to Example (1) in Section 3.1.1 where monthly cdfs of failure incidents are superimposed on a seasonal distribution of mean failure incidents.

3.2 Transformation of a Random Variable

3.2.1 Linear Transformation

The *transformation* of one probabilistic variable into another is used to test whether the transformed variable has a known and easily usable probability distribution. A well known example is the use of the logarithms of flow data instead of the real data because the logarithms of the flow data may fit a normal distribution whereas the real data do not. A transformation from a variable X into a variable Y can be accomplished by any monotonically rising or falling function. Functions that qualify are linear or nonlinear functions, such as $y = a + bx$, $y = \log x$, $y = e^x$, and so on. These functions do not reverse the sign of their gradient, i.e., they do not traverse a maximum or a minimum in the variable range.

The simplest transformation is a linear transformation. Let X be a population of temperature observations in degrees Fahrenheit that is to be transformed into a population Y of degrees Celsius. The transformation function is

$$Y_x = \frac{5}{9}(X - 32)$$ (3.2-1)

where Y_x is a random outcome produced by a random input $X = x$. It is strictly incorrect to drop the index x on Y, as Y is not a random outcome once a choice has been made for X. Using lower case letters means that specific values of x and y are meant, while X and Y stand for any realization in the population of random numbers.

Assume X is normally distributed. Then the parameters required to establish the pdf $f_X(x)$ are $\mu = E(X)$ and $\sigma_X = [Var(X)]^{1/2}$. With these parameters, the variable X is normalized as

$$Z_x = \frac{X - E(X)}{\sigma_X}.$$ (3.2-2)

With Equation (3.2-2), $f(z)$ and $F(z)$ can be calculated for any desired x using the normal pdf.

The X-variable is transformed into the Y-variable by the linear transformation

$$Y = a + bX$$ (3.2-3)

which means that for a collective of random numbers X a collective Y will be produced. The parameters of the X-distribution must now be replaced by parameters of the Y-distribution. The expectation of Y for a linear relation is (Section 2.4.1)

$$E(Y) = a + bE(X).$$ (3.2-4)

The variance for Y was derived in Section 2.4.2 as

$$Var(Y) = \sigma_Y^2 = E(Y^2) - [E(Y)]^2.$$ (3.2-5)

Substituting Equation (3.2-3) into Equation (3.2-5) results in

$$\sigma_Y^2 = E[(a + bX)^2] - [a + bE(X)]^2.$$ (3.2-6)

Evaluating the squares and taking expectations produces (see Section 2.4.2)

$$\sigma_Y^2 = b^2 \sigma_X^2,$$ (3.2-7)

and

$$\sigma_Y = b\ \sigma_X .$$ (3.2-8)

The normalized variables are

$$z_x = \frac{x - E(X)}{\sigma_X},$$ (3.2-9)

and

$$z_y = \frac{y - E(Y)}{\sigma_Y}.$$ (3.2-10)

Here the indices of z serve to distinguish the normalized variable z for the two variable x and y. The pdf's for the two normally distributed variables, x and y, can now be written as

$$f_X(x) = \frac{1}{\sqrt{2\pi}\ \sigma_X} e^{-0.5z_x^2}$$ (3.2-11)

and

$$f_Y(y) = \frac{1}{\sqrt{2\pi}\ b\sigma_X} e^{-0.5z_y^2} .$$ (3.2-12)

In Equation (3.2-12), σ_Y has been substituted by making use of Equation (3.2-8). Since the probabilities of the corresponding x-value and y-value must be the same, one can write

$$f_X(x)dx = f_Y(y)dy .$$ (3.2-13)

Differentiation of the normalized variables, z_x and z_y, Equations (3.2-9) and (3.2-10), versus x and y, respectively, results in

$$dz_x = \frac{dx}{\sigma_X} , \text{ and } dz_y = \frac{dy}{b\sigma_X} .$$ (3.2-14)

Since the normalized variable increment, dz, is the same in both cases, the following relation is obtained:

$$\frac{dx}{\sigma_X} = \frac{dy}{b\sigma_X}, \tag{3.2-15}$$

so that $dx/dy = 1/b$. The transformed pdf becomes

$$f_Y(y) = |\frac{dx}{dy}|f_X(x) = |\frac{1}{b}|f_X(x). \tag{3.2-16}$$

The absolute value, $|1/b|$, ensures a positive value for $f_Y(y)$, if dx/dy is negative, as in the case of a decreasing function $y = f(x)$ (Benjamin and Cornell, 1971, p. 109).

Example: A temperature distribution, given in Fahrenheit, has a normal pdf. Transform the °F-distribution into a °C-distribution. Solution: The quantities in °F are X and the quantities in °C are Y. The corresponding pdf's are $f_X(x)$ and $f_Y(y)$. The monotonic transformation function, here a linear function, is

$$y = \frac{5}{9}(x - 32) = -17.778 + \frac{5}{9}x. \tag{3.2-17}$$

Graphically, the transformation function states that in an x,y-coordinate system, the original data (in °F) are given on the x-axis and the transformed data (in °C) are given on the y-axis. The conversion from x to y is accomplished by Equation (3.2-17). A normal pdf can be visualized over a data range on the x-axis data and the transformed pdf can be visualized over the y-axis data. An area slice under the pdf of the x-range, $f_X(x)\,dx$, is the probability of an x-value. Similarly, an area slice under the pdf of the y-range is the probability of the corresponding y-value. The two probabilities of the pair (x, y) must be equal, as expressed by Equation (3.2-13). Differentiating the transformation function, Equation (3.2-17), produces $dy/dx = 5/9$. The pdf for Y is obtained from Equation (3.2-13) as

$$f_Y(y) = |\frac{dx}{dy}|f_X(x) = \frac{9}{5}f_X(x) = 1.8f_X(x). \tag{3.2-18}$$

For example, for $x = 68$ °F, the ordinate of the pdf that corresponds to that value, $f_X(68)$, is multiplied by 1.8. The width of the slice of the Celsius-distribution is $dy = (5/9)dx$. This means that the narrower slices of the Y-distribution have ordinates 9/5 times those of the X-distribution, so that the probabilities stay the same in fulfillment of Equation (3.2-13).

The shapes of the original pdf and the transformed pdf are shown in Figure 3-

4. The upper part of the figure shows the original °F-distribution

$$f_X(z_x)dz_x = 0.3989e^{-0.5z_x^2}dz_x \qquad\qquad (3.2\text{-}19)$$

where $dz_x = dx/\sigma_X$. The lower part of Figure 3-4 shows the transformed °C-distribution

$$f_Y(z_y)dz_y = 0.3989e^{-0.5z_y^2}dz_y \qquad\qquad (3.2\text{-}20)$$

where

$$dz_y = 1.8\,dy/\sigma_X . \qquad\qquad (3.2\text{-}20\text{a})$$

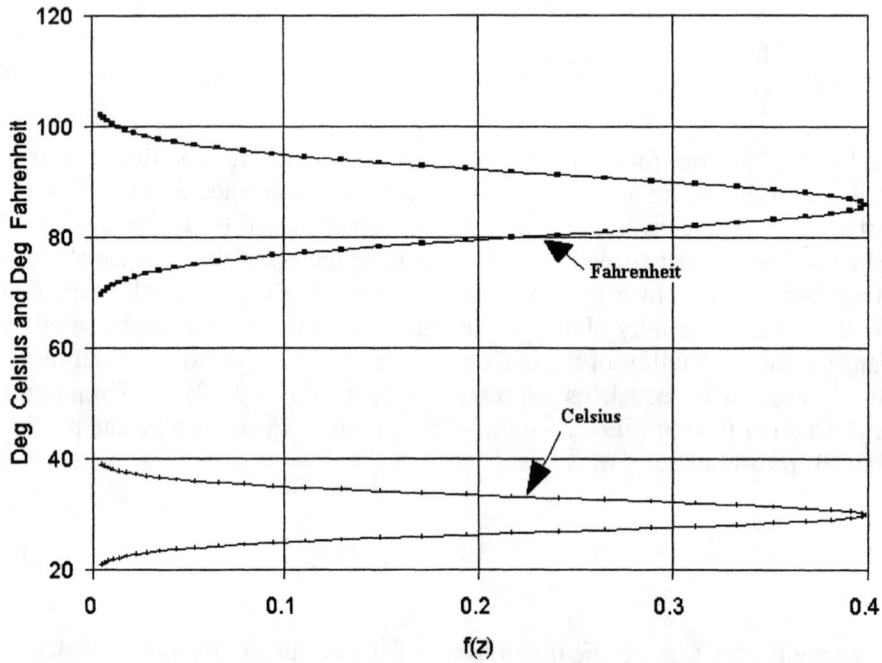

Figure 3-4: Normalized probability density function for temperature in degrees Fahrenheit and the transformed function for degrees Celsius (This plot is the result of a plotting software shortcoming that can handle two y-axes, but not two x-axes in one plot).

In Figure 3-4, $f_X(z_x)$ and $f_Y(z_y)$ have the same maximum, 0.3989, but because of the smaller standard deviation, $\sigma_X/1.8$, used in z_y, $f_Y(z_y)$ has a narrower bell shape. This was explained in Section 2.6.3 in connection with Figure 2-17. Also, $f_X(z_x)$ and $f_Y(z_y)$ do not enclose an area 1, only $f_X(z_x)\,dz_x$ and $f_Y(z_y)\,dz_y$ do. Equations (3.2-19) and (3.2-20) both have the same maximum ordinate, $(2\pi)^{-1/2} = 0.39894$, as shown in Figure 3-4. This is the result of the transfer of σ_X and $b\,\sigma_X$ from the denominator of the factor in front of Equations (3.2-11) and (3.2-12), into dz, as shown, for example, by Equation (3.2-20a). This makes both $f(z)$ and dz dimensionless and the central ordinate of $f(z)$ a constant.

3.2.2 Logarithmic Transformation

A stochastic variable can sometimes be simulated as the product of several random variables. Taking the logarithms of such a multifactor product becomes a sum of random variables. It is a general property of sums of random variables to be normally distributed. This property of random variables is further discussed in the context of the *central limit theorem* (Section 3.4.2). This property also explains why logarithms of random variables may be normally distributed. Many random variables are not symmetrically distributed or can assume only positive values. If the range of a random variable is bounded by zero and infinity, then the logarithms range from $-\infty$ to $+\infty$, which corresponds to the range of the normal distribution. Repair times have been found to exhibit log-normal distributions. In hydrology, daily average flows, annual maximum floods, annual precipitation, grain sizes of ground rock, and natural sand mixtures have been found to display such characteristics (Yevjevich, 1972, pp. 140 and 141).

If the logarithms, $\ln x$, of random realizations x are normally distributed, then the distribution of the real quantities, x, also can be easily found. This log-normal transformation produces a rather complex pdf for x that could not otherwise easily be found. The data fit thus obtained allows more reliable interpolation and extrapolation of probabilities associated with data points of a limited data set. This is especially important for the tails (extreme values) of distributions where data usually are scarce.

Suppose the realizations of random variable, X, are transformed into their logarithms to test whether the logarithms can be represented by a normal pdf. The transformed variable is

$$y_x = \ln x \tag{3.2-21}$$

where y_x is a random realization y that is produced by a random realization x via the deterministic logarithmic function. The normalized logarithmic random variable becomes

$$z_y = \frac{y_x - \mu_Y}{\sigma_Y} \tag{3.2-22}$$

where μ_Y and σ_Y are mean and standard deviation, respectively, of the transformed variable y_x. The mean μ_Y is computed by

$$\mu_Y = \frac{1}{n}[\ln x_1 + \ln x_2 + \ldots + \ln x_n] \tag{3.2-23}$$

which can be written in compact notation as

$$\mu_Y = \ln(\sqrt[n]{x_1 x_2 \ldots x_n}) \tag{3.2-23a}$$

where the nth root is the geometric mean of x; n is the number of elements. The standard deviation of the logarithms is

$$\sigma_Y = \sqrt{\frac{(\ln x_i - \mu_Y)^2}{n-1}} \tag{3.2-23b}$$

where μ_Y is from Equation (3.2-23a).

The normal distribution function of the logarithms of x is

$$f(z_y) = \frac{1}{\sqrt{2\pi}} e^{-0.5 z_y^2} \tag{3.2-24}$$

where $f(z_y)$ is the normal pdf of the normalized variable; z_y is computed by Equation (3.2-22).

Functions other than logarithms could be used as variable transformations in conjunction with the normal distribution. The use of log x with the normal distribution is also referred to as *Galton distribution* (Remenieras, 1960, p. 287). Here only the natural logarithm and the decimal logarithm will be used.

The probability of the original variable, x, and of the transformed variable, $\ln x$, remains the same as long as the transformation function is monotonical (see Section 3.2.1). Hence, one can equate their probabilities (Yevjevich, 1971, p. 134)

$$f(z_y)\,dz_y = f_X(x)\,dx = f_Y(\ln x)\,d(\ln x) \tag{3.2-25}$$

where $f_X(x)\,dx$ is the pdf for the original variable X and $f_Y(\ln x)$ is the pdf for the transformed variable Y_x. Since $d(\ln x) = dx/x$, and

$$dz_y = \frac{dy_x}{\sigma_Y} = \frac{d(\ln x)}{\sigma_Y} = \frac{dx}{x\sigma_Y} , \tag{3.2-25a}$$

the relation between the original function and the transformed function is

$$f_X(x) = f(z_y)\frac{dz_y}{dx} = \frac{f(z_y)}{x\,\sigma_Y} , \tag{3.2-25b}$$

and finally, with $1/(2\pi)^{0.5} = 0.39894$, the transformed pdf becomes

$$f_X(x) = \frac{0.39894}{x\,\sigma_Y} e^{-0.5z_y^2} . \tag{3.2-26}$$

The parameters of the log-normal pdf, μ and σ, are obtained by the method of moments (Yevjevich, 1971, p. 135). For $y = \ln x$, they are

$$\mu = e^{\mu_Y + 0.5\sigma_Y^2} , \tag{3.2-27}$$

$$\sigma = \mu\sqrt{e^{\sigma_Y^2} - 1} \;, \text{ and} \tag{3.2-28}$$

$$C_v = \frac{\sigma}{\mu} = \sqrt{e^{\sigma_Y^2} - 1} \tag{3.2-29}$$

where μ_Y and σ_Y are the distribution parameters based on the logarithms of the original data calculated by Equations (3.2-23a) and (3.2-23b).

If decimal (base-10) logarithms are used, as usually is the case for plotting purposes, the original variable x is transformed into logarithms by $y_x = \log x$, and the normalized variable becomes

$$z_y = \frac{y_x - \mu_{Yd}}{\sigma_{Yd}} \tag{3.2-30}$$

where μ_{Yd} and σ_{Yd} are calculated by equations similar to Equations (3.2-23a) and (3.2-23b) only with $\ln x$ replaced by $\log x$; the index "d" stands for "decimal" to distinguish variables and parameters using the decimal log from those using the natural log. The differential of Equation (3.2-30) becomes

$$dz_y = \frac{dy_x}{\sigma_{Yd}} = 0.4343 \frac{dx}{x \, \sigma_{Yd}} \tag{3.2-30a}$$

and the transformed pdf using $\log x$ becomes

$$f_X(x) = f_y(z_y)\frac{dz}{dx} = 0.4343\frac{f_y(z_y)}{x \, \sigma_{Yd}}. \tag{3.2-31}$$

The normalized variable, z_y, is given by Equation (3.2-30). Actually, z_y is the same for natural and decimal logarithms, because the conversion factor, $\log x / \ln x = 0.4343$, can be extracted from all quantities and cancels out. Then Equation (3.2-31), for base-10 logarithms, becomes

$$f_X(x) = 0.4343\frac{0.39894}{x \, \sigma_{Yd}}e^{-0.5\,z_y^2}. \tag{3.2-31a}$$

The parameters of the log-normal transform using base-10 logarithms are derived from Equations (3.2-27) and (3.2-28) by substituting the ln-based parameters by those derived from base-10 logarithms (index "d"):

$$\mu_{Yd} = 0.4343\,\mu_Y \tag{3.2-32}$$

and

$$\sigma_{Yd} = 0.4343\,\sigma_Y. \tag{3.2-33}$$

The parameters of the base-10 log-normal transform are

$$\mu = 10^{\mu_{Yd}+1.1513\sigma_{Yd}^2} \tag{3.2-34}$$

$$\sigma = \mu\sqrt{10^{2.3026\sigma_{Yd}^2}-1} \tag{3.2-35}$$

$$C_v = \sqrt{10^{2.3026\sigma_{Yd}^2}-1} \tag{3.2-36}$$

where μ_{Yd} and σ_{Yd} are mean and standard deviation, respectively, of base-10 logarithms of the original variable X (see also Stedinger et al., 1993, p. 18.15).

By multiplying $f_X(x)$ of Equation (3.2-31) by the increment of the untransformed variable, dx, and integrating, the non-exceedance probability, $F(x)$, is obtained:

$$F(x) = \int_0^x f_X(x)dx \ . \tag{3.2-37}$$

The integration of the log-transform is confined to the positive branch of the x-axis.

The successful use of the log-transform is affected by the variability of the data. For $C_v < 0.3$, the log-transform approaches the normal distribution (Stedinger et al., 1993, p. 18.14). A reduced range of X (by dropping outliers) results in a reduced σ, which in turn reduces the coefficient of variation, which in turn leads to a better approximation of the normal distribution by the log-transformed data.

Examples: (1) A sample of 86 annual maximum floods, from 1869 to 1954 (Remenieras, 1960, p. 403) is hypothesized to have a log-normal distribution. In the same way other quantities, such as the number of routine maintenance hours for doing specific tasks, the life lengths of similar items, the costs of specific activities, and so on, could be analyzed. Determine $f(x)$ and $F(x)$ of the real data by using the log-normal transform. Solution: The logarithms of the data are tested for normal distribution. If this is the case, the distribution of the real data is also known in the form of the log-normal transform. The tabular calculations are omitted here, but the data set is given so that the analysis can be repeated. The 86 values representing maximum annual floods of the Rhine River at Rheinfelden are given in rows of consecutive events with the ending value of a row being followed by the beginning value of the next row:

2169, 2759, 2160, 3525, 1936, 2740, 2264, 5530, 2827, 2984, 2352, 3482, 4764, 4371, 2506, 1640, 2034, 1801, 1880, 3239, 2001, 3259, 2721, 2299, 1452, 1840, 2018, 2836, 3428, 2433, 2626, 1896, 2797, 2768, 2589, 2325, 2143, 3085, 2273, 2229, 2827, 4040, 1952, 2570, 2160, 2469, 2135, 2579, 1912, 3623, 2696, 2617, 1435, 2535, 2243, 2682, 1990, 3121, 2699, 2414, 1798, 3063, 2508, 2835, 2372, 2108, 2713, 2430, 2208, 2526, 2834, 2913, 2244, 2080, 1939, 3258, 2640, 2655, 1883, 2997, 1361, 2543, 2574, 2575, 3515, 2588.

The data mean is $\mu_x = 2579.5$; the standard deviation is $\sigma_x = 685$; and the units are m^3/s. The data are made dimensionless by dividing them by μ_x. The mean of the ratios, x_i/μ_x, is 1, and the standard deviation of the ratios, as derived from the standard deviation of the original data, is $685/2579.5 = 0.265$. The ratios are sorted in descending order and the logarithms are taken of the sorted data, $y_{xi} = \log(x_i/\mu_x)$. The parameters μ_Y and σ_Y are calculated from the logarithms, y_x (the index "d" is omitted here as only decimal logarithms are used). The normalized variable, z_y is calculated by Equation (3.2-30) using $\log(x_i/\mu_x)$ instead of $\log x_i$. The normal pdf, $f(z_y)$, is calculated by Equation (3.2-24). This function has the typical normal shape with a maximum $f(0) = 0.39834$ at $z_y = 0$. The pdf of X based on $y_x = \log(x/\mu_x)$ is

$$f_X(x) = \frac{0.4343}{\frac{x}{\mu_x}\sigma_Y} f(z_y)$$

where $f(z_y)$ is the normal distribution of the logarithms.

The function $f_X(x)$ is shown in Figure 3-5. Its shape is close to normal, but its mean is shifted away from $z = 0$. Its mean and standard deviation are obtained from Equations (3.2-34) and (3.2-35), respectively: for $\mu_{Yd} = -0.01035$, $\sigma_{Yd} = 0.1073$, and $\sigma_{Yd}^2 = 0.0115$,

$$\mu = 10^{(-0.0135 + 1.1513 \cdot 0.0115)} = 1.000$$

$$\sigma = \mu[10^{(2.3026 \cdot 0.0115)} - 1]^{1/2} = 1.000 \cdot 0.251 = 0.251$$

$$C_v = 0.251/1.000 = 0.251.$$

For $C_v < 0.3$, the log-normal pdf approaches the normal pdf, as shown in Figure 3-5. The cdf, $F(x)$, of the log-normal distribution is obtained by integrating $f_X(x)$. As a computational check, if $f_X(x)$ is a pdf, the integration over the range of the random variable must produce 1:

$$F(x) = \int_{-\infty}^{+\infty} f_X(x)dx = 1 .$$

By definition, the original variable probabilities and the transformed variable probabilities remain the same so that

$$f_X(x) \, dx = f_Y(z_y) \, dz_y.$$

Given $f_Y(z_y)$ is the normal pdf computed by Equation (3.2-24), it should integrate to 1. The numerical integration consists of summing all increments $f_X(x) \, \Delta x$ over the variable range using the finite difference form

$$F(x) = \frac{0.4343}{2\sigma_Y} \sum_{i=1}^{n} [\frac{f(z_{yi})}{x_i/\mu_X} + \frac{f(z_{yi+1})}{x_{i+1}/\mu_X}](\frac{x_{i+1}}{\mu_X} - \frac{x_i}{\mu_X})$$

where $n = 85$. Proceeding in ascending order, $x_{i+1}/\mu_x \geq x_i/\mu_x$, and $\Delta(x/\mu_x) = x_{i+1}/\mu_x - x_i/\mu_x > 0$. The summation produces $F(x) = 0.998$, a value close to 1.

(2) Without knowing the empirical distribution of the data, $F(z)$ and z_y can be calculated from the mean and standard deviation of a chosen distribution, here the log-normal distribution. Check how closely $F(z)$ derived from the ranking of the data resembles the calculated $F(z)$ based on the data parameters of the log-normal distribution. <u>Solution</u>: The nondimensional variable, x_i/μ_x, is ranked in descending order, i.e., rank $m = 1$ is given to the largest value. From the rank numbers, the non-exceedance probability is computed by

$$F(z) = \frac{n+1-m}{n+1}$$

where n is the total number of elements and m is the rank (see Equation (3.7-4)). This formula gives the largest value of the sample an $F(z) < 1$, which accounts for the fact that larger values are still possible. It is a commonly used formula for the calculation of empirical cdf's (see Section 3.7.1). Inverting the empirical $F(z)$ values produces the argument, z_F (see Section 2.6.4)

$$z_F = I[F(z)].$$

When calculating z_F, the inversion formula must be changed in mid-column to observe different expressions for $F(z) > 0.5$ and $F(z) < 0.5$. The z_F are plotted against $z_y = [\log(x_i/\mu_x) - \mu_Y]/\sigma_Y$, with $\mu_Y = -0.0135$, and $\sigma_Y = 0.1073$, based on $\log(x_i/\mu_x)$, where $\mu_x = 2580$. A plot of z_F versus z_y is shown in Figure 3-6. The z_F extend from -2.274 to $+2.274$, and the z_y from -2.462 to $+3.213$. The trendline computed as a linear best fit of the data is

$$z_F \approx 0.947 \, z_y,$$

with a very small intercept omitted. This means that the trendline practically passes through the origin of the z_F-z_y graph and approximates the diagonal $z_F = z_y$. Figure 3-6 shows that the data fit by the normal distribution is good, with $R^2 = 0.97$, meaning that 97 % of the scatter about the mean is explained by the regression.

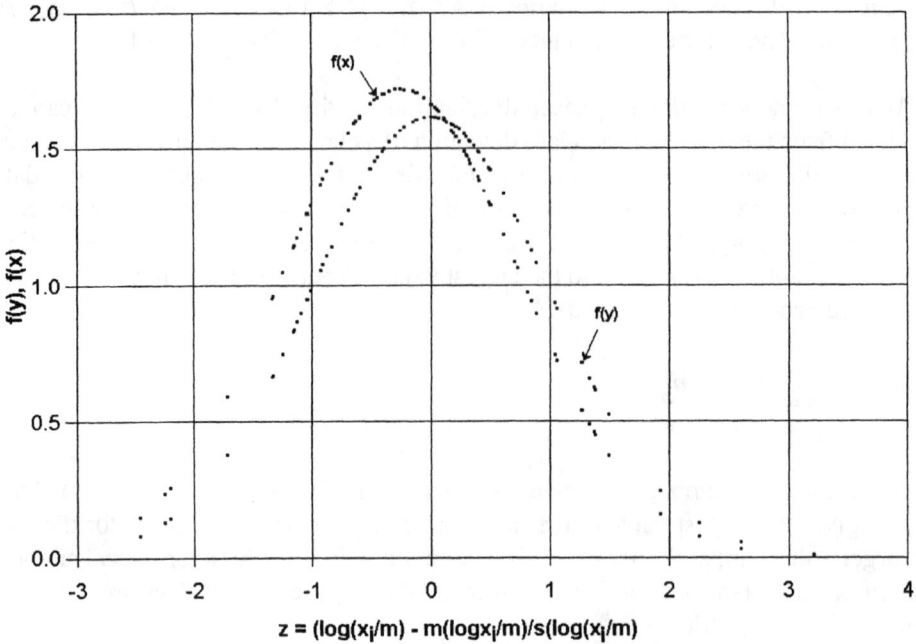

Figure 3-5: Normal pdf, $f(y)$, with $y = \log(x)$, and log-normal transform, $f(x)$. The normal pdf has its maximum at $z = 0$, whereas the transform is shifted somewhat to the left, but is still quite similar to the normal pdf.

The fit is good in the center but poor at the low and high ends. If high end values are of interest it would be wise not to rely on the trendline beyond $z = 2$ for estimating the probability of large x. If a better fit is needed, other distributions, such as the double exponential function or Gumbel function, could be tested (Stedinger et al., 1993, p. 18.16; Yevjevich, 1972, p. 149).

For the acceptable range, $-2.46 \leq z \leq 2.0$, one may proceed with finding the probability of x_i/μ_x by first calculating z_y. Then one enters the z-axis at z_y and reads z_F, or one uses the derived regression equation to compute z_F, whereupon the non-exceedance probability $F(z_F)$ is determined. For example, $x = 4100$ produces $z_y = 2$. The trendline gives $z_F = 1.89$. A table lookup produces $F(z) = 0.971$. Instead of the table lookup, the polynomial approximation $z = f[F(z)]$ can be used (see Section 2.6.4). Using the inversion of $F(z)$ to inspect the data fit does not require *probability paper* and is computationally efficient. One should never just rely on a high value of R^2. The empirical function $z_F = f(z_y)$ should always be examined by inspection.

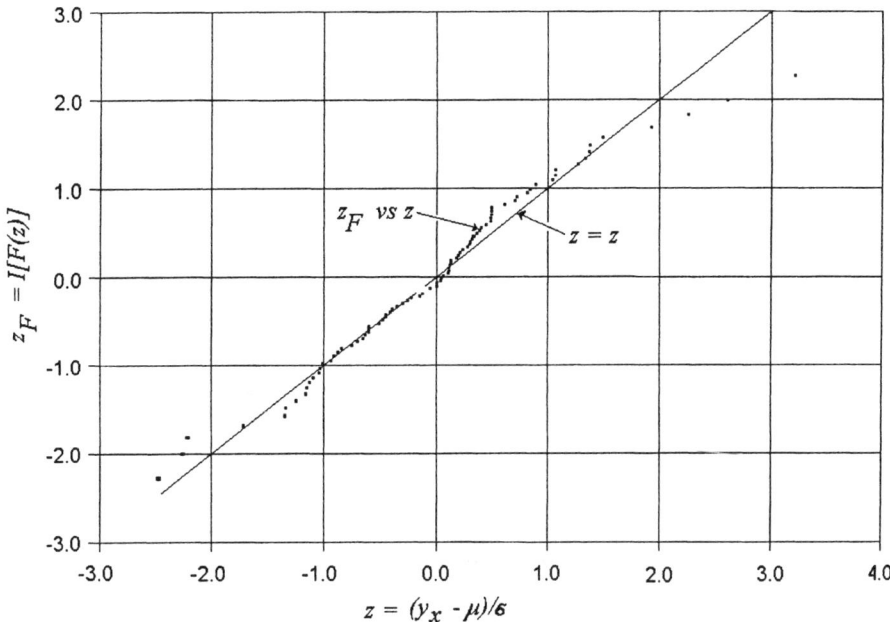

Figure 3-6: The inversion, $z_F = I[F(z)]$, is plotted against $z = (y_x - \mu)/\sigma_Y$. The regression yields $z_F \approx 0.947\,z$. The diagonal $z_F = z$ represents the perfect normal fit. The points z_F represent the departure of the empirical individual values of $F(z)$ from the normal $F(z)$ calculated from the data parameters via $f(z)$. The goodness of fit of the regression is $R^2 = 0.97$.

3.3 Arithmetic with Random Variables

3.3.1 Sum and Difference of Random Variables

Adding or subtracting random variables produces a sum or difference that is again a random variable, i.e., a variable with a range and a pdf. This is a more complicated process than adding or subtracting deterministic variables, which have no range and no pdf. The arithmetical operations with random variables must be carried out, in principle, for all realizations of the random variables. Calculus provides an organized approach to such operations. Very similar approaches are used for calculating the sum and difference of random variables.

(a) The sum of two random variables X and Y is

$$Z^* = X + Y \tag{3.3-1}$$

where Z^* denotes a random variable, not a normalized variable, as in previous sections. Implementing Equation (3.3-1) amounts to adding a realization each of X and Y, which produces a new realization z^*,

$$z^* = x + y \tag{3.3-2}$$

where z^* is the sum of the two random realizations , x and y. Since both x and y have ranges, numerous z^* are produced with a new range. The population of all possible z^* makes up the collective Z^* which has its own probability distribution. The probability of a particular z^* is the joint probabilities of the random variables x and y that produce it:

$$P(Z^* = z^*) = P(X \cap Y) = P(X)\, P(Y), \tag{3.3-3}$$

where the product of probabilities is applicable if x and y are independent. The total probability of one item of the sum, z^*, is the sum of all joint probabilities of possible realizations of the constituents x and y. For continuous variables, the total probability of z^* in finite difference form is

$$f(z^*)\Delta x = \sum [f(x)\Delta x \cdot f(y)\Delta x] \tag{3.3-3a}$$

where Δx is a small increment of the range of the functions $f(x)$ and $f(y)$; and $f(z^*)$ is the ordinate of the pdf of z^*. The sum is taken over all possible combinations of x and y that add to a particular z^*. For dimensional correctness, all products in Equation (3.3-3a), the one on the left and the two under the sum, are probabilities and as such must be dimensionless. If x has a dimension, then z^*, y, and Δx must have the same dimension, and the pdf's $f(x)$, $f(y)$, and $f(z^*)$ must have the reciprocal dimension of x. To avoid dimensional errors, one can make random variables dimensionless, as was done in Section 3.2.2.

The evaluation of a sum of random variables proceeds with choosing a z^* and a random selection of x. Then a combination of x and y that produces z^* is obtained by

$$y = z^* - x. \tag{3.3-4}$$

Substituting Equation (3.3-4) into (3.3-3a), and integrating leads to

$$f(z^*) = \int_{-\infty}^{+\infty} f(x) f(z^* - x)\, dx \tag{3.3-5}$$

where $f(z^*)$ is the pdf of the sum of X and Y if x and y are independent (see also Benjamin and Cornell, 1970, p. 119). The integration range in the general form $(-\infty, +\infty)$ is meant to include all possible selections of x and y, but $f(z^*)$ exists only where the integrand exists (see Example 1). Equation (3.3-5) is known as the *convolution* integral. An explanation of this integral without making use of another advanced concept and of its integration limits is given by Brigham (1974, pp. 50–54).

In practical applications one should be careful about what is being evaluated. If Equation (3.3-3a) is evaluated, the probability of z^* is obtained. If Equation (3.3-5) is evaluated in its finite difference form,

$$f(z^*) = \Delta x \sum [f(x) \cdot f(z^* - x)] \tag{3.3-3b}$$

the ordinate of the pdf $f(z^*)$ is obtained. If $\Delta x = 1$ is chosen, Equations (3.3-3a) and (3.3-3b) give the same numerical value, but one is a probability and the other is a pdf ordinate. This could cause a misinterpretation of the result if $\Delta x \neq 1$.

(b) The difference of two random numbers is

$$z^* = x - y. \tag{3.3-6}$$

For a chosen z^* and a random selection of x the choice of y is

$$y = x - z^*. \tag{3.3-6a}$$

Substituting Equation (3.3-6a) into Equation (3.3-5) gives the pdf of the difference

$$f(z^*) = \int_{-\infty}^{+\infty} f(x) f(x - z^*) dx \ . \tag{3.3-7}$$

The numerical integration of the general form of a bivariate pdf, $f(x,y)$, is obtained by the Euler integration formula (trapezoidal formula)

$$f(z^*) = \Delta x [0.5 f(x,y)_0 + \sum f(x,y)_i + 0.5 f(x,y)_n] \tag{3.3-8}$$

where Δx is the increment at which x is chosen; $f(x,y)_0$ is the first feasible function value of the pdf and $f(x,y)_n$ is the last one; $\sum f(x,y)_i$ is the sum of all function values in between. Strictly speaking, the functions dealt with are pdf's and should carry an identifying index or indices. For independent variables, x and y, the bivariate function simplifies to a product of single variable functions: $f_{XY}(x,y) = f_X(x) f_Y(y)$. The sum is taken over all feasible combinations of x and y that give a specified z^*. To keep the computations simple, Equation (3.3-8) is evaluated at equal increments Δx.

Examples: (1) Add two uniformly distributed random variables, each with the range (3,13). Solution: For each z^*, the feasible range of x is scanned for feasible y. Only the x that produce a feasible y for a selected z^* are included. The increment is $\Delta x = 1$. For example, for $z^* = 8$, $x = 3, 4, 5$, and $y = 5, 4, 3$, respectively, need to be evaluated. The probability densities of the uniform distribution are the same for all x and y: $f_1(3) = 0.1, f_2(5) = 0.1$. Then, for $x = 3$, $f(8) = f_X(3) \cdot f_Y(5) \Delta x = 0.01$. If $x = 4$, $y = 4$, the probability density is again $f(8) = 0.01$, and so on. For $x = 6$, $y = 2$, y is outside its range (3,13), and therefore this selection does not contribute to the total pdf $f(8)$. The numerical integration of $f(z^*)$ for $z^* = 8$ and $z^* = 16$ is demonstrated in Table 3-3(a), column 7. For $z^* = 8$, the pdf ordinate $f(8)$ is obtained by the Euler integration formula

$$f(8) = \Delta x \{0.5 f_X(3) f_Y(5) + f_X(4) f_Y(4) + 0.5 f_X(5) f_Y(3)\}$$

$$= 1.0 [0.5 \cdot 0.1^2 + 0.1^2 + 0.5 \cdot 0.1^2] = 0.02.$$

This result is shown in column 8.

For $z^* = 16$, the entire range of each variable, $3 \le (x, y) \le 13$, contributes to the pdf ordinate $f(16)$. The calculation is also shown in Table 3-3(a). X is tabulated from 3 to 13 and Y from 13 to 3 in columns 1 and 4, respectively. Adding the 11 contributing ordinates in column 7 gives the total pdf ordinate $f(16) = 0.1$ in column 8. The range of Z^* extends from a lower bound that is the minimum sum $z^* = x + y = 3 + 3 = 6$, to an upper bound that is the maximum sum, $z^* = 13 + 13 = 26$. Calculations similar to the two sample calculations must be carried out for all z^* in the range (6,26). Additional details are given in Example (3). The summary of the calculations is given in Table 3-3(c), column 2, which represents the pdf $f(z^*)$. It is shown on the right-hand side of Figure 3-7 and has a triangular distribution in contrast to x and y which have a uniform distribution.

(2) Calculate the difference, $Z^* = X - Y$ with the same range and probability distribution as in Example (1). Solution: The calculation of the pdf for the difference of two random variables follows a similar pattern with $y = x - z^*$. Two evaluations for $z^* = -6$ and $z^* = 8$ are given in Table 3-3(b). Similar calculations must be carried out for all z^* that result from the ranges of X and Y. The resulting difference z^* is found to have the range $-10 \le Z^* \le +10$. The result of all calculations that produce pdf ordinates $f(z^*)$ is given in Table 3-3(c). A graph of $f(z^*)$ is shown on the left-hand side of Figure 3-7.

Table 3-3: Adding and Subtracting Two Uniformly Distributed Random Variables

(a) Adding: $z^* = x + y$. Range $3 \leq x, y \leq 13$. Evaluation for $z^* = 8$ and $z^* = 16$.

x	$f_X(x)$	z^*	$z^* - x$	$f_Y(z^* - x)$	Δx	$f_X(x) f_Y(z^* - x) \Delta x$	$f(z^*)$
1	2	3	4	5	6	7	8
3	0.1	8	5	0.1	1	0.01	0.02
4	0.1	8	4	0.1	1	0.01	
5	0.1	8	3	0.1	1	0.01	
3	0.1	16	13	0.1	1	0.01	0.10
4	0.1	16	12	0.1	1	0.01	
5	0.1	16	11	0.1	1	0.01	
6	0.1	16	10	0.1	1	0.01	
7	0.1	16	9	0.1	1	0.01	
8	0.1	16	8	0.1	1	0.01	
9	0.1	16	7	0.1	1	0.01	
10	0.1	16	6	0.1	1	0.01	
11	0.1	16	5	0.1	1	0.01	
12	0.1	16	4	0.1	1	0.01	
13	0.1	16	3	0.1	1	0.01	

(b) Subtracting: $z^* = x - y$. Range $3 \leq x, y \leq 13$. Evaluation for $z^* = -6$ and $z^* = 8$.

x	$f_X(x)$	z^*	$x - z^*$	$f_Y(x - z^*)$	Δx	$f_X(x) f_Y(x - z^*) \Delta x$	$f(z^*)$
1	2	3	4	5	6	7	8
3	0.1	-6	9	0.1	1	0.01	0.04
4	0.1	-6	10	0.1	1	0.01	
5	0.1	-6	11	0.1	1	0.01	
6	0.1	-6	12	0.1	1	0.01	
7	0.1	-6	13	0.1	1	0.01	
11	0.1	8	3	0.1	1	0.01	0.02
12	0.1	8	4	0.1	1	0.01	
13	0.1	8	5	0.1	1	0.01	

(c) Resulting pdf $f(z^*)$ for Addition and Subtraction

z^*	Addition $f(z^*)$	Subtraction $f(z^*)$
– 10	0	0.00
– 8	0	0.02
– 6	0	0.04
– 4	0	0.06
– 2	0	0.08
0	0	0.10
2	0	0.08
4	0	0.06
6	0.00	0.04
8	0.02	0.02
10	0.04	0.00
12	0.06	0
14	0.08	0
16	0.10	0
18	0.08	0
20	0.06	0
22	0.04	0
24	0.02	0
26	0.00	0
Total	0.50	0.50

Figure 3-7 shows that the pdf's of sum and difference of two uniformly distributed random variables take on a triangular shape. This phenomenon and its implications will be further discussed in connection with the central limit theorem (Section 3.4.2). It is recalled here that the sum z^* which is also the abscissa of Figure 3-7 and has the dimension of x and y, is not the dimensionless normal variable z. The sum of the pdf ordinates in Table 3-3(c) multiplied by $\Delta x = 2$, which is the increment between the pdf ordinates, gives 1, which is the value of the integral of a pdf.

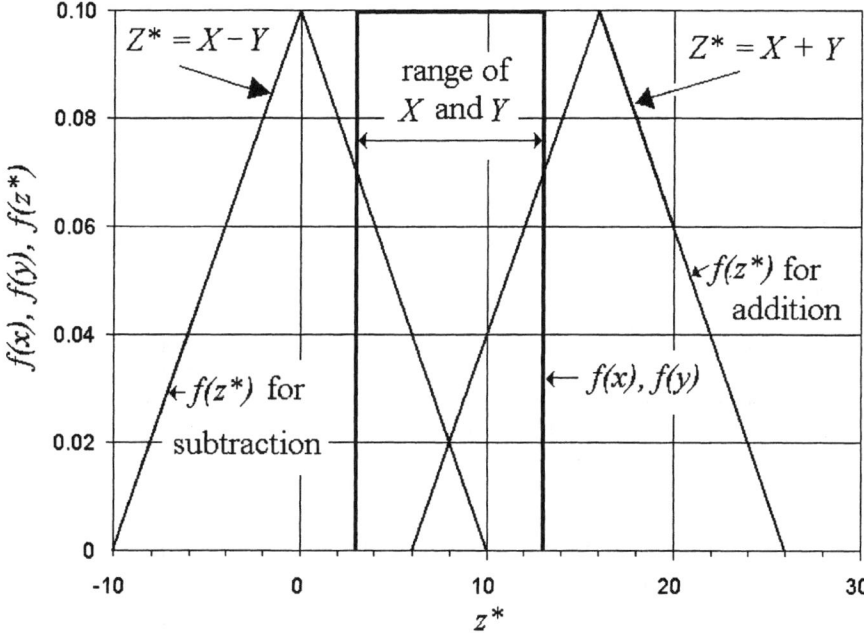

Figure 3-7: Addition, $z^* = x + y$, and subtraction, $z^* = x - y$, of two uniformly distributed random variables, X and Y. The pdf's of the original random variables are represented by the rectangle with range $3 \leq (x, y) \leq 13$. The pdf's of sum and difference are represented by triangular distributions with ranges $6 \leq z^* \leq 26$ and $-10 \leq z^* \leq +10$, respectively.

(3) Examples (1) and (2) only calculate two examples of $f(z^*)$ in Table 3-3(a) and (b) for addition and subtraction. The addition and subtraction of two uniformly distributed random variables each with the range (3,13) requires many more evaluations of $f(z^*)$ ordinates. Z^* exists only if X and Y exist, in other words, the selections must come from both ranges. For example, if $x = 3$, y cannot be 0, 1, or 2, because these are values outside the range $3 \leq y \leq 13$. The first possible value in the x-range as well as in the y-range is 3, so that there is only one possible combination for $z^* = 6$, which is $x = 3$ and $y = 3$. As x is evaluated over its range, $3 \leq x \leq 13$, y can vary only over its range, $3 \leq y \leq 13$. For $z^* = 26$, the only possible combination is $x = 13$ and $y = 13$. For z^* in the range $6 < z^* < 26$, there is a varying number of possible (x, y) combinations that produce a specific z^*. The number of these combinations also determines the frequency of z^*. Examples of such combinations for three ranges of z^*, $6 \leq z^* \leq 10$, $14 \leq z^* \leq 18$, and $22 \leq z^* \leq 26$ follow.

Tabular Evaluation of Example (3)

z^*	x	z^*	x	z^*	x
	y		y		y
6	3	14	3, 4, 5, 6, 7, 8, 9, 10, 11	22	9, 10, 11, 12, 13
	3		11, 10, 9, 8, 7, 6, 5, 4, 3		13, 12, 11, 10, 9
7	3, 4	15	3, 4, 5, 6, 7, 8, 9, 10, 11, 12	23	10, 11, 12, 13
	4, 3		12, 11, 10, 9, 8, 7, 6, 5, 4, 3		13, 12, 11, 10
8	3, 4, 5	16	3, 4, 5, 6, 7, 8, 9, 10, 11, 12, 13	24	11, 12, 13
	5, 4, 3		13, 12, 11, 10, 9, 8, 7, 6, 5, 4, 3		13, 12, 11
9	3, 4, 5, 6	17	4, 5, 6, 7, 8, 9, 10, 11, 12, 13	25	12, 13
	6, 5, 4, 3		13, 12, 11, 10, 9, 8, 7, 6, 5, 4		13, 12
10	3, 4, 5, 6, 7	18	5, 6, 7, 8, 9, 10, 11, 12, 13	26	13
	7, 6, 5, 4, 3		13, 12, 11, 10, 9, 8, 7, 6, 5		13

The table shows that between $z^* = 6$ and $z^* = 26$ for which there is only one combination each: $x = y = 3$ and $x = y = 13$, respectively, the number of combinations increases to a maximum of 11 for $z^* = 16$. This also is the most likely value of Z^*, represented by the peak of the triangle in Figure 3-7. The integration by Equation (3.3-8) results in $f(16) = 0.1$, as shown in Table 3-3(a), column 8.

(4) The pdf of the sum is a continuous function, because values between the integers used in Table 3-3(a) also can be combined to produce a specific z^*. If the increment Δx is made very small, many more combinations must be evaluated. As an additional example, one may want to carry out an evaluation of $z^* = 8$ using $\Delta x = 0.1$ in the format of Table 3-3(a). For the evaluation of $z^* = 8$, x would assume values from 3 to 5 by increments $\Delta x = 0.1$, whereas $z^* - x$ would assume values from 5 to 3 by decrements of 0.1. This would yield a total of 21 rows, each producing a $\Delta x\, f(x)\, f(y)$

= 0.001. Summing over the 21 values using Equation (3.3-8) gives $f(z^*) = 0.5 \cdot 0.001$ + $19 \cdot 0.001 + 0.5 \cdot 0.001 = 0.02$, the same $f(z^*)$ that was obtained for $\Delta x = 1$ in Table 3-3(a).

(5) Calculate (a) the mean and (b) the variance of the sum and difference , Z^*, for Examples (1) and (2). <u>Solution</u>: The mean and variance of Z^* can be calculated from the individual data, as they are listed in a spreadsheet. But they also can be calculated by the formulas derived in Section 2.4.1, where it was shown that the expectation of a sum is the sum of the expectations, and the expectation of a difference is the difference of the expectations.

(a) The expectation of a sum, with $E(X) = E(Y) = 8$, is

$$E(Z^*) = E(X) + E(Y) = 16.$$

The expectation of the difference, with $E(X) = E(Y) = 8$, is

$$E(Z^*) = E(X) - E(Y) = 0.$$

Figure 3-7 shows the calculated triangular pdf's of Z^* for the sum and difference. They are centered on their means, $E(Z^*) = 16$ and 0, respectively.

(b) The variance of a difference and of a sum of two independent random variables was shown in Section 2.4.3 to be equal to the *sum* of the variances. Hence, one can write for the variance of a sum or difference,

$$Var(Z^*) = Var(X) + Var(Y).$$

According to Section 2.7.3, the variance of the individual variables which have uniform distribution with a range (3,13) is

$$\sigma^2 = \frac{(b-a)^2}{12} = \frac{(13-3)^2}{12} = 8.33 \ .$$

Hence, the variance of Z^* is the sum of the variances of X and Y, or $Var(Z^*) = 8.33$ + $8.33 = 16.67$, which is twice the variance of the original variables. Examples (1) and (2) showed that the sum and the difference of two uniformly distributed random variables have a triangular distribution whose variance is (Section 2.7.4)

$$\sigma^2 = \frac{(b-a)^2}{24}$$

where (a,b) is the range of the triangular distribution. The range of both sum and variance in the examples is 20 so that $\sigma^2 = 20^2/24 = 16.67$. This shows that the sum or difference of random variables have a greater spread or uncertainty associated with them than the original random variables. This problem usually is completely ignored when deterministic values are used without any allowance for ranges.

3.3.2 Sum and Difference of Normal Random Variables

Random variables can be relatively easily handled analytically if they are described by simple analytical functions. This is demonstrated by calculating the sum and difference of two normally distributed random variables. The sum of the two independent random variables is as in the previous section

$$Z^* = X + Y. \tag{3.3-9}$$

If X and Y are normally distributed random variables, their pdf's are analytically described by

$$f_X(x) = \frac{1}{\sqrt{2\pi}\,\sigma_X} e^{-0.5(\frac{x-\mu_X}{\sigma_X})^2} \tag{3.3-10a}$$

and

$$f_Y(y) = \frac{1}{\sqrt{2\pi}\sigma_Y} e^{-0.5(\frac{y-\mu_Y}{\sigma_Y})^2}. \tag{3.3-10b}$$

Since the mean of a sum of random variables is the sum of their means and the variance of a sum of random variables is the sum of their variances, knowing these two parameters one can immediately write for the pdf of the *sum*

$$f(z^*) = \frac{1}{\sqrt{2\pi(\sigma_X^2 + \sigma_Y^2)}} e^{-0.5\frac{(z^*-\mu_X-\mu_Y)^2}{\sigma_X^2+\sigma_Y^2}}. \tag{3.3-11}$$

where z^* denotes a realization of the sum of two random variables, not a normalized variable. By temporarily using the symbol ϕ for the normalized variable of the sum, one gets

$$\phi = \frac{z^* - (\mu_X + \mu_Y)}{\sqrt{\sigma_X^2 + \sigma_Y^2}}. \tag{3.3-12}$$

The pdf for the *difference* of two independent normally distributed random variables is

$$f(z^*) = \frac{1}{\sqrt{2\pi(\sigma_X^2 + \sigma_Y^2)}} e^{-0.5\phi'^2} \tag{3.3-13}$$

with

$$\phi' = \frac{z^* - (\mu_X - \mu_Y)}{\sqrt{\sigma_X^2 + \sigma_Y^2}} \tag{3.3-14}$$

where ϕ' is the normalized variable of the difference of two normally distributed independent random variables. If X and Y are correlated, then the correlation coefficient, ρ_{XY}, is greater than zero and ϕ' becomes

$$\phi' = \frac{z^* - (\mu_X - \mu_Y)}{\sqrt{\sigma_X^2 + 2\rho_{XY}\sigma_X\sigma_Y + \sigma_Y^2}} \ . \tag{3.3-14a}$$

The expressions for $f(z^*)$ for sum and difference, Equations (3.3-11) and (3.3-13), respectively, show that the sum and difference of two normally distributed random variables are again normally distributed random variables. This result holds true for adding and subtracting any number of normally distributed variables, not just two. In the notation used for normally distributed variables, one can write for the sum of two independent random variables ($\rho_{XY} = 0$):

$$N[\mu_X, \sigma_X^2] + N[\mu_Y, \sigma_Y^2] = N[\mu_X + \mu_Y, \sqrt{\sigma_X^2 + \sigma_Y^2}] \ , \tag{3.3-15}$$

and for the difference of two independent random variables:

$$N[\mu_X, \sigma_X^2] - N[\mu_Y, \sigma_Y^2] = N[\mu_X - \mu_Y, \sqrt{\sigma_X^2 + \sigma_Y^2}] \ . \tag{3.3-16}$$

The resulting distributions have increased variances as was discussed in Section 3.3.1.

Examples: (1) Calculate the addition of two random variables that are exponentially distributed (after Shooman, 1968, p.78). Solution: Suppose the pdf's of the variables X and Y are

$$f(x) = \lambda e^{-\lambda x}$$

with a range $0 \le x \le +\infty$, and

$$f(y) = \lambda e^{-\lambda y}$$

with a range $0 \le y \le +\infty$. Then the pdf ordinate for the sum of the two functions, $f(z^*)$ is, with the integration limits at first in general form, then adjusted to the range of positive x and positive z^*:

$$f(z^*) = \int_{-\infty}^{+\infty} f(x) f(z^* - x) dx = \lambda^2 \int_0^{z^*} e^{-\lambda x} e^{-\lambda(z^*-x)} dx$$

Since $e^{-\lambda z^*}$ is a constant factor for an assumed z^*, the integral becomes

$$f(z^*) = \lambda^2 e^{-\lambda z^*} \int_0^{z^*} dx = \lambda^2 z^* e^{-\lambda z^*} .$$

(2) Calculate the sum and the difference of two normally distributed random variables with the parameters $\mu_X = \mu_Y = 8$, and $\sigma_X = \sigma_Y = 1.67$. <u>Solution</u>: The parameters for the sum are $\mu_{Z^*} = \mu_X + \mu_Y = 16$, $\sigma_{Z^*} = (\sigma_X^2 + \sigma_Y^2)^{1/2} = 2.36$. The parameters for the difference are $\mu_{Z^*} = \mu_X - \mu_Y = 0$, and $\sigma_{Z^*} = 2.36$. The normal pdf's for sum and difference are obtained by using μ_{Z^*} and σ_{Z^*}, as calculated for the sum or difference in Equations (3.3-11) and (3.3-13). The result of the addition and subtraction of two normal random variables is given in Table 3-4 and displayed in Figure 3-8. The original variables, X and Y (column 4 of Table 3-4), are distributed around $\mu_X = \mu_Y = 8$. The sum is distributed around $\mu_{Z^*} = 16$, and the difference is distributed around $\mu_{Z^*} = 0$. The standard deviation of both the sum and the difference is 2.36, an increase of 41 % over 1.67 for X and Y. The increased dispersion of $f(z^*)$ compared with $f(x)$ and $f(y)$ is clearly recognizable by the increased width of the sum and difference distributions in Figure 3-8. The two identical normal random variables, X and Y, are represented by the pdf in the center. To the right of it is the pdf of the sum, and to the left is the pdf of the difference; both are again normal pdf's.

Table 3-4: Adding and Subtracting Two Independent, Normally Distributed Random Variables. Assumed Range is 6 σ = 10; $\sigma_X = \sigma_Y = 1.667$; $\mu_X = \mu_Y = 8$; $\rho_{XY} = 0$.

x, y	$f(x), f(y)$	Sum $f(z^*)$	Difference $f(z^*)$	x, y	$f(x), f(y)$	Sum $f(z^*)$	Difference $f(z^*)$
-10	0.0000	0.0000	0.0000	9	0.0021	0.0001	0.1999
-9	0.0000	0.0001	0.0000	10	0.0066	0.0000	0.1165
-8	0.0000	0.0005	0.0000	11	0.0179	0.0000	0.0474
-7	0.0000	0.0021	0.0000	12	0.0401	0.0000	0.0134
-6	0.0000	0.0066	0.0000	13	0.0753	0.0000	0.0027
-5	0.0000	0.0179	0.0000	14	0.1181	0.0000	0.0004
-4	0.0000	0.0401	0.0000	15	0.1547	0.0000	0.0000
-3	0.0000	0.0753	0.0000	16	0.1692	0.0000	0.0000
-2	0.0000	0.1181	0.0000	17	0.1547	0.0000	0.0000
-1	0.0000	0.1547	0.0000	18	0.1181	0.0000	0.0000
0	0.0000	0.1692	0.0000	19	0.0753	0.0000	0.0000
1	0.0000	0.1547	0.0000	20	0.0401	0.0000	0.0000
2	0.0000	0.1181	0.0004	21	0.0179	0.0000	0.0000
3	0.0000	0.0753	0.0027	22	0.0066	0.0000	0.0000
4	0.0000	0.0401	0.0134	23	0.0021	0.0000	0.0000
5	0.0000	0.0179	0.0474	24	0.0005	0.0000	0.0000
6	0.0000	0.0066	0.1165	25	0.0001	0.0000	0.0000
7	0.0001	0.0021	0.1999	26	0.0000	0.0000	0.0000
8	0.0005	0.0005	0.2393				

Notes: column 1: variables x, y, and z^*, from -10 to 26.

column 2: pdf $f(x)$ is calculated by $f(x) = (0.398942/\sigma_X) \exp\{-0.5[(x-\mu_X)/\sigma_X]^2\}$; the same formula is used for $f(y)$, with $\mu_X = \mu_Y = 8$, and $\sigma_X = \sigma_Y = 1.667$.

column 3: the random variable of the sum is $Z^* = X + Y$; the pdf of the sum is $f(z^*) = (0.398942/\sigma_Z^*) \exp\{-0.5[(z^*-\mu_Z^*)/\sigma_Z^*]^2\}$, where $\mu_Z^* = \mu_X + \mu_Y = 16$; $\sigma_Z^* = (\sigma_X^2 + \sigma_Y^2)^{1/2} = (2 \cdot 1.667^2)^{1/2} = 2.3575$; z^* is from column 1.

column 4: the random variable of the difference is $Z^* = X - Y$; the pdf of the difference is $f(z^*) = (0.398942/\sigma_Z^*) \exp\{-0.5[(z^*-\mu_Z^*)/\sigma_Z^*]^2\}$, where $\mu_Z^* = \mu_X - \mu_Y = 0$, and $\sigma_{Z^*} = 2.3575$; z^* is from column 1.

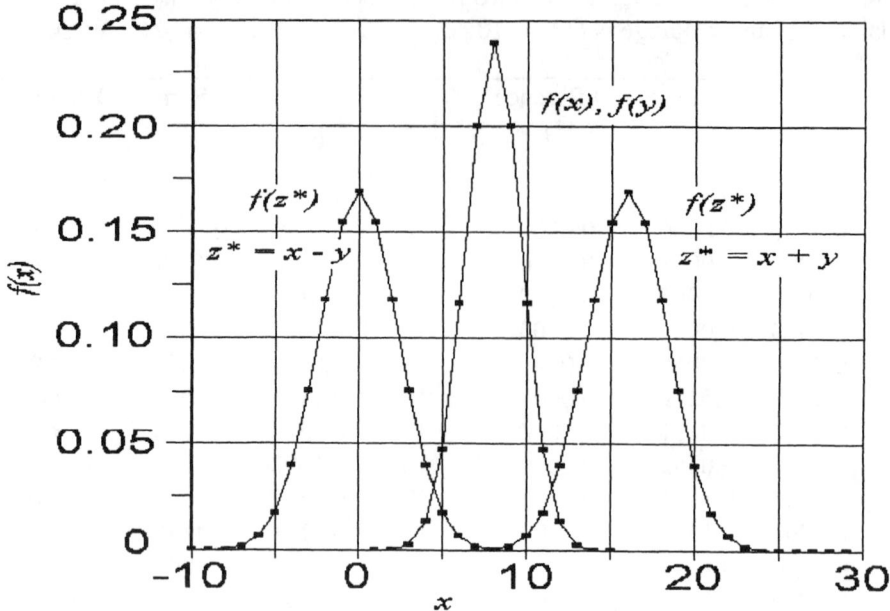

Figure 3-8: Probability density functions of two identical, normal random variables, X and Y, both with $\mu = 8$, $\sigma = 1.667$ and their sum and difference. The abscissa accommodates the variables x, y, and z^*, which have the same dimension. The ordinate accommodates $f(x)$, $f(y)$, and $f(z^*)$, which have the same dimension (reciprocal of the x-dimension). The probability density functions of the sum (*right*) and the difference (*left*), are centered on their respective means, $\mu = 16$ and $\mu = 0$, with standard deviations $\sigma = 2.36$. The variable z^* represents the sum or difference of x and y, not the normalized variable.

3.3.3 Product of Random Variables

The pdf of the product of two independent random variables, $Z^* = X \cdot Y$, is approached in a similar way as for additions and subtractions. The substitution $x = z^*/y$ is used to tie an x to a y in order to evaluate a particular z^*. In this way, the total probability of a particular z^* is determined, and so forth, for all z^* (Bronstein and Semendjajew, 1984, p. 674; Benjamin and Cornell, 1970, p. 208). Another approach is to use the logarithmic transform and add the logarithms of random variables. Here only some calculations with expectations are given.

(a) The *expectation of a product* of two random variables, $Z^* = X \cdot Y$, was obtained in connection with calculating the variance of a difference in Section 2.4.3:

$$E(X \cdot Y) = E(X)E(Y) + Cov(X,Y) \qquad (3.3\text{-}17)$$

where $E(X \cdot Y)$ is the expectation of the product of the random variables X and Y, and $Cov(X,Y)$ is the covariance of the X and Y. If X and Y are independent, then $Cov(X,Y) = 0$, and the expectation of the product is equal to the product of expectations,

$$E(X \cdot Y) = E(X)E(Y) . \qquad (3.3\text{-}18)$$

For independent random variables, Equation (3.3-18) can be expanded to the expectation of the product of multiple random variables:

$$E(X_1 \cdot X_2 \cdot \ldots \cdot X_n) = E(X_1) \cdot E(X_2) \cdot \ldots \cdot E(X_n) . \qquad (3.3\text{-}19)$$

(b) The *variance of a product* is derived by expanding the formula for the variance, Equation (2.3-9) of Section 2.3.2:

$$Var(Z) = E(Z^2) - [E(Z)]^2 . \qquad (3.3\text{-}20)$$

Substituting for Z the product $Z^* = X \cdot Y$ leads to

$$Var(X \cdot Y) = E[(X \cdot Y)^2] - [E(X \cdot Y)]^2 . \qquad (3.3\text{-}21)$$

The products on the right-hand side can be further processed by using Equation (3.3-17):

$$E(X^2 \cdot Y^2) = E(X^2)E(Y^2) + Cov(X^2,Y^2)$$

and

$$[E(X \cdot Y)]^2 = [E(X)E(Y) + Cov(X,Y)]^2 .$$

For independent variables X and Y, the covariances are zero and Equation (3.3-21) becomes

$$Var(X \cdot Y) = E(X^2)E(Y^2) - [E(X)]^2[E(Y)]^2 . \qquad (3.3\text{-}22)$$

If the variances and the expectations of X and Y are known, the variance of the product can be obtained by substituting

$$E(X^2) = Var(X) + [E(X)]^2$$

and

$$E(Y^2) = Var(Y) + [E(Y)]^2$$

into Equation (3.3-22) and multiplying out, which gives

$$Var(X \cdot Y) = Var(X)Var(Y) + Var(X)[E(Y)]^2 + Var(Y)[E(X)]^2 . \qquad (3.3-23)$$

For checking purposes, note that all terms of the sum of products must have the same units. For example, if X and Y are in meters, then all terms in Equation (3.3-23) must be in m^4.

Of interest may also be the coefficient of variation of the product of two independent random variables. It is obtained from Equation (2.3-11) of Section 2.3.3 as the ratio of the standard deviation and the mean,

$$C_v = \frac{\sqrt{Var(X \cdot Y)}}{E(X \cdot Y)} \qquad (3.3-24)$$

with C_v being a dimensionless number.

Example: For two random variables with mean $E(X) = E(Y) = 8$, and $Var(X) = Var(Y) = 2.78$, calculate (a) the variance of the product; (b) calculate the coefficient of variation of the product and compare the results with those of X and Y. Solution: (a) The variance of the product is by Equation (3.3-23)

$$Var(X \cdot Y) = 2.78^2 + 2.78 \cdot 64 + 2.78 \cdot 64 = 363.6.$$

Hence the variance of the product is 130 times the variance of the original variables.

(b) The coefficient of variation of the product calculated by Equation (3.3-24) is $C_v = \sqrt{363.6} / 64 = 0.298$, whereas the coefficient of variation of the original variables is $C_v = \sqrt{2.78} / 8 = 0.208$. This amounts to a 43 % increase of the product C_v over the original variable C_v.

3.3.4 Quotient of Random Variables

The quotient of two random variables, $Z^* = X/Y$, occurs frequently. Examples are the benefit-cost ratio of random benefits and costs, and the safety factor of random resistance and load. The pdf of the quotient can be found by direct integration of the quotient pdf (Bronstein and Semendjajew, 1984, p. 674; Benjamin and Cornell, 1970, p. 115). Another approach is by logarithmic transformation of X and Y and calculation of the difference of the transformed variables (Section 3.3.1). A numerical approach uses repeated sampling of the random variables X and Y, and calculates an empirical pdf of X/Y. This procedure is known as *Monte Carlo method*. Instead of pursuing these somewhat involved methods only some calculations with expectations are discussed.

(a) The *expectation of a ratio* can be found as the expectation of a product. Suppose $B_C = B/C$ is the benefit-cost ratio, where B is the benefit, and C is the cost, both being independent random variables. The ratio can be written as a product, $B_C = B \cdot (1/C)$. Its expectation is obtained as

$$E(B_C) = E(B/C) = E(B) \cdot E(\frac{1}{C}) \ . \tag{3.3-25}$$

(b) The *variance of a ratio* is found as the variance of the product from Equation (3.3-23) as

$$Var(B_C) = Var(B)Var(\frac{1}{C}) + Var(B)[E(\frac{1}{C})]^2 + Var(\frac{1}{C})[E(B)]^2 \ . \tag{3.3-26}$$

<u>Example</u>: A benefit-cost analysis has produced $E(B) = 16$, $Var(B) = 4$, $E(1/C) = 0.1$, and $Var(1/C) = 0.01$. What is the expected benefit-cost ratio, its variance and standard deviation? <u>Solution</u>: The expected benefit-cost ratio is according to Equation (3.3-25)

$$E(B_C) = E(B) \cdot E(\frac{1}{C}) = 16 \cdot 0.1 = 1.6 \ .$$

The variance of the benefit-cost ratio is according to Equation (3.3-26):

$$Var(B_C) = 4 \cdot 0.01 + 4 \cdot 0.1^2 + 0.01 \cdot 16^2 = 2.64.$$

The standard deviation of B_C is $\sigma_{BC} = [Var(B_C)]^{1/2} = 1.62$.

3.4 Sampling Distributions

3.4.1 Distribution of the Sample Mean

It is important to understand the difference between population parameters and sample parameters. Samples are used to statistically represent populations. If the sample size includes the entire population, then sample and population parameters are identical. The sample size usually is relatively small compared with the population size. The question then arises how well does a randomly drawn sample or a collective of random samples represent a population? A normal distribution is sufficiently described by two parameters, the mean and the standard deviation. The representativeness of the sample is therefore expressed by how closely its parameters approach the population parameters.

When random samples are taken from a normal distribution that has the population parameters μ and σ, then the means of these samples, $\overline{X}$, which are sums of random variables and therefore themselves random variables, are again normally distributed with mean μ and standard deviation of the sample mean $\sigma_{\overline{X}} = \sigma / \sqrt{n}$, where n is the sample size (see Section 2.4.4). Given the mean and variance of the population, the standard normal variable of the sample mean, $\overline{X}$, is

$$z_i = \frac{\overline{X}_i - E(\overline{X})}{\sigma_{\overline{X}}}$$
(3.4-1)

and with $E(\overline{X}) = \mu$,

$$z_i = \frac{\overline{X}_i - \mu}{\sigma / \sqrt{n}} = \sqrt{n}\, \frac{\overline{X}_i - \mu}{\sigma} .$$
(3.4-1a)

Equation (3.4-1) expresses the normalized variable of the sample mean, z_i in terms of sample parameters, and Equation (3.4-1a) expresses z_i in terms of population parameters. Both equations show in somewhat different ways that with increasing sample size, n, the z_i for the sample mean $\overline{X}_i$ are $\sqrt{n}$ times larger than for a regular population element x_i, given both are of the same numerical size. Since the normal pdf, $f(z)$, decreases rapidly with increasing z, the pdf for the sample mean declines much more rapidly than the $f(z)$ of the population distribution. This makes the pdf of the sample mean very spiky around the population mean as n gets large (Example (1)).

The knowledge that the distribution of the sample mean is normal can be used to sharpen the accuracy by which the population mean can be estimated from the sample mean. The remainder of this section is a retake of confidence intervals of Section 2.4.5 with emphasis on probability estimates for the sample mean and its representativeness of the population mean.

The probability of the sample mean being within a desirable or tolerable distance from the population mean in terms of a fraction or a multiple of $\sigma_{\bar{X}}$ is

$$P[-z\sigma_{\bar{X}} \le \bar{X} - \mu \le +z\sigma_{\bar{X}}] = F(z) - F(-z) \qquad (3.4\text{-}2)$$

where $F(z) - F(-z)$ represents the probability that the difference between $\bar{X}$ and μ does not exceed $\pm z\sigma_{\bar{X}}$. The probability, which is assumed normal in all cases here, can be evaluated by $F(z) - F(-z) = 2 F_0(z)$, where $F_0(z)$ is the integral of the normal pdf for positive z that is usually given in table lookups.

The probability of $\bar{X}$ being within a specified band around the population mean is obtained from Equation (3.4-2) by adding μ to all three terms in the brackets:

$$P[\mu - z\sigma_{\bar{X}} \le \bar{X} \le \mu + z\sigma_{\bar{X}}] = F(z) - F(-z) \quad . \qquad (3.4\text{-}3)$$

Finally, the subtraction of $\bar{X}$ from all terms in the brackets of Equation (3.4-3), multiplying through by -1, which reverses the inequalities, and adding μ to all terms produces

$$P[\bar{X} - z\sigma_{\bar{X}} \le \mu \le \bar{X} + z\sigma_{\bar{X}}] = F(z) - F(-z) \quad . \qquad (3.4\text{-}4)$$

Equation (3.4-4) is the *interval estimate for the population mean* (Lapin, 1978, p. 243). Each $\bar{X}$ calculated from a sample of n x_i's, is a random variable and, therefore, always has a different value so that the interval has a random size and location, whereas μ is in a fixed location. Equations (3.4-2) to (3.4-4), depending on the parameters known, $(\bar{X}, \sigma_{\bar{X}})$ or (μ, σ), allow an unknown quantity to be bracketed by known quantities.

Example: (1) The difference between the distribution of the sample mean and the population elements can be demonstrated analytically. Suppose the normalized variable of the pdf of $\bar{X}$ is z' and the normalized variable of a population element x is z. Then, according to Equation (3.4-1a), for the same numerical values of $\bar{X}$ and x, $z' = \sqrt{n} z$, $f(z') = 0.3989 \cdot \exp[-n z^2/2]$ and $f(z) = 0.3989 \cdot \exp[-z^2/2]$. The ratio of the two functions is $f(z') / f(z) = \exp[-(n-1) z^2/2]$, where n is the sample size. Suppose $f(z)$ is evaluated at $z = 1$, then $f(1) = 0.242$. For a modest sample size of $n = 5$, $f(z') / f(z) = \exp[-4/2] = 0.135$. This means that $f(z')$ is about 13 % of $f(z)$ for the assumed numerical value x. This shows that a distribution of the sample mean of size n is very spiky around the population mean even if the sample means are based on a relatively small n. The usually spiky shape of the distribution of the sample mean is, of course, also expressed by the use of $\sigma/\sqrt{n}$ instead of σ.

(2) For a population with known parameters μ and σ, determine the probability that $\overline{X}$ lies within a band of ± 1 σ from the population mean, μ, given the sample size is $n = 10$. <u>Solution</u>: The band size is $\mu \pm 1 \ \sigma = \mu \pm \sqrt{10} \ \sigma_{\overline{x}} = \mu \pm 3.16 \ \sigma_{\overline{x}}$. According to Equation (3.4-3),

$$P[\mu - 3.16\sigma_{\overline{x}} \leq \overline{X} \leq \mu + 3.16\sigma_{\overline{x}}] = F(3.16) - F(-3.16) = 2F_0(3.16) \quad .$$

A table lookup provides $F_0(3.16) = 0.49921$, so that $2 F_0(3.16) = 0.998$. The result shows that the sample distribution is almost 100 % within $\mu \pm 1$ σ, whereas only 68 % of the elements of the normal distribution are within $\mu \pm 1$ σ.

3.4.2 The Central Limit Theorem

In the previous section the point was made that sample means taken from a normal distribution are again normally distributed. When a population has mean μ and a standard deviation σ, and its pdf is *not* normal, the sample means, $\overline{X}$, taken from it are still approximately normally distributed with mean μ and standard deviation $\sigma_{\overline{x}} = \sigma / \sqrt{n}$ if some conditions are met. The *central limit theorem* (CLT) states that *sample means taken from <u>any</u> distribution approach the normal distribution as the sample size on which they are based goes to infinity*. Based on formulations by Parzen (1965, p. 19), Hoel (1971, p. 125), and Bronstein and Semendjajew (1984, p. 677), the CLT can be explained as follows. Assume a sample of random variables of size n, $X_1, X_2, \dots , X_n$ taken from a population with mean μ and standard deviation σ. Using the results of Section 2.4.4, the sample parameters are

$$n \overline{X} = X_1 + \dots + X_n = \sum_{i=1}^{n} X_i = Y_n \tag{3.4-5a}$$

where Y_n is the sum of random realizations. Given all X_i are from the same distribution, taking expectations gives

$$n E(\overline{X}) = E(X_1) + \dots + E(X_n) = n \mu \tag{3.4-5b}$$

and

$$n\sigma_{\overline{x}} = \sqrt{Var(X_1)+...+Var(X_n)} = \sqrt{n\sigma^2} \ . \tag{3.4-5c}$$

Each sample mean is again a random variable which can be expressed as a normalized variable by

$$Z = \frac{\overline{X} - \mu}{\sigma / \sqrt{n}} \tag{3.4-6}$$

which is the standard normalized variable of the sample mean; index n indicates the number of random realizations that are used to calculate the sample mean by Equation (3.4-5a).

A formulation based on sample parameters uses the centered sum of random variables (Bronstein and Semendjajew, 1984, p. 677)

$$Z = \frac{X_1 - E(X_1)+...+ X_n - E(X_n)}{\sqrt{Var(X_1) +...+ Var(X_n)}} \tag{3.4-7a}$$

or

$$Z = \frac{\sum[X_i - E(X_i)]}{\sqrt{\sum Var(X_i)}} \tag{3.4-7b}$$

where n denotes the sample size and the sums are taken over the n items of the sample. If all X_i are from the same population then by using Equations (3.4-5a) through (3.4.-5c), Equation (3.4-7b) becomes

$$Z = \frac{Y_n - nE(\overline{X})}{n\sigma_{\overline{x}}} \tag{3.4-8a}$$

or

$$Z = \frac{Y_n - n\mu}{\sqrt{n}\,\sigma} \tag{3.4-8b}$$

where n indicates the number of original random selections X_i, $i = 1, ... , n$, used to calculate Z. Each Z is obtained from a sample of n random selections. Dividing the numerator and denominator of Equation (3.4-8a) by n and substituting $Y_n/n = \overline{X}$, gives

$$Z = \frac{\overline{X} - E(\overline{X})}{\sigma_{\overline{X}}}$$ (3.4-8c)

which is the normalized variable of the sample mean.

All the expressions for Z derived here are equivalent. Each is based on a sample of n random variables X_i. Once the sample size n is chosen, each sample of n independent random variables (sampling with replacement), $X_1, \dots, X_n$, creates one normalized variable Z for the new variable, $\overline{X}$ or Y_n. It is the sample size n that is used for the calculation of $\overline{X}$ or Y_n that causes Z to become normally distributed, which is guaranteed only for $n \rightarrow \infty$. The number of samples or of Z-values, however, creates a data set that establishes the empirical pdf, $f(z)$, which can be expected to resemble the normal distribution if n is sufficiently large. This function, $f(z)$, based on sampling is called the *sampling distribution* in contrast to the *population distribution* from which the random variables X_i are drawn.

The CLT makes the following probability statement on Z: Given individual items that make up Z, such as a term of the sum of Equation (3.4-7b),

$$\frac{X_i - E(X_i)}{\sqrt{Var(X_1) + \dots + Var(X_n)}},$$ (3.4-9)

are homogeneous in size and small, the cdf of Z approaches the normal cdf as n approaches infinity. This is expressed by (Bronstein and Semendjajew (1984, p. 677)

$$\lim_{n \to \infty} P(Z \le u) = F(u) = 0.39894 \int_{-\infty}^{u} e^{-0.5z^2} dz$$ (3.4-10)

where $F(u)$ is the non-exceedance probability of the normal distribution, a function of its upper integration limit u.; n determines Z and its realizations z; therefore Z is sometimes denoted Z_n. Equation (3.4-10) holds regardless of the shape of the distribution from which the elements x_i in Z have been sampled.

The practical importance of Equation (3.4-10) is that the probability of any Z can immediately be stated. Once a population has been sampled and an $\overline{X}$ or Y_n have been calculated, its probability immediately can be found by a lookup of the normal cdf, $F(z)$. Furthermore, the probability distribution of the sample mean approaches the normal distribution regardless of the distribution of the random selection X_i. This property of being a limiting distribution gives the normal distribution its dominant importance among all distributions.

Equation (3.4-10) strictly holds only in the limit $n \to \infty$, and for means or sums that are made up of similarly small individual items. But the CLT becomes applicable long before the strict conditions are met. This holds especially for sample sizes which can be much less than infinity. Thirty items is considered a reasonable number of contributions to satisfy the requirement of large n. According to Hoel (1971, p. 126), $n > 50$ produces about the same normal shape regardless of population distribution.

The population distribution from which the random variables are sampled can be exponential, uniform, triangular, binomial, skewed, or even multimodal (with two or more frequency peaks). The CLT holds the better the larger the sample size n, and convergence toward the normal distribution is the faster the more the sampled distributions resembles the normal distribution. This holds for symmetric distributions, such as the symmetric triangular distribution and the uniform distributions. If the population distribution is normal to start with, then the sampling distribution is exactly normal, not just in the limit, but for any number of random variables in the sample.

An example of the distribution of the sum of two random variables that are drawn from the same uniform distribution and produce a triangular distribution is discussed in Section 3.3.1 and illustrated in Figure 3-7. If four or more random variables had been added in that example, a distribution resembling a bell shape would have been produced.

A demonstration of the CLT for different underlying distributions is given in Figure 3-9 (data after Myers and Wunderlich, 1983, pp. 18–19). The left column shows four distributions of quarterly flows which have different ranges and shapes because of the different (meteorologic) processes that generate them. Adding them produces the sums of these distributions in the right column. The uppermost distribution is the 1. Quarter by itself, the same as in the left column. The second, third, and fourth distribution in the right column are sums of the random combinations of two, three and four quarterly flows, respectively. The last diagram on the lower right, which represents sums of four elements drawn from different distributions, shows a central tendency that is predicted by the CLT. Examples of the tendency toward the limiting normal distribution are given by Lapin (1978, pp. 216–217).

The property of the normal distribution of being the limiting distribution of all sampling distributions, and the fact that the normal pdf is approximated by only a relatively small number of elements in the sample sums instead of the theoretical requirement of an infinite number, is of great practical importance. The CLT explains why in many cases random variables may be assumed to be normally distributed. For example, errors in tests can be assumed to be made up of many small additive random contributions, so that the average error tends to be normally distributed. A project cost that consists of many small independent cost items may be found to have a normal distribution. The CLT also justifies the practice of simulating the sampling of a normal distribution by actually drawing random samples from a uniform distribution (see Example (3) of this section and Section 3.4.4).

The random impacts of molecules on small particles in a fluid (fat particles in milk or dust particles in air) cause the particle to assume a random location $\{X(t), Y(t), Z(t)\}$ at time t. This phenomenon is known as *Brownian motion* (Parzen, 1965, p. 27). The displacements of a particle during a period $(0, t)$ are made up of many small displacements at time increments much shorter than the observation period, $(0,t)$, with 0 meaning the beginning of each observation period and t indicating the end. According to the CLT, one can expect the resulting (x,y,z)-path lengths measured during many periods of equal length, $(0, t)$, to be normally distributed, given the process is stationary.

Examples: (1) Write the parameters of a sampling distribution for four random selections from a population distribution. Solution: Using Equations (3.4-5a) through (3.4-5c) gives the following sample parameters:

$$\overline{X} = \frac{1}{4}(X_1 + X_2 + X_3 + X_4)$$

$$E(\overline{X}) = \frac{1}{4}[E(X_1) + E(X_2) + E(X_3) + E(X_4)]$$

$$\sigma_{\overline{X}} = \frac{1}{4}\sqrt{Var(X_1) + Var(X_2) + Var(X_3) + Var(X_4)}$$

where $X_1, \dots, X_4$ are random selections from the population, $E(X_i)$ are the expectations of these random selections, and $Var(X_i)$ are their variances, which for selections from the same populations are all μ and σ, respectively. Hence, for the selection of four random variables from the same population distribution one obtains the normalized sampling variable according to Equation (3.4-6), (3.4-8b), or (3.4-8c) for $n = 4$:

$$Z = \frac{\overline{X} - \mu}{\sigma / \sqrt{4}} = \frac{Y_4 - 4\mu}{\sigma\sqrt{4}} = \frac{\overline{X} - E(\overline{X})}{\sigma_{\overline{X}}}$$

where $\overline{X}$ and Y_4 are the alternative sampling variables, mean and sum; $E(\overline{X})$ and $\sigma_{\overline{X}}$ are the sample parameters, and μ and σ are the population parameters. If the sample size were increased from $n = 4$ to $n = 10$, a better approximation of the normal distribution by the Z's could be expected. See Example (3) for the creation of a sampling distribution.

(2) If a process is a product of many variables each having a relatively small effect on the overall result, then the output from this process may be normally distributed. For

the Potomac River basin, a regression formula was suggested of the form (Benson and Matalas, as cited by Fiering and Jackson, 1971, p. 37)

$$Y = a \; A^b \; S^c \; S_t^d \; p^e \; S_n^f \; F^g$$

where Y is a statistical parameter, i.e., annual average flow; A is the drainage area, in square miles; S is slope of the main channel, in feet per miles; S_t is surface storage in lakes and pond, in percent of the total drainage area, S_n is annual snowfall, in inches; p is annual precipitation, in inches; F is forested area in percent of total area; the coefficient a and the exponents b through g are empirical numbers. Taking logarithms gives

$$\log Y = \log a + b \log A + c \log S + d \log S_t + e \log p + f \log S_n + g \log F$$

A regression analysis provides the intercept $\log a$ and the coefficients b through g. Some of the variables are random variables which makes this expression a sum of random variables that may be normally distributed. This approach is mentioned here not because it is a recommendable procedure but because it explains why annual flows have been found to be close to normally or log-normally distributed. Yevjevich (1972, p. 141) gives a sample of 150 years (1808-1957) of annual Rhine River flows at Basle whose logarithms line up almost perfectly in a log-normal plot. Another Rhine River data set was used in Section 3.2.2 to explain the log-normal transform.

(3) A uniform distribution with the parameters $a = 3$, $b = 13$, $\mu = (a + b)/2 = 8$, $\sigma = [(b - a)^2/12]^{1/2} = 2.887$, is sampled four times. The four random numbers drawn are added to produce the sum of random realizations, y_4. This sampling process is done 100 times to create a data set for an empirical pdf $f(z)$ and its integral $F(z)$. Compare the sampling distribution of the random sums generated with the normal distribution. Solution: A random number generator (a spreadsheet function) is used to pick four random numbers from the range (3,13). Then the sum, $y_4 = x_1 + x_2 + x_3 + x_4$ is calculated, which is again a random variable. This calculation of y_4 is repeated 100 times. The resulting 100 values of y_4 are sorted in descending order. A histogram is constructed with a class width of $\Delta y_4 = 1$ (the calculation could proceed without histogram, but this was the approach selected at the time). All samples are allocated to classes, as shown in columns 1 and 2 of Table 3-5. In class (19, 19.99) there is one occurrence, in class (20, 20.99) there are two occurrences, and so on. These counts are given in column 3. The cumulation starts at the lower limit of the first class, and the cumulated sum occurs at the upper limit of each class. At the upper limit of the second class, at sum value 20.99, or for practical purposes 21, the number of occurrences has reached 3. This cumulation starts at the bottom of column 3 and continues to the top, where it reaches 100 (the cumulation is not shown in Table 3-5).

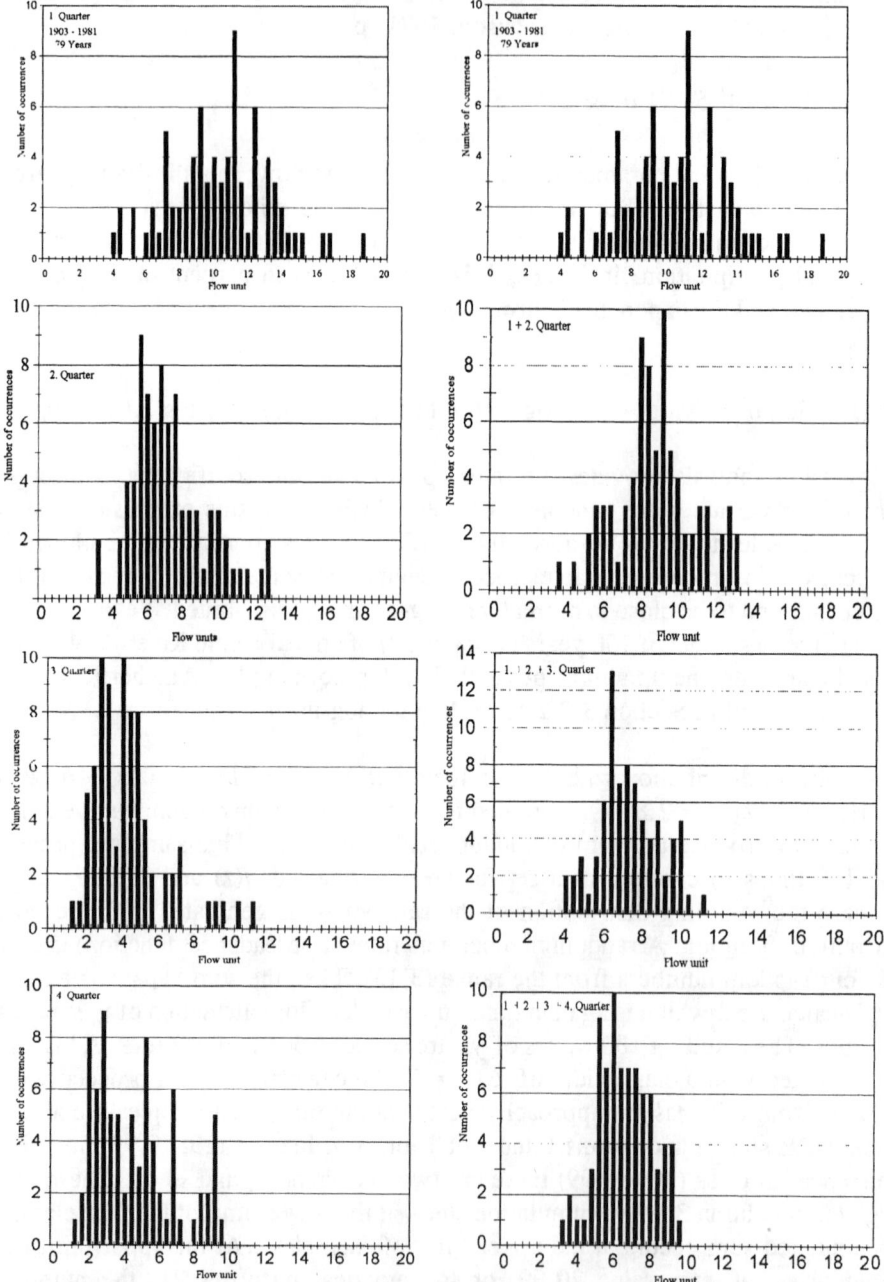

Figure 3–9: Four samples of elements with different pdf's (left column); the pdf's of sums of 1, 2, 3, and 4 elements (right column) indicate a central tendency.

The possible minimum of y_4 is $4 \cdot 3 = 12$, and the possible maximum is $4 \cdot 13 = 52$. Neither of these extremes occurred in the drawn random samples. The sample sums ranged from 19 to 46. For example, sums in the class (31,32), i.e., $31 \leq y_4 < 32$, occurred 10 times (column 3). Dividing these numbers by $N = 100$ and cumulating them from the lowest to the highest class produces the   *empirical* cumulative distribution function (cdf), $F(y_4)$, in column 7; each function value occurs and is plotted at the upper limit of each class. The result is shown in Figure 3-10.

The next step is to compare the empirical distribution with the *limiting distribution*. For this purpose, the normalized variable z is calculated using Equation (3.4-8b)

$$z = \frac{y_4 - 4\mu}{\sigma\sqrt{4}}$$

for all upper limits of the class intervals. The parameters μ and σ refer to the population distribution with the parameters $\mu = 8$, and $\sigma = 2.887$. The mean of y_4 is $\mu = 4\mu = 4 \cdot 8 = 32$, and the standard deviation is $\sigma = \sigma\sqrt{4} = 2 \cdot 2.887 = 5.77$. The 100 samples of random sums in Table 3-5 produce $\mu = 31.4$ and $\sigma = 5.76$. The normalized variable using parameters derived from sampling is $z = (y_4 - 31.4)/5.76$, whereas the normalized variable using population parameters is $z = (y_4 - 32)/5.774$. Using the latter the normal pdf is calculated by

$$f(z) = 0.39894e^{-0.5z^2}$$

where z is taken from column 4 and $f(z)$ is given in column 5 of Table 3-5.

Table 3-5: Calculation of the Distribution of the Sum of Four Uniformly Distributed Random Variables and Its Comparison with the Normal Distribution.
Parameters of x: $a = 3$, $b = 13$; $\mu = 8$; $f(x) = 0.1 = $ constant (uniform pdf); $\sigma = 2.887$. Random variable generated: $y_4 = x_1 + x_2 + x_3 + x_4$.

y_4-Class limits (lower) (1)	(upper) (2)	Count of y_4's (3)	z (4)	pdf $f(z)$ (5)	Calculated $F(z)$ (6)	Experimental $F(y_4)$ (7)	Difference col.7 - col. 6 (8)
46	47.0	1	2.5981	0.0137	0.9951	1.0000	0.0049
45	46.0	0	2.4249	0.0211	0.9921	0.9900	-0.0021
44	45.0	1	2.2517	0.0316	0.9875	0.9900	0.0025
43	44.0	0	2.0785	0.0460	0.9808	0.9800	-0.0008
42	43.0	2	1.9053	0.0650	0.9712	0.9800	0.0088
41	42.0	3	1.7321	0.0890	0.9578	0.9600	0.0022
40	41.0	2	1.5588	0.1184	0.9399	0.9300	-0.0099
39	40.0	2	1.3856	0.1528	0.9164	0.9100	-0.0064
38	39.0	1	1.2124	0.1913	0.8866	0.8900	0.0034

37	38.0	6	1.0392	0.2325	0.8499	0.8800	0.0301
36	37.0	4	0.8660	0.2742	0.8060	0.8200	0.0140
35	36.0	5	0.6928	0.3138	0.7551	0.7800	0.0249
34	35.0	3	0.5961	0.3486	0.6977	0.7300	0.0323
33	34.0	4	0.3464	0.3757	0.6350	0.7000	0.0650
32	33.0	6	0.1732	0.3930	0.5684	0.6600	0.0916
31	32.0	10	0.0000	0.3989	0.4998	0.6000	0.1002
30	31.0	9	−0.1732	0.3930	0.4313	0.5000	0.0687
29	30.0	8	−0.3464	0.3757	0.3647	0.4100	0.0453
28	29.0	5	−0.5196	0.3486	0.3020	0.3300	0.0280
27	28.0	6	−0.6928	0.3138	0.2446	0.2800	0.0354
26	27.0	2	−0.8660	0.2742	0.1937	0.2200	0.0263
25	26.0	6	−1.0392	0.2325	0.1498	0.2000	0.0502
24	25.0	5	−1.2124	0.1913	0.1131	0.1400	0.0269
23	24.0	2	−1.3856	0.1528	0.0833	0.0900	0.0067
22	23.0	1	−1.5588	0.1184	0.0598	0.0700	0.0102
21	22.0	3	−1.7321	0.0890	0.0419	0.0600	0.0181
20	21.0	2	−1.9053	0.0650	0.0285	0.0300	0.0015
19	20.0	1	−2.0785	0.0460	0.0189	0.0100	-0.0089
	19	100	−2.2517	0.0316	0.0122	0.0000	-0.0122

Notes: columns 1 and 2: lower and upper class limits of histogram of experimental y_4's; the range of y_4 is $4\,a \leq y_4 \leq 4\,b$, or (12,52); no experimental y_4's occurred at or near the range limits.

column 3: 100 sums of y_4's generated by adding four random realizations x drawn from a uniform distribution; they are distributed according to their size to the classes specified by columns 1 and 2; the distribution thus obtained is random and cannot be repeated.

column 4: normalized variable z of y_4 using mean and standard deviation of the population distribution and the upper bound of the class of y_4 in column 2.

column 5: normal pdf $f(z)$ for z of column 4.

column 6: normal cdf $F(z)$ by integrating $f(z)$ from the bottom up to the upper limit of each class; the ending value should be close to 1.

column 7: $F(y_4)$ is the experimental cdf obtained by cumulating column 3 after dividing each increment by 100. The column starts with 0 and ends with 1 because all sampled elements are within this range.

column 8: difference between the empirical cdf in column 7 and the calculated normal cdf in column 6.

The pdf $f(z)$ is numerically integrated from the bottom up. The starting value of the integral is the value of the integral from $z = -\infty$ to the lower limit of the lowest interval, (19,20), $z_{19} = -2.252$. It is calculated using symmetry $F(-z) = 0.5 - F_0(z)$:

$$F(-2.252) = 0.5 - \int_0^{2.252} f(z)dz = 0.5 - 0.4878 = 0.0122 \quad .$$

Once the starting value $F(z_{19}) = F(-2.252)$ is found, the rest of the integration proceeds by cumulation:

$$F(z_{20}) = F(z_{19}) + \frac{1}{2}[f(z_{19}) + f(z_{20})]dz$$

where $dz = dy_4/\sqrt{4}\,\sigma$, with $dy_4 = 1$, and $\sigma = 2.887$. The numerical value of the integral at the upper interval limit, z_{20}, is

$$F(z_{20}) = 0.0122 + \frac{1}{2}(0.0316 + 0.0460)\frac{1.0}{5.774} = 0.0189$$

where $f(z_{19})$ and $f(z_{20})$ are the values at the bottom of column 5, Table 3-5. The integration continues from the bottom up with the result given in column 6. The top value must be 1 or close to it, which serves as a check for the accuracy of the numerical integration.

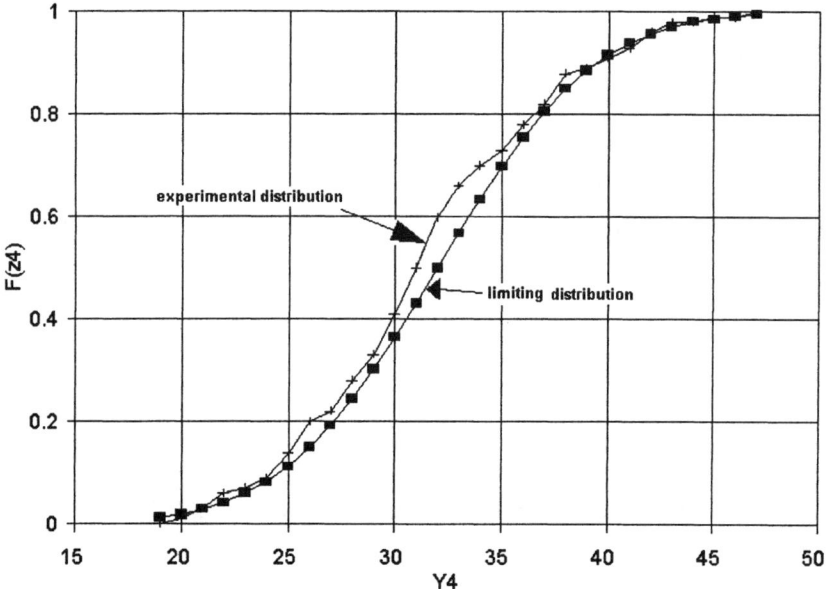

Figure 3-10: The experimental distribution is based on a sampling frequency distribution (histogram) of sums of four uniformly distributed random variables. The limiting (normal) distribution is calculated using the normal pdf, $f(z)$, with z being based on the mean and standard deviation of the random sums.

Column 7 is the "experimental" cdf, which is derived from the histogram in column 3. Column 8 gives the difference between the two $F(z)$, columns 6 and 7. The two cdf's are shown in Figure 3-11. The experimental distribution is similar to the Gaussian distribution, but there are some discrepancies. After all, the random sums are based on only four population elements ($n = 4$), far from $n \rightarrow \infty$, as required for the CLT to hold, and the elements are sampled from a uniform distribution that has little resemblance with a normal distribution.

3.4.3 Sampling the Uniform Distribution

The CLT justifies the practice of sampling a normal random variable by drawing about 12 random numbers from a uniform distribution (Hammersley and Hanscomb, 1964 and 1967, p. 39). Nowadays spreadsheets usually have various *random number generators* as built-in functions that can produce uniform random numbers or normal random numbers. The method is discussed here as another example of the power of the CLT. The approach makes use of the fact that the uniform distribution is more easily sampled than the normal distribution; for this reason, random number tables based on the uniform distribution can be found in textbooks. A random variable with a known normal distribution can then be used in probabilistic models, such as a *Monte Carlo* model. The fact that a normal variable can be thought of as a sum of uniformly distributed random variables was demonstrated in Example (3) of the previous section. The approach is acceptable if the problem to be modeled does not depend on the tails of the normal distribution (Hammersley and Hanscomb, 1967, p. 40). If the parameters of the normal variable are known, then the variable that is supposed to be normally distributed must in the limit have the parameters

$$\mu_N = \mu_u \tag{3.4-11a}$$

and

$$\sigma_N = \sigma_u \tag{3.4-11b}$$

where index N refers to the normal distribution and u refers to the uniform distribution.

The parameters of the uniform distribution are (Section 2.7.3)

$$\mu_u = \frac{a+b}{2} \tag{3.4-12}$$

and

$$\sigma_u = \frac{b-a}{\sqrt{12}} \tag{3.4-13}$$

where a and b are the limits of the uniform distribution. The range of the uniform variable follows from Equation (3.4-13) as

$$b - a = \sqrt{12}\ \sigma_u . \tag{3.4-14}$$

From the combination of Equations (3.4-12) and (3.4-13) one obtains for the lower bound

$$a = \mu_u - \sqrt{3}\ \sigma_u , \tag{3.4-15}$$

and for the upper bound

$$b = \mu_u + \sqrt{3}\ \sigma_u . \tag{3.4-16}$$

A typical random number generator has the form

$$rv = a + rand(b - a) \tag{3.4-17}$$

where rv is the random number, a is the lower bound of the range to which is added $rand(b - a)$, a random fraction of the range.

Examples: (1) A normally distributed variable with $\mu_N = 80$ and $\sigma_N = 9.13$ is to be sampled using a uniform distribution. The number of sample elements is $n = 12$. Determine the parameters of the uniform distribution to be sampled. Solution: The parent distribution (*population*) is the uniform distribution, and the sampling distribution should approach the normal distribution. The sample parameters obtained from sampling the uniform distribution will be compared with those of the normal distribution. The lower limit and upper limit of the range of the uniform distribution are calculated from Equations (3.4-15) and (3.4-16) using the parameters given by Equations (3.4-11a) and (3.4-11b): $\mu_u = \mu_N = 80$, and $\sigma_u = \sigma_N = 9.13$. The range limits of the uniform distribution are

$$a = \mu_u - \sigma_u \sqrt{3} = 80 - 9.13 \cdot \sqrt{3} = 64.19$$

and

$$b = \mu_u + \sigma_u \sqrt{3} = 80 + 9.13 \cdot \sqrt{3} = 95.81.$$

The random number generator can be set up according to Equation (3.4-17) and a number of samples with 12 elements each can be generated for further processing.

(2) Take three samples with 12 elements each from the uniform distribution using the bounds calculated in Example (1), and compare the sample parameters with those of the normal distribution. Solution: A random number generator is used for a uniform distribution with the standard range (0,1). It returns a random fraction of the range $0 \le rv_i \le 1$. The random numbers, rv_i, are shown in columns 2, 5, and 8 of the tabular evaluation that follows. They are adapted to the specified range computed in Example (1) by

$$X_i = a + rv_i (b - a)$$

where $a = 64.19$ (63.89 was inadvertently used), and $b = 95.81$. Successive random observations are denoted X_i, $i = 1, \dots, n$. In independent sampling (sampling with replacement), two random observations, X_1 and X_2, can have the same realization, x_j. The observations for three samples are given in columns 3, 6, and 9 of the tabular evaluation.

Tabular Evaluation of Example (2)

No.		$k = 1$				$k = 2$				$k = 3$	
	rv_i	X_i	$(X_i - \overline{X})^2$	rv_i	X_i	$(X_i - \overline{X})^2$	rv_i	X_i	$(X_i - \overline{X})^2$
(1)	(2)	(3)	(4)	(5)	(6)	(7)	(8)	(9)	(10)
1	0.32	74.2	104.5	0.27	72.5	53.0	0.74	87.5	35.1
2	0.40	76.7	59.7	0.11	67.4	153.2	0.62	83.8	5.1
3	0.46	78.5	34.5	0.55	81.3	2.4	0.72	87.0	29.4
4	0.39	76.3	65.2	0.34	74.7	25.2	0.99	95.6	198.0
5	0.92	93.2	77.5	0.28	72.7	49.7	0.18	69.6	143.3
6	0.61	83.3	1.2	0.97	95.0	231.9	0.73	87.3	33.2
7	0.79	89.2	22.6	0.58	82.5	7.4	0.15	68.6	168.0
8	0.92	93.4	79.8	0.84	90.8	121.3	0.46	78.6	8.4
9	0.59	82.9	2.5	0.72	86.9	51.2	0.75	87.7	38.2
10	0.89	92.2	59.9	0.64	84.3	20.6	0.72	86.9	28.5
11	0.52	80.4	16.1	0.64	84.2	20.1	0.37	75.6	34.9
12	0.90	92.8	69.5	0.03	64.8	224.0	0.20	70.3	126.1
$\overline{X}_k$ and σ_k^2		84.42	53.91		79.74	87.28		81.55	77.12
σ_k		7.34			9.34			8.78	
$\sigma_{\overline{X}}$		2.06 (computed from the three sample means)							
$	\overline{X} - \mu_N	$		4.42			0.26			1.55	

The sums of columns 3, 6, and 9 divided by $n = 12$ give the sample means, $\overline{X}$. The sums of columns 4, 7, and 10 divided by $n - 1 = 11$ produce the sample variances according to

$$\sigma_k^{\,2} = \frac{1}{11} \sum_{i=1}^{12} (X_i - \overline{X}_k)^2$$

where $k = 1, 2, 3$ is the index of the sample.
The three sample means are $\overline{X}_1 = 84.42$, $\overline{X}_2 = 79.74$, and $\overline{X}_3 = 81.55$. The expectation of the sample mean is

$$E(\overline{X}) = (\overline{X}_1 + \overline{X}_2 + \overline{X}_3)/3 = (84.42 + 79.74 + 81.55)/3 = 81.90.$$

Since all random samples are drawn from the same population, the expectation of their means should approach the population mean, $\mu_u = 80$. The sample standard deviations are $\sigma_1 = 7.34$, $\sigma_2 = 9.34$, and $\sigma_3 = 8.78$. The variance of the sample mean is

$$\sigma_{\overline{X}}^2 = E(\overline{X}^2) - [E(\overline{X})]^2$$

$$= (84.42^2 + 79.74^2 + 81.55^2)/3 - 81.90^2 = 4.259.$$

This gives a standard deviation of the sample mean $\sigma_{\overline{X}} = 2.06$. The population standard deviation is $\sigma = 9.13$. The standard deviation of the sample mean, based on the fact that all random variables have the variance of the population distribution, is

$$\sigma_{\overline{X}} = \sigma/\sqrt{12} = 9.13/\sqrt{12} = 2.63.$$

The standard error of the expectation of the sample mean is

$$|E(\overline{X}) - \mu| = 81.9 - 80.0 = 1.91,$$

a 2.4% departure from the population mean. A characteristic of the normal pdf is that 68.26% of all sample elements fall into a band of width $\pm 1\,\sigma$ centered on the mean. With $\sigma_{\overline{X}} = 2.63$, the boundaries of this interval around the mean are

$$77.37 \le \mu \le 82.63.$$

Of the 3 generated sample means 2 fall within this interval, or 66 %.

(3) A single sample with $n = 36$ elements can be constructed using the rv_i's of the three samples of Example (2). The new mean is the mean of all 36 elements which is the same as the average of the three means of Example (2), $\overline{X} = 81.91$. In this case, it is the sample mean of a single sample with 36 elements, not the expectation of three sample means. The sample variance based on the new mean $\overline{X} = 81.91$, and all 36 X_i in columns 3, 4 and 9 of Example (2), is 72.43, and the standard deviation is σ $= \sqrt{72.43} = 8.51$, which can be taken as an estimate of the population standard deviation which is actually 9.13. The standard deviation of the sample mean is estimated as $\sigma_{\overline{x}} \approx 9.13/\sqrt{36} = 1.52$, a 42 % reduction with respect to the standard deviation of the sample mean for the smaller samples of Example (2).

(4) Use a random number generator based on the normal distribution (as available in a spreadsheet) to create three samples with 12 elements and compare the sample parameters with those of the normal distribution.

Tabular Evaluation of Example (4)

No.	x_i	$(X_i - \overline{X})^2$	x_i	$(X_i - \overline{X})^2$	x_i	$(X_i - \overline{X})^2$
		$k = 1$		$k = 2$		$k = 3$
(1)	(2)	(3)	(4)	(5)	(6)	(7)
1	61.9	312.6	93.1	126.5	88.4	116.5
2	83.5	15.3	87.3	29.7	84.0	40.4
3	71.4	67.2	73.8	64.8	77.3	0.1
4	59.0	421.5	84.1	5.2	77.3	0.1
5	86.7	50.8	95.0	174.3	67.2	108.8
6	78.2	1.7	77.7	17.1	85.7	64.6
7	91.7	148.5	72.4	89.8	81.4	14.4
8	84.4	23.4	92.8	120.0	66.0	136.3
9	81.5	3.7	75.2	44.1	80.6	8.7
10	86.7	51.4	77.7	17.5	78.6	0.9
11	80.4	0.7	71.1	114.5	70.9	45.6
12	89.3	94.3	81.9	0.0	74.3	11.0
$\overline{X}_k$ and σ_k^2	79.56	108.28	81.83	73.03	77.63	49.77
σ_k		10.41		8.55		7.06
$\sigma_{\overline{x}}$		1.86 (computed from the three sample means)				
$|\overline{X} - \mu_N|$		0.44		1.83		2.37

Solution: It is assumed that the spreadsheet normal random number generator does not actually use a uniform distribution; but even if it does one may assume that it uses a sample size of $n > 30$ elements, substantially larger than the sample size of $n = 12$ used here. For $\mu_N = 80$ and $\sigma_N = 9.13$, the random numbers generated for three samples with 12 elements each and their analysis are summarized in the tabular evaluation. The sample means are used to calculate the expectation of the sample mean:

$$E(\overline{X}) = (\overline{X}_1 + \overline{X}_2 + \overline{X}_3)/3 \ = (79.56 + 81.83 + 77.63)/3 = 79.67$$

The normal population mean is $\mu = 80$. The standard error of the expectation of the sample mean is $|E(\overline{X}) - \mu_N| = |79.67 - 80| = 0.33$, or 0.4 %. The standard deviation of the sample mean based on the fact that each random selection has the same variance as the population distribution is

$$\sigma_{\overline{X}} = \sigma_N/\sqrt{12} \ = 9.13/\sqrt{12} = 2.63.$$

It is the same for each sample. The variance of the sample mean computed from the three sample means is

$$\sigma_{\overline{X}}^2 = E(\overline{X}^2) - [E(\overline{X})]^2 \ = (79.56^2 + 81.83^2 + 77.63^2)/3 - 79.67^2 = 3.48,$$

and the standard deviation of the sample mean is $\sigma_{\overline{X}} = 1.865$, an approximation.

The $\pm 1 \ \sigma_N$ confidence band of the normal pdf contains 68.26 %. With $\sigma_N = 9.13$, this interval is bounded by $70.87 \leq X \leq 89.13$. Of the 36 generated elements, 27 elements are within this interval, or 75 %, instead of 68.26 % that the normal distribution predicts. The $\pm 1 \ \sigma_{\overline{X}}$ confidence band of the sampling distribution should contain 68.26 % of the sample means. With $\sigma_{\overline{X}} = 2.63$, its band width is bounded by $77.37 \leq \mu \leq 82.63$. All three sample means are within this interval with one just barely making it. A single sample with $n = 36$ elements drawn from the normal distribution (the random variables of the three samples together) has the mean $\overline{X} = 79.67$. The sample variance based on this mean is 75.67, and the sample standard deviation is $\sqrt{75.67} = 8.70$ whereas the population standard deviation is $\sigma_N = 9.13$. The standard deviation of the sample mean is $\sigma_{\overline{X}} = \sigma_N/\sqrt{36} \ = 9.13/6 = 1.52$, a reduction of 42 % with respect to the 12-element sample. This is an indication that the sampling distribution for $n = 36$ is spikier than the sampling distribution for $n = 12$. In summary the information obtained from the two samples of Example (2), which is supposed to be a simulation of drawing from a normal distribution, are very similar to those of Example (4), which represents a sample drawn directly from a normal distribution (using a spreadsheet routine).

3.4.4 Testing the Sampling Distribution

In the context of approximating the normal distribution by sampling of other distributions, such as the uniform distribution in Section 3.4.3, the question arises of how good the approximation is. Usually there are differences between the empirical distribution and the limiting distribution, as shown in Figure 3-10 and in column 8 of Table 3-5. Tests have been developed to quantitatively assess the importance of these differences and decide whether the approximation of the limiting distribution by the empirical distribution is acceptable.

The *Kolmogorov–Smirnov statistic* is used to evaluate such differences (Lapin, 1978, p. 642). The statistic is the vertical difference in the ordinates $F(x)$ between two cumulative distribution functions (cdf):

$$D = \max|F_a(x) - F_e(x)| \tag{3.4-18}$$

where D is the absolute value of the maximum difference between the ordinates of the *actual* (a) and the *expected* (e) cumulative distribution functions, $F_a(x)$ and $F_e(x)$, respectively, that occurs at some point x. The actual cdf results from integrating an experimental histogram or pdf, and the expected cdf is the limiting distribution. D can be positive or negative, but only absolute values are considered in the test. A sample of such differences is given in column 8 of Table 3-5 with a maximum $D = 0.1002$.

The *decision rule* for rejecting or not rejecting the approximation to the limiting distribution, based on the absolute maximum difference, is as follows (Lapin, 1978, p. 643):

Null-hypothesis H_0: <u>normal distribution applies</u>:

not reject H_0:
normal distribution <u>applies</u>, it cannot be rejected: $D \le D_\alpha$,   (3.4-19)

reject H_0:
normal distribution <u>does not apply</u>: $D > D_\alpha$.   (3.4-19a)

According to the test, as long as D is less than a critical difference, D_α, for a selected *significance level*, α, the expected distribution, here the normal distribution, cannot be rejected as a fit of the empirical distribution. If D exceeds D_α, then H_0 is rejected at the selected significance level, and some other distribution applies. The tests should be regarded as a means for detecting significant differences. The significance level, α, is the exceedance probability of D_α so that the probability of D exceeding D_α, or the probability of rejecting the null hypothesis is

$$P(D \ge D_a) = \alpha . \tag{3.4-20}$$

Critical differences, D_α, as a function of α and sample size n are given in Table 3-6. The sample size refers to the number of function values that constitute the empirical cdf. Some examples of critical values, D_α, for $n = 25$ and exceedance probabilities α follow:

$$P(D \geq 0.208) = 0.1$$
$$P(D \geq 0.238) = 0.05$$
$$P(D \geq 0.317) = 0.005$$

Table 3-6: Critical Differences, D_α, for the Kolmogorov–Smirnov Maximum Deviation Test as Function of Sample Size n and Significance Level α (values taken from Lapin, 1978, Table L, pp. A-38 to A-40)

	$P(D \geq D_\alpha] = \alpha$				
n	$\alpha = 0.1$	$\alpha = 0.05$	$\alpha = 0.025$	$\alpha = 0.01$	$\alpha = 0.005$
1	0.90000	0.95000	0.97500	0.99000	0.99500
5	0.44698	0.50945	0.56328	0.62718	0.66853
10	0.32260	0.36866	0.40925	0.45662	0.48893
15	0.26588	0.30397	0.33760	0.37713	0.40420
20	0.23156	0.26473	0.29408	0.32866	0.35241
25	0.20790	0.23768	0.26404	0.29516	0.31657
30	0.19032	0.21756	0.24170	0.27023	0.28987
35	0.17659	0.20185	0.22425	0.25073	0.26897
40	0.16547	0.18913	0.21012	0.23494	0.25205
45	0.15623	0.17856	0.19837	0.22181	0.23798
50	0.14840	0.16959	0.18841	0.21068	0.22604
60	0.13573	0.15511	0.17231	0.19267	0.20673
70	0.12586	0.14381	0.15975	0.17863	0.19167
80	0.11787	0.13467	0.14960	0.16728	0.17949
90	0.11125	0.12709	0.14117	0.15786	0.16938
100	0.10563	0.12067	0.13403	0.14987	0.16081

Notes: column 1: n is the sample size, the number of points that make up the sampling distribution.

columns 2 to 6: α is the probability that D exceeds the critical difference D_α; the number field of the table represents the maximum difference D_α between the cdf of a sampling distribution and the cdf of a normal distribution that must not be exceeded for a given n and a selected α.

The examples show that the probability of the difference D exceeding a small critical difference, such as $D_{0.1}$, is higher than the probability of exceeding a larger difference. The complementary probability further explains this criterion:

$$P(D \leq D_\alpha) = 1 - \alpha \,. \tag{3.4-20a}$$

Equation (3.4-20a) states that if D is confined by a narrow limit D_α or a narrow confidence band, there is a smaller probability that D will not exceed it. According to the previous examples, the probability that $D \leq D_{0.1} = 0.208$ is 90 %, and the probability that D is smaller than $D_{0.05} = 0.238$ and $D_{0.005} = 0.317$, is 95 % and 99.5 %, respectively. Examples: (1) In the example given in Table 3-5, column 8, for $n = 29$, or about 30, and $D = 0.1002$, Table 3-6 gives $D_{0.1} = 0.190$, so that $D \leq D_{0.1}$. Therefore, H_0, the hypothesis that the sampling distribution is normal cannot be rejected at the 10 % significance level.

(2) The splitting tensile strengths of 100 concrete samples are shown in the histogram of Figure 2-4. Check whether the distribution can be approximated by a normal distribution. Solution: The data are grouped in classes with class width 0.1 MPa. The class frequencies are given in Table 3-7, column 3 (the data are by D. L. Ivey, as reported by Kreyszig, 1979, p. 840). Their range is 2 MPa $\leq x \leq$ 3.1 MPa. The mean is $\mu = 2.51$ MPa, and the standard deviation is $\sigma = 0.185$ MPa. The coefficient of variation is $0.185/2.51 = 0.074 < 0.3$. The lower limit of the range is 2 MPa. The corresponding normalized variable is $z = (2 - 2.51)/0.185 = -2.757$. There are no data points below $x = 2$ so that the experimental non-exceedance probability $P[X \leq 2] = 0$, whereas the normal non-exceedance probability of $x = 2$ is

$$F(-2.757) = 0.5 - F_0(2.757) = 0.5 - 0.49708 = 0.00292.$$

The cumulation of the empirical frequencies for the selected classes in column 3 produces the actual (experimental) cdf, $F(z)$, in column 4. The normalized variable is computed for the upper interval limits x (column 2) by $z = (x - 2.51)/0.185$ in column 5. The normal pdf, $f(z)$, in column 6 is calculated by

$$f(z) = 0.39894 e^{-0.5z^2} \,.$$

The integral of $f(z)$ is calculated by numerical integration in direction of ascending z, from the top of column 7 down. The starting value for the integral $F(z)$ is the non-exceedance probability of $x = 2$, $F(-2.757) = 0.00292$. The increment is $dz \approx \Delta z = 0.1/0.185 = 0.54$. The final value of the integral is $F(3.189) = 0.9985$, as the normal cdf does not reach 1 until $z = +\infty$. The empirical cdf, in contrast, runs from 0 to 1. This property of the normal distribution can be tolerated in applications where the tails of the distribution are not critical. The maximum departure between the empirical cdf and the normal cdf is $D = 0.0516$, as shown in column 8 of Table 3-7. The Kolmogorov–Smirnov test is carried out at the significance level, $\alpha = 0.1$. While there are 100 data elements, the shape of the experimental curve is a function of only 12 points. According to Table 3-6, for $n = 12$, $D_{0.1}$ is about the average between the values for $n = 10$ and $n = 15$, or $D_{0.1} = 0.296$. Since $D < D_{0.1}$, the null hypothesis, H_0, that the sampling distribution is normal, cannot be rejected at the 10 % significance level. A comparison of the experimental (actual) and the limiting (expected) distribution in Figure 3-11 shows the normal distribution in comparison to the data distribution.

Table 3-7: Testing the Normality of Concrete Splitting Tensile Strength Data. The experimental distribution parameters $\mu = 2.51$ MPa and $\sigma = 0.185$ MPa are based on 100 data points (data are by D. L. Ivey as reported by Kreyszig, 1979, p. 840)

x_l (1)	x_u (2)	n_i/N (3)	$F(x)$ (4)	z (5)	$f(z)$ (6)	$F(z)$ (7)	D (8)	z_F (9)
	2.0	0.00	0.00	-2.757	0.009	0.0029	0.0029	$-\infty$
2	2.1	0.02	0.02	-2.216	0.034	0.0146	0.0054	-2.054
2.1	2.2	0.00	0.02	-1.676	0.098	0.0503	0.0303	-2.054
2.2	2.3	0.10	0.12	-1.135	0.209	0.1334	0.0134	-1.175
2.3	2.4	0.11	0.23	-0.595	0.334	0.2804	0.0504	-0.739
2.4	2.5	0.30	0.53	-0.054	0.398	0.4784	0.0516	0.075
2.5	2.6	0.15	0.68	0.486	0.354	0.6818	0.0018	0.468
2.6	2.7	0.18	0.86	1.027	0.235	0.8412	0.0188	1.080
2.7	2.8	0.08	0.94	1.568	0.117	0.9364	0.0036	1.555
2.8	2.9	0.02	0.96	2.108	0.043	0.9797	0.0197	1.751
2.9	3.0	0.03	0.99	2.649	0.012	0.9946	0.0046	2.326
3	3.1	0.01	1.00	3.189	0.002	0.9985	0.0015	$+\infty$
						max D:	0.0516	

Notes: columns 1 and 2: lower and upper class boundaries.

column 3: empirical frequencies; n_i is the number of elements in class i and N is the total number of elements, $N = 100$.

column 4: cumulation of frequencies of column 3.

column 5: normalized variable z calculated from x in column 2 and population parameters $\mu = 2.51$ MPa and $\sigma = 0.185$ MPa.

column 6: normal pdf $f(z)$ using z of column 5.

column 7: numerical integral of $F(z)$ of column 6.

column 8: absolute difference of columns 4 and 7.

column 9: inverse of the empirical cdf of column 4.

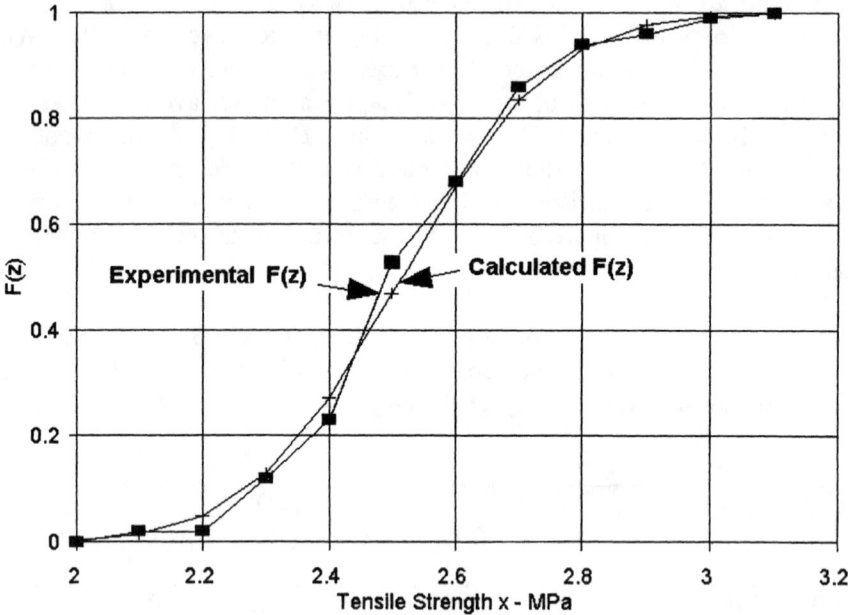

Figure 3-11: Comparing the cumulative distribution function of splitting tensile strength test data based on the experimental histogram (actual cdf) with the normal cumulative distribution function (expected cdf) which was calculated from the mean and standard deviation of the data sample (see Table 3-7; data are from D. L. Ivey as reported by Kreyszig, 1979, p. 840). The maximum difference in the ordinates of the cdf's is $D = 0.0516$ and occurs at $x = 2.5$ MPa.

3.4.5 Testing the Normal Approximation by Regression

A test that includes every point of the distribution is to compare the z obtained from the inverse of the experimental $F(z)$, denoted z_F, with the z obtained from the data parameters. Such a comparison includes all data pairs, (z_F, z), in a criterion in the form of the regression coefficient (Section 3.1.2). The actual $F(z)$ is obtained from data ranking or from cumulating a histogram. It is also referred to as "experimental." The z_F-values obtained from this cdf by inversion (Section 2.6.4) are plotted against the standard normal variable calculated from the data parameters, $z = (x - \mu)/\sigma$. The fit is deemed satisfactory if the empirical z_F plotted versus z results in a straight line close to the diagonal $z_F = z$. This kind of test has been historically used to check a data sample's approach to an expected (hypothesized) distribution by using probability paper designed for the expected distribution.

The goodness-of-fit test is the regression of z_F on z. The regression coefficient, Equation (3.1-33a) is based on all usable data pairs consisting of the fitted function ordinate, $\hat{y}$, and the data points

$$R^2 = 1 - \frac{\sum (z_F - z)^2}{\sum [z_F - E(Z_F)]^2} \qquad (3.4\text{-}21)$$

where R^2 is the sample coefficient of determination; $E(Z_F)$ is the mean of the usable z_F-values of column 9, Table 3-7; z is from column 5 of Table 3-7; $\hat{y}$ in Equation (3.1-33a) is replaced here by z; and the sums are taken over the number of data pairs, (z_F, z). Figure 3-12 displays the z_F versus z plot in which the diagonal $z_F = z$ serves as an indicator of the closeness of approach to the normal distribution.

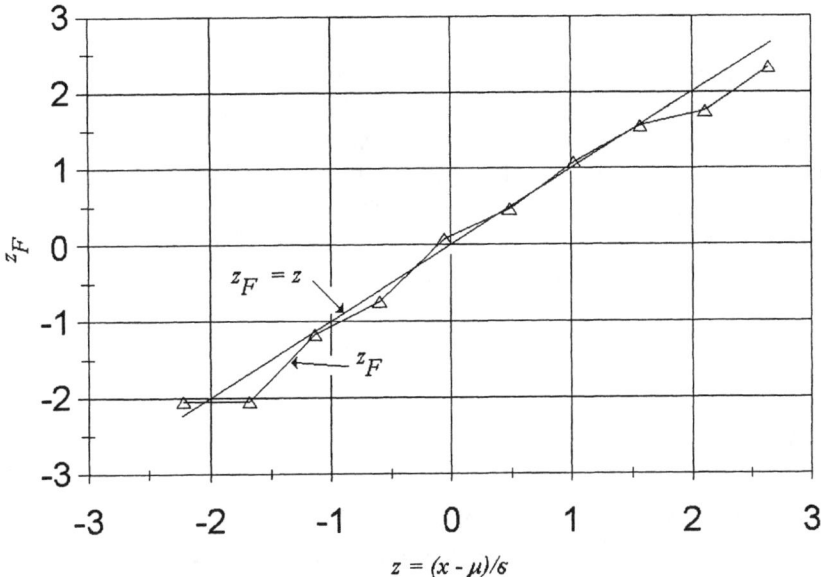

Figure 3-12: The z_F obtained from the inverse of the empirical (actual) cumulative distribution function $F(z)$ is plotted against the z obtained from $z = (x - \mu)/\sigma$. The diagonal $z_F = z$ is used to visualize how well z_F and z agree with each other.

The empirical function, $z_F = g(z)$, according to the data fit, is a straight line $g(z) = \zeta z + \epsilon$ whose coefficients ζ and ϵ are determined by the regression. Inspection shows that the linear fit is quite good with $\zeta \approx 1$ and $\epsilon \approx 0$. The sample coefficient of determination, calculated from 10 pairs of usable data pairs, (z_F, z), is $R^2 = 0.98$. As is often the case when using the normal distribution, the fit at the tails of the distribution is not very reliable.

3.5 Reliability

3.5.1 Reliability and Unreliability

A component or system is reliable if it performs as expected. If a component is scheduled or designed to function for a desired period without failure and experience shows that this is usually the case then this component is considered reliable, even if failure cannot be ruled out. *Reliability* is therefore a probability of success (Moan, 1982, p. 4-16). If t_w is a warranty period, a period in which the new or repaired component should not fail, then the probability that there is no failure during t_w is

$$P(T > t_w) = R(t_w) \qquad\qquad (3.5\text{-}1)$$

where T is a random variable of time periods between failures; and t_w is the *warranty time*; $R(t_w)$ is the exceedance probability of t_w, or the probability of no failure during t_w. $R(t)$ is known as *reliability or survival probability* with t representing a period from a starting time 0 (or t_0) to a time t. High reliability means that $R(t)$ is close to 1. High reliability usually is guaranteed only for relatively short periods, or some fraction of the expected life of the component or system. It is a task of maintenance to achieve high reliability during and beyond the warranty period.

The cumulative distribution function (cdf) of T, $F(t)$, is the probability that a realization t of T is less than or at most equal to a desired time, say t_w. Then $F(t_w)$ is the complementary probability of $R(t_w)$

$$P(T < t_w) = F(t_w) = 1 - R(t_w) . \qquad\qquad (3.5\text{-}2)$$

Equation (3.5-2) gives the probability of t not reaching or just barely reaching the success event t_w. $F(t_w)$ is called the *unreliability*. For a pdf of survival times t, $F(t)$ is the sum of mutually exclusive probabilities that one of the survival times t in the range from t_o to t_w will be realized. For a continuous pdf the probability that t_w may be reached but not exceeded is

$$P(T \leq t_w) = F(t_w) = \int_{t_0}^{t_w} f(t)dt \quad . \tag{3.5-3}$$

The integral of Equation (3.5-3) can be interpreted as the cumulation of many small increments

$$dF(t) = f(t)dt \quad , \tag{3.5-4}$$

each being the probability of a random realization t. The warranty period ends with the completion of t_w. Hence, the failure range is defined as $t_o \leq t \leq t_w$ and success range is defined as $t > t_w$.

The reliability may also be specified as the probability that a variable assumes a value in a desired range,

$$R(t_w) = P(t_l \leq t_w \leq t_u) = F(t_l) - F(t_u) . \tag{3.5-5}$$

where t_w is the variable that is desired to be in the range $t_l \leq t_w \leq t_u$, with t_l and t_u being lower and upper bounds, respectively, of the range (Moan, 1982, p. 4-16). This reliability statement defines $R(t_w)$ as the difference of two non-exceedance probabilities of specified upper and lower limits of the desired range. This means that in this case only the sum of probabilities in the range from t_l to t_u is included in the calculation of $R(t_w)$. For a continuous distribution, as $\Delta t = t_u - t_l$ approaches dt, $R(t_w) = f(t_w)dt$ becomes infinitesimally small. This means that the probability of a specific abscissa value of a continuous pdf is practically zero.

In a k-out-of-n system, the probability of k successes in n trials is a reliability statement for the system (Moan, 1982, p. 4-17; see also Section 2.7.2 and Section 4.3).

Example: Two reliability functions, $R(t) = 1 - F(t)$, for two types of lamps, incandescent (i), and fluorescent (u), are shown in Figure 3-13 (after Morrow, 1957, p. 7-232, source: General Electric). The abscissa is the ratio $t/E(t)$, in percent, where t is elapsed life, and $E(t)$ is the expected survival time or expected lifetime. The shapes of the reliability functions are similar, even if $E(t)$ may not be the same, as is usually the case for these lamp types. For $t/E(t) < 40 \%$, the reliability or survival probability is very high for both types. This percentage of the lifetime could qualify as warranty period for this product.

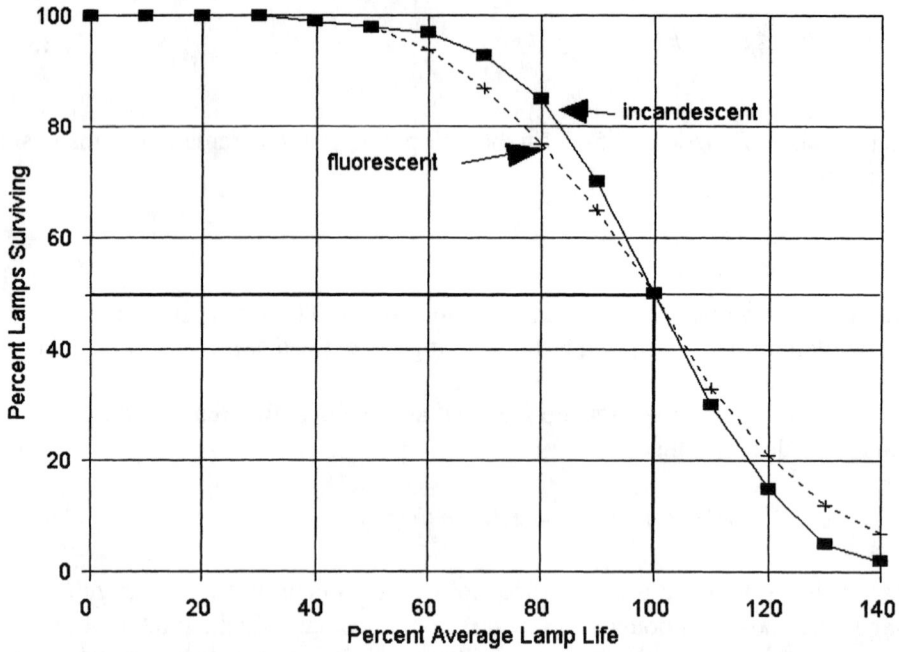

Figure 3-13: Reliability as a function of time (after Morrow, 1957, p. 7-232. Data source: General Electric Co.).

3.5.2 Reliability Criteria

Reliability analysis is concerned with a component's resistance to failure under an imposed load. If the resistance exceeds the load by a "comfortable margin," the element very likely will perform as expected and will not fail. If the component is frequently stressed to the limit or its resistance is exceeded by the load, the component is likely to fail. The causes of failure may not be obvious. Engineering calculations usually are based on simplifying assumptions, as the behavior of the material under multidimensional stress patterns, heat flows, and other loads may be difficult to assess, and failure mechanisms may not be well known. Therefore, some contingency usually is included in the design of components in the form of a *safety margin* or a *safety factor*. The definitions of these quantities follow:

(a) <u>Safety margin</u>:

$$SM = R - L > 0 \tag{3.5-6}$$

(b) Safety factor:

$$SF = \frac{R}{L} > 1 \tag{3.5-7}$$

where R is resistance and L is load. The safety margin has the dimension of the difference of two quantities, whereas the safety factor is a dimensionless number. If the safety margin is divided by the load, one obtains

$$\frac{SM}{L} = \frac{R-L}{L} = SF - 1 \tag{3.5-8}$$

where SM/L is the *relative safety margin*, the amount by which the safety factor exceeds 1. If FS and SM are given, R and L can be calculated from Equation (3.5-8). If SF is greater than 1, as it is expected to be, SM/L is a positive quantity. Risk management programs usually require some minimum size for SF and SM. If R and L are quantities with ranges, as they usually are, and if the can be construed as random variables, then SM and SF are differences and quotients of random variables, similar to net benefits, and benefit-cost ratios (see Section 3.3).

The load-resistance comparison by safety margin and safety factors can be extended to various engineering sectors, such as structural engineering, power system analysis, and benefit-cost analysis. The meanings of the terms in these categories are explained in Table 3-8.

Table 3-8: Load-Resistance Analogies in Various Engineering Sectors

Load (L)	Resistance (R)	$R - L$	R/L	$(R - L)/L$
structural load	structural resistance	safety margin	safety factor	relative safety margin
power system demand (load)	power system capacity	safety margin	safety factor	relative safety margin
project cost	project benefit	net benefit	benefit-cost ratio	net benefit-cost ratio

The probabilistic assessment of safety margins and safety factors is not just a computational embellishment of traditional deterministic analysis. *The deterministic analysis may be deficient, nonconservative, and misleading*. This has been demonstrated by the sums and differences of random variables in Sections 3.3.1 and 3.3.2. For example, Figure 3-8 shows that whereas the difference of the means of the two variables is zero, the fact that the variables have ranges leads to differences ranging from about $+10$ to -10. Deterministic analysis may be deficient because it evaluates the problem only partially by selecting one or a few special cases. It may be nonconservative because it may miss the worst case or neglect the fact that there usually is a non-zero probability of failure, given the possible combinations of variables. It may be misleading because it replaces a complex problem by an oversimplistic and a possibly overoptimistic scenario. It may be risky because the decision maker may adopt an alternative that he would reject if the associated probability of failure were known to him.

An example that illustrates this possibility is the evaluation of an available power capacity (or resistance) that is called upon to meet a load. According to Table 3-8, the variables can be translated into analogous variables of other problems, such as the resistance of a structure under its load or a planned project's costs and benefits. Suppose the average capacity of the system is $R = 12,000$ MW, and the average load is $L = 10,500$ MW. The average safety factor is $SF = R/L = 12,000/10,500 = 1.143$, and the average safety margin is $SM = (SF - 1) L = 0.143 \cdot 10,500 = 1,500$ MW.

In a realistic situation, the load can be construed as the sum of a large number of random demands. Similarly, the system capacity is a sum of generating capacities of many units that are either on or off. In other words, both R and L are sums of random numbers. The CTL has shown (Sections 3.4.2 and 3.4.4) that sums of random variables, whatever their original distribution, tend toward a normal distribution. In Figure 3-14, the two pdf's of load, $f(x)$, and capacity (resistance), $f(y)$, are shown to overlap. This means that if loads and capacities can occur independently of each other, there are instances in which load exceeds capacity. This means that there is a probability that the capacity fails to meet the load despite the average safety factor exceeding 1 and the safety margin being positive. Average SF and SM are obviously not sufficient criteria for meeting the load. The question arises of how big SF or SM must be to prevent failure. Building and maintaining excess capacity that provides a large safety margin may be uneconomical. But a supply shortfall due to an insufficient safety margin may also be costly. The probabilistic approach combined with an economic analysis can provide an answer to an optimal trade-off between system reliability and system investment cost.

The more accurately a variable can be assessed, the smaller its range will be. Small ranges tend to reduce or eliminate the intersection of the variable ranges. It is clear from Figure 3-14 that there should be a reasonably sized SM, because both load and capacity are not precisely known and also are not under complete control by the decision maker or by the system operators. Equipment maintenance is a control measure that can increase the probability of units being available when scheduled to operate. If the standard deviation of a system's total capacity can be kept small by maintaining all units at high availability, then the required safety margin in the form of excess capacity can be kept small. This means that if *system maintenance* can achieve high system reliability, then investment in excess or redundant capacity can be minimized. Furthermore, high reliability of efficient units reduces the occurrences when high cost energy and capacity must be used or purchased as replacement for unexpected capacity loss. On the load side, good load prediction provides the basis for reliable scheduling of system capacity to meet the load at the lowest possible cost. Hence, system maintenance combined with load prediction are key ingredients to efficient system operation management.

3.5.3 Reliability Assessment

In general, a probabilistic analysis requires information on plausible ranges for the variables and the pdf's over these ranges. The information may be obtained from test data, subjective judgments based on past experience, data records collected over the years, and other sources (Yen et al., 1986, pp. 3–8). In any case, the aim is to establish plausible pdf's for the probabilistic variables. If data are missing, it often is possible to assume a reasonable variable range and a frequency distribution over it as a result of an expert's judgment or a consensus of experts. If load and resistance (capacity) are independent random variables, the reliability assessment can be carried out as follows.

Similar to the addition and subtraction of two random variables in Section 3.3.1, the probability is assessed of the capacity meeting the load. The success event is maintaining a positive difference $Y - X$, where Y stands for resistance or capacity and X stands for load, both are assumed to be independent random variables. The probability of $Y - X$ is the reliability of the system. The probability of a specific load, x_1, is

$$P(X = x_1) = f(x_1)dx \tag{3.5-9}$$

where X represents the random variable *load*, and $f(x_1)$ dx is a special value of the pdf of X, $f(x)$. The probability that the resistance Y is greater than the load x_1 is the integral over the resistance pdf, $f(y)$, from x_1 to infinity which means over the remaining range of Y to the right of x_1.

$$P(Y > x_1) = R(x_1) = \int_{x_1}^{+\infty} f(y)dy \ . \tag{3.5-10}$$

where infinity generally stands for the end of the variable range wherever it may be; y is measured on the same axis as x and in the same units, here MW. $R(x_1)$ is the exceedance probability of x_1 by capacity Y.

The joint probability of x_1 occurring and of the system having a resistance greater than x_1 is the product $R(x_1) f(x_1)$ dx. The sum of all joint probabilities, $R(x_1)$ $f(x_1)$ $dx + R(x_2) f(x_2)$ $dx + ...$, produces the total probability of all mutually exclusive events that the load x_1 is exceeded by a capacity, Y,

$$P(Y \geq X) = \int_0^{+\infty} R(x)f(x)dx \qquad\qquad (3.5\text{-}11)$$

where the integral from 0 to $+\infty$ represents the sum of joint probabilities, $R(x)f(x)\,dx$, in the range in which both functions exist, which is the overlapping portion of $f(x)$ and $f(y)$; outside this range the integral either has reached its maximum value or does not exist; $R(x)$ is a precomputed integral for all x in the common range of both variables, (x,y).

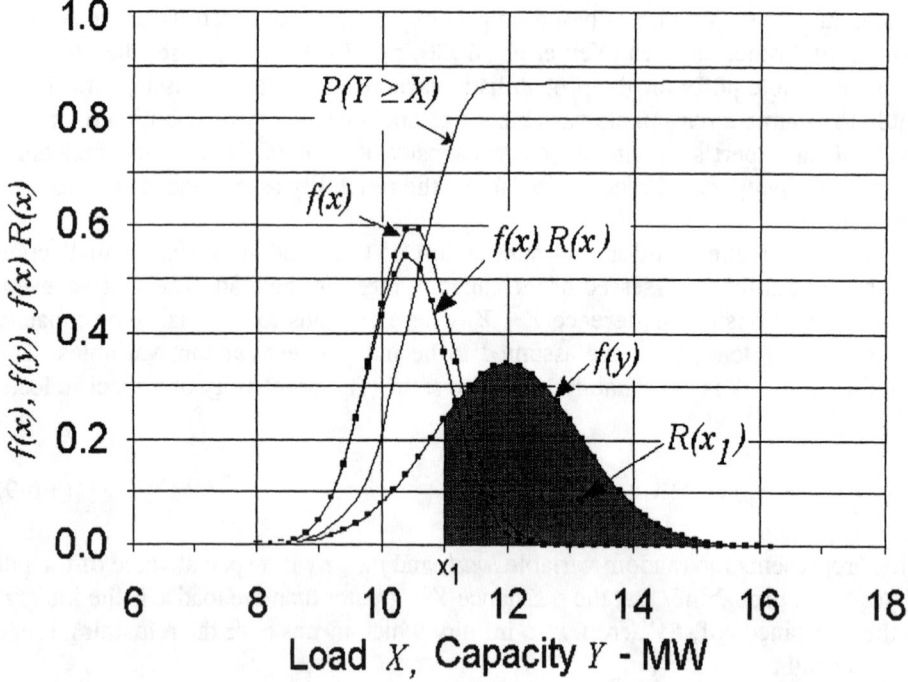

Figure 3-14: Reliability assessment for an electrical system's generating capacity, Y (or resistance to load) satisfying a load X. The pdf of the load is $f(x)$, and the pdf of capacity is $f(y)$. The area under $f(y)$ to the right of the load realization x_1 is $R(x_1)$, which is the probability that the capacity Y for all practical purposes meets the load x_1 and also exceeds it. The product $f(x)\,R(x_1)$ is the joint probability of the load x_1 occurring and being meet by a capacity greater than x_1. The integral of the pdf, $R(x)f(x)dx$, is the probability $P(Y \geq X)$ which is the probability that the capacity meets the load, also called system reliability.

The probability $P(Y \geq X)$, as computed by Equation (3.5-11), is the total probability of capacity (or resistance) Y exceeding the load X or the *reliability of the system*. The calculation is carried out in tabular form, in a spreadsheet. A numerical evaluation of an example is given in Table 3-9 and the results are illustrated in Figure 3-14. The calculations show that the actual range of integration is limited to the rather small range in which the product $[R(x) f(x) dx]$ exists. The result of this integration or summation is the success probability $P(Y \geq X)$, or the reliability of the system. Understanding this calculation goes a long way toward *demystifying* probability calculus and holds the key for understanding many similar reliability calculations. A reliability assessment of corroding vertical lift gates based on this principle was described by Bryant and Mlakar (1987). Also, the load and resistance factor design (LRFD) method is based on this principle (Galambos, 1981).

The reliability assessment explained in this section is a probabilistic assessment of SM, the difference between all possible capacity and load realizations. The fact that $P(Y \geq X) = 0.87 < 1$ indicates that not all possible loads can be met by the available capacity and that there is a probability of failure of 13 %. The decision maker then has to decide if he wants to accept this risk or not. The assumption of capacity and load being independent random variables simplifies the calculations but does not limit the method. A formulation for dependent variables is given by Yen et al. (1986, p. 3).

<u>Examples</u>: (1) A system has a load with a mean $m_X = 10.5$ MW, and a standard deviation $\sigma_X = 0.667$ MW. The capacity has a mean $m_Y = 12$ MW, and $\sigma_Y = 1.1667$ MW. If the total capacity can be assumed to be composed of many comparably sized contributions, such as relatively small, randomly on and off power units, and if the total demand can be assumed to be the sum of many small random demands on the system then, by the CLT, capacity and load can be assumed to be normally distributed. Hence, the use of the normal pdf for both capacity and load is acceptable for a demonstration. <u>Solution</u>: Load and capacity are represented by normally distributed random variables X and Y, respectively. The calculations are summarized in Table 3-9. Column 1 contains the ranges of X and Y. For the numerical evaluation, the theoretical range of the normal variable is limited here to 6 σ. Column 2 contains the normalized variable, z_X, and column 3 contains $f(x) = f(z_X)/\sigma_X$, as computed by the normal pdf. The smaller standard deviation for the load means a narrower bell-shape for the load than for the capacity (resistance). The pdf's in the form of $f(x)$ and $f(y)$ produce ordinates which summed and multiplied by dx and dy, respectively, produce 1 for each pdf. It is important to be aware of units. For example, $f(x)$ has the dimension of $1/x$, because of the standard deviation in the denominator, as shown in Equation (3.5-13). The same is true for $f(y)$. The sums of columns 3 and 5 multiplied by $dx \approx \Delta x = 0.2$ MW are practically 1, which indicates that the numerical integration accuracy is satisfactory. Column 6 shows the area under $f(y)$ to the right of the specified load value, x_1, in Figure 3-14. This integral,

$$R(x_k) = \int_{x_k}^{\infty} f(y)dy \; , \tag{3.5-12}$$

represents the probability that capacity Y exceeds a specific load x_k. It is the reliability for that load level, $R(x)$, and is integrated numerically by the Euler integration formula applied to $f(y)$. The integration starts at the selected value x_k (lower bound), which is the k-th row of the spreadsheet and continues to the end of the range of y,

$$R(x_k) = \Delta x[\frac{1}{2}(E_k + E_n) + \sum_{i=k+1}^{n-1} E_i] \tag{3.5-13}$$

where E is an abbreviation for $f(y)$ with the units MW^{-1}, $\Delta x = 0.2$ MW. Proper use of units will make $R(x)$ a dimensionless number. For $k = n - 2$, there is only one value, E_{n-1}, left under the sum, and for $k = n - 1$, there is none, so that

$$R(x_{n-1}) = \frac{\Delta x}{2}(E_{n-1} + E_n) . \tag{3.5-13a}$$

Column 7 contains the ordinates $[f(x)\ R(x)]$. They are plotted in Figure 3-14 next to the ordinates of $f(x)$. The product $[f(x)\ R(x)]$ is dominated by $f(x)$ because $R(x) \leq 1$. The sum of column 7 is the sum of the ordinates $[f(x)\ R(x)]$. Multiplied by $\Delta x = 0.2$ MW it produces $P(Y \geq X)$, the probability of meeting the load. The same value is obtained in the last row of column 8 which represents the numerical integration of the pdf $f[(x)\ R(x)]$ by

$$H_k = H_{k-1} + \frac{\Delta x}{2}(G_{k-1} + G_k) \tag{3.5-14}$$

where H_k is the abbreviation for the kth function value of $F(x)$ in Table 3-9, column 8. G_k is the abbreviation for the product $f(x)\ R(x)$ of column 7, with units of MW^{-1}, and $\Delta x = 0.2$ MW. Since the pdf value at the beginning of the range, $x = 6$ MW, is very small, the starting value, $F(6)$, is assumed to be zero. In the example, this causes no error in the first four digits of $F(x)$. The complete cumulation of the ordinates of the pdf in column 7 using $\Delta x = 0.2$ MW produces $F(17) = 0.8674$. Hence, the total probability of the available capacity covering the load is $P(Y \geq X) = 0.8674$.

Table 3-9: Calculation of the Reliability for Normally Distributed, Independent Random Variables of Load and Capacity

Load: $m_X = 10.5$ MW $\sigma_X = 0.667$ MW $\Delta x = 0.2$ MW

Capacity: $m_Y = 12$ MW $\sigma_Y = 1.1667$ MW

x	z_X	$f(x)$	z_Y	$f(y)$	$R(x)$	$f(x)R(x)$	$F(x)$
(1)	(2)	(3)	(4)	(5)	(6)	(7)	(8)
6	−6.7466	0.0000	−5.1429	0.0000	1.0000	0.0000	0.0000
6.2	−6.4468	0.0000	−4.9714	0.0000	1.0000	0.0000	0.0000
6.4	−6.1469	0.0000	−4.8000	0.0000	1.0000	0.0000	0.0000
6.6	−5.8471	0.0000	−4.6286	0.0000	1.0000	0.0000	0.0000
6.8	−5.5472	0.0000	−4.4571	0.0000	1.0000	0.0000	0.0000
7	−5.2474	0.0000	−4.2857	0.0000	1.0000	0.0000	0.0000
7.2	−4.9475	0.0000	−4.1143	0.0001	1.0000	0.0000	0.0000
7.4	−4.6477	0.0000	−3.9429	0.0001	1.0000	0.0000	0.0000
7.6	−4.3478	0.0000	−3.7714	0.0003	0.9999	0.0000	0.0000
7.8	−4.0480	0.0002	−3.6000	0.0005	0.9998	0.0002	0.0000
8	−3.7481	0.0005	−3.4286	0.0010	0.9997	0.0005	0.0001
8.2	−3.4483	0.0016	−3.2571	0.0017	0.9994	0.0016	0.0003
8.4	−3.1484	0.0042	−3.0857	0.0029	0.9990	0.0042	0.0009
8.6	−2.8486	0.0103	−2.9143	0.0049	0.9982	0.0103	0.0023
8.8	−2.5487	0.0232	−2.7429	0.0079	0.9969	0.0232	0.0057
9	−2.2489	0.0477	−2.5714	0.0125	0.9948	0.0475	0.0128
9.2	−1.9490	0.0895	−2.4000	0.0192	0.9917	0.0888	0.0264
9.4	−1.6492	0.1535	−2.2286	0.0285	0.9869	0.1515	0.0504
9.6	−1.3493	0.2407	−2.0571	0.0412	0.9799	0.2358	0.0891
9.8	−1.0495	0.3448	−1.8857	0.0578	0.9700	0.3345	0.1462
10	−0.7496	0.4516	−1.7143	0.0787	0.9564	0.4319	0.2228
10.2	−0.4498	0.5406	−1.5429	0.1040	0.9381	0.5071	0.3167
10.4	−0.1499	0.5914	−1.3714	0.1335	0.9144	0.5408	0.4215
10.6	0.1499	0.5914	−1.2000	0.1664	0.8844	0.5230	0.5279
10.8	0.4498	0.5406	−1.0286	0.2015	0.8476	0.4582	0.6260
11	0.7496	0.4516	−0.8571	0.2368	0.8037	0.3630	0.7081
11.2	1.0495	0.3448	−0.6857	0.2703	0.7530	0.2597	0.7704
11.4	1.3493	0.2407	−0.5143	0.2996	0.6960	0.1675	0.8131
11.6	1.6492	0.1535	−0.3429	0.3224	0.6338	0.0973	0.8396
11.8	1.9490	0.0895	−0.1714	0.3370	0.5679	0.0508	0.8544
12	2.2489	0.0477	−0.0000	0.3420	0.5000	0.0239	0.8619
12.2	2.5487	0.0232	0.1714	0.3370	0.4321	0.0100	0.8653
12.4	2.8486	0.0103	0.3429	0.3224	0.3662	0.0038	0.8666
12.6	3.1484	0.0042	0.5143	0.2996	0.3040	0.0013	0.8671
12.8	3.4483	0.0016	0.6857	0.2703	0.2470	0.0004	0.8673

13	3.7481	0.0005	0.8571	0.2368	0.1963	0.0001	0.8674
13.2	4.0480	0.0002	1.0286	0.2015	0.1524	0.0000	0.8674
13.4	4.3478	0.0000	1.2000	0.1664	0.1156	0.0000	0.8674
13.6	4.6477	0.0000	1.3714	0.1335	0.0856	0.0000	0.8674
13.8	4.9475	0.0000	1.5429	0.1040	0.0619	0.0000	0.8674
14	5.2474	0.0000	1.7143	0.0787	0.0436	0.0000	0.8674
14.2	5.5472	0.0000	1.8857	0.0578	0.0300	0.0000	0.8674
14.4	5.8471	0.0000	2.0571	0.0412	0.0201	0.0000	0.8674
14.6	6.1469	0.0000	2.2286	0.0285	0.0131	0.0000	0.8674
14.8	6.4468	0.0000	2.4000	0.0192	0.0083	0.0000	0.8674
15	6.7466	0.0000	2.5714	0.0125	0.0052	0.0000	0.8674
15.2	7.0465	0.0000	2.7429	0.0079	0.0031	0.0000	0.8674
15.4	7.3463	0.0000	2.9143	0.0049	0.0018	0.0000	0.8674
15.6	7.6462	0.0000	3.0857	0.0029	0.0010	0.0000	0.8674
15.8	7.9460	0.0000	3.2571	0.0017	0.0006	0.0000	0.8674
16	8.2459	0.0000	3.4286	0.0010	0.0003	0.0000	0.8674
16.2	8.5457	0.0000	3.6000	0.0005	0.0002	0.0000	0.8674
16.4	8.8456	0.0000	3.7714	0.0003	0.0001	0.0000	0.8674
16.6	9.1454	0.0000	3.9429	0.0001	0.0000	0.0000	0.8674
16.8	9.4453	0.0000	4.1143	0.0001	0.0000	0.0000	0.8674
17	9.7451	0.0000	4.2857	0.0000	0.0000	0.0000	0.8674
	Sum	4.9996		4.9996		4.3369	

Notes: column 1: load realizations x.

column 2: normalized variable x (load), $z_X = (x - m_X)/\sigma_X$

column 3: $f(x) = f(z_X)/\sigma_X = (0.39894/\sigma_X) \exp(-0.5\, z_X^2)$

column 4: normalized variable y (capacity), $z_Y = (y - m_Y)/\sigma_Y$

column 5: $f(y) = f(z_Y)/\sigma_Y = (0.39894/\sigma_Y) \exp(-0.5\, z_Y^2)$

column 6: $R(x_k)$ by numerical integration of $f(y)$ using Equations (3.5-13) and (3.5-13a).   column 7: $f(x_i)\, R(x_i)$, for $i = 1, \dots, n$, with $f(x_i)$ from the ith row of column 3 and $R(x_i)$ from the ith row of column 6.

column 8: cumulation of increments $dF(x) = f(x_i)\, R(x_i)\, \Delta x$; which are the values of column 7 multiplied by Δx using Equation (3.5-14).

bottom row: column sums multiplied by $\Delta x = 0.2$ give values of area integrals; $4.3369/4.9996 = 0.8674$.

(2) An additional calculation was carried out using $m_X = 9.5$ MW, with all other data remaining the same. The resulting probability of the capacity covering the load is $P(Y \geq X) = 0.97$. This means that the same capacity can cover a smaller load with average 9.5 MW and the same dispersion as before with 97 % probability. The safety margin in this case is 2,500 MW, and $SF = 1.25$. Carrying out this analysis for several loads and/or system capacities establishes a relation between the safety factor, the system capacity, and the reliability of meeting the load. The decision maker then selects a comfort level in the form of an acceptable reliability and augments the capacity or reduces the capacity standard deviation until this level is reached. There also may be

a possibility of managing (influencing) the load with the intention of achieving changes in load and its standard deviation. This approach of recognizing the random nature of the input variables is more meaningful than an *a priori* setting of a safety factor and a safety margin without knowledge of load and capacity variability. Figure 3-14 shows that as $R(x) \rightarrow 1$, the function $f(x)R(x) \rightarrow f(x)$ which integrates to 1. This happens when $f(x)$ and $f(y)$ do not overlap. Then each $f(x)$ would be multiplied by $R(x) = 1$. In other words, there is sufficient capacity to cover the range of the load completely.

3.6   Hazard Function and Related Functions

3.6.1   Derivation of the Hazard Function

The *hazard function* assesses the failure rate of a population relative to the momentary number of surviving elements. The function is therefore also called the *instantaneous failure rate distribution* (Moan, 1982, p. 4-17; Siddall, 1972, p. 289). The probability of a system failing by time t is $dF(t) = f(t) \, dt$, where $f(t)$ is the pdf of the system lifetime. The integral of $dF(t)$ is $F(t)$, the probability that failure occurs at any time up to time t. This probability can be expressed in a discrete form by the ratio of the cumulated number of failures, N_f, up to time t, divided by the initial number of elements, N_0,

$$F(t) = \frac{N_f}{N_0} .$$

(3.6-1)

The complement of $F(t)$ is the probability that the system survives beyond time t, also called reliability,

$$R(t) = 1 - F(t)$$

(3.6-2)

An instantaneous failure rate is defined as the rate of surviving elements failing in the subsequent period dt:

$$h(t)N(t) = \frac{dN_f}{dt}$$

(3.6-3)

where $h(t)$ is the rate of failure of elements $N(t)$ that have survived by time t, and dN_f/dt is the number of failures in the upcoming period dt. The reliability $R(t)$ can also be expressed by the ratio of surviving elements, $(N_0 - N_f)$ to initial total elements, N_0:

$$R(t) = \frac{N_0 - N_f}{N_0} = 1 - \frac{N_f}{N_0} = \frac{N(t)}{N_0} \quad . \tag{3.6-4}$$

Taking the derivative of the reliability leads to

$$\frac{dR(t)}{dt} = -\frac{dN_f}{dt} \frac{1}{N_0} \quad . \tag{3.6-5}$$

An expression for $h(t)$ can be written that is entirely expressed by functions of the total distribution:

$$h(t) = -\frac{dR(t)}{dt} \frac{1}{R(t)} . \tag{3.6-6}$$

Since

$$\frac{dR(t)}{dt} = -\frac{dF(t)}{dt} = -f(t) ,$$

the *instantaneous failure rate*, $h(t)$, or *hazard function*, becomes

$$h(t) = \frac{f(t)}{R(t)} = \frac{f(t)}{1 - F(t)} \tag{3.6-7}$$

where $f(t)$ is the pdf of the system life, and $F(t)$ is the cdf of system life.

By its definition, $h(t)$ must have the same units as $f(t)$, $1/t$. If the argument were a distance, x, as in the case of failures per distance, the units of $h(x)$ would be $1/x$. The instantaneous failures per period dt, or distance dx, $h(t)\, dt$, or $h(x)\, dx$, respectively, is not a probability, such as $f(x)\, dx$ or $R(x)$, because it has a changing reference base, i.e., the surviving elements at t. It is a probability defined on the surviving population at time t. Its integral over t or x does not cumulate to 1, as the rate may reach 1 in a single period in which the remaining population dies. Due to the declining reference base at each time t that includes only items that have not yet failed, $h(t)$ is either equal to or greater than $f(t)$. This is seen from Equation (3.6-7) where $h(t)$ is obtained by dividing $f(t)$ by a probability that is a number less than 1.

Examples: (1) Out of a population of 10 items, 4 have failed by time t, and 2 more fail during the time increment dt. What are the failure probability, the survival probability, and the hazard rate? Solution: With $N_0 = 10$, $N(t) = 6$, $N_f = 4$, the failure probability is $F(t) = 4/10 = 0.4$, the incremental failure probability is $dF(t) = 0.2$, the survival probability is $R(t) = 6/10 = 0.6$, and the hazard rate is $h(t)\ dt = dF(t)/[1 - F(t)] = 0.2/(1 - 0.4) = 0.33$.

(2) Calculate $h(t)$ when $f(t)$ is a normal distribution. What is the value of $h(t)$ at the maximum value of $f(t)$? <u>Solution</u>: Note the role of the variable t in $f(t)$ and $R(t)$ as running variable and as integration boundary, respectively. In the second case, as a distinction, x is substituted as the running variable under the integral but it retains the meaning of t. With

$$f(t) = \frac{0.39894}{\sigma} e^{-0.5(\frac{t-\mu}{\sigma})^2}$$

and

$$R(t) = \frac{0.39894}{\sigma} \int_t^\infty e^{-0.5(\frac{x-\mu}{\sigma})^2} dx$$

one obtains

$$h(t) = \frac{f(t)}{R(t)} = \frac{e^{-0.5(\frac{t-\mu}{\sigma})^2}}{\int_t^\infty e^{-0.5(\frac{x-\mu}{\sigma})^2} dx} \cdot \qquad (3.6\text{-}7a)$$

In Equation (3.6-7a), the numerator is evaluated at abscissa value t, whereas the integral in the denominator is evaluated over the remaining variable range from t to infinity. $R(t)$ is said to be a function of its lower boundë The maximum value of a normal pdf occurs at $z = 0$, which corresponds to $t = \mu$, so that $f(\mu) = 0.39894/\sigma$. At this point, $R(\mu) = 0.5$, so that $h(\mu) = 0.7979/\sigma$. The ordinate of $h(t)$ at $t = \mu$ is not the maximum of $h(t)$, as will be seen in the more complete example given in Table 3-12 and in Figure 3-19.

3.6.2 Relations Among Probability Functions

The failure probability is frequently used with time as the argument. If $h(t)\ dt$ is given by observations, other characteristics of the (data) population can be calculated by the relation between $h(t)$, $f(t)$, and $R(t)$ given by Equation (3.6-7). With $f(t) = dF(t)/dt$, and $R(t) = 1 - F(t)$, the integration of Equation (3.6-7) gives

$$H(t_u) = \int_0^{t_u} h(t)dt = \int_0^{t_u} \frac{dF(t)}{1 - F(t)}$$

(3.6-8)

$$= -\ln[1 - F(t)]_0^{t_u} = -[\ln R(t_u) - R(0)]$$

where $H(t_u)$ is the cumulated number of failures over period t_u, the elapsed time from $t = 0$ to the time t_u when the evaluation of reliability is made. The reliability, or probability of survival, at time t_u follows from Equation (3.6-8) as

$$R(t_u) = R(0)e^{-H(t_u)}$$

(3.6-9)

where $R(0)$ is the reliability at starting time, here $t = 0$. It usually is assumed that $R(0) = 1$.

In analogy to the reliability of an exponential pdf (Section 2.7.5), one can construe the exponent, $H(t_u)$, as the product of an average rate multiplied by a time, so that the reliability becomes

$$R(t_u) = R(0)e^{-\lambda_u t_u}$$

(3.6-10)

where

$$\lambda_u = \frac{H(t_u)}{t_u}$$

(3.6-10a)

with λ_u being an average failure rate over the period t_u. Equation (3.6-10a) tends to overestimate failure rates, if t_u is long, as a linearization does not take into account the usually very low rates early in the life of a component.

Since $R(t_u) = 1 - F(t_u)$, and $R(0) = 1 - F(0)$, the cumulative distribution function of failure time, or the probability of failure during time t_u, is

$$F(t_u) = 1 - [1 - F(0)]e^{-\lambda_u \cdot t_u} .$$

(3.6-11)

With $F(0) = 0$, one obtains

$$F(t_u) = 1 - e^{-\lambda_u \cdot t_u} .$$

(3.6-11a)

Differentiation of the general form of Equation (3.6.11a), with index u omitted, gives $dF(t)/dt = f(t)$, or

$$f(t) = \lambda \, e^{-\lambda t} \; . \tag{3.6-12}$$

Equation (3.6-12) is the pdf of the exponential distribution (see Section 2.7.5). By substituting Equation (3.6-12) into Equation (3.6-7), setting $h(t) = \lambda$, and solving for $R(t)$, the reliability or survival probability for an exponentially distributed t is

$$R(t) = e^{-\lambda t} \; . \tag{3.6-13}$$

Equation (3.6-13) assumes a constant failure rate over period $(0,t)$. Equations (3.6-11), (3.6-12), and (3.6-13) are all functions of the exponential distribution, as given in Section 2.7.5. No specific assumptions have been made here for this distribution to appear except for the basic definition of the hazard rate as being the number of failures out of the momentarily existing population. This assumption brings forth the exponential decay function and its associated functions, such as the exponential pdf, the non-exceedance probability, and reliability.

3.6.3 Mortality Rates

A typical example of $h(t)$ is the mortality rate, the die-off rate of an existing population of animate or inanimate entities. Human mortality rates, for ages from 0 to 99 years in incidents per 1,000 of the momentarily existing population are given in Table 3-10, column 2 (Commissioners Standard Ordinary Mortality Table [1958], from Greene, 1977, p. 412). They were recorded by insurance companies during the years 1950 to 1954. These rates are updated from time to time (the latest version dates from 1980). According to the definition of mortality rates, they represent deaths per 1,000 of population for an age group and a short period at this age, usually one year. When multiplied by that time increment, they become $h(t) \, dt$, the number of failures for the population group that had survived to the beginning of that time increment. These failure frequencies are also called the *instantaneous failure probability* for that specific age group. The instantaneous failure probability is a function of age and adds to 1 only with its complement in the age group, i.e., the survivor probability. If $dt = 1$ year, the numerical values of $h(t)$ and $h(t) \, dt$ may be the same, but not their dimensions. The rates have all been reduced to a common denominator, a reference survivor population of 1,000 at the beginning of dt. Adding the rates for a period t_u produces the total failure rate for the period t_u, which is $H(t_u)$, as expressed by Equation (3.6-8) in Section 3.6.2.

Table 3-10: Instantaneous Failure Rates $h(t)$; Reliability $R(t)$; Life Length pdf $f(t)$; and Expected Remaining Life, $E[L_R(t_a)]$. The instantaneous failure rates of column 2 are based on Commissioners Standard Ordinary Mortality Table (CSO, 1958; Greene, 1977, Table 19-1, p. 412).

Age in Years	Failure Rate	Number of Failures	Reliability	Life pdf	Reliability (alt.)	Probability of	Numerical Integral of	Remaining Life
	$h(t)$	$H(t)$	$R(t)$	$f(t)$	$R(t)$	t	$t\,f(t)\,dt$	$E[L_R(t_a)]$
1	2	3	4	5	6	7	8	9
0	0.00708	0.0000	1.0000	0.0071	1.0000	0.0000	68.1801	68.18
1	0.00176	0.0044	0.9956	0.0018	0.9956	0.0018	68.1792	67.48
2	0.00152	0.0061	0.9940	0.0015	0.9940	0.0030	68.1768	66.59
3	0.00146	0.0076	0.9925	0.0014	0.9925	0.0043	68.1732	65.69
4	0.00140	0.0090	0.9911	0.0014	0.9911	0.0055	68.1682	64.78
5	0.00135	0.0104	0.9897	0.0013	0.9897	0.0067	68.1621	63.87
6	0.00130	0.0117	0.9884	0.0013	0.9884	0.0077	68.1549	62.96
7	0.00126	0.0130	0.9871	0.0012	0.9871	0.0087	68.1467	62.04
8	0.00123	0.0142	0.9859	0.0012	0.9859	0.0097	68.1375	61.11
9	0.00121	0.0154	0.9847	0.0012	0.9847	0.0107	68.1273	60.19
10	0.00121	0.0166	0.9835	0.0012	0.9835	0.0119	68.1160	59.26
11	0.00123	0.0179	0.9823	0.0012	0.9823	0.0133	68.1034	58.33
12	0.00126	0.0191	0.9811	0.0012	0.9811	0.0148	68.0893	57.40
13	0.00132	0.0204	0.9798	0.0013	0.9798	0.0168	68.0735	56.48
14	0.00139	0.0217	0.9785	0.0014	0.9785	0.0190	68.0556	55.55
15	0.00146	0.0232	0.9771	0.0014	0.9771	0.0214	68.0353	54.63
16	0.00154	0.0247	0.9756	0.0015	0.9756	0.0240	68.0126	53.71
17	0.00162	0.0262	0.9741	0.0016	0.9741	0.0268	67.9872	52.80
18	0.00169	0.0279	0.9725	0.0016	0.9725	0.0296	67.9590	51.88
19	0.00174	0.0296	0.9708	0.0017	0.9708	0.0321	67.9281	50.97
20	0.00179	0.0314	0.9691	0.0017	0.9691	0.0347	67.8948	50.06
21	0.00183	0.0332	0.9673	0.0018	0.9673	0.0372	67.8588	49.15
22	0.00186	0.0350	0.9656	0.0018	0.9656	0.0395	67.8205	48.24
23	0.00189	0.0369	0.9638	0.0018	0.9638	0.0419	67.7798	47.33
24	0.00191	0.0388	0.9619	0.0018	0.9619	0.0441	67.7368	46.42
25	0.00193	0.0407	0.9601	0.0019	0.9601	0.0463	67.6916	45.51
26	0.00196	0.0427	0.9582	0.0019	0.9582	0.0488	67.6440	44.59
27	0.00199	0.0447	0.9563	0.0019	0.9563	0.0514	67.5939	43.68
28	0.00203	0.0467	0.9544	0.0019	0.9544	0.0542	67.5411	42.77
29	0.00208	0.0487	0.9524	0.0020	0.9524	0.0575	67.4852	41.85
30	0.00213	0.0508	0.9504	0.0020	0.9504	0.0607	67.4261	40.94
31	0.00219	0.0530	0.9484	0.0021	0.9484	0.0644	67.3636	40.03
32	0.00225	0.0552	0.9463	0.0021	0.9463	0.0681	67.2973	39.12

33	0.00232	0.0575	0.9441	0.0022	0.9441	0.0723	67.2271	38.21
34	0.00240	0.0599	0.9419	0.0023	0.9419	0.0769	67.1525	37.29
35	0.00251	0.0623	0.9396	0.0024	0.9396	0.0825	67.0728	36.39
36	0.00264	0.0649	0.9372	0.0025	0.9372	0.0891	66.9870	35.48
37	0.00280	0.0676	0.9346	0.0026	0.9346	0.0968	66.8941	34.57
38	0.00301	0.0705	0.9319	0.0028	0.9319	0.1066	66.7924	33.67
39	0.00325	0.0736	0.9290	0.0030	0.9290	0.1178	66.6802	32.78
40	0.00353	0.0770	0.9259	0.0033	0.9259	0.1307	66.5559	31.89
41	0.00384	0.0807	0.9225	0.0035	0.9225	0.1452	66.4180	31.00
42	0.00417	0.0847	0.9188	0.0038	0.9188	0.1609	66.2649	30.12
43	0.00453	0.0891	0.9148	0.0041	0.9148	0.1782	66.0953	29.25
44	0.00492	0.0938	0.9105	0.0045	0.9105	0.1971	65.9077	28.39
45	0.00535	0.0989	0.9058	0.0048	0.9058	0.2181	65.7001	27.53
46	0.00583	0.1045	0.9008	0.0053	0.9008	0.2416	65.4703	26.68
47	0.00636	0.1106	0.8953	0.0057	0.8953	0.2676	65.2157	25.84
48	0.00695	0.1173	0.8893	0.0062	0.8893	0.2967	64.9335	25.01
49	0.00760	0.1245	0.8829	0.0067	0.8829	0.3288	64.6208	24.19
50	0.00832	0.1325	0.8759	0.0073	0.8759	0.3644	64.2742	23.38
51	0.00911	0.1412	0.8683	0.0079	0.8683	0.4034	63.8903	22.58
52	0.00996	0.1508	0.8601	0.0086	0.8601	0.4454	63.4659	21.79
53	0.01089	0.1612	0.8511	0.0093	0.8511	0.4913	62.9975	21.01
54	0.01190	0.1726	0.8415	0.0100	0.8415	0.5407	62.4815	20.25
55	0.01300	0.1850	0.8311	0.0108	0.8311	0.5942	61.9140	19.50
56	0.01421	0.1986	0.8199	0.0117	0.8199	0.6524	61.2907	18.76
57	0.01554	0.2135	0.8078	0.0126	0.8078	0.7155	60.6068	18.03
58	0.01700	0.2298	0.7947	0.0135	0.7947	0.7836	59.8572	17.32
59	0.01859	0.2476	0.7807	0.0145	0.7807	0.8563	59.0373	16.62
60	0.02034	0.2670	0.7657	0.0156	0.7657	0.9344	58.1420	15.93
61	0.02224	0.2883	0.7495	0.0167	0.7496	1.0168	57.1663	15.27
62	0.02431	0.3116	0.7323	0.0178	0.7323	1.1037	56.1061	14.61
63	0.02657	0.3370	0.7139	0.0190	0.7139	1.1950	54.9567	13.98
64	0.02904	0.3648	0.6943	0.0202	0.6944	1.2904	53.7140	13.36
65	0.03175	0.3952	0.6735	0.0214	0.6736	1.3900	52.3738	12.75
66	0.03474	0.4285	0.6515	0.0226	0.6516	1.4938	50.9320	12.17
67	0.03804	0.4649	0.6282	0.0239	0.6283	1.6011	49.3845	11.60
68	0.04168	0.5047	0.6037	0.0252	0.6038	1.7109	47.7285	11.05
69	0.04561	0.5484	0.5779	0.0264	0.5780	1.8187	45.9637	10.52
70	0.04979	0.5961	0.5510	0.0274	0.5511	1.9203	44.0942	10.00
71	0.05415	0.6480	0.5231	0.0283	0.5233	2.0110	42.1285	9.51
72	0.05865	0.7044	0.4944	0.0290	0.4946	2.0877	40.0792	9.03
73	0.06326	0.7654	0.4651	0.0294	0.4654	2.1480	37.9613	8.57
74	0.06812	0.8311	0.4356	0.0297	0.4358	2.1957	35.7895	8.11

75	0.07337	0.9018	0.4058	0.0298	0.4061	2.2331	33.5751	7.67
76	0.07918	0.9781	0.3760	0.0298	0.3764	2.2628	31.3271	7.24
77	0.08570	1.0605	0.3463	0.0297	0.3466	2.2850	29.0532	6.82
78	0.09306	1.1499	0.3167	0.0295	0.3171	2.2985	26.7615	6.41
79	0.10119	1.2471	0.2873	0.0291	0.2878	2.2971	24.4637	6.01
80	0.10998	1.3526	0.2586	0.0284	0.2590	2.2749	22.1777	5.62
81	0.11935	1.4673	0.2305	0.0275	0.2311	2.2288	19.9259	
82	0.12917	1.5916	0.2036	0.0263	0.2041	2.1566	17.7332	
83	0.13938	1.7258	0.1780	0.0248	0.1786	2.0595	15.6252	
84	0.15001	1.8705	0.1540	0.0231	0.1546	1.9410	13.6249	
85	0.16114	2.0261	0.1318	0.0212	0.1325	1.8059	11.7514	4.45
86	0.17282	2.1931	0.1116	0.0193	0.1122	1.6582	10.0194	
87	0.18513	2.3721	0.0933	0.0173	0.0939	1.5025	8.4390	
88	0.19825	2.5638	0.0770	0.0153	0.0776	1.3436	7.0159	
89	0.21246	2.7691	0.0627	0.0133	0.0633	1.1859	5.7511	
90	0.22814	2.9894	0.0503	0.0115	0.0509	1.0331	4.6416	3.12
91	0.24577	3.2264	0.0397	0.0098	0.0403	0.8879	3.6811	
92	0.26593	3.4822	0.0307	0.0082	0.0314	0.7521	2.8611	
93	0.28930	3.7598	0.0233	0.0067	0.0239	0.6266	2.1718	
94	0.31666	4.0628	0.0172	0.0054	0.0178	0.5120	1.6025	
95	0.35124	4.3968	0.0123	0.0043	0.0129	0.4110	1.1410	1.64
96	0.40056	4.7727	0.0085	0.0034	0.0091	0.3252	0.7729	
97	0.48842	5.2171	0.0054	0.0026	0.0060	0.2569	0.4818	
98	0.66815	5.7954	0.0030	0.0020	0.0037	0.1991	0.2538	
99	0.98000	6.6195	0.0013	0.0013	0.0020	0.1294	0.0895	0.30
100	1	7.6095	0.0005	0.0005	0.0011	0.0496	0.0000	
				0.9980		68.1801		

Notes: column 1: age t.

column 2: instantaneous failure rates, $h(t)$, which are the fraction of the surviving age group; at age 99, 98 % of the surviving population that enters the year dies; multiplying the column values by 1,000 gives the number of deaths per thousand at age t; at age 99, 980 of 1,000 die during the time increment $dt \approx \Delta t = 1$ year. The failure rates are from the CSO (1958).

column 3: integration of failure rate from age 0 to age t by $H(t) = \sum h(t)\Delta t$

column 4: reliability calculated as $R(t) = R(0) \exp[-H(t)]$, where $R(0) = 1$.

column 5: life pdf calculated by $f(t) = h(t) R(t)$.

column 6: alternative calculation of $R(t)$ by $1 - F(t)$, where $F(t) = \int f(t) \, dt$, with integration from 0 to t (from the bottom up). The numerical integral is calculated by the Euler integration formula (trapezoidal formula).

column 7: probabilities of t by $t f(t) \, dt$; the column sum is the expectation of t, $E(t) = 68.18$ years.

column 8: numerical integration of $\int t f(t) \, dt$ by cumulating the values in column 7 (from the bottom up).

column 9: expected remaining life, $E[L_R(t_a)]$ for a given age t_a; the calculation from $t_a = 85$ years on is explained in Section 3.6.4. The CSO (1958) gives values slightly higher than those in column 9.

The instantaneous failure probability, $h(t)$ dt, of Table 3-10 is plotted in Figure 3-15. It starts with a relatively high rate of 0.0071, or about 7 deaths per 1,000, at age 0. This means that in the first year of life, about 7 of each 1,000 entrants into life die. Then the rate drops sharply to a minimum of 0.0012 at age 10. The minimum is rather flat from 7 to 12 years, but then the rate rises more and more rapidly with age. It exceeds the death rate at birth at age 48, reaches twice that rate by age 56, and about 0.0500 at age 70. By age 80 the rate has climbed to almost 0.1100, and at age 99, $h(t)$ $dt = 1$. This means that all members of this age group die in the upcoming year. The instantaneous failure probability, $h(t)$ dt cannot exceed 1, as the number of deaths cannot exceed the group population.

The mortality rates for any population, human or other, are subject to updating from time to time as new data become available. Environmental conditions or technological innovation may change the overall shape of the instantaneous failure rate. For example, human survival beyond 100 years has increased in frequency. But for age 100 and beyond, the mortality rate will remain quite high. For example, if it were 0.99, the population would be reduced exponentially at this rate until there are no survivors.

The human mortality rates in Figure 3-15 follow the typical shape of a *bathtub curve* (see also Section 2.7.6, Figure 2-23, and Section 6.3.2, Figure 6-3). Figure 3-15 is not a probability distribution but the distribution of the quantity "death rate" over age. The walls of the bathtub on both sides represent the highest rate changes with time at the beginning and end of the "system's" life, with a minimum rate change in between. Until about age 10, the rate change is negative, which means the failure rate actually decreases. At age 10, the rate change is zero for a short period whereupon, it becomes positive and increases with increasing age due to wearout failures. It is interesting to note the parallels between biological and technical systems. Both have relatively high failure rates at the start, due to flaws and errors in design, construction, and operation, the latter caused by lack of experience with system operation. After the start-up difficulties are overcome, and the system is "debugged" it runs for a while with minimum failure. The bottom of the bathtub curve is marked by a failure rate that is almost constant with time, which means it has a horizontal tangent. During this period the rate may reflect random external causes, such as accidents, that are unrelated to age. With increasing age, however, wear-out failures start to increase. They about double every 10 years between 30 and 90, and from 90 to 100 they grow five-fold from about 0.2 to 1.0, the maximum possible instantaneous failure rate. It is one of the tasks of maintenance to keep down the rise of the failure rate, thereby extending the life of the system.

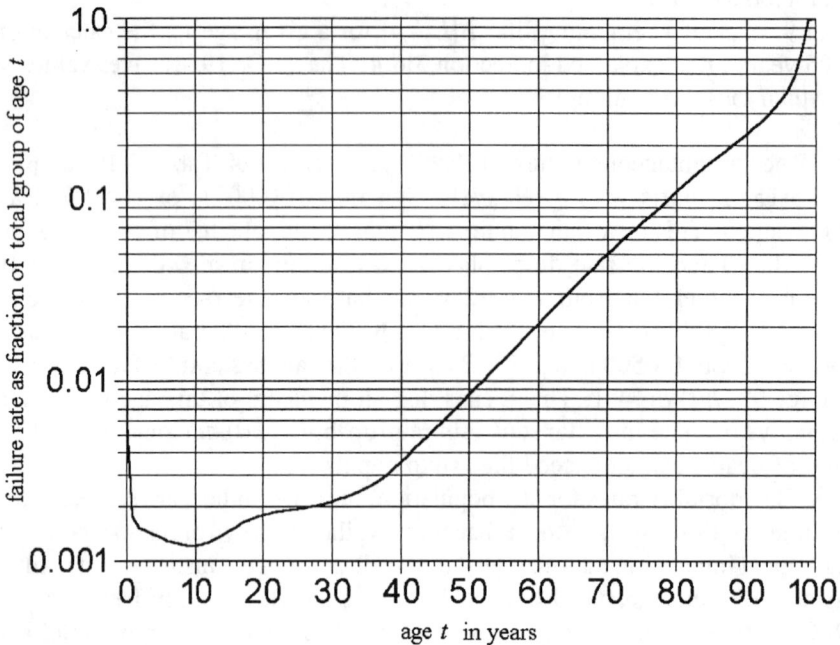

Figure 3-15: Human mortality (or death) rates from Table 3-10, column 2, after CSO (1958). Mortality rates are the fraction of the population group at age t that dies off during age increment $\Delta t = 1$ year. For example, at age 100, the total population dies in the year ahead which makes the mortality rate 1, or 1,000 death per 1,000 members of the population group at age 100. This, of course, is not quite true, as these rates are averages.

$H(t)$ in column 3 of Table 3-10 is obtained by integration of $h(t)$ using Equation (3.6-8). If $dt = 1$ year, it amounts to adding the $h(t)$ values. The survival probability, $R(t)$, is computed by using Equation (3.6-9) with $H(t)$ as exponent. The result is given in column 4 and plotted in Figure 3-16. The reliability decline from age 40 to age 90 goes hand in hand with a more than 70-fold increase in mortality rate during this period. The pdf of life length, $f(t)$, is calculated by Equation (3.6-7), which is solved for $f(t)$, given $h(t)$,

$$f(t) = h(t)R(t) \ . \tag{3.6-7b}$$

The $f(t)$ mark the frequency of life ending at age t. The calculated values are shown in column 5, and are graphically represented in Figure 3-17. The pdf has an extended minimum from about 2 to 30 years. This is the period in which the system survives with minimum age-related failures. After age 30, $f(t)$ starts to increase rapidly and reaches its maximum at about 75 years, which means the frequency of lives ending is highest at this age. Column 6 shows an alternative calculation of $R(t)$ based on Equation (3.6-2). Both values of $R(t)$ in columns 4 and 6 are, of course, the same.

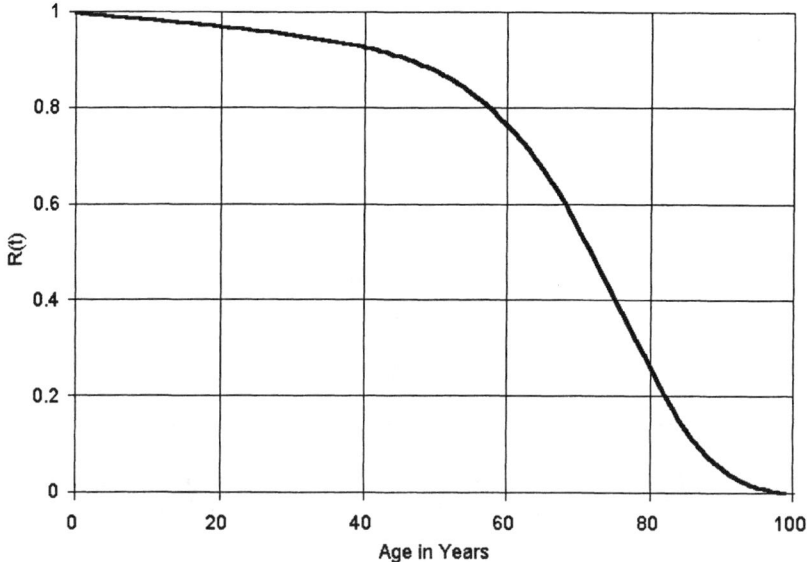

Figure 3-16: Reliability of a "human system," $R(t)$, based on column 4 of Table 3-10.

Examples: (1) A population of six motors 1 MW and above of a California pumping plant worked for 25 years when they showed signs of a need for rewinds. The performance record is shown in the table that is given with this example (the data are taken from DOE/BUREC, 1989, pp. A-66 and A-137). Calculate the reliability of the population. Solution: The initial population of six units is depleted over time by retirement and by taking units out for rewinds. For the purpose of demonstrating the method, units taken out for rewind are assumed to be reductions of the aging population, which thus dwindles to zero. The first retirement or failure occurs at age 26 when the first unit is taken out. The failure rate is $1/6 = 0.1667$.

Performance Record for Example (1)

Age in Years	Retirement or Failure	Survivor Population	Retirement Rate	$H(t)$	$R(t)$
(1)	(2)	(3)	(4)	(5)	(6)
0	0	6	0	0	1
1	0	6	0	0	1
2	0	6	0	0	1
—	—	—	—	—	—
24	0	6	0	0	1
25	0	6	0	0	1
26	1	6	0.1667	0.1667	0.8465
27	0	5	0.0000	0.1667	0.8465
28	0	5	0.0000	0.1667	0.8465
29	3	5	0.6000	0.7667	0.4646
30	1	2	0.5000	1.2667	0.2818
31	1	1	1.0000	2.2667	0.1037

Then, in year 29, three out of five units are taken out, which produces a failure rate of $3/5 = 0.6$. In year 31, the last unit is taken out of the remaining population of 1, which reduces the survivor population to 0 and produces a failure rate of 1.0. Column 5 shows the integral of the failure rate after Equation (3.6-8), and column 6 shows the reliability or survival probability after Equation (3.6-9). Three important characteristics of the hazard function are demonstrated: (1) The retirement rates in column 4 can be construed as probabilities that refer to the dwindling survivor population in column 3, and not to the total population. (2) The vertical summation of failure rates in column 4 usually exceeds 1.0 when the population die-off is completed, as the last $h(t)$ by itself is already 1.0 (see also Table 3-10). (3) The reduction of reliability in column 6 reaches asymptotically zero because of Equation (3.6-9) being an exponential function. In the aforementioned case, this would happen if $H(t)$ had reached a higher value, as for example in Table 3-10. Because of the few steps used, one should assume that $R(t)$ reaches 0 in the next time step (year 32), as a very small virtual die-off of a zero population would send $h(t)$ and $H(t)$ to infinity, and thus $R(t)$ to zero. A fifty percent survivor probability is reached at 29 years. The simplicity of the example should not mislead the reader to believe that this computation can, in general, be replaced by simple averaging.

(2) A population of 290 units holds up at high reliability for about 25 years. Then units need to be taken out for repair and replacement. The data resemble a case of refurbishing pump motors in a power system (the data are taken from DOE/BUREC,

1989, pp. A-66 and A-137). Calculate the reliability of the population. Solution: The initial population of 290 units is depleted over time by taking units out for replacements or rewinds without replacements in the aging population. The population and its reduction over time are shown in the table.

Performance Record for Example (2)

Age in Years	Retirement or Failure	Survivor Population	Retirement Rate	$H(t)$	$R(t)$
(1)	(2)	(3)	(4)	(5)	(6)
0	0	290	0	0	1
1	0	290	0	0	1
2	0	290	0	0	1
—	—	—	—	—	—
24	0	290	0	0	1
25	0	290	0	0	1
26	1	290	0.0034	0.0034	0.9966
27	0	289	0.0000	0.0034	0.9966
28	0	289	0.0000	0.0034	0.9966
29	3	289	0.0104	0.0138	0.9863
30	1	286	0.0035	0.0173	0.9828
31	1	285	0.0035	0.0208	0.9794

The failure rates in column 4 are much smaller than those in Example (1). They start with $1/290 = 0.0035$ in year 26 and reach maximally 0.0104 in year 29. Also, the summation $H(t)$, according to Equation (3.6-8), remains very small, so that $R(t)$, calculated by Equation (3.6-9), remains close to 1. In year 31, the survivor population is still very large, and the probability of survival, $R(31) = 97.94$ %, as shown at the bottom of column 6. With six units removed out of 290, the fraction surviving is $(290 - 6)/290 = 0.9793$. An important point is made by this example: For small failure rates, the population remains almost intact as a constant reference base, so that the failure rates calculated for the surviving population are almost the same as the failure probabilities of the total population. This impression is given when comparing the numbers of columns 5 and 6. That this is only a special case is clearly shown when comparing these numbers with those of Example (1).

3.6.4 Failure and Survival Probabilities and Expectations

For maintenance and insurance purposes, it may be important to know the *probable length of life* of a component, the *average length of life*, and the *remaining life*. These quantities can all be computed from previously derived relations.

(a) The *probability of system failure*, or the probability of life to end at any time t between $t = 0$ and $t = t_e$, is

$$P(t \leq t_e) = F(t_e) = \int_0^{t_e} f(t)dt \tag{3.6-14}$$

where $f(t)$ is the pdf of life length. The probability of life ending at any particular time t, for a continuous function, can only be expressed for an infinitesimally short period, dt, at time t, to obtain a probability greater than zero:

$$dF(t) = f(t)dt \tag{3.6-15}$$

If the infinitesimally short period is expanded to a finite period, e.g., a year, from time t_1 to t_2, then the probability of failure in this period is

$$P(t_1 \leq t \leq t_2) = F(t_2) - F(t_1) \tag{3.6-16}$$

where t_1 and t_2 could be numbers of years, say 40 and 41. The probability of no failure during this period is

$$1 - P(t_1 \leq t \leq t_2) = 1 - [F(t_2) - F(t_1)]. \tag{3.6-17}$$

Equation (3.6-17) does not represent a reliability because it does not exclude failures prior to t_1 (Siddall, 1972, p. 294). The survival probability, or reliability, for period $(0,t_1)$ is

$$R(t_1) = 1 - F(t_1). \tag{3.6-18}$$

The probability of survival up to t_1, and also in the subsequent period, from t_1 to t_2, is the conditional probability of survival from t_1 to t_2, given survival has occurred until t_1 (see Section 2.1.4):

$$P(t_2|t_1) = \frac{P(t_1 \cap t_2)}{P(t_1)} \tag{3.6-19}$$

where $P(t_1 \cap t_2)$ is the joint probability that there is no failure until t_1 and also none from t_1 until t_2. With

$$P(t_1 \cap t_2) = 1 - F(t_2), \text{ and } P(t_1) = 1 - F(t_1)$$

Equation (3.6-19) becomes

$$P(t_2|t_1) = \frac{1 - F(t_2)}{1 - F(t_1)} = \frac{R(t_2)}{R(t_1)}, \tag{3.6-20}$$

which is the probability of no failure during period $t_2 - t_1$, given there is no failure in period $(0, t_1)$. The probability of no failure until t_2 without the additional condition is

$$R(t_2) = P(t_2|t_1)R(t_1) \tag{3.6-20a}$$

This shows that $P(t_2|t_1) > R(t_2)$, because if there is no failure until t_1, the probability is increased that there also is no failure during $t_2 - t_1$.

(b) The *expected life* is the expectation of t, which is the sum of all probability-weighted life lengths, $t f(t) dt$. The life length pdf, $f(t)$, can be calculated from the mortality rates by Equation (3.6-7):

$$f(t) = h(t)R(t) . \tag{3.6-7b}$$

This calculation is carried out in Table 3-10, column 5. Figure 3-17 shows the pdf of life length $f(t)$. It is a rather irregular curve. No analytical fit was attempted as it is not needed for numerical integration.

If T is the collective of random realizations t, then the expectation of T is the first moment of t (see Section 2.5.3),

$$E(T) = \int_0^\infty tf(t)dt . \tag{3.6-21}$$

where the integration is over the entire range of t, theoretically from $0 \le t \le \infty$, but for the numerical integration, the range is from $0 \le t \le 100$. Table 3-10, column 5, gives $f(t)$, but with $\Delta t = 1$, the values also are numerically equal to $f(t) dt$, the probabilities of t, $P(T = f(t) dt)$, where T is the random variable of life length. The probability-weighted t's, $t f(t) dt$, are given in column 7 and their numerical integration is given in column 8. The result is the *expected life length*, $E(T) = 68.18$ years, which is valid for the CSO [1958]. The present life expectancy of the U.S. population is about 76 years according to newer mortality tables. Here Table 3-10 serves only as a data source and is not a presently valid mortality table.

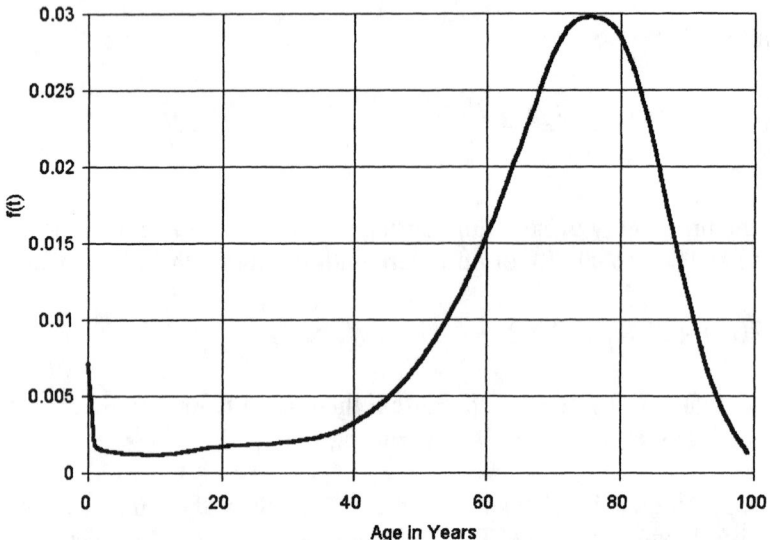

Figure 3-17: Human life length probability density function, $f(t)$, derived from mortality rates in Table 3-10.

(c) The *expected remaining life*. After a system has survived to age $T = t_a$, the actual or attained age, there is a probability that a higher age $t > t_a$ is reached. The life that remains after a life t_a is the *remaining life*, $T_r = t - t_a$. It is the difference of two random variables and therefore also a random variable. At birth, $t_a = 0$, and the expected remaining life is equal to the expected life, $E(T_r) = E(T)$, here 68.2 years. As the system advances in age, the range of T_r shrinks and the possible remaining life length get shorter. Hence $E(T_r)$ is a function of the actual life already reached. Figure 3-17shows that if one disregards the long left tail, $f(t)$ is almost symmetric around the mean, with most life lengths ranging from about 50 to 90, with the mean life length at around 70. If the actual age is $t_a = 50$, there is a good chance that the remaining life, $T_r = T - t_a = 20$ years, or more.

The probability of the remaining life span, $T - t_a$, where $T > t_a$ is the conditional probability that a time $t > t_a$ will be reached, given that t_a has been reached,

$$P[(T - t_a) \cap t_a] = P(T - t_a | t_a) P(t_a) \qquad (3.6\text{-}22)$$

where T is the random variable of age, and t_a is a reference age, the actual age, for which the remaining life is to be determined. The remaining life at every age t_a has a probability distribution of possible remaining life lengths. This distribution is a function of the parameter t_a. The probability of exceeding t_a is $R(t_a)$, and the probability of not failing before t is $R(t)$. Hence, similar to Equation (3.6-20), Equation (3.6-22) can be written as

$$P(t - t_a | t_a) = \frac{R(t)}{R(t_a)}.$$

$(3.6\text{-}23)$

Weighing all possible periods $t - t_a$ by their probabilities given by the life length pdf $f(t)\,dt$, from t_a to infinity gives

$$E[L_R(t_a)] = \frac{1}{R(t_a)} \int_{t_a}^{\infty} (t - t_a) f(t) dt$$

$(3.6\text{-}24)$

where $E[L_R(t_a)]$ is the *expected remaining life* as a function of age t_a; $R(t_a)$ is the probability that no failure occurs until time t_a. The integral of Equation (3.6-24) can be split into two parts:

$$E[L_R(t_a)] = \frac{1}{R(t_a)} \int_{t_a}^{\infty} t f(t) dt - \frac{1}{R(t_a)} \int_{t_a}^{\infty} t_a(t) dt$$

and t_a can be pulled in front of the second integral whereupon the second integral is seen to equal $R(t_a)$ so that the equation simplifies to

$$E[L_R(t_a)] = \frac{1}{R(t_a)} \int_{t_a}^{\infty} t f(t) dt - t_a$$

$(3.6\text{-}25)$

where $E[L_R(t_a)]$ is the expectation of the function of remaining life evaluated for a selected t_a; the integral is the sum of the probability-weighted life lengths $t > t_a$; $R(t_a)$ is the sum of all probabilities of $t > t_a$. $R(t_a)$ for selected t_a and the integral of $t\,f(t)$ are given in Table 3-10, columns 6 and 8, respectively. The remaining expected life is given in column 9. The result is illustrated in Figure 3-18. The often quoted *life after fifty-five* is $E[L_R(55)] = 20$ years. At the expected age of 68 years, $E[L_R(68)] = 10$ years, and at 99 years one can expect $E[L_R(99)] = 0.3$ years (depending on computation accuracy). It must be emphasized that $E[L_R(t_a)]$ is an expectation. There are probability distributions around the expected values similar to those discussed in Section 3.1.6 and shown in Figure 3-3. They would have ranges anywhere from 0 to

68 for the range of t_a from 0 to 100 years, with the mean of each range being $E[L_R(t_a)]$.

The first term on the right side of Equation (3.6-25) is always greater than t_a, so that $E[L_R(t_a)] > 0$. In the example used, the integration $\int (t - t_a) f(t)\, dt$ requires a somewhat more elaborate procedure to prevent $E[L_R(t_a)] < 0$ from occurring. An example for $t_a = 85$ years is given in Table 3-11. Column 4 contains the possible remaining life periods from 0 to 15. The column values for $f(t)$ and $R(t_a)$ are taken from columns 5 and 6, respectively, of Table 3-10. The integration in column 6 is carried out by applying Euler's integration formula to the values in column 5 and summing from the bottom up. $E[L_R(t_a)]$ is obtained by dividing the top value of column 6 by $R(t_a)$. For better accuracy, t should continue beyond 100 until $f(t)$ is zero, to improve the accuracy of the calculation.

Calculations similar to those in Table 3-11 were carried out for $t_a = 90$ years and $t_a = 95$ years. The results are given in Table 3-10, column 9.

Table 3-11: Calculation of Expected Remaining Life for Attained Age $t_a = 85$ years

t	$f(t)$	$R(t_a)$	$t - t_a$	$f(t)(t - t_a)$	$\int (t - t_a) f(t) dt$	$E[L_R(t_a)]$
(1)	(2)	(3)	(4)	(5)	(6)	(7)
85	0.0212	0.1325	0	0	0.5898	4.45
86	0.0193		1	0.0193	0.5801	
87	0.0173		2	0.0345	0.5532	
88	0.0153		3	0.0458	0.5130	
89	0.0133		4	0.0533	0.4635	
90	0.0115		5	0.0574	0.4081	
91	0.0098		6	0.0585	0.3502	
92	0.0082		7	0.0572	0.2923	
93	0.0067		8	0.0539	0.2367	
94	0.0054		9	0.0490	0.1853	
95	0.0043		10	0.0433	0.1391	
96	0.0034		11	0.0373	0.0989	
97	0.0026		12	0.0319	0.0643	
98	0.0020		13	0.0264	0.0352	
99	0.0013		14	0.0183	0.0129	
100	0.0005		15	0.0075	0	

Examples: (1) Using the exponential pdf, $f(t)$, by Equation (3.6-12), the exponential non-exceedance probability, $F(t)$, by Equation (3.6-11a), and the exponential reliability, $R(t)$, by Equation (3.6-13), demonstrate the following functional relations: (a) Show the difference between the probability of nonfailure and reliability; (b) Calculate the conditional probability that no failure occurs in period (t_1, t_2), given there was none in period $(0, t_1)$; (c) Calculate the reliability for different failure rates in the two periods, and for the same failure rate. <u>Solution</u>: (a) The probability that failure occurs during period (t_1, t_2) is $F(t_2) - F(t_1)$. The complementary probability of nonfailure is according to Equation (3.6-17)

$$1 - [F(t_2) - F(t_1)] = 1 - [(1 - e^{-\lambda t_2}) - (1 - e^{-\lambda t_1})] = 1 + e^{-\lambda t_2} - e^{-\lambda t_1}.$$

Any probability of a success event can be construed as a reliability. Hence Equation (3.6-17) is a reliability, namely the nonfailure probability for the interval $t_2 - t_1$ (see Section 3.5.1, Equation 3.5-5).
(b) According to Equation (3.6-20), the probability that there will be no failure in period (t_1, t_2), given there was none in period $(0, t_1)$, is

$$P(t_2 | t_1) = \frac{1 - F(t_2)}{1 - F(t_1)} = e^{-\lambda(t_2 - t_1)}.$$

The reliability of no failure in the entire period $(0, t_2)$ is according to Equation (3.6-20)

$$R(t_2) = 1 - F(t_2) = e^{-\lambda t_2} < e^{-\lambda(t_2 - t_1)},$$

which means $R(t_2) < P(t_2 | t_1)$, or the reliability of the whole period is smaller than the reliability of its parts.
(c) With $R(t_2) = 1 - F(t_2)$ and $R(t_1) = 1 - F(t_1)$, the probability of no failure in both periods with failure rates λ_1 and λ_2, respectively, is

$$R(t_2) = P(t_2 | t_1) R(t_1) = e^{-\lambda_2(t_2 - t_1)} e^{-\lambda_1 t_1}$$

and for the same failure rate in both periods, $\lambda = \lambda_2 = \lambda_1$, $R(t_2) = e^{-\lambda t_2}$.

(2) Calculate the expected remaining life for $t_a = 99$ years. <u>Solution</u>: As in Table 3-11, $f(t)$ and $R(t_a)$ are taken from Table 3-10. A small increment of $f(t)$ is added for year 101. The calculation of the expected remaining life is shown in tabular form.

Tabular Evaluation of Example (2)

t	$f(t)$	$R(t_a)$	$t - t_a$	$(t - t_a)f(t)$	$\int (t - t_a)f(t)dt$	$E[L_R(t_a)]$
99	0.0013	0.0020	0	0	0.00065	0.3
100	0.0005		1	0.0005	0.00045	
101	0.0002		2	0.0004	0	

The result is $E[L_R(t_a)] = 0.3$ years. If $f(t)$ for $t = 101$ had been set to zero, $E[L_R(t_a)]$ $= 0.1$ years would have been obtained. Numerical accuracy becomes important when the calculation involves the tail of the distribution.

(3) Calculate the hazard function and its integral for a normal pdf of life length t of a mechanical component. Similar to the life length pdf shown in Figure 3-17, the life of components ends after they have been in operation for a time t. The random variable t and its distribution parameters have the units years. Assume the life length distribution is normal with $\mu = 2.5$ years and $\sigma = 0.19$ years. <u>Solution</u>: Based on the given parameters of the normal distribution, one finds $f(t)$, $R(t)$, then $h(t)$, and finally $H(t)$. The calculations are carried out in Table 3-12. The normal variable is $z = (t - \mu)/\sigma$, its derivative is $dz = dt/\sigma = 0.05/0.19 = 0.263$. The normal pdf in column 3 is

$$f(t) = \frac{0.39894}{\sigma} e^{-0.5z^2}$$

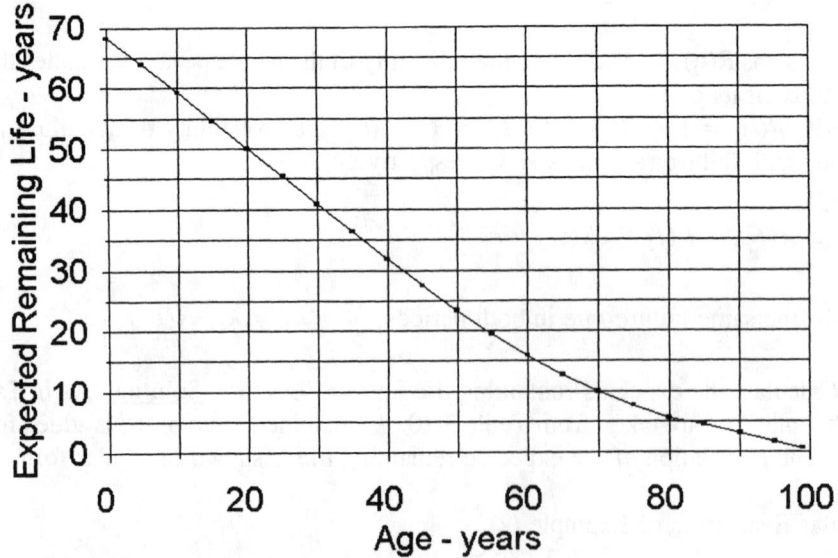

Figure 3-18: Expected remaining life, $E[L_R(t_a)]$, as function of age, derived from the human life probability density function, $f(t)$, which in turn was derived from the hazard function, $h(t)$. The calculations are based on the mortality rates or instantaneous failure rates of Table 3-10, column 2.

In column 4, use is made of $f(t)\, dt = f(z)\, dz$. The functions derived from $f(z)$ are (see also Example (2) of Section 3.6.1):

$$R(z) = 1 - F(z) = 1 - \int_{-\infty}^{z} f(z)dz \ ,$$

$$h(z) = \frac{f(z)}{R(z)}, \tag{3.6-7c}$$

and

$$H(z) = \ln[\frac{R(0)}{R(z)}].$$

The pdf $f(t)$ is rather spiky around the mean with a maximum value at $t = \mu$ of $0.39894/0.19 = 2.1$. Practically all life lengths are within $1.5 \le t \le 3.5$. This is seen by the very large (absolute) z-values in column 2 for $t = 1.5$ and $t = 3.5$. For these values, $f(t) \approx 0$, and the probability of z, $f(z)dz \approx 0$, as shown in column 4. Given the very small probabilities, also $F(z)$ in column 5 is zero in this low range of t, whereas $R(z)$ in column 6 remains 1 indicating 100 % survival. Column 8 lists the calculated hazard function, $h(z)$. According to Equation (3.6-7c), $h(z) \ge f(z)$, as shown in column 7, because $R(z) \le 1$. The integral of $h(z)$, which is $H(z)$, given in column 9, is calculated using $R(0) = 1$. It exceeds 1, whereas the integral of $f(z)$, which is represented by $F(z)$ in column 5, is less than or maximally equal to 1. The instantaneous failure rate, $h(z)\, dz$, relates failures to a survivor population at time t. Therefore it is a failure probability only for the population at t, in contrast to the probability $f(z)dz$, which relates to the total population. The hazard function derived from the normal pdf is shown in Figure 3-19. Its maximum occurs at $z = 4.21$, which corresponds to $t = 3.3$. At this point, $h(z)\, dz = 3.813 \cdot 0.263 = 1.003 \approx 1$. This means all population elements die off and the hazard function should end at that point and be zero thereafter. The normal pdf, $f(z)$, and the reliability $R(z)$, the latter being the integral of $f(z)$ beyond a given z, continue to form a ratio $f(z)/R(z) > 0$. Thus $h(z)$ remains a positive quantity that approaches zero because $f(z)$ approaches zero faster than $R(z)$. This is shown by the decline of $h(t)$ in Figure 3-19 that starts at $t = 3.3$. In a practical application of the normal distribution, t is given a finite range, usually $6\ \sigma$ or $8\ \sigma$. In such a case, $f(z)$ and all the derived functions end at the upper limit of the range. In this example, with $8\ \sigma = 1.52$, the upper range limit would be $t_u = 2.5 + 1.52/2 = 3.26 \approx 3.3$, which corresponds to $z_u = (3.26 - 2.5)/0.19 = 4$.

Table 3-12: Evaluation of Instantaneous Failure Rate Function or Hazard Function
$h(z)$ and Its Integral $H(z)$ for a Normal pdf
$\mu = 2.5$ years, $\sigma = 0.19$ years, $dt = 0.05$ years, $dz = dt/\sigma = 0.263$

t	$z(t)$	$f(t)$	$f(z)\,dz$	$F(z)$	$R(z)$	$f(z)$	$h(z)$	$H(z)$
(1)	(2)	(3)	(4)	(5)	(6)	(7)	(8)	(9)
1.50	-5.2632	0.0000	0.0000	0.0000	1.0000	0.0000	0.0000	0.0000
1.55	-5.0000	0.0000	0.0000	0.0000	1.0000	0.0000	0.0000	0.0000
1.60	-4.7368	0.0000	0.0000	0.0000	1.0000	0.0000	0.0000	0.0000
1.65	-4.4737	0.0001	0.0000	0.0000	1.0000	0.0000	0.0000	0.0000
1.70	-4.2105	0.0003	0.0000	0.0000	1.0000	0.0001	0.0001	0.0000
1.75	-3.9474	0.0009	0.0000	0.0000	1.0000	0.0002	0.0002	0.0000
1.80	-3.6842	0.0024	0.0001	0.0001	0.9999	0.0005	0.0005	0.0001
1.85	-3.4211	0.0060	0.0003	0.0003	0.9997	0.0011	0.0011	0.0003
1.90	-3.1579	0.0143	0.0007	0.0008	0.9992	0.0027	0.0027	0.0008
1.95	-2.8947	0.0318	0.0016	0.0020	0.9980	0.0060	0.0061	0.0020
2.00	-2.6316	0.0658	0.0033	0.0044	0.9956	0.0125	0.0126	0.0044
2.05	-2.3684	0.1271	0.0064	0.0093	0.9907	0.0241	0.0244	0.0093
2.10	-2.1053	0.2289	0.0114	0.0182	0.9818	0.0435	0.0443	0.0183
2.15	-1.8421	0.3849	0.0192	0.0335	0.9665	0.0731	0.0757	0.0341
2.20	-1.5789	0.6037	0.0302	0.0582	0.9418	0.1147	0.1218	0.0600
2.25	-1.3158	0.8835	0.0442	0.0954	0.9046	0.1679	0.1856	0.1000
3.30	-1.0526	1.2066	0.0603	0.1477	0.8523	0.2292	0.2690	0.1598
2.35	-0.7895	1.5375	0.0769	0.2163	0.7837	0.2921	0.3727	0.2437
2.40	-0.5263	1.8281	0.0914	0.3004	0.6996	0.3473	0.4965	0.3572
2.45	-0.2632	2.0282	0.1014	0.3968	0.6032	0.3854	0.6389	0.5055
2.50	-0.0000	2.0997	0.1050	0.5000	0.5000	0.3989	0.7979	0.6931
2.55	0.2632	2.0282	0.1014	0.6032	0.3968	0.3854	0.9712	0.9243
2.60	0.5263	1.8281	0.0914	0.6996	0.3004	0.3473	1.1563	1.2027
2.65	0.7895	1.5375	0.0769	0.7837	0.2163	0.2921	1.3508	1.5313
2.70	1.0526	1.2066	0.0603	0.8523	0.1477	0.2292	1.5526	1.9129
2.75	1.3158	0.8835	0.0442	0.9046	0.0954	0.1679	1.7596	2.3497
2.80	1.5789	0.6037	0.0302	0.9418	0.0582	0.1147	1.9700	2.8435
2.85	1.8421	0.3849	0.0192	0.9665	0.0335	0.0731	2.1823	3.3960
2.90	2.1053	0.2289	0.0114	0.9818	0.0182	0.0435	2.3950	4.0084
2.95	2.3684	0.1271	0.0064	0.9907	0.0093	0.0241	2.6070	4.6819
3.00	2.6316	0.0658	0.0033	0.9956	0.0044	0.0125	2.8172	5.4173
3.05	2.8947	0.0318	0.0016	0.9980	0.0020	0.0060	3.0244	6.2154
3.10	3.1579	0.0143	0.0007	0.9992	0.0008	0.0027	3.2272	7.0767
3.15	3.4211	0.0060	0.0003	0.9997	0.0003	0.0011	3.4233	8.0013
3.20	3.6842	0.0024	0.0001	0.9999	0.0001	0.0005	3.6061	8.9883
3.25	3.9474	0.0009	0.0000	1.0000	0.0000	0.0002	3.7564	10.0333
3.30	4.2105	0.0003	0.0000	1.0000	0.0000	0.0001	3.8130	11.1216
3.35	4.4737	0.0001	0.0000	1.0000	0.0000	0.0000	3.5956	12.2056
3.40	4.7368	0.0000	0.0000	1.0000	0.0000	0.0000	2.7718	13.1573
3.45	5.0000	0.0000	0.0000	1.0000	0.0000	0.0000	1.4415	13.7846
3.50	5.2632	0.0000	0.0000	1.0000	0.0000	0.0000	0.4907	14.0575

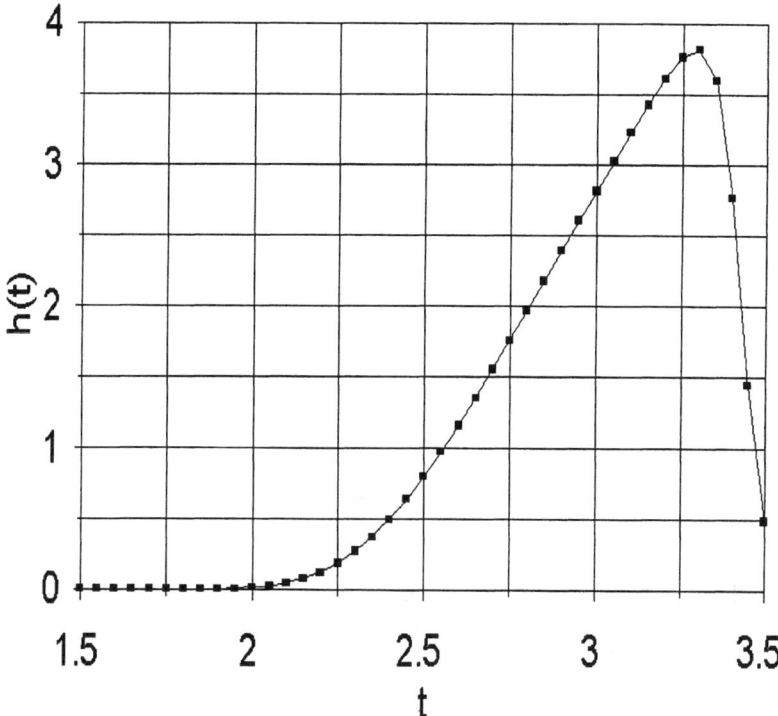

Figure 3-19: Hazard function, $h(t)$, derived from a normal pdf in Table 3-12 with parameters $\mu = 2.5$ and $\sigma = 0.19$. If a practical range of 8 σ were chosen, $h(t)$ would end at about $t = 3.3$ where $h(z)\, dz \approx 1$. The instantaneous failure rate would be zero beyond this point. The normal distribution with its extension to infinity continues to produce an $h(t)$ which decreases asymptotically toward zero as t goes to infinity.

(4) Given are four instantaneous failure rates similar to the last three rates in Table 3-10: 400/1,000; 500/1,000; 600/1,000; and 1,000/1,000. How is an initial population of 1,000 reduced by these rates and how would it be reduced by a constant rate of an exponential function? <u>Solution</u>: The population reduction for given $h(z)$ and $dz = 1$ is given in tabular form.

Tabular Evaluation of Example (4)

Period x	$h(z)$	Population at Beginning of Period	Failures During Period	Remaining Population at End of Period	Exponential Function $1000\,e^{-0.625\,x}$
1	0.4	1000	400	600	535
2	0.5	600	300	300	286
3	0.6	300	180	120	153
4	1.0	120	120	0	82

If an average rate, $(0.4 + 0.5 + 0.6 + 1.0)/4 = 0.625$ were applied in each period to the surviving population, the remaining population would decay exponentially according to the exponential function in the last column of the table, where x is the number of periods. It would approach zero asymptotically, as $x \to \infty$. In contrast to the exponential die-off at an average rate, the increasing rates $h(z)$ cause a progressive die-off.

(5) Derive the hazard function based on the Weibull pdf. <u>Solution</u>: For the special case (1) of the Weibull distribution, for which $c = 0$, one obtains (see Section 2.7.8, Equations (2.7-47) and (2.7-48))

$$f(x) = (\frac{k}{v})(\frac{x}{v})^{k-1} e^{-(\frac{x}{v})^k}$$

and

$$F(x) = 1 - e^{-(\frac{x}{v})^k}.$$

The hazard function follows from the combination of these two formulas (Moan, 1982, p. 4-48):

$$h(x) = \frac{f(x)}{1 - F(x)} = (\frac{k}{v})(\frac{x}{v})^{k-1}. \tag{3.6-26}$$

Special cases: (1): $k = 1$, $h(x) = 1/v$, a constant. (2): $k = 2$, $h(x) = (2/v^2)\,x$, a straight line. (3): $k < 1$, $h(x)$ is a hyperbola.

3.7 Probability and Return Periods

3.7.1 Probability Assignment by Data Ranking

In predicting rare events, the use of the return period is common. It gives a more concrete meaning to probability to say that an event occurs "once in 100 years on average" instead of "with probability 0.01." It will be shown that using the return period instead of probability distorts the upper end of the probability scale so that the record can be more easily extrapolated to higher values.

If a series of events has been recorded over a period, there usually is a largest and a smallest event. This may be true for costs, repair times, flood peaks, and so on. The events on record, if they are independent of each other, can be construed as realizations of a random variable produced by a stochastic process. If the process is stationary it can be expected to randomly repeat its realizations so that the already observed event population represents a range of possible future realizations.

Ordering the random realizations according to size lines them up in descending order over the observed range of events. An example of independent random events would be the largest daily flow rate of each year for n years. To be independent they must be produced by individual weather processes. Traditionally, weather processes were thought of as being stationary. This means that while random, each event would be produced by an overall similar process. In other words, a historic event could repeat itself in the future at any time without an underlying trend toward stronger or weaker processes over the long run. If any of the events of an n-year record can recur in the future, without preference for one or the other, then each event would have the same probability of occurrence, $1/n$, regardless of their size.

The largest or top-ranking event is given rank $m = 1$. It equals or exceeds all other events. The second event in descending order is given rank $m = 2$. It equals or exceeds all but one event, in other words, only one event is bigger than the event rank $m = 2$. The third event from the top is ranked $m = 3$. It equals or exceeds all but two events. The event x ranked m equals or exceeds $n - m + 1$ events. For example, event x_5 ranked $m = 5$ in descending order exceeds events 6 to 10 and equals itself, so that $n - m + 1 = 6$. Since each of the events x was given equal probability $1/n$, the sum of the $n - m + 1$ probabilities, from the lowest ranked event up to and including the kth event, is the probability of event x_k being maximally equaled but not exceeded, which is the non-exceedance probability $F(x_k)$,

$$P(X \le x_k) = F(x_k) = \sum_{i=1}^{k} f(x_i) = \frac{n - m + 1}{n} \tag{3.7-1}$$

where $f(x_i)$ are the ordinates of the discrete pdf summed up to and including $f(x_k)$, here all $f(x_i)$ are equal to $1/n$; k is the index of the element whose non-exceedance probability is to be calculated and which becomes the upper integration (summation) limit for $F(x)$,

$$k = n - m + 1 .$$ (3.7-1a)

The exceedance probability is obtained as the complementary probability,

$$P(X > x_k) = 1 - F(x_k) = 1 - \sum_{i=1}^{k} f(x_i) = \frac{m-1}{n} .$$ (3.7-2)

The ratio $(m - 1)/n$ represents a discrete exceedance probability with the highest-ranked event being assigned an exceedance probability of zero. This makes sense for a discrete sample that is known to end at this limit, as no event is possible beyond it. But this is not the case for data records of limited length of ongoing processes where the largest and the smallest events may not have occurred during the usually limited period of observation. Assuming that the largest event in a 10-year record will never be exceeded would almost certainly be a nonconservative assumption. According to experience, if the observations continue only long enough, new maxima and minima will be recorded. To allow for a small exceedance probability, one could modify Equation (3.7-2) to

$$1 - F(x) = \frac{m}{n} ,$$ (3.7-3)

which gives the rank-1 event an exceedance probability of $1/n$. As the record length increases, the exceedance probability automatically becomes smaller with increasing n. Various formulas have been proposed for the calculation of $1 - F(x_k)$. The Weibull formula is an example (Stedinger et al., 1993, p. 18.24):

$$1 - F(x) = \frac{m}{n+1} .$$ (3.7-4)

It increases both numerator and denominator over what is used in Equation (3.7-2). Table 3-13 compares $F(x)$ and $1 - F(x)$ according to Equations (3.7-1) through (3.7-4). Column 6 gives the exceedance probabilities by the Weibull formula. The differences between column 6 and columns 4 and 5 are greatest at the boundaries and relatively small in the interior. For large n, the differences are even smaller. Unfortunately, the values at the boundaries usually are critical.

Table 3-13: Comparison of Exceedance Probabilities and Return Periods for a Discrete Series of $n = 10$ Elements with Uniform pdf

x_i	m	$F(x) =$ $(n - m + 1)$ $/n$	$1 - F(x)$ $= (m$ $- 1)/n$	$1 - F(x)$ $= m/n$	$1 - F(x)$ $=$ $m/(n+1)$	$T =$ $n/(m- 1)$	$T =$ n/m	$T_w =$ $(n +1)$ $/m$
(1)	(2)	(3)	(4)	(5)	(6)	(7)	(8)	(9)
x_{10}	1	1.0	0.0	0.1	0.09	∞	10	11.0
x_9	2	0.9	0.1	0.2	0.18	10	5	5.5
x_8	3	0.8	0.2	0.3	0.27	5	3.3	3.7
x_7	4	0.7	0.3	0.4	0.36	3.3	2.5	2.8
x_6	5	0.6	0.4	0.5	0.45	2.5	2.0	2.2
x_5	6	0.5	0.5	0.6	0.55	2.0	1.7	1.8
x_4	7	0.4	0.6	0.7	0.64	1.7	1.4	1.6
x_3	8	0.3	0.7	0.8	0.73	1.4	1.3	1.4
x_2	9	0.2	0.8	0.9	0.82	1.3	1.1	1.2
x_1	10	0.1	0.9	1.0	0.91	1.1	1.0	1.1

Notes: column 3: Equation (3.7-1); column 4: Equation (3.7-2); column 5: Equation (3.7-3); column 6: Equation (3.7-4); column 7: Equation (3.7-2a); column 8: Equation (3.7-5); column 9: Equation (3.7-7).

3.7.2 Return Periods

The ratio m/n can be construed as an occurrence rate "m times in n years," or a frequency. If m/n is written as $1/(n/m)$ it can be construed as "normalized" to "1 time in (n/m) years." The period (n/m)-years is a mean time between events of rank m in a series of n events, or a *mean interarrival time*, or *return period*, T, for an event of rank m:

$$T = \frac{n}{m} .$$

(3.7-5)

For example, the biggest event in the series, ranked $m = 1$, has a return period $T = n$, where n has the unit of a number, here a number of years. Since m/n is a frequency, n/m is a time, but with some caveats, as will be seen next. The relation between return period and probability is

$$T = \frac{n}{m} \approx \frac{1}{P(X > x)} = \frac{1}{1 - F(x)} \tag{3.7-6}$$

where T is seen to be the inverse of the exceedance probability, Equation (3.7-3). Since a probability has no dimension, "time" must be understood in the sense of "number of trials" or "number of opportunities." For example, a flood ranked $m = 1$, the highest on record in a series of 38 annual floods, according to Equation (3.7-6), has a return period of $T = n = 38$ years, or in other words, it takes 38 trials on average to encounter one outcome of that special kind. If the trials occur once each year, then T is measured in years. Similarly, if the trials occur at fractions of a year, than T is measured in these fractions.

The return periods vary according to the formulas used. For a sample of 10 elements, the results of three formulas for T, are given in columns 7 through 9 of Table 3-13. T of column 7 is based on the reciprocal of Equation (3.7-2),

$$T = \frac{n}{m-1} . \tag{3.7-2a}$$

This formula has a singularity for the most important rank, $m = 1$, for which $T \to \infty$; it is therefore unacceptable. Unfortunately this formula is the one that is based on the correct formula for the non-exceedance probability. The reciprocal of the modified formula, Equation (3.7-3), is Equation (3.7-5). It is given in column 8. It has no singularity and is acceptable. The formula known as Weibull formula, Equation (3.7-4), gives

$$T_w = \frac{n+1}{m} . \tag{3.7-7}$$

Its values are given in column 9 of Table 3-13. They are only slightly bigger than those of Equation (3.7-5). Equation (3.7-7) is widely used in hydrologic return period calculations (Linsley et al., 1958, p. 249; Yevjevich, 1972, p. 90).

Example: Compare the return period formulas, $T_2 = n/m$ after Equation (3.7-5) and the Weibull formula, $T_w = (n + 1)/m$, Equation (3.7-7). Solution: The ratio of the formulas is

$$\frac{T_w}{T_2} = \frac{n+1}{n} = 1 + \frac{1}{n} .$$

The ratio converges to 1 as n becomes large, in other words, the two formulas give the same result for a reasonable size of n. The Weibull formula has the edge over n/m because it better simulates the probabilities at the limits of the data range. Within the range, both formulas are satisfactory for reasonably sized n, say $n > 30$.

3.7.3 Partial-Duration Series and Return Period

Probabilistic methods make a point of recommending the use of all available information. In flood analysis, one distinguishes between an *annual maximum series* and a *partial-duration series*. The annual maximum series uses only the maximum event for each year, whereas the partial-duration series uses all events above a certain magnitude that can be considered independent. For example, the annual maximum event is assumed to be independent of similar events in the years before and after. If two or more floods occur in rather short succession, the first flood may influence the second flood by filling surface and groundwater storage, thus causing increased surface runoff, and by riding on a high basin depletion flow of the first event. Floods in short succession could be caused by independent events. For example, in September/October 2002, two hurricanes affected southeastern drainage basins of the United States about one week apart. In general, there are many flood events large and small, during the year, which are independent, but are not counted in an annual maximum series. The annual maximum series tries to capture the largest event that can produce a design flood at a site. For maintenance, not only the size but also the time of occurrence and the interarrival time between successive events become variables of importance.

Usually runoff is not equally distributed over the year. A year usually has flood seasons and dry periods. For maintenance or repair work the *interarrival time* of high flow becomes a random variable of interest. Table 3-14 lists 120 flood events in 38 years. The arrival dates for the event are given in columns 2 and 3. The interarrival times as differences between dates are given in column 4. The interarrival times sorted in descending order (detached from the chronological order of column 4) are given in column 5. They range from 381 days to 3 days. This large spread is caused by the concentration of flood events in a relatively short period of the year from day 100 to day 180, so that the interarrival time between the last event of the current season and the first event of the next season can be 285 days. Interarrival times are much shorter within the 80-day flood season. Such spacing between events is obviously of importance for maintenance planning.

There are numerous analogies between floods and accident events. The size of floods and accidents are random variables as are their interarrival times; cleanup costs and repair costs, which can be construed as stochastic processes. The flood events, like accidents, may trigger a string of additional accidents and repairs. Flood events can be controlled by providing storage space, and accident occurrences can be controlled, at least to some extent, by preventive maintenance. Thus it is not surprising that similar planning methods can be used in both hydrology and maintenance.

Table 3-14: Events, Interarrival Times, and Return Periods for the Clearwater River at Kamiah, Idaho for 1911–1948 (after U.S. Geological Survey, as reported by Linsley et al., 1958, pp. 254–256). For the annual maximum series contained in this table, μ = 1,502 m^3/s, σ = 467 m^3/s. For all reported data (partial series), μ = 1,250 m^3/s.

No.	Year	Day	Interarrival time-days	Interarrival time-sorted	Q m^3/s	Q sorted	T_p years
(1)	(2)	(3)	(4)	(5)	(6)	(7)	(8)
1	1911	126	–	381	980	2803	38.00
2		137	11	377	833	2449	19.00
3		155	18	364	1017	2305	12.67
4		164	9	363	1119	2169	9.50
5	1912	141	342	360	1563	2042	7.60
6		150	9	360	1753	2042	6.33
7		172	22	357	1076	2016	5.43
8	1913	110	303	355	833	1996	4.75
9		117	7	353	869	1979	4.22
10		131	14	352	1297	1974	3.80
11		146	15	350	2169	1974	3.45
12	1914	138	357	344	1195	1943	3.17
13		143	5	342	1175	1860	2.92
14		154	11	342	869	1818	2.71
15	1915	139	350	341	799	1801	2.53
16	1916	118	344	339	850	1795	2.38
17		127	9	336	1257	1790	2.24
18		156	29	336	1036	1767	2.11
19		170	14	335	1586	1761	2.00
20		180	10	334	1036	1753	1.90
21	1917	135	320	329	1801	1722	1.81
22		150	15	328	1974	1716	1.73
23		160	10	326	1608	1693	1.65
24		169	9	326	1996	1668	1.58
25		363	194	324	1056	1608	1.52
26	1918	125	127	323	1495	1586	1.46
27		135	10	320	997	1563	1.41
28		161	26	320	1495	1495	1.36
29	1919	119	323	318	869	1495	1.31

No.	Year	Day	Interarrival time-days	Interarrival time-sorted	Q m^3/s	Q sorted	T_p years
30		143	24	317	1472	1492	1.27
31	1920	138	360	313	1235	1478	1.23
32		167	29	311	1215	1475	1.19
33	1921	113	311	303	997	1472	1.15
34		140	27	196	1974	1433	1.12
35	1922	139	364	194	1716	1410	1.09
36		146	7	194	1475	1405	1.06
37		157	11	164	1767	1314	1.03
38	1923	128	336	150	1099	1314	1.00
39		146	18	144	1405	1300	0.97
40		163	17	127	1223	1297	0.95
41	1924	124	326	97	1291	1291	0.93
42		133	9	38	1668	1269	0.90
43	1925	107	339	35	1184	1257	0.88
44		127	20	30	1269	1257	0.86
45		140	13	29	1693	1246	0.84
46	1926	109	334	29	1017	1243	0.83
47		121	12	28	1017	1235	0.81
48		141	20	27	917	1235	0.79
49	1927	118	342	26	1314	1223	0.78
50		137	19	25	1818	1223	0.76
51		159	22	24	1943	1223	0.75
52		309	150	24	1243	1215	0.73
53		330	21	23	827	1195	0.72
54	1928	129	164	22	1860	1184	0.70
55		146	17	22	2042	1175	0.69
56	1929	144	363	21	1492	1155	0.68
57		152	8	20	807	1136	0.67
58		160	8	20	1014	1119	0.66
59	1930	115	320	19	878	1116	0.64
60	1931	127	377	19	1155	1099	0.63
61		134	7	18	1034	1087	0.62
62	1932	104	335	18	807	1076	0.61
63		134	30	17	2042	1070	0.60

No.	Year	Day	Interarrival time-days	Interarrival time-sorted	Q m^3/s	Q sorted	T_p years
64		141	7	17	1761	1065	0.59
65		164	23	16	994	1056	0.58
66	1933	117	318	16	1014	1051	0.58
67		155	38	16	2016	1051	0.57
68		161	6	15	2305	1051	0.56
69		357	196	15	1235	1036	0.55
70	1934	89	97	15	915	1036	0.54
71		104	15	14	1070	1036	0.54
72		115	11	14	1300	1034	0.53
73		128	13	14	971	1031	0.52
74	1935	144	381	13	1246	1022	0.51
75		151	7	13	974	1017	0.51
76		157	6	13	847	1017	0.50
77	1936	109	317	13	1433	1017	0.49
78		125	16	13	1410	1014	0.49
79		135	10	13	1790	1014	0.48
80		148	13	13	971	997	0.48
81		152	4	13	932	997	0.47
82	1937	139	352	12	971	994	0.46
83		148	9	12	912	980	0.46
84	1938	109	326	11	1795	974	0.45
85		121	12	11	1116	971	0.45
86		137	16	11	892	971	0.44
87		148	11	11	1722	971	0.44
88	1939	124	341	11	1314	968	0.43
89		137	13	11	1031	960	0.43
90	1940	132	360	10	1051	957	0.42
91		145	13	10	838	954	0.42
92	1941	133	353	10	818	943	0.41
93	1942	104	336	10	818	932	0.41
94		111	7	10	818	923	0.40
95		146	35	9	1051	917	0.40
96	1943	110	329	9	1223	915	0.40
97		121	11	9	838	912	0.39

No.	Year	Day	Interarrival time-days	Interarrival time-sorted	Q m³/s	Q sorted	T_p years
98		149	28	9	1478	892	0.39
99		162	13	9	1051	883	0.38
100		170	8	9	1223	878	0.38
101		173	3	9	1136	869	0.38
102	1944	136	328	8	968	869	0.37
103	1945	126	355	8	1257	869	0.37
104		151	25	8	1087	850	0.37
105	1946	110	324	7	943	850	0.36
106		116	6	7	954	847	0.36
107		126	10	7	1036	838	0.36
108		139	13	7	850	838	0.35
109		148	9	7	1022	838	0.35
110		155	7	7	801	833	0.35
111		349	194	7	960	833	0.34
112	1947	128	144	7	1979	833	0.34
113		147	19	6	1065	827	0.34
114		160	13	6	883	818	0.33
115	1948	108	313	6	833	818	0.33
116		112	4	5	923	818	0.33
117		128	16	4	957	807	0.32
118		142	14	4	2449	807	0.32
119		149	7	3	2803	801	0.32
120		173	24	-	838	799	0.32

Notes: column 1: number given to event in chronological order.
column 2: year of occurrence.
column 3: dates of occurrence in Julian days[1], which is the current number of the day

---

[1]The sequential day numbers of the year are often referred to as Julian Days. The Julian Day calendar is a day count that starts with January 1, 4713 B.C. This makes January 1, 1980 Julian Day 2,444,240. This Julian Day (long) count is used to calculate time differences between dates. The Julian Day lasts from noon Greenwich Mean Time to the same time the next day. The Roman day count of Julius Caesar's time was more complicated with days being counted forward and backward from three days: Kalendae (1st day of month), Nonae (5th day of month) and Idus (13th day of month; in March, Idus was the 15th day). One innovation of lasting value that the Julian calendar introduced was the 365-day year with one additional day every four

counted from January 1 as 1 to December 31 as 365 without leap day.
column 4: interarrival times are calculated as the difference between day numbers.
column 5: interarrival times *sorted* in descending order.
column 6: flow rates corresponding to dates in columns 2 and 3.
column 7: flow rates *sorted* in descending order.
column 8: partial series return periods calculated by Equation (3.7-9): $T_p = n/m_p$.

The inclusion of all floods above a minimum peak flow rate for a period of record results in a population usually significantly larger than the number of years, here 120 events in 38 years. If the concept of one event per period is used, then there is one event assigned to each of $n_p = 120$ periods, with one period being $n/n_p = 38/120 = 0.317$ years. Each event is actually recorded for a particular date, which may not coincide with the period assigned to it by this averaging process. The time of most likely occurrence must be found by a separate analysis (see Section 3.7.4).

The partial series return period, in analogy to the annual return period, is

$$T_p = \frac{n_p}{m_p} \approx \frac{1}{P(X > x)} = \frac{1}{1 - F(x_p)} \tag{3.7-8}$$

where T_p is in units of partitions of the record period, which is about 0.32 years; $n_p = 120$, is the number of partitions of the total record period, one for each event of the period; and $m_p = 1,\dots,120$, is the rank number for the $n_p = 120$ events; the index p refers to *partial series*. $F(x_p) = F(Q_p)$ is the non-exceedance probability of Q_p, which is the sum of $f(x_p) = 1/n_p = 1/120$, up to and including Q_p. This means that each event of the partial series is given the same (uniform) probability. The top event has rank $m_p = 1$, which is the same as for the annual maximum series, but its return period is $T_p = 120/1 = 120$ "partial-year return periods," each about 0.32 years long.

In order for the partial series return period to have the units of full years, Equation (3.7-8) is multiplied by $n/n_p = 38/120$, which gives

$$T_p = \frac{n}{m_p} \tag{3.7-9}$$

where T_p is the partial series return period in years; n is the length of the record in years; and m_p is the rank number of the partial-year events. The return periods for the partial-duration series calculated by Equation (3.7-9) are given in column 8 of Table 3-14. Since m_p exceeds n, events with return periods less than one year are included.

---

years. It would be more appropriate to say "sequential day number of the year."

The partial-duration flow series is shown in Figure 3-20.

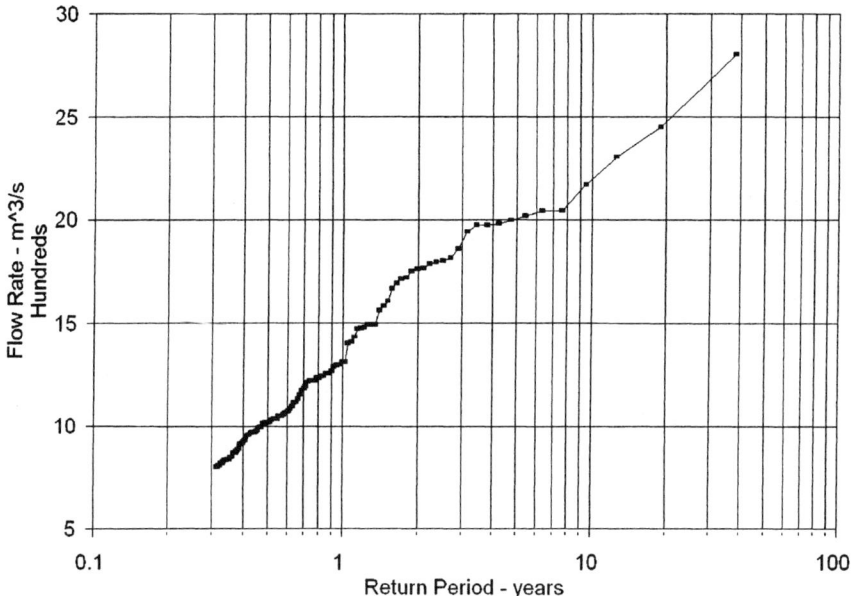

Figure 3-20: Partial-duration series of floods of the Clearwater River, Idaho, for 1911–1948. The alignment of the flow rates Q versus the logarithms of the return periods suggests an underlying exponential distribution of Q. The data are from Table 3-14 (from Linsley et al., 1958, pp. 254–256). In the annual maximum series, all events with less than 1-year return period are suppressed.

The selection of suitable scales for Q and T is an empirical way of finding a probability distribution for Q. In Figure 3-20 Q versus $\log\{1/[1 - F(Q_p)]\}$ or Q versus $\log(T_p)$ is such a choice. The log-functions are expressions of the probability assigned to Q. If the assumed function simulates the probability distribution of the random variable Q, then the probability for any Q inside or outside the data range can be calculated by the derived function. Whereas this assumption is often made, it cannot be proved, usually because of insufficient sample size of Q. In such cases, there is not much that can be done to infuse more information into the data set than there is, even with the use of sophisticated mathematical analysis (Klemes, 2000, pp. 159–162). Hence there is more or less uncertainty associated with using an empirical function based on a small data set to calculate the probability of Q inside as well as outside the data range. The justification for applying the analysis is squeezing the available data for information and then using judgment while being fully aware of the limitations of the analysis.

The data in Figure 3-20 seem to be approximated reasonably well by a straight line in the chosen coordinate system,

$$Q = a + b' \log T \tag{3.7-10}$$

where Q is the flood flow rate in m^3/s; a and b' are empirical coefficients; b' is the coefficient associated with the decimal logarithm; and T is the return period in years. The index p is dropped here, but a and b are preferably derived from the partial series, which includes all data. This may be advantageous for establishing the trend (slope) of the curve.

Usually what is needed for decision making is Q for a desired return period, say for 100 years. To also establish the relation between the empirical fit, Equation (3.7-10), and the probability of Q, one converts the decimal log to the natural log by $\log T = 0.4343 \ln T$ and takes the antilog of Equation (3.7-10),

$$T = e^{\frac{Q-a}{b}} \tag{3.7-11}$$

where $b = 0.4343\, b'$. Since the e-function is dimensionless, T must also be a dimensionless number; here it is the number of trials until the specified Q can be expected to occur; for an annual event, this is one trial per year; for a subannual period, it is one trail per subperiod. The coefficients a and b are found by inspection or by regression in an appropriate coordinate system, here the semi-log plot of Q versus $\log T$ in Figure 3-20. The relation between T and $F(Q)$ is obtained by making use of Equation (3.7-8),

$$F(Q) = 1 - e^{-\frac{Q-a}{b}} = 1 - \frac{1}{T} \tag{3.7-12}$$

where $F(Q)$ is the non-exceedance probability of Q. Equation (3.7-12) is the cdf of the exponential distribution (see Equation (2.7-28b) of Section 2.7.5).

A partial series includes all events of a selected range of magnitude regardless of their ranking within subdivisions of the period of record, such as a year. It should therefore be more representative of the general event frequency than the annual series. But this augmentation of the data sample of a short record still cannot be representative of all events that can happen in a longer record. Therefore, caveats are always appropriate when making extrapolations beyond a data range.

Examples: (1) For an event x ranked $m = 100$ in a partial-duration series that stretches over 40 years, calculate $F(x)$, T_p, and T. Solution: The discrete uniform probability for any event in the partial-duration series is $f(x) = 1/120$. The cumulation of the probabilities from the smallest event, ranked $m = 120$, to event $m = 100$ gives $F(x)$ $= 20 \cdot 1/120 = 0.167$. The exceedance probability is $1 - F(x) = 1 - 0.167 = 0.833$. The reciprocal is the return period of the event in subperiods, $T_p = 1/0.833 = 1.2$ partial-duration units. The return period in years is $T = 1.2 \cdot 40/120 = 0.4$ years.

(2) Fit the annual series contained in the data of Table 3-14 by the exponential distribution. (a) Determine the empirical coefficients. (b) Estimate the return period of the 100-year flood. Solution: (a) Mean and standard deviation of the *annual series* are given in Table 3-14 as $\mu = 1502$ m^3/s, $\sigma = 467$ m^3/s. The parameters of the exponential cdf are according to Section 2.7.5:

$$a = \mu - b = 1502 - 467 = 1035 \text{ m}^3\text{/s, and } b = \sigma = 467 \text{ m}^3\text{/s.}$$

The exponential distribution for these parameters is

$$F(Q) = 1 - e^{-\frac{Q-a}{b}} = 1 - e^{-\frac{Q-1035}{467}}. \tag{3.7-12a}$$

$F(Q)$ is the cdf of the annual Q-series based on the annual data parameters μ and σ. The formula is not necessarily in complete agreement with the data if the data distribution is not exactly exponential. For example, the lower bound on flow is 799 m^3/s. $F(Q)$ should become zero for this flow. $F(Q) = 0$ means $T = 1$; it would be associated with an event that can be expected annually, if T is in years. The exponent of Equation (3.7-12a) shows, however, that this happens for 1035 m^3/s.
(b) From Equation (3.7-12) follows

$$T = e^{\frac{Q-1035}{467}}.$$

Solving for Q, gives

$$Q = 1035 + 467 \ln T,$$

and in decimal logarithms,

$$Q = 1035 + 1075 \log T. \tag{3.7-10a}$$

Equation (3.7-10a) is calculated for the parameters of the *annual series* with its coefficients based on μ and σ assuming the distribution is exponential. Since the data do not line up in a straight line the distribution is not truly exponential and Equation (3.7-10a) is only an approximation of the exponential distribution. The regression based on the annual data using $T = n/m$ gives

$$Q = 1021.12 + 1192.98 \log T \tag{3.7-10b}$$

with a coefficient of determination $R^2 = 0.934$. Also this equation is based on the exponential distribution assumption but uses the empirical data plot or experimental data distribution instead of exponential distribution parameters. It also is an approximation of the data but possibly a better one as it accounts for the location of each data point in the plot.

(3) Compare the annual series adjustment with the partial series adjustment in Figure 3-20. Solution: The partial series in Figure 3-20 can be fitted with a straight line by eye or by a regression. A fit by eye (Linsley et al., 1958, Figure 11-4, p. 258) of the *partial series* data in Figure 3-20 can be represented by the empirical formula

$$Q = 1302 + 1029 \log(T) . \tag{3.7-10c}$$

A comparison of Q calculated by using the function Q versus T based on the annual series and on the partial series is given in tabular form below.

Tabular Evaluation of Example (3)

T Years	Annual Series Q in m^3/s	Annual Series Q in m^3/s	Partial Series Q in m^3/s
(1)	(2)	(3)	(4)
1	1035	1021	1302
2	1358	1380	1612
5	1786	1855	2021
10	2110	2214	2321
20	2434	2573	2641
50	2861	3048	3050
100	3186	3407	3360

Notes: column 2: Equation (3.7-10a) based on exponential distribution parameters. column 3: Equation (3.7-10b) based on regression involving the data distribution. column 4: Equation (3.7-10c) fitted by eye.

The regression on the annual data and the partial series give about the same result, whereas the annual series that relies on the exponential distribution parameters μ and σ (column 2) gives smaller values. The partial series uses a larger data set than the annual series and therefore is considered more reliable than the others. The greatest differences between the two series reach about 20 % at the lower end of the Q-range with respect to the partial series; they are only from 1% to 5% at the extrapolated upper end. (4) What is the probability that the flow rate of 3,000 m^3/s will be exceeded and what is its return period? From Equation (3.7-12a) one obtains the exceedance probability as

$$1 - F(Q) = 1 - [1 - e^{-\frac{3000-1035}{467}}] = e^{-\frac{3000-1035}{467}} = 0.015 .$$

The inverse of this exceedance probability is the return period, $T = 1/0.015 = 67$ years.

3.7.4 Seasonal Distribution

So far, the random events were assumed to occur within some time period, such as a year or some subperiod of a year, without identifying a particular time or season. A year is a rather long time, and planning for protection against a major event may require more precise information. For example, if maintenance work needs to be done on components of a project, one wants to know the period that is most favorable for such work. Here, the year is divided into 10-day periods, x_j, and the 120 events on record, as given in Table 3-14, are allocated to these periods. The result is shown in Table 3-15. The frequencies of these events per period, $n_j/120$, are given in column 5 of the table, where n_j is the number of events falling into period j. Cumulating these frequencies gives the function shown in Figure 3-21. The resulting S-shaped curve shows that about 1 % of events occur during the first 10 periods of the year, 95 % of all events occur during periods 11 through 17, or from day 101 through day 180 (April 10 through June 29). No events occur during the 22 periods from day 180 through day 300, and 4 % of events occur during periods 33 through 36. Obviously the summer periods, 19 through 30, June 30 through October 27, are uniquely suited for work that requires the absence of floods.

The reader should note that the event distribution over time of Figure 3-21 while has the shape of a cdf is not a probability distribution. It is the distribution of the cumulative number of events over time. The x-axis is not a random variable whose values can occur independently of each other; and the frequencies over the x-axis do not relate to frequencies of the x-variable but to a different variable, namely the cumulated number of events at time x (see Section 2.2.1). Hence, not every curve that looks like a probability distribution is a probability distribution. A pdf is a relationship between a quantity and its frequency, for example, a life length (survival time), x, and its frequency of occurrence, $f_X(x)$, and the cdf is its cumulation or integral. The reader is referred to Figure 3-1 where the top left graph is a pipe failure distribution over time whereas the remaining graphs are cdfs of pipe failure incidents for selected months.

Table 3-15: Event Distribution of 120 Flood Events for a 38-Year Period of the Clearwater River, Idaho (data from Linsley et al., 1958, pp. 254–256)

Period j	From Day	To Day	Number of Events in Period	Event Frequencies $n_j/120$	Cumulation of Event Frequencies
(1)	(2)	(3)	(4)	(5)	(6)
9	81	90	1	0.008	0.008
10	91	100	0	0.000	0.008
11	101	110	11	0.092	0.10
12	111	120	11	0.092	0.19
13	121	130	18	0.150	0.34
14	131	140	20	0.167	0.51
15	141	150	23	0.192	0.70
16	151	160	14	0.117	0.82
17	161	170	11	0.092	0.91
18	171	180	6	0.050	0.96
19-30	181	300	0	0	0.96
31	301	310	1	0.008	0.97
32	311	320	0	0	0.97
33	321	330	1	0.008	0.98
34	331	340	0	0	0.98
35	341	350	1	0.008	0.98
36	351	365	2	0.017	1.00

Notes: columns 2 and 3: day count for 10-day periods, starting with day 1 for January 1, for the events and dates given in Table 3-14.

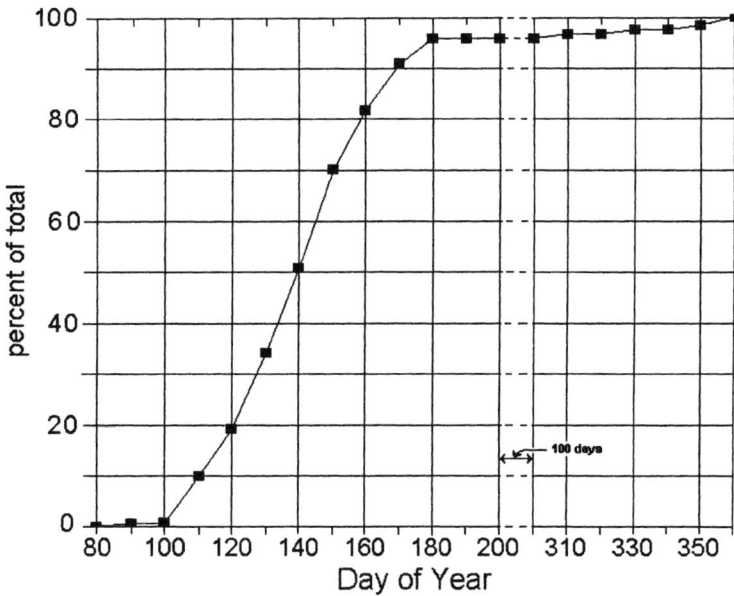

Figure 3-21: Cumulative distribution of 120 flood events in excess of 800 m³/s of a 38-year period for the Clearwater River, Idaho (data from Table 11-5 of Linsley et al., 1958, pp. 254–256). By day 100, the percentage of the total number of events having occurred is less than 1 %, and by day 180, it is 96 %. From day 200 to day 300, no qualifying event occurs. The remaining 3 % of events occur during the last 65 days of the year. The almost linear increase of the cumulation indicates that the events are nearly uniformly distributed over the flood season. This plot is a seasonal event distribution, not a probability distribution. Chronological time is not a random variable and the frequencies do not relate to this time but to frequencies of floods at a given time.

3.7.5 Interarrival Times

The interarrival times of 120 flood events are listed in column 4 of Table 3-14. They are the time intervals between selected flood events, in days. Column 5 gives the sorted sequence of the observed interarrival times. They range from 3 days to 381 days. The average interarrival time for 119 events (the first event does not have one) in 38 years is 38 · 365/119 = 116 days. Column 5 of Table 3-14 shows no value in this range. This is due to the fact that the events are not randomly distributed over the period of record. Whereas 96 % of all events are bunched together in an annual flood season of about 80-day length, the return periods of the annual flood seasons themselves are 300 days and longer. Thus the interarrival times of floods fall into two distinct groups: interarrival times within the flood season from 3 to 38 days for 78 events, and interarrival times for events belonging to different flood seasons that include the flood season return periods from 303 to 381 days for 33 events, and interarrival times from 97 to 196 days for eight events. The frequency distribution for the interarrival times within the flood season is given in Table 3-16. A frequency peak occurs for interarrival times from 7 to 14 days. The estimated probability of an interarrival time, T_I, from 7 days to 14 days based on the data is $P(7 \text{ days} \leq T_I \leq 14 \text{ days}) = 42/78 = 54 \%$.

Table 3-16: Frequencies of Flood Interarrival Times During the 80-Day Annual Flood Season of the Clearwater River, Idaho, 1911–1948 (from Table 3-14, column 5, which is based on data from Linsley et al., 1958, Table 11-5, pp. 254–256)

Interarr. Time Days	Absol. Freq. n	Interarr. Time Days	Absol. Freq. n	Interarr. Time Days	Absol. Freq. $n.$	Interarr. Time Days	Absol. Freq. n
(1)	(2)	(3)	(4)	(5)	(6)	(7)	(8)
1	–	11	6	21	1	31	0
2	–	12	2	22	2	32	0
3	1	13	8	23	1	33	0
4	2	14	3	24	2	34	0
5	1	15	3	25	1	35	1
6	3	16	3	26	1	36	0
7	8	17	2	27	1	37	0
8	3	18	2	28	1	38	1
9	7	19	2	29	2	39	0
10	5	20	2	30	1	40	0

Notes: column 1: interarrival time in days between floods during the 80-day annual flood season; column 2: the number of interarrival times of the length given in column 1 in the 38-year period (absolute frequency of interarrival times); column pairs 3 & 4, 5 & 6, and 7 & 8 are continuations of the column pair 1 & 2. The total number of events in the 38 flood seasons is 78.

3.7.6 Earthquake Probabilities and Return Periods

Earthquakes, like floods, are external random events. Such events may affect a structure without having any relation to the existence or operation of the structure. Earthquake magnitude is usually measured by magnitude on the Richter scale[2], which is the logarithm of the ratio of the maximum observed earthquake amplitude to a reference amplitude. In frequency analysis, this magnitude, M, is commonly related to the number of earthquake events in a given region and for a given period by the Gutenberg-Richter relation,

$$\log N = a - b M \tag{3.7-13}$$

where $\log N$ is the decimal logarithm of the number of earthquakes of magnitude M that occur during a specified period in some defined region; a and b are empirical coefficients.

Table 3-17: Earthquake Magnitude (Richter Scale) and Numbers of Occurrence in a 10-Year Period. Data Collected by the Seismological Institute of Uppsala (data by S. J. Duda, as reported by Bath, 1979, p. 154, Table 12, for the magnitude range $7.0 \leq M \leq 8.9$ and class width $\Delta M = 0.5$)

Magnitude M	Center of Class	N for a 10-Year Period
8.5 – 8.9	8.7	3 (2.5)
8.0 – 8.4	8.2	11 (9.3)
7.5 – 7.9	7.7	31 (35)
7.0 – 7.4	7.2	149 (132)
extrapolated		
6.5 – 6.9	6.7	560 (495)
6.0 – 6.4	6.2	2100 (1862)

Notes: column 3: extrapolations in column 3 are those given by Bath (1979, p. 154, Table 12); numbers in parentheses are calculated by Equation (3.7-13a) for M at center of class.

---

[2]There is a considerable unit insecurity in the earthquake science field as almost every researcher has his own preference for earthquake measures.

The meaning of N in Equation (3.7-13) is not always the same. Kanamori and Brodsky (2001, p. 39) explain N as the average number of earthquakes per year of a magnitude of at least M. This makes N a sort of rank number. Bath (1979, p. 154), however, gives N as the number of earthquakes of magnitude M, with M being in a class of width $\Delta M = 0.5$, as shown in Table 3-17. This makes N the number of elements in a set. This definition is in line with the definition of (absolute or relative) frequencies in histograms and will be used here. An empirical fit of Equation (3.7-13) to the earthquake data of Table 3-17 given by Bath (1979, p. 155) is

$$\log N = 10.40 - 1.15 M \qquad (3.7\text{-}13a)$$

where N is the absolute frequency of a class of earthquakes of magnitude M during the 10-year period.

Equation (3.7-13a) is now converted into a pdf by at first transforming it into an e-function:

$$N = e^{2.3026(10.4 - 1.15 M)} \qquad (3.7\text{-}14)$$

Plotting the data of Table 3-17 in a histogram suggests that the function $N = f(M)$ resembles indeed an exponential ddistribution. N is converted into a relative frequency by dividing it by the total number of elements of the sample. This number is the sum of the elements in the first four rows of column 3 of Table 3-17, $N_t = 194$, so that Equation (3.7-14) becomes

$$f(M) = c \frac{N}{N_t} = \frac{c}{194} e^{(23.947 - 2.648 M)} = c\, e^{-(2.648 M - 18.679)} \qquad (3.7\text{-}15)$$

where $f(M)$ is the pdf of M; and c is a scale factor to be determined by integrating $f(M)$ over its range so that the integral meets the requirement of being equal to 1 (see Section 2.7.5). Introducing the auxiliary variable $u = -2.648 M + 18.679$ and its derivative $du = -2.648\, dM$, the integral gives

$$1 = -\frac{c}{2.648} \int_7^{8.9} e^u du = -\frac{c}{2.648} [e^{-(2.648 \cdot 8.9 - 18.679)} - e^{-(2.648 \cdot 7 - 18.679)}] = 0.433c$$

From $1 = 0.433\, c$ follows $c = 2.310$. Incorporating this factor into the exponent of Equation (3.7-15) by adding $\ln 2.310 = 0.837$ to it gives the pdf of M,

$$f(M) = e^{-(2.648 M - 19.516)} . \qquad (3.7\text{-}16)$$

Equation (3.7-16) shows that M has an exponential pdf in the range $7 \leq M \leq 8.9$.

The non-exceedance probability of a design earthquake, M_d, can be obtained from Equation (3.7-16) by

$$P(M \leq M_d) = \int_7^{M_d} f(M)dM \qquad (3.7\text{-}17)$$

Using the substitution $u = 2.648\ M - 19.516$, and $du = 2.648\ dM$, the integral becomes

$$P(M \leq M_d) = -\frac{1}{2.648}[e^{-(2.648\ M_d - 19.516)} - e^{-(2.648 \cdot 7 - 19.516)}] \quad.$$

By incorporating the factor $1/2.648$ into the exponent and disregarding some numerical inaccuracy, i.e., substituting 1 for 1.006, one obtains

$$P(M \leq M_d) = F(M_d) = 1 - e^{-(2.648\ M_d - 18.542)} \quad, \qquad (3.7\text{-}18)$$

which is the exponential non-exceedance probability. The return period is the inverse of the exceedance probability (see Section 3.7.2):

$$T = \frac{1}{1 - P(M \leq M_d)} = \frac{1}{1 - F(M_d)} \qquad (3.7\text{-}19)$$

As was pointed out in Section 3.7.3, the return period is a dimensionless number, i.e., the expected number of trials until a specified event occurs. Suppose 194 trials occur over a 10 year period. Then each trial is allocated to a subperiod so that there are 194 subperiods each $10/194 = 0.052$ years in length. Equation (3.7-18) gives the number of occurrences out of the total number of occurrences of M that are less than or maximally equal to M_d. Hence, T of Equation (3.7-19) measures the return period by the number of trials or by the number of sub-periods of 0.052 years.

Suppose $M_d = 8$. Then Equation (3.7-18) gives

$$P(M \leq 8) = 1 - e^{-(2.648 \cdot 8 - 18.542)} = 0.929$$

and Equation (3.7-19) gives

$$T = \frac{1}{1 - 0.929} = 14.1$$

where T is in subperiods of 0.052 years; hence, $T = 14.1 \cdot 0.052 = 0.73$ years, which means that an earthquake up to $M = 8$ can occur about every year.

Example: Assume the event population includes the magnitude range $6.0 \leq M \leq 8.9$. According to data provided (Bath, 1979, p. 154, Table 12), this increases the population by 2,660 to a total of 2,854 (sum of elements in column 3 of Table 3-17). The exponential pdf is assumed to also hold for the range $7.0 \leq M \leq 8.9$. Calculate the return period for $M_d = 8$. Solution: With the new total population of $N_t = 2,854$, Equation (3.7-14) becomes

$$f(M) = c\frac{N}{N_t} = \frac{c}{2854} e^{(23.947 - 2.648 M)} = c\, e^{-(2.648 M - 15.990)} \qquad (3.7\text{-}15a)$$

The scale factor is obtained by integrating Equation (3.7-15a):

$$1 = -\frac{c}{2.648} \int_6^{8.9} e^u du = -\frac{c}{2.648} (e^{-(2.648 \cdot 8.9 - 15.990)} - e^{-(2.648 \cdot 6 - 15.990)}) = 0.418c$$

and $c = 2.392$. The pdf for the expanded populations is obtained from Equation (3.7-15a) by incorporating the factor c into the exponent by adding $\ln 2.392 = 0.872$ to the constant term:

$$f(M) = e^{-(2.648 M - 16.862)} \qquad (3.7\text{-}16a)$$

The non-exceedance probability of $M_d = 8$ is obtained by integration of Equation (3.7-16a):

$$P(M \leq 8) = -\frac{1}{2.648} [e^{-(2.648 \cdot 8 - 16.862)} - e^{-(2.648 \cdot 6 - 16.862)}] = 1 - 0.005 = 0.995.$$

The return period according to Equation (3.7-19) becomes

$$T = \frac{1}{1 - 0.995} = 200$$

where T is the number of periods. Conversion into units of years by the factor 10/2854 gives $200 \cdot 10/2854 = 0.70$ years. This result is similar to 0.73 years found for the smaller population.

In the case of earthquakes, frequent small events actually may reduce the frequency of large events or prevent them altogether. Also, events in one area may trigger events in other areas, in other parts of the same continent, or in other parts of the world. Hence, the assumption of event independence may not strictly hold for earthquakes.

The probabilistic analysis of magnitude was chosen here as one example of the probabilistic nature of earthquake engineering. There are empirical relations that connect magnitude to other technically relevant parameters, such as seismic moment, which is the product of fault slip distance, fault slip area, and crustal modulus of elasticity (Kanamori and Brodsky, 2001, p. 36). Other parameters relate to the earthquake resistance of structures, such as lateral and vertical accelerations, lateral velocity, and lateral and vertical ground displacement. Whereas they all occur simultaneously during the 10 to 15 seconds of a shock, their maxima occur at different times with different effects on structures (Bolt, 1978, pp. 955 to 957).

Probabilistic aspects of earthquake engineering are associated with four problem areas: (1) The earthquake probabilities and the associated exposure to harm in a region. This includes the probability of repeated crustal stress accumulation and release that may resemble a probabilistic sawtooth pattern with a random stress release level and a random interarrival time of stress release (the intervals between stress releases may range from decades to centuries, and the stress release level may vary from event to event due to changing fault activity). (2) The resistance of structures that are required to maintain a high probability of survival, such as dams, bridges, high rise buildings, (nuclear) power plants, industrial, commercial, and housing structures, hospitals, public buildings, and so on, to ground motion and shaking. (3) The earthquake–underground interactions that propagate, amplify or dampen waves, ground and foundation subsidence, slope collapse, seiches (tsunamis), and soil liquefaction. (4) The direct measurement of, or empirical relationships between various measures of earthquake power, such as magnitude, energy release rate, and shock, shear, pressure and tension forces on various parts of structures.

Many processes that may seem random may be actually quite predictable were it not for the lack of knowledge of such processes. Earthquake—structure interactions are a good example. There is generally only a short record of seismic observations. The possible seismic activity in a given area is often not well understood, which makes the prediction of earthquake occurrence and magnitude difficult if not impossible. Also there remains a residual possibility of failure of high risk engineering structures under earthquake forces. The reduction of residual failure probabilities often require expensive fail-safe construction and/or rehabilitation measures.

******

4. Probabilistic Approach to Systems

4.1 The System as an Assembly of Components

4.1.1 Probabilistic Nature of Systems

A *system* is defined here as an assembly of components that work in a coordinated way to transform an input into a desired output. For a system to be practical it must have a sufficient degree of functional reliability. Unreliability can reside in all areas contributing to the system's functioning, such as the supply of input, the workings of the system's individual components, and the coordinated functioning of all components together. The more complex a system is, the more likely it is that something fails. The joint probability that all components work as expected is the reliability of the system. In order for the reliability of a system to reach a high level the reliability of its components must even be much higher. In addition, a system may be subject to disturbances of its functioning or threats to its integrity by external events that are independent of the internal functioning of the system. The fact that a system is usually not fully controllable by its operators gives a system and its performance a probabilistic character.

Dependent on how the system is structured, all components or at least a characteristic number of them must be operational at the same time for the system to perform as expected. Maintenance in its various forms, such as repair when damage has occurred and anticipative repair before damage or failure occurs, is used to maximize system reliability. Yet major breakdowns have occurred, sometimes immediately after maintenance had been completed. Failures have occurred in systems where none are permissible because of the certainty of catastrophic consequences. It is typical of the probabilistic character of systems that total control remains elusive. The probability of malfunction or failure can be reduced by operation and maintenance policies. But as long as forced outages cannot be completely eliminated and thus remain possible, the system remains a probabilistic system with a reliability less than 1. The causes that prevent achieving full control include lack of complete knowledge of the production process and of how the system functions, design and construction flaws, detection and elimination of hazards in their incipient states, human errors, and economic infeasibility of achieving perfection in all areas of a system and its operation.

The most amazing phenomenon in probability theory, the central limit theorem (Section 3.4.2), explains why a large system that uses the random contributions of many components is more predictable than a single contributor. For example, an individual power unit of a power system may be on or off at random making its output a rather unreliable quantity. If the system has a sizable number of such units, the resulting sum of random contributions has a normal distribution regardless of the distributions of the original contributors, thus providing a degree of reliability of some level of system capacity within a reasonably small range around the mean capacity.

After planning, design, and construction have created a system, operation and maintenance (O&M) go hand in hand to achieve its optimal performance. Starting with the planning phase and continuing through the productive life of a system, there are many manifestations of a system's behavior that are indicative of its probabilistic nature and of O&M actions that can mitigate or amplify them. Some examples follow.

- design flaws (errors and oversights in the planning and design phase)
- construction flaws (errors and omissions in the construction phase)
- equipment shortcomings or failures (manufacturing flaws, lack of quality control, material flaws, installation errors)
- unexpected operational wear and tear (concrete deterioration, leakage, steel corrosion; machine vibration; abrasion, cavitation)
- operator error (lack of experience, lack of communication, surprise effects, panic)
- surveillance and monitoring shortfalls (lack of inspection, lack of routine and planned maintenance, insufficient scope of surveillance, lack of analysis and use of past findings)
- environmental exposure (freeze/thaw, acid rain, corrosion by polluted air and water)
- external natural events (floods, earthquakes, landslides)
- unintentional and intentional human events (accidents, fires, explosions, vandalism, terrorist attack, war)

A prudent approach to system control by O&M tries to anticipate consequences and weaknesses of the system through competent operation, continued surveillance, competent diagnoses of symptoms, and preventive maintenance action. Hidden flaws may be built into the system through errors in design and construction and may not be discovered until they reveal themselves by unexpected damage and failure symptoms, or actual failure. The more components a system has, the greater is the probability of hidden flaws in one or several of them, manifesting themselves either in isolated incidences involving independent components or in joint incidences or chains of failure events. The internal structure of a system, here meaning the arrangement of the components with respect to each other to perform the intended process, is a major factor in shaping the probability of partial or complete failure as a consequence of one or more component failures.

The uncertainty encountered in the performance of a system is of two types: *system uncertainty* and *input/output uncertainty*. *System uncertainty* relates to the probability of all components functioning as expected, individually and as an ensemble, whereas *input/output uncertainty* relates to externally imposed probabilistic input and output conditions that are related to uncertainty about supply and demand. If just one aspect of the system, including its input and output areas, has probabilistic characteristics, then the system and its output have probabilistic characteristics (see also Section 2.2.4). The usual policy of system operation is to reduce uncertainty. In the hydro sector or the power sector in general, input uncertainty is reduced by a water reservoir or a fossil fuel stockpile. Output uncertainty is reduced by a network as large as possible that has many individual consumers. Each individual consumption is a term in a sum of random variables that makes up the total demand for a given period and thus tends to be normally distributed around a reasonably predictable mean. System uncertainty is reduced by anticipating production unit failures through preventive maintenance.

4.1.2 System Disaggregation

Each system is unique for various reasons. This is also true for systems that perform the same production process, such as power production, using the same input and producing the same output. Despite these similarities there are differences in overall size, in the location and emplacement of subsystems, in type and number of components used, in the arrangement of the components, in type and size of equipment used, and so on. Thus it is not possible to statistically predict system performance or reliability based on a large number of identical systems. An alternative approach is to build a conceptual model of system reliability. This approach consists of two major steps. A first phase of system analysis analyzes the behavior of individual components. The second phase of system synthesis determines system reliability. The first phase that takes the system apart into its components is also called *system disaggregation*. After each component's function is analyzed for its contribution, *system aggregation* synthesizes the system components back into a whole. The entire system's performance is then predicted as the result of the integrated probabilistic functioning of all components. The functioning of a system can be construed as a series of joint events that are independent or dependent on each other.

A system may be a unique entity, but it still consists of basic components that are used in many other systems. Thus, performance data on availability and reliability can be collected and analyzed on a component basis. For example, time to failure and repair time may be available for basic components, such as motors, pumps, switches, valves, and so on. Components are called "similar" if they look alike, have the same purpose, and act alike, but still their performance is different. For example, three light bulbs of the same kind, with the same power output, and used in the same location for the same purpose usually do not fail all at the same time. One may fail before its average lifetime, the second may fail after it, and the third may fail somewhere in between. Their life length is a random variable with a range and a pdf, a perfect example of what is understood here as a system component with probabilistic characteristics.

A water project, for example, may have multiple purpose functions, such as hydropower generation, flood control, drinking water supply, irrigation, recreation, aquatic habitat enhancement, and navigation. These functions require a system with several subsystems to provide the services and outputs expected from the system. For simplicity, only the hydropower aspect will be pursued. The major system components and their function are:

- The *dam* impounds the water thereby providing head and water reserve.
- The *intakes* provide the input to the production process.
- The *spillway and other outlets* protect the structures against floods.
- The *conduits* and/or penstocks convey the water, from the collection area to the use area.
- The *valves* or gates control water flow for production, protect the structure, and make components accessible for maintenance.
- The *turbine/generator units* extract mechanical energy from the water and convert it into electrical energy.
- The *draft tubes* contribute to energy extraction and dispose of the water.
- The *switchyard* connects the electrical power and energy output to the electrical energy distribution network.

This component listing is a first level of *disaggregation* of the system "hydroproject." Several additional disaggregation sublevels are needed to dissect the system into its basic components that are not project- or site-dependent. A four-level disaggregation is shown in Figure 4-1. The first level subsystems, or components, are disaggregated into second level subsystems, which are disaggregated further into still lower level subsystems and finally into individual components. The disaggregation goes to the lowest meaningful level at which general information on the probabilistic behavior of a subsystem or component can be gained.

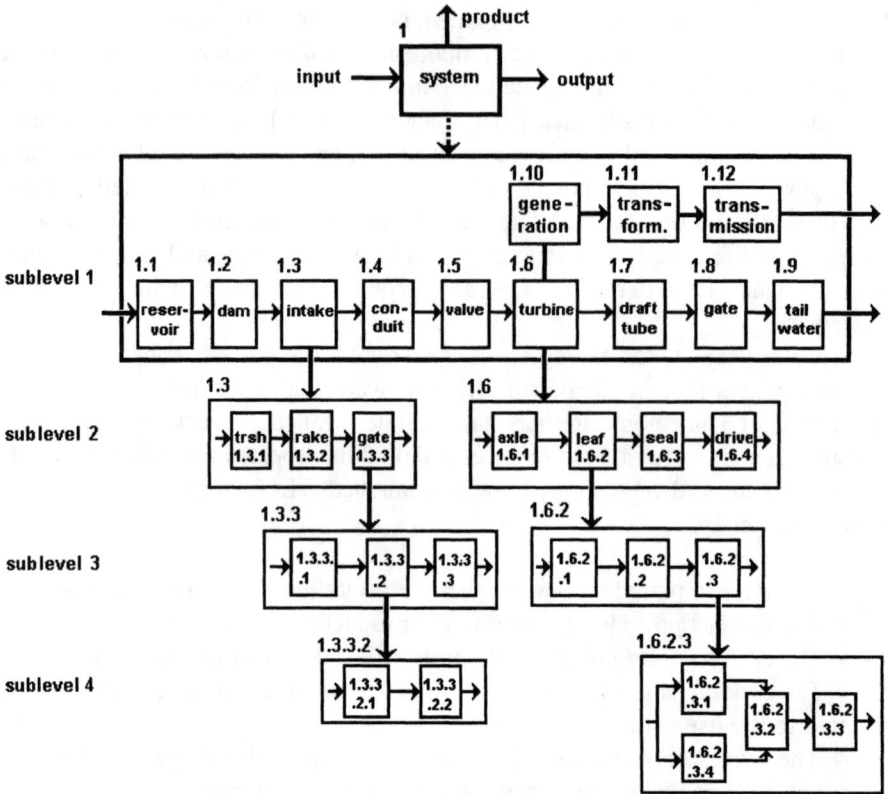

Figure 4-1: Disaggregation of a system (hydroproject) into several sublevels of subsystems and components. The aggregated system is represented by the top box marked "1." The system at this level is disaggregated into sublevel 1 subsystems 1.1 through 1.12. The numbers on the subsystems at the lower disaggregation levels connect these subsystems to the sublevel 1 components of the aggregated system.

    By looking at a hydroproject in this way, it becomes clear that certain components must be operable for the system to be fully or at least partially functioning. These components lined up in a series that connects input and output form a *critical path*. Components in the critical path must have high maintenance priority.

For example, if all turbines are supplied by one water conduit, then this conduit is in the critical path. If parallel power units have their own water conduits, various levels of partial functioning are possible, and each assembly has its own critical path. These paths usually emanate from a common input source (the dam and reservoir) and converge toward a common output delivery point, here to the tailwater and to the switchyard, which are usually common also in systems with parallel production subsystems. All systems, however complex, can be reduced to serial, parallel, or mixed serial/parallel subsystems, as will be discussed in Section 4.2.

4.1.3 Combinations and Permutations

A system is *not* a random combination or permutation of elements. However, a brief discussion of the subject is useful because it provides insight and some quantitative tools for assessing the probabilistic character of systems. The binomial coefficient, $B(n, k)$ has been encountered with the binomial distribution in Section 2.7.2. Here,

$$Comb(n, k) = B(n, k) = \frac{n!}{k!(n-k)!} \qquad (4.1\text{-}1)$$

represents the number of *combinations* of k items in a total of n items without repetition of items. This means the same item must not occur more than once in a combination. Given the definition of $0! = 1$, one obtains for $k = 0$,

$$B(n,0) = \frac{n!}{0!\, n!} = 1 ,$$

which means that zero items can be combined in n items one time. $B(n, n) = 1$ means that $k = n$ items can be combined in n items one time. This is true as long as there is no distinction being made on the serial arrangement of the elements.

If combinations of two are taken from three elements, a, b, c, then with $n = 3$, and $k = 2$, $B(3, 2) = 3 \cdot 2/(1 \cdot 2) = 3$. The three combinations are ab; ac; and bc. Repetitions such as aa are excluded.

With repetitions, the number of combinations is (Kreyszig, 1979, p. 860)

$$Comb_r(n, k) = B(n + k - 1, k) = \frac{(n+k-1)!}{k!\, (n-1)!} \qquad (4.1\text{-}2)$$

where the index r refers to repetitions. Equation (4.1-2) produces a substantially larger number of combinations than Equation (4.1-1). For $n = 3$, $k = 2$, $B(4, 2) = 6$; the number of combinations is ab; ac; bc; aa; bb; and cc.

Permutations are even more numerous than combinations with repetitions, as they also count combinations that are distinguished only by the serial arrangement of the elements. The number of permutations for given n and k without repetitions is

$$Per(n, k) = \frac{n!}{(n-k)!} = B(n, k)k!, \tag{4.1-3}$$

which shows that the number of permutations is k-factorial-times the number of combinations. This produces very large numbers of permutations when k becomes large. For example, the elements a, b, c can be permutated into acb, bca, bac, cab, cba, so that together with abc a total of six permutations is produced whereas only one combination of three elements is possible in three elements.

For permutations with repetitions the number of permutations becomes

$$Per_r(n, k) = n^k \tag{4.1-4}$$

where $Per_r(n, k)$ is the number of permutations with repetitions of n items taken k at a time.

In many problems, n is large and the factorial produces very large numbers. Stirling's formula can be used to approximate the n-factorial (Kreyszig, 1979, p. 861):

$$n! = (\frac{n}{e})^n \sqrt{2\pi n} \tag{4.1-5}$$

with $e \approx 2.71828$, and $\pi \approx 3.14159$. This formula gives only approximate values.

When calculating the binomial coefficient, a reduction in faculty sizes can be achieved by canceling a common factor in numerator and denominator of Equation (4.1-1), which gives (see also Section 2.7.2)

$$B(n, k) = \frac{n(n-1)(n-2)\cdot...\cdot(n-k+2)(n-k+1)}{k!} \tag{4.1-6}$$

Equation (4.1-6) is useful for hand calculations when n is large and k is small.

Examples: (1) Let $n = 5$ and $k = 2$. What is the number of combinations of two items without repetitions? Give an example of such combinations. Solution: Using Equation

(4.1-6), $B(5, 2) = 5 \cdot 4/(1 \cdot 2) = 10$. If the five elements are the numbers 1, 2, 3, 4, 5, then the combinations are: 12, 13, 14, 15, 23, 24, 25, 34, 35, 45, where the digits represent the elements of each combination.

(2) Let $n = 5$ and $k = 2$. What is the number of combinations of two items with repetitions? Give an example of the added combinations. Solution: Using Equation (4.1-2) and the short form of the binomial coefficient, Equation (4.1-6), one obtains $B(n + k - 1, k) = B(6, 2) = 6 \cdot 5/(1 \cdot 2) = 15$. Without repetitions there are 10 combinations, as shown by Example (1). With repetitions there are five more: 11, 22, 33, 44, and 55.

(3) Calculate the number of combinations of three items in a total of five items, without and with repetition. Solution: Without repetition, the number of combinations is $B(5, 3) = 5 \cdot 4 \cdot 3/(1 \cdot 2 \cdot 3) = 10$. For the elements 1, 2, 3, 4, 5, the combinations without repetitions are 123, 124, 125, 134, 135, 145, 234, 235, 245, 345. With repetitions, the total number of combinations is $B(7, 3) = 7 \cdot 6 \cdot 5/(1 \cdot 2 \cdot 3) = 35$. The repetitions include combinations, such as 111, 112, 113, 114, 115, 221, 222, 223, 224, 225, 331, 332, ..., and so on, altogether 25 combinations in addition to the ten without repetition.

(4) From three different elements, a, b, c, two are taken at a time. What is the number of permutations without and with repetition? Solution: The number of permutations without repetitions is $B(3, 2) \cdot 2! = [3!/(2! \, 1!)] \, 2! = 6$. The permutations are ab, ac, bc, ba, ca, cb. With repetitions, the number of permutations is $3^2 = 9$: ab, ac, bc, ba, ca, cb, aa, bb, cc.

(5) Calculate $B(n, k)$ and $Per(n, k)$ for $n = 12$ and $k = 3$ (without repetitions). Solution: Using Equation (4.1-6) one obtains $B(12, 3) = (12 \cdot 11 \cdot 10)/(1 \cdot 2 \cdot 3) = 220$. Equation (4.1-3) gives $Per(12, 3) = B(12, 3) \cdot 3! = 220 \cdot 1 \cdot 2 \cdot 3 = 1320$, which amounts to six times more permutations than combinations.

(6) An example that goes back to Laplace (von Mises, 1981, p. 19) is to compute the probability that the 14 letters of the word *constantinople* in that sequence are drawn from an urn that holds the 26 letters of the alphabet. Suppose each of the 26 letters is written on a tag and placed in the urn. The tag is replaced after each draw. (a) What is the probability of drawing 14 letters that constitute the word *constantinople*? (b) What is the probability of just drawing the letters in any sequence? Solution: (a) The word *constantinople* has 3 n's and 2 t's in it. Therefore, the word is a permutation with repetitions. The number of permutations with repetitions is $n^k = 26^{14} = 6.45 \cdot 10^{19}$. Each combination has the same probability. Hence the probability of this word to be drawn is the same as the probability of any other combination or $(1/26)^{14} = 1.55 \cdot 10^{-20}$ The probability of this word is also found as the joint probability of drawing

the correct letter 14 times in a row. Drawing the correct letter, or any special letter, has the probability 1/26. Drawing 14 correct letters has the joint probability $(1/26)^{14}$. (b) The number of combinations of 14 letters in 26 letters with repetitions is $(26 + 14 - 1)!/(14! \, 25!) = 1.51 \cdot 10^{10}$. The probability of drawing this combination is $0.66 \cdot 10^{-10}$.

(7) 69! is the usual limit of hand calculators. The result is $1.711 \cdot 10^{98}$. Applications can easily exceed this number. The Sterling formula, Equation (4.1-5), gives $69! = (69/e)^{69} (2 \, \pi \, 69)^{1/2} = 1.709 \cdot 10^{98}$.

4.1.4 The k-out-of-n System

For systems with many components, all with the same probability of functioning, p, the binomial distribution provides a means of analyzing the system for its probability of functioning. The binomial formula, in slightly different notation than that used in Section 2.7.2, states:

$$P(n,k,p) = B(n,k)\, p^k \, (1-p)^{n-k} \tag{4.1-7}$$

where $P(n, k, p)$ is used for $f(k, n, p)$ in Section 2.7.2. Equation (4.1-7) gives the probability of k special events in n trials, given the probability of the special event at any trial is p. The binomial pdf is a discrete distribution which makes $f(k, n, p)$ a probability. If k is the number of failed elements out of a total of n elements, then $P(n, k, p) = P(p_k = k/n)$ is the probability of k failing elements in a total of n elements.

Special cases: (1) For $k = n$, and with p a success probability, the n-out-of-n system is one in which all n components are successful. With $B(n, n) = 1$, and $p^n (1 - p)^0 = p^n$, Equation (4.1-7) becomes

$$P(n, n, p) = p^n . \tag{4.1-8}$$

Equation (4.1-8) gives the probability of n successes in n independent trials (Section 2.1.3). In a binomial process such as the one shown in Figure 2-5, p is the branching probability, here toward the successful outcome. If a unit succeeds with probability p during every successive hour, then each hour represents a new trial. If the unit succeeds in n successive hours then it follows the uppermost path in Figure 2-5. The probability that this happens is p^n, the joint probability of n successful events. This probability can also be construed as the probability of n links of a chain to hold each with a probability p, so that the entire chain holds.

A system of units concatenated in a row is a *serial* system (see Section 4.2.1). In real-world systems, p may be different for each component of the system, so that the joint probability of n units being successful is

$$P(n) = \prod_{i=1}^{n} p_i \qquad (4.1-9)$$

where p_i are the individual probabilities of the i components, and $P(n)$ is the product of the n individual probabilities. Equation (4.1-9) holds if the units are independent which means that the success or failure of one unit is not related to the success or failure of another unit (see Section 2.1.4).

(2) For the case of $p = 0.5$, Equation (4.1-7) becomes

$$P(n, k, 0.5) = B(n, k) \, 0.5^n . \qquad (4.1-10)$$

This is the probability distribution of the heads or tails problem. For a number of trials, n, and k being designated as heads, the number of heads is a function of the binomial coefficient only. For $n = 5$, $k = 0, 1, \cdots , 5$, and $0.5^5 = (1/2)^5 = 1/32$, the distribution of k follows.

Tabular Evaluation of Special Case (2)

k	0	1	2	3	4	5	Sum
$B(5, k)$	1	5	10	10	5	1	32
$P(5, k, 0.5)$	1/32	5/32	10/32	10/32	5/32	1/32	1

In n coin tosses, k is the number of heads (H), and $n - k$ is the number of tails (T). Keeping track of the outcome of each toss, one may record outcomes of the five tosses like HHHHH, HHTTH, and so on. Five consecutive branchings into two branches produce a total of $2^5 = 32$ outcomes. If all possible outcomes are represented graphically, a tree structure similar to the one of Figure 2-5 results. If the number of H are counted in each outcome, one will count maximally five and minimally zero. These are the k's in the top row of the result table. The binomial coefficient $B(n, k)$ in row 2 gives the number of k's that *can* occur in each outcome, and row 3 gives the theoretical frequency of k. The factor 0.5^n is the reciprocal of the population size so that $B(n, k) \, 0.5^n$ is the frequency of k in all possible outcomes, as required by Equation (4.1-10).

Example: (1) A decision maker has a record of making decisions that are right or wrong with 50 % probability. Now he has to make five decisions in a row. What is the chance that he is always right or that he is always wrong?   Solution: According to Equation (4.1-10) and the result table of special case (2), the probability of always being right ($k = 5$) or always being wrong ($k = 0$) is the same, namely, 1/32.

(2) Given is a system of 100 components with a probability of failure of $p = 0.0001$. (a) What is the probability that one fails? (b) What is the probability that 99 succeed? (c) Show the general relationship between (a) and (b). Solution: (a) By Equation (4.1-7), the probability of one out of 100 failing is

$$P(100,1, 0.0001) = B(100, 1) \cdot 0.0001 \cdot 0.9999^{99} = 0.0099 \approx 0.01,$$

or about 1 %. This is also the probability of 100 independent, mutually exclusive failure events, which is $0.0001 \cdot 100 = 0.01$.

(b) the success probability of one unit is $p = 1 - 0.0001 = 0.9999$. The probability of 99 successes is

$$P(100,99, 0.9999) = B(100, 99) \cdot 0.9999^{99} \cdot 0.0001 = 0.0099 \approx 0.01.$$

The probability of 1 out of 100 failing is equal to the probability of 99 out of 100 succeeding.

(c) The result of Example (2b) can be generalized. The ratio of the probability of k out of n failing, $P(n, k, p)$, and the probability of $n - k$ out of n succeeding, $P(n, n - k, 1 - p)$ is

$$\frac{P(n, k, p)}{P(n, n - k, 1 - p)} = \frac{B(n, k)p^k(1 - p)^{n-k}}{B(n, n - k)(1 - p)^{n-k}p^k} = 1$$

with $B(n, k) = B(n, n - k)$ by symmetry. For example, for $n = 20$ and $k = 17$, $B(20, 17) = B(20,3) = 1140$ (Abramowitz and Stegun, 1970, pp. 828–830).

4.2 System Structure

4.2.1 Serial and Parallel Systems

The components of a system are arranged in the sequence that is required for accomplishing the intended purpose of a system. The *structure function* expresses this arrangement of components in a quantitative way to allow an assessment of the availability of various levels of system performance, subject to the availability of individual components. There are two basic component arrangements, *serial* and *parallel*, as shown in Figure 4.2. The functioning of a system relies on the joint functioning (intersection) or the exclusive functioning (union) of contributing components (events). In a serial system that is composed of three subsystems in sequence, *A*, *B*, *and* *C*, all three subsystems must function for the system as a whole to function, as there is no partial functioning. In a parallel system that is composed of three parallel subsystems, at least one subsystem out of *A*, *B*, *or* *C* must function for the system to function at least partly. The conjunctions *and* and *or* distinguish between joint and mutually exclusive events where *joint* can mean one event being dependent on another or occurring simultaneously and independently with another or others, whereas *mutually exclusive* means that one event or another may occur but not both at the same time. These characteristics and their effect on systems are discussed in some detail.

(1) *Serial system*: The system in Figure 4-2a consists of two sequential components with the second taking the output from the first as input to produce the system output. Obviously, no output is obtained if either *A* or *B* fails. Hence, for the system to function, both components, *A and B*, must function. The probability of functioning or the reliability of the system is the joint success probability of *A* and *B*,

$$R = P(A \cap B).$$ (4.2-1)

The joint probability of two ordinary events, events for which independence is not invoked, is

$$P(A \cap B) = P(A|B)P(B) = P(B|A)P(A) .$$ (4.2-2)

Equation (4.2-2) is the general form of the joint probability that assumes that one event is conditioned on the other. The joint probability of *n* events is (see also Section 2.1.4, Equation (2.1-23))

$$P(E_1 \cap E_2 \cap E_3 \cap \ldots \cap E_n) = P(E_1)P(E_2|E_1)P(E_3|E_1 \cap E_2)\ldots$$

(4.2-3)

$$\ldots P(E_n|E_1 \cap E_2 \cap E_3 \cap \ldots \cap E_{n-1})$$

Equation (4.2-3) states that once event E_1 has occurred, all subsequent events depend on the previous or jointly occurring events. A simple example of this process is drawing a tag from an urn without replacement. All future draws are influenced by the fact that the drawn tags are no more in the game.

If the n events are independent, then the subsequent events are not conditioned on any previous event, and the joint probability of n events is just the product of all n probabilities:

$$P(E_1 \cap E_2 \cap E_3 \ldots \cap E_n) = P(E_1)P(E_2)P(E_3)\ldots P(E_n) \cdot \qquad (4.2\text{-}3a)$$

The subject of success probability or reliability of serial systems is further discussed in Section 4.3.3.

(2) *Parallel system*: The system of two parallel components or subsystems, A and B, in Figure 4-2b functions, if A or B functions. This is expressed by the probability of the union of two ordinary events "A and B functioning." The reliability of the system is

$$R = P(A \cup B). \qquad (4.2\text{-}4)$$

A parallel system with two or more components can function partially. This property of "partial functioning" makes the system more robust against total failure, a property that is attractive and sometimes vital for practical applications (see Section 4.4.1).

The probability of functioning of a two-component parallel system, such as the one in Figure 4-2b, is

$$P(A \cup B) = P(A) + P(B) - P(A \cap B) \cdot \qquad (4.2\text{-}5)$$

Equation (4.2-5) gives the probability of either A or B is functioning, but not both. If A and B are mutually exclusive, then they cannot function at the same time, and $P(A \cap B) = 0$. In this case, the probability of A or B functioning is the sum of their probabilities

$$P(A \cup B) = P(A) + P(B) \cdot \qquad (4.2\text{-}5a)$$

As $P(A)$ and $P(B)$ become large, events A and B are likely to occur also jointly. This becomes obvious when Equation (4.2-5a) produces probabilities in excess of 1. The subtractions of joint probabilities must be observed, however, in all cases where the probability of A or B is wanted (see Section 2.1.3). If $P(A) = P(B) = 0.9$, then joint events must occur and Equation (4.2-5) must be used. However, if A and B are ordinary sets and joint events are possible, then Equation (4.2-5) also applies even if $P(A) + P(B) < 1$.

(a) serial: $A \cap B$

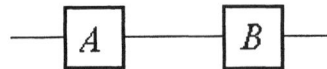

(b) parallel: $A \cup B$

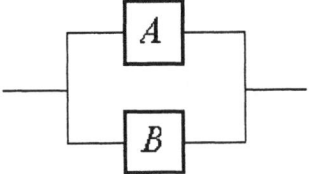

(c) mixed: $A \cap B \cap C$

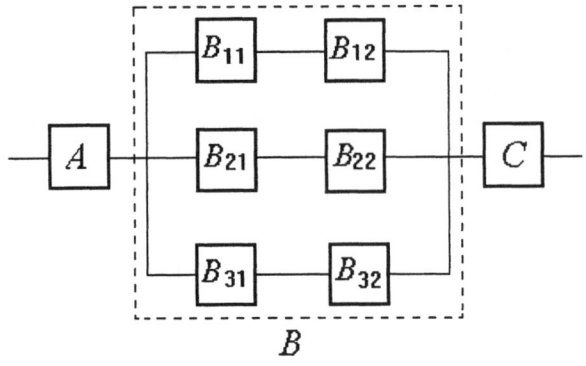

Figure 4-2: Serial, parallel, and mixed (serial-parallel) systems

(3) _Mixed system_: The system in Figure 4-2c consists of a serial system A-B-C, with subsystem B having three serial subsystems in parallel. The system functions if A and B and C function. The reliability of the system is

$$R = P(A \cap B \cap C). \tag{4.2-6}$$

The probability of functioning of the mixed system, such as Figure 4-2c, is the joint probability of each subsystem functioning:

$$P(A \cap B \cap C) = P(A)\,P(B|A)\,P(C|A \cap B) \; . \tag{4.2-7}$$

Equation (4.2-7) is the probability statement for the ordinary case where the subsystems are dependent on the functioning of the subsystem upstream. If A, B, and C are independent of each other, then

$$P(A \cap B \cap C) = P(A)\,P(B)\,P(C) \; . \tag{4.2-8}$$

If the probability of functioning of each system is 0.9, then the joint probability of A, B, and C functioning is $R = P(A \cap B \cap C) = 0.9^3 = 0.73$. Note that the reliability of the serial system is smaller than, or at best equal to, the lowest component reliability.

(4) *Partially functioning total system*: A partial functioning system requires that at least one or two subsystems of B function. To make this statement for the subsystem in Figure 4-2c, some simplified notation is introduced. Subsystem B consists of three secondary subsystems: B_1, B_2, and B_3, each consisting of two components that must function for the subsystem B_1, etc., to function: $B_1 = B_{11} \cap B_{12}$, $B_2 = B_{21} \cap B_{22}$, and $B_3 = B_{31} \cap B_{32}$. Then the probability of at least partial functioning of the total system is

$$P(A \cap B \cap C) = P[A \cap (B_1 \cup B_2 \cup B_3) \cap C] \; . \tag{4.2-9}$$

Equation (4.2-9) states that components A and C must function and, in addition, either B_1 or B_2 or B_3 must function for the system to function at least partially. The probability of the union of B_1, B_2, and B_3 is

$$P(B_1 \cup B_2 \cup B_3) = P(B_1) + P(B_2) + P(B_3)$$

$$\tag{4.2-10}$$

$$- P(B_1 \cap B_2) - P(B_1 \cap B_3) - P(B_2 \cap B_3) + P(B_1 \cap B_2 \cap B_3)$$

If the B_{ij} of each serial system in B are independent of each other, then their joint probabilities of functioning are

$$P(B_1) = P(B_{11} \cap B_{12}) = P(B_{11})\,P(B_{12}),$$

$$P(B_2) = P(B_{21} \cap B_{22}) = P(B_{21})\,P(B_{22}),$$

$$P(B_3) = P(B_{31} \cap B_{32}) = P(B_{31}) \, P(B_{32}).$$

$$P(B_1 \cap B_2) = P[(B_{11} \cap B_{12}) \cap (B_{21} \cap B_{22})] = P(B_{11}) \, P(B_{12}) \, P(B_{21}) \, P(B_{22}),$$

$$P(B_1 \cap B_3) = P(B_{11}) \, P(B_{12}) \, P(B_{31}) \, P(B_{32}),$$

$$P(B_2 \cap B_3) = P(B_{21}) \, P(B_{22}) \, P(B_{31}) \, P(B_{32}),$$

$$P(B_1 \cap B_2 \cap B_3) = P[(B_{11} \cap B_{12}) \cap (B_{21} \cap B_{22}) \cap (B_{31} \cap B_{32})]$$

$$= P(B_{11}) \, P(B_{12}) \, P(B_{21}) \, P(B_{22}) \, P(B_{31}) \, P(B_{32})$$

Substituting these expressions into Equation (4.2-10) gives the probability of one of the three subsystems functioning as

$$P(B_1 \cup B_2 \cup B_3) = P(B_{11}) \, P(B_{12}) + P(B_{21}) \, P(B_{22}) + P(B_{31}) \, P(B_{32})$$

$$- P(B_{11}) \, P(B_{12}) \, P(B_{21}) \, P(B_{22}) - P(B_{11}) \, P(B_{12}) \, P(B_{31}) \, P(B_{32})$$

$$- P(B_{21}) \, P(B_{22}) \, P(B_{31}) \, P(B_{32})$$

$$+ P(B_{11}) \, P(B_{12}) \, P(B_{21}) \, P(B_{22}) \, P(B_{31}) \, P(B_{32}). \qquad (4.2\text{-}11)$$

Assuming the component probabilities are the same for all components, $P(B_{ij}) = 0.9$, one obtains from Equation (4.2-11) for the success probability of subsystem B

$$P(B_1 \cup B_2 \cup B_3) = 0.9^2 + 0.9^2 + 0.9^2 - 0.9^4 - 0.9^4 - 0.9^4 + 0.9^6$$

$$= 2.43 - 1.97 + 0.53 = 0.99.$$

The numerical evaluation shows that if the probability were calculated as if the B_i were three mutually exclusive sets, one would obtain $3 \cdot 0.9^2 = 2.43 > 1$, which obviously is an error. This error is caused by double counting of elements in the intersecting portions of the sets. With high set probabilities (when one or several sets occupy a large portion of the sample space) it becomes obvious that intersections of sets exist and the probability of the union of intersecting sets must be calculated by Equation (4.2-11) to eliminate double counting of the intersecting parts. Intersections are less obvious for small sets and erroneous probabilities may be calculated. Therefore, one must make sure that sets are mutually exclusive with no intersections before calculating the probability of their unions by simply adding up their probabilities.

The probability of the partially functioning total system $A \cap B \cap C$ is

$$P(A \cap B \cap C) = P(A)\ P(B)\ P(C) = 0.9 \cdot 0.99 \cdot 0.9 = 0.80.$$

(5) *Full functioning total system*: In this case, subsystems A, C, and all subsystems of B must fully function. The probability of this event is

$P(A \cap B \cap C)$

$$= P[A \cap (B_{11} \cap B_{12}) \cap (B_{21} \cap B_{22}) \cap (B_{31} \cap B_{32}) \cap C] \qquad (4.2\text{-}12)$$

If all components and subsystems are independent of each other and the parentheses inside the brackets are dropped, the probability of the total system functioning is the product of all component probabilities as if they were all lined up in series. In other words, in a fully functioning system all components and subsystems must function regardless of whether they are in serial arrangement or parallel arrangement. If all component probabilities of functioning are 0.9, then the probability of the entire system functioning is $0.9^8 = 0.43$. The fully functioning case requires all components to function which is less likely than a partially functioning system and thus has a lower probability.

4.2.2 Structure Functions

System B of Figure 4-2c is a k-out-of-n system. Its n components can be represented by the component vector $\mathbf{x} = (x_1, x_2, \dots , x_n)$. Each component is given a binary state variable $x_i = (0,1)$, where

$x_i = 1$ the functioning state
$x_i = 0$ the shutdown state
$k =$ number of functioning components required for the system to function

The *structure function* is an index tagged to a system that marks it as functioning or not functioning. Depending on the system's structure a number k or all n components of the system must work for the system to work.

$$\text{SF}(\mathbf{x}) = 1,\ \text{the system works, if } \sum_{i=1}^{n} x_i \geq k\ . \qquad (4.2\text{-}13)$$

$$\text{SF}(\mathbf{x}) = 0,\ \text{system does not work, if } \sum_{i=1}^{n} x_i < k\ . \qquad (4.2\text{-}14)$$

For subsystem B in Figure 4-2c, the smallest number of components required for the system to function with a probability greater than zero is $k = 2$. This is the minimum number required for at least one subsystem to work. But arbitrary arrangement of the functioning components, for example, one in two different subsystems, can still result in system failure. For the system to function at least partially with probability 1, the number of functioning components required is $k = 4$, which guarantees two components in series functioning in at least one subsystem.

A serial system, Figure 4-2a, is a special case of a k-out-of-n system. It requires the number of functioning components to be $k = n$. A parallel system that consists of m subsystems with $n = 1$ component each, Figure 4-2b, requires $k = 1$ component for partial functioning. A mixed system that consists of m parallel subsystems, with n serial components each, Figure 4-2c, requires one subsystem to function, $k = n$, for a probability greater than zero of partial functioning. But it requires at least

$$k = m(n - 1) + 1$$

components to partially function with probability 1. This means that the matrix of system components, $m \times n$, must be filled with functioning components with the exception of the last column in which only 1 functioning component is required, where m and n are the rows the columns, respectively, of the matrix.

Figure 4-3 illustrates an application of this concept to a three-unit pumped-storage plant. In order for the parallel portion of the system to function, three devices in series need to function for partial operation. But at least $3(3 - 1) + 1 = 7$ components out of 9 must function to fill enough slots so that at least one serial subsystem is functioning for sure.

The index values of the structure functions of serial and parallel systems are obtained as follows.

Serial system:

$$SF_s = \min(x_1, x_2, \dots, x_n) = \prod_{i=1}^{n} x_i \qquad (4.2\text{-}15)$$

where SF_s is the structure function for a serial system. The value of the function is the minimum value of all x_i that can take on values of 0 or 1. This value is found by taking the product of all x_i. This product determines the minimum of the structure function, which is either 0 or 1. If any $x_i = 0$, $SF_s = 0$, which means the system fails. If all components have $x_i = 1$, $SF_s = 1$ and the serial system works

Parallel system:

$$SF_p = \max(x_1, x_2, \ldots, x_n) = 1 - \prod_{i=1}^{n}(1 - x_i) \qquad (4.2\text{-}16)$$

where SF_p is the structure function of the parallel system; $\max(x_1, x_2, \ldots)$ means taking the largest number of the x_i and set SF_p equal to it. This is accomplished by the product formula. With the x_i being either 0 or 1, one single $x_i = 1$ makes $SF_p = 1$, and the total system works. This means that an x_i must represent an entire parallel subsystem. If such a parallel subsystem contains serial components, the serial subsystem is tested first, and the result is then entered into the test of the next higher level aggregated system.

Examples: (1) The parallel-serial subsystem components of system B in Figure 4-2c have the following state numbers: $B_{11} = B_{32} = 0$, all others have state numbers 1. Test the functioning of B. Solution: The structure function for serial systems, Equation (4.2-15) is applied to each subsystem B_1, B_2, and B_3.

$$
\begin{aligned}
B_1: \quad & SF_s(B_1) = \min(B_{11}, B_{12}) = \min(0, 1) = 0 \\
B_2: \quad & SF_s(B_2) = \min(B_{21}, B_{22}) = \min(1, 1) = 1 \\
B_3: \quad & SF_s(B_3) = \min(B_{31}, B_{32}) = \min(1, 0) = 0.
\end{aligned}
$$

The same result is obtained by applying the second right-side term of Equation (4.2-15) to each subsystem.

$$
\begin{aligned}
B_1: \quad & SF_s(B_1) = B_{11} \cdot B_{12} = 0 \cdot 1 = 0 \\
B_2: \quad & SF_s(B_2) = B_{21} \cdot B_{22} = 1 \cdot 1 = 1 \\
B_3: \quad & SF_s(B_3) = B_{31} \cdot B_{32} = 1 \cdot 0 = 0.
\end{aligned}
$$

The structure function for a parallel system, Equation (4.1-16), gives for B:

$$SF_p(B) = \max(B_1, B_2, B_3) = \max(0, 1, 0) = 1.$$

The same result is obtained from the second right-side term of Equation (4.2-16):

$$SF_p(B) = 1 - (1 - 0)(1 - 1)(1 - 0) = 1.$$

This means the parallel-serial subsystem B in Figure 4-2c works under the assumed conditions. While the first and third serial subsystems are down, the central subsystem is operational. This makes subsystem B partially functioning.

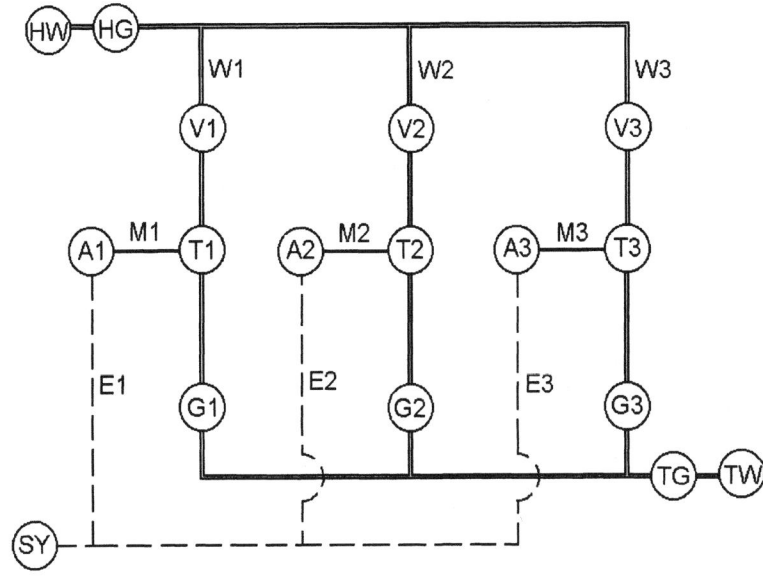

Figure 4-3: A pumped storage plant is a system that can store electric surplus energy by impounding water in a hilltop reservoir and produce energy by discharging the water from the reservoir. The rate at which energy storage and release occur is determined by the installed power. The electromechanical system consists of two double-duty components pump/turbine and motor/alternator, which convert electrical energy into stored energy (pumping mode), or stored energy into electrical energy (generating mode). The system components are the hilltop reservoir, or headwater (HW); the head gate (HG) that controls reservoir inflow and outflow mainly as a safety device; three parallel water conveyance systems (W1, W2, W3), each individually controllable by valves (V); three energy conversion units consisting of reversible pump-turbines (T); mechanical couplings (M) provide energy transfer to/from the motor/alternator (A), which consume/produce electrical energy (E) that is provided/delivered from/to the switchyard (SY), which in turn is connected to the electrical distribution system. Downstream gates (G) and the tailgate (TG) in the water conveyance system are used for maintenance purposes to isolate parallel subsystems that are being serviced from operating subsystems. The tailwater (TW) is the water entrance/exit of the system.

4.2.3 Hydrosystems as Mixed Systems

Hydroplants are typically *serial* or *mixed* serial-parallel systems. Water and head is supplied by a serial subsystem consisting of a reservoir and a dam; the power plant consists of one serial subsystem for each installed unit in parallel arrangement; the power output and the water discharge are again delivered to serial output systems. Parallel systems have the advantage of higher reliability. However, the nature of the process, such as the single fuel source, structural requirements, and project economy, often limit the extent to which parallel systems can be implemented.

A mixed serial-parallel system of a pumped storage plant is shown in Figure 4-3. The headwater reservoir, main conduit, and tailwater are the serial components that serve the three parallel generation units. Each generation unit can be cut off from the upstream and downstream serial subsystems by valves so that one or several units can operate while others are shut down for operational or maintenance reasons. Thus, the system remains at least partially operational as long as the serial components at the front end *and* back end remain operational. Emergency gates at the front end and the back end of the parallel production system permit the shutdown of the entire system. Installation of control gates and valves is costly, but is usually justified or required to meet safety, reliability, and availability requirements of the system.

A reliability analysis of a hydroplant begins with disaggregating it into its subsystems and components. The spillway is one of these subsystems. Suppose it has four bays each being controlled by a gate. Consider two alternative designs:

1. The gates have no individual drive mechanism and are served by a mobile crane.
2. Each gate has individual lifting equipment.

The mobile crane has a self-propulsion motor, a winch, cables, a grabbing mechanism, and a release mechanism. The device moves on track along the crest of the dam, takes position over the gate, grabs the gate and lifts it, and then releases it in the desired position and moves on. The individual gate-lifting device consists of a stationary motor next to the spillway bay, a winch, and a drive mechanism for each gate. The two systems are illustrated in Figures 4-4 and 4-5. They have the same task: to pass water from the reservoir directly into the tailwater. Both systems consist of mixed serial-parallel systems. The individual drive system is an arrangement of parallel subsystems with a high probability of at least partial functioning. The mobile crane system consists of a serial subsystem that controls the functioning of the parallel gate system. Usually cost is the constraint that imposes limits on the installation of parallel systems. A probabilistic benefit/cost analysis would be in order to weigh the pros and cons of parallel and serial designs.

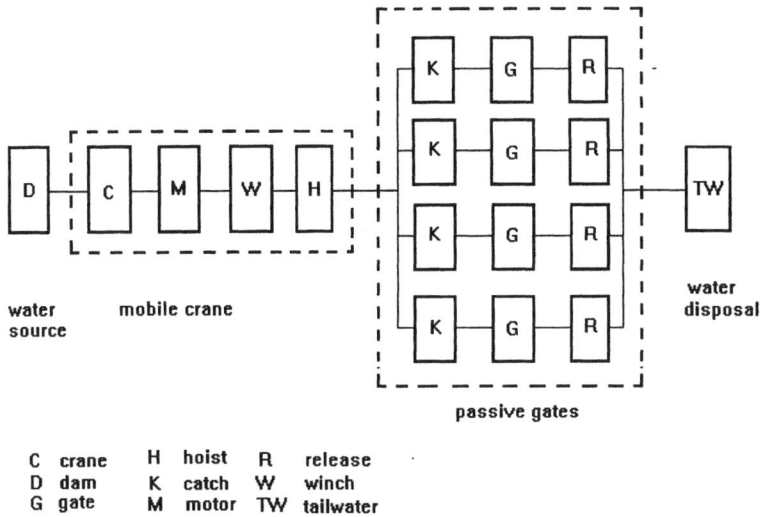

C crane   H hoist   R release
D dam   K catch   W winch
G gate   M motor   TW tailwater

Figure 4-4: A serial system for gate operation: one mobile crane services several gates.

If x_i's are assigned to each component and the serial structure function is applied to each subsystem, any zero encountered in the **x** vectors of a serial subsystem will return $SF_s = 0$ for this serial subsystem. For the mobile crane subsystem, this means that *all* gates fail to open. In the individual hoist subsystems, it means that the respective gate lift fails and that *one* gate fails to open. A joint probability of several simultaneous gate failures exists but it should be small and become smaller for each additional gate, so that the probability of all gates failing is very small. Hence, the mobile crane subsystem must be a focus of preventive maintenance because of the serious consequences of failure of this subsystem.

Examples: (1) The system in Figure 4-4 represents a mobile crane that services four gates. Formulate the probability that the system works. Solution: The system works, given the traveling crane works. The crane opens one gate at a time, so the crane must be operational not just once but four times to open all four gates. The probability that the components of the mobile crane work is

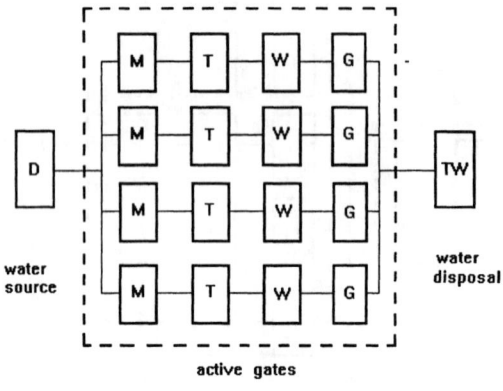

active gates

D dam   M motor   W   winch
G gate   T transmission TW tailwater

Figure 4-5: A parallel system for gate operation: each gate is equipped with an individual hoist mechanism.

$$P(MC) = P(C \cap M \cap W \cap H)$$

where $P(MC)$ is the probability of the mobile crane working; C is the crane, M is the motor, W is the winch, and H is the hooking device. The probability that the crane works on the first gate and then three more times in a row, is

$$P(MC_1 \cap MC_2 \cap MC_3 \cap MC_4) =$$

$$P(MC_1) \, P(MC_2|MC_1) \, P(MC_3|MC_1 \cap MC_2) \, P(MC_4|MC_1 \cap MC_2 \cap MC_3)$$

where $P(MC_1)$ is the probability of the first successful crane operation; MC_i, $i = 1,...,$ 4, are the four crane operations in sequence. The probability of the first crane operation is the unconditional success probability that the crane system works. The second and following operations depend on the successful completion of the previous operation, and their probabilities are conditioned on the previous successful operation or operations (Section 2.1.4).

The total probability that all gates will open is conditioned on the MC working.

$$P(G) = P\{[(K_1 \cap G_1 \cap R_1)|MC_1] \cap [(K_2 \cap G_2 \cap R_2)|MC_2]$$

$$\cap [(K_3 \cap G_3 \cap R_3)|MC_3] \cap [(K_4 \cap G_4 \cap R_4)|MC_4]\}$$

where K_i, G_i, and R_i are the serial components of each gate, $i = 1, ..., 4$. If the abbreviation $E_i = \{K_i \cap G_i \cap R_i\}$ is used, and if $P(E_i) = 0.9$, and all events are assumed independent, then one obtains for the probability of subsequent crane operations $P(MC_i) = 0.9^i$, and the probability of successful gate operation with the mobile crane is

$$P(G) = P(E_1)P(MC_1)P(E_2)P(MC_2)P(E_3)P(MC_3)P(E_4)P(MC_4)$$

$$= 0.9 \cdot 0.9 \cdot 0.9 \cdot 0.9^2 \cdot 0.9 \cdot 0.9^3 \cdot 0.9 \cdot 0.9^4 = 0.9^{14} = 0.23.$$

This rather low probability shows that, given there are many components in a system, the probability of functioning of each component must be very high to achieve a sufficiently high system functioning probability.

(2) Formulate the probability for at least two spillway gates functioning. Solution: The probability of at least two gates functioning out of four is the probability that one of the pairs out of all possible pairs of gates functions. The number of pairs of gates is the number of combinations of two gates in four without repetitions. According to Equation (4.1-1), $B(4, 2) = 6$. The possible gate combinations are 1 and 2, 1 and 3, 1 and 4, 2 and 3, 2 and 4, and 3 and 4. To keep the expressions simple, the combinations are designated $A = 1$ and 2, $B = 1$ and 3, $C = 1$ and 4, $D = 2$ and 3, $E = 2$ and 4, and $F = 3$ and 4. The probability of the union of six ordinary events is according to Equation (2.1-14) of Section 2.1.3:

$P(A \cup B \cup C \cup D \cup E \cup F)$

$= P(A) + P(B) + P(C) + P(D) + P(E) + P(F)$

$- [P(AB) + P(AC) + P(AD) + P(AE) + P(AF) + P(BC) + P(BD) + P(BE)$

$+ P(BF) + P(CD) + P(CE) + P(CF) + P(DE) + P(DF) + P(EF)]$

$+ [P(ABC) + P(ABD) + P(ABE) + P(ABF) + P(ACD) + P(ACE)$

$$+ P(ACF) + P(ADE) + P(ADF) + P(AEF) + P(BCD) + P(BCE) + P(BCF)$$

$$+ P(BDE) + P(BDF) + P(BEF) + P(CDE) + P(CDF) + P(CEF) + P(DEF)]$$

$$- [P(ABCD) + P(ABCE) + P(ABCF) + P(ABDE) + P(ABDF)$$

$$+ P(ABEF) + P(ACDE) + P(ACDF) + P(ACEF) + P(ADEF)$$

$$+ P(BCDE) + P(BCDF) + P(BCEF) + P(BDEF) + P(CDEF)]$$

$$+ [P(ABCDE) + P(ABCDF) + P(ABCEF) + P(ABDEF) + P(ACDEF)$$

$$+ P(BCDEF)] - P(ABCDEF).$$

The intersection symbols between the letters were omitted. Probability statements for the union of ordinary sets become cumbersome to evaluate as all set intersection probabilities must be evaluated. The number of set intersections is also the number of set combinations and thus the number is given by the binary coefficients for these combinations. The first series are the six probabilities, $P(A) + P(B)$, and so on. Their number is given by $B(6, 1) = 6$, where $B(n, k)$ is the binomial coefficient. This sum of mutually exclusive event probabilities has to be corrected for an ordinary set by eliminating double counting due to intersections. The number of two-set intersection probabilities, $P(AB)$, $P(AC)$, and so on, is $B(6, 2) = 15$. The number of three-set intersection probabilities is $B(6, 3) = 20$. The number of four-set intersection probabilities is $B(6, 4) = 15$. The number of five-set intersection probabilities is $B(6, 5) = 6$. There is one six-set intersection probability, $B(6, 6) = 1$. The brackets in the equation contain the groups of 2-, 3-, 4-, and 5-set intersection probabilities. The signs for the groups are those given by Equation (2.1-14) of Section 2.1-3. For ordinary events, the joint probabilities, according to Equation (2.1-17), are $P(AB) = P(A \cap B) = P(A) P(B|A)$. If the events are independent, $P(AB) = P(A) P(B)$. The probabilities for more than two events are expanded according to Equations (2.1-22) and (2.1-23).

4.3. System Reliability Functions

4.3.1 System Reliability

Reliability is defined as the probability of success. A success occurs when the realizations x of the random variable X meet a stated criterion, as reliability may have different meanings.

(a) Reliability as exceedance probability of some minimum threshold $x = a$:

$$R(a) = P(X > a) = 1 - F(a) \tag{4.3-1}$$

where $1 - F(a)$ is the exceedance probability of a. A threshold a could be a standard material strength, and the requirement could be that 80 % of samples taken exceed a. Then $R(a) = 0.8$.

(b) Reliability as non-exceedance probability of some upper limit, $x = b$:

$$R(b) = P(X \le b) = F(b) \tag{4.3-2}$$

where $F(b)$ could be the non-exceedance probability of a load on a structural component. The requirement could be that with 90 % probability a load b is not exceeded. Then $R(b) = 0.9$.

(c) Reliability as probability of x being within a range, $a \le X \le b$ (see Sections 2.4.5 and 3.5.1):

$$R(a,b) = P(a \le X \le b) = F(b) - F(a) \tag{4.3-3}$$

where a and b represent an acceptable range of concrete strength samples.

4.3.2 k-out-of-n Reliability

The binomial probability density function, $P(n, k, p)$ was given by Equation (4.1-7). Being a discrete pdf, its cumulative distribution function is (see Section 2.7.2, Equation 2.7-8)

$$F(x) = \sum_{i=0}^{x} B(n, i) p^i (1 - p)^{n-i} \tag{4.3-4}$$

where $F(x)$ is the non-exceedance probability of x; x is used here to denote any number of occurrences of a special event, i.e., a success or a failure, in n trials whereas k is used as a critical or desirable number, but both are always the upper limit of an integral; if $p_x = x/n$ is the proportion of special events in n trials, then p is the expectation of p_x, $E(p_x) = p$ (see Section 2.7.2, Equation (2.7-5)). Since n is a constant, the proportion x/n has the same probability as x. Hence, $F(x)$ can also be written as

$$F(x) = P(p_x \le \frac{x}{n}) . \tag{4.3-5}$$

If the exceedance probability of x is a success event, such as defined by Equation (4.3-1), then $1 - F(x)$ is a success probability or reliability that can be written as $R(x + 1) = 1 - F(x)$. This avoids double counting of x and fulfills the complementary probability requirement $F(x) + R(x + 1) = 1$. Examples are given in Table 4-2 of Example (2).

The k-out-of-n reliability is the probability that k or more components out of n are successful. Using Equation (4.3-4), but with the summation starting a k, the reliability is

$$R(n, k, p) = \sum_{i=k}^{n} B(n, i) p^{i} (1 - p)^{n-i} \tag{4.3-6}$$

where $R(n, k, p)$ is the exceedance probability of $k-1$, or the complementary probability of $F(n, k - 1, p)$, so that one can also write

$$R(n, k, p) = 1 - F(n, k - 1, p) . \tag{4.3-6a}$$

The reliability or probability of having k or more components operating depends on the total number of components, n, the necessary number of success events k, and the probability of the successful event or component reliability, p. Reliability is seen to depend on n and k, which are determined by system design and load, whereas p can be influenced by maintenance.

If a probabilistic system is expected to produce an overall reliability R, its components must have a sufficiently high component reliability p. For a specified k-out-of-n reliability, R, one solves Equation (4.3-6) for p (Moan, 1982, p. 4-21):

$$\sum_{i=k}^{n} B(n,i) p^{i} (1 - p)^{n-i} = R . \tag{4.3-7}$$

Expanding Equation (4.3-7) into a sum produces

$$B(n, k) p^{k} (1 - p)^{n-k} + B(n, k + 1) p^{k+1} (1 - p)^{n-k-1} + \ldots + p^{n} = R . \tag{4.3-8}$$

Equation (4.3-8) is a polynomial in p and $1 - p$. It is solved by trial and error by assuming values for p until one is found that matches a given R.

Sometimes a large number of binomial terms must be calculated and recalculated, for example, in the solution of Equation (4.3-8). Given the structure of the terms in Equation (4.3-8), a recursive formula can be derived for computing the next term from the previous term. By comparing the two left-most terms in Equation

(4.3-8) and by using Equation (4.1-3) for the binomial coefficients, one can make both terms equal by adding coefficients to the left-hand side:

$$\frac{n!}{k!(n-k)!}p^k(1-p)^{n-k}\frac{p}{1-p}\frac{n-k}{k+1} = \frac{n!}{(k+1)!(n-k-1)!}p^{k+1}(1-p)^{n-k-1}.$$

By using the general notation x for k and turning the equation around one obtains

$$P(n, x+1, p) = P(n, x, p)\, f_x \qquad\qquad (4.3\text{-}9)$$

where

$$f_x = \frac{n-x}{x+1}\frac{p}{1-p}. \qquad\qquad (4.3\text{-}9a)$$

$P(n, x, p)$ is the present value; $P(n, x+1, p)$ is the next value of the binomial probability; and f_x is the probability adjustment factor. A starting value $P(n, x, p)$ can be selected as an easily computable term, for example, for $x_0 = 0$, or $x_0 = 1$. The computation is carried out for all x from x_0 to $x = k$. The discrete non-exceedance probability

$$F(k) = \sum_{i=0}^{k} B(n,i)p^i(1-p)^{n-i} \qquad\qquad (4.3\text{-}10)$$

is then converted to the reliability by

$$R(k+1) = 1 - F(k). \qquad\qquad (4.3\text{-}10a)$$

The exceedance probability, $1 - F(k)$, is the sum from $k + 1$ to n,

$$R(k+1) = \sum_{i=k+1}^{n} B(n,i)p^i(1-p)^{n-i}. \qquad\qquad (4.3\text{-}11)$$

As discussed in Section 4.3.1, reliability is a success probability. As such it can be an exceedance or a nonexceedance probability. Here reliability is assumed to be defined by Equation (4.3-10a).

Attention is drawn to avoiding double counting of elements which would produce wrong probabilities. For example, if $F(x)$ is integrated from $x_0 = 0$ to $x = k$, then $R(x)$ must be integrated from $x = k+1$ to n, so that

$$F(k) + R(k + 1) = 1. \tag{4.3-11a}$$

The same is true for $F(k - 1) + R(k) = 1$ (see Example (2) for details).

Special case: For $k = n$, and p defined as success probability, all units are functioning, and Equation (4.3-8) reduces to one term,

$$p^n = R, \tag{4.3-12}$$

and

$$p = R^{\frac{1}{n}}. \tag{4.3-13}$$

Table 4-1: Component Success Probabilities p that are Required to Achieve a Specified System Reliability R as a Function of the Number of Components n

n	R	p	$p/R - 1$	R	p	$p/R - 1$
(1)	(2)	(3)	(4)	(5)	(6)	(7)
10	0.9	0.9895	0.0994	0.99	0.9990	0.0090
20	0.9	0.9947	0.1052	0.99	0.9995	0.0096
100	0.9	0.9989	0.1099	0.99	0.9999	0.0100

Notes: column 1: number of component; column 2: desired system reliability; column 3: necessary component reliability calculated by Equation (4.3-13); column 4: excess of p in percent of R by which p must exceed R to achieve a desired R for a given number of elements n, $100 (p - R)/R$; columns 6 and 7 repeat the calculations for $R = 0.99$.

If R and p are given, then n follows from Equation (4.3-12) as

$$n = \frac{\log R}{\log p}. \tag{4.3-14}$$

Some numerical values of Equation (4.3-13) are given in Table 4-1. They show that for a system to meet a required R, the component success p must be larger than R. For two selected system reliabilities, $R = 0.9$ and $R = 0.99$, and different numbers of components, the component success p must exceed system reliabilities by about 10 % and 1 %, respectively.

Examples: (1) A power plant has four units with 10 MW each. (a)What is the probability of at least two units operating and producing 20 MW or more, given all units are independent of each other and have the same functioning probability $p = 0.9$? (b) What is the probability of all units operating? (c) What is the probability of less than full power? (d) What is the probability of all units down? (e) How much power can the system offer if a reliability of 99 % is required? Solution: (a) Equation (4.3-6) gives the probability of two or more units out of four operating:

$$R(2) = \sum_{i=2}^{4} B(n, i) p^i (1-p)^{n-i}$$

$$= B(4, 2) \cdot 0.9^2 \cdot 0.1^2 + B(4, 3) \cdot 0.9^3 \cdot 0.1 + B(4, 4) \cdot 0.9^4$$

$$= 0.0486 + 0.2916 + 0.6561 = 0.9963$$

(b) The probability of all units operating, according to Equation (4.3-12), is $p^4 = 0.656$. This is the probability of the full output of 40 MW. (c) The probability of less than full power is $1 - p^4 = 0.344$. This result is also obtained as the total probability of the mutually exclusive events 0, 1, 2, or 3 units operating:

$$B(4, 0) \cdot 0.9^0 \cdot 0.1^4 + B(4, 1) \cdot 0.9^1 \cdot 0.1^3 + B(4, 2) \cdot 0.9^2 \cdot 0.1^2$$

$$+ B(4, 3) \cdot 0.9^3 \cdot 0.1 = 0.344.$$

(d) The probability of all units down is $(1 - p)^4 = 0.1^4 = 0.0001$. (e) From the calculation under (a), two or more units can be expected to operate with probability 0.9963 whereas three or more units operate with a probability of $0.2916 + 0.6561 = 0.9477$. Hence, if a reliability of 99 % is required, the system only can offer the output of two units, 20 MW, as *reliable* output.

(2) Explore the effect of different success probabilities, $p = 0.9, 0.5, 0.1$, on the probability of the units being operable and on the reliability of the four-unit system of the previous example. Solution: For $n = 4$, $k = 0, 1, 2, 3, 4$, Equation (4.1-7) for $P(n, k, p)$, Equations (4.3-4) and (4.3-6a) for $F(n, k, p)$, and Equation (4.3-6) for $R(n, k, p)$ are evaluated in Table 4-2. Column 5 shows the binomial probabilities of k-out-of-n units, $P(n, k, p)$. These probabilities are cumulated to give $F(n, k, p)$ in column 6. The complementary probability, $R(n, k, p)$, is given in column 7, but offset according

Table 4-2: System Reliability as Function of Component Success Probability

(a) $n = 4$; $p = 0.9$

k	$B(4, k)$	0.9^k	0.1^{n-k}	$P(4, k, 0.9)$	$F(4, k, 0.9)$	$R(4, k, 0.9)$
(1)	(2)	(3)	(4)	(5)	(6)	(7)
0	1	1	0.0001	0.0001	0.0001	1
1	4	0.9	0.001	0.0036	0.0037	0.9999
2	6	0.81	0.01	0.0486	0.0523	0.9963
3	4	0.729	0.1	0.2916	0.3439	0.9477
4	1	0.6561	1	0.6561	1	0.6561

(b) $n = 4$; $p = 0.5$

k	$B(4, k)$	0.5^k	0.5^{n-k}	$P(4, k, 0.5)$	$F(4, k, 0.5)$	$R(4, k, 0.5)$
0	1	1	0.0625	0.0625	0.0625	1
1	4	0.5	0.125	0.2500	0.3125	0.9375
2	6	0.25	0.25	0.3750	0.6875	0.6875
3	4	0.125	0.5	0.2500	0.9375	0.3125
4	1	0.0625	1	0.0625	1	0.0625

(c) $n = 4$; $p = 0.1$

k	$B(4, k)$	0.1^k	0.9^{n-k}	$P(4, k, 0.1)$	$F(4, k, 0.1)$	$R(4, k, 0.1)$
0	1	1	0.6561	0.6561	0.6561	1
1	4	0.1	0.729	0.2916	0.9477	0.3439
2	6	0.01	0.81	0.0486	0.9963	0.0523
3	4	0.001	0.9	0.0036	0.9999	0.0037
4	1	0.0001	1	0.0001	1	0.0001

Notes: Column 1: number of special events; column 2: binomial coefficients for $n = 4$, and $k = 1, ..., 4$; column 3 and 4: factors of the terms making up Equation (4.3-4); column 5: binomial probabilities for Equation (4.3-10); column 6: cumulation of column 5; column 7: complement of column 6 according to Equation (4.3-10a); for $p = 0.1$ and $k = 2$, $1 - F(4, 2, 0.1) = R(4, 3, 0.1) = 1 - 0.9963 = 0.0037$.

to Equation (4.3-10a). The complementarity is met by these probabilities according to Equation (4.3-11a). The relation $P(n, k, p) = P(n, n - k, 1 - p)$ is shown by several pairs of the same numbers in column 5 of Table 4-2(a) and 4-2(c). For example, $P(4, 4, 0.9) = P(4, 0, 0.1) = 0.656$. With increasing k, the probabilities in column 5 increase for $p = 0.9$, decrease for $p = 0.1$, and form a symmetric bell-shape for $p = 0.5$. For $n \to \infty$, the binomial distribution approaches the normal distribution (see Section 2.7.2).

(3) What would the operating probability of each unit have to be to achieve a 40 MW output with reliability of 0.996 with four 10 MW units? Solution: Equation (4.3-13) gives $p = 0.996^{1/4} = 0.999$. The rather high unit reliabilities may not be realistically achievable. Then additional units (reserve) are needed to achieve the 40 MW output with an achievable reliability, similar to the previous example, where two times 10 MW was achieved with a reliability of 0.996 from four 10 MW units each with $p = 0.9$.

(4) Consider a spillway with 10 gates that is to be equipped either with individual hoist mechanisms for each gate or with two cranes that serve two sets of five gates. Assume the probability of proper functioning is $P(\text{gate}) = 0.9$ for the case of individual gates, and $P(\text{crane}) = 0.9$ for each of the two-crane systems. (a) What is the probability that five or more of the independent gates work? (b) What is the probability that one of the two crane-serviced systems works? Solution: (a) The probability of at least five individual gates functioning is the total probability of five gates, or six gates, or seven gates, or eight gates, or nine gates, or ten gates functioning. This total probability is obtained as the sum of these mutually exclusive probabilities by evaluating $R(10, 5, 0.9)$:

$$R = B(10, 5) \cdot 0.9^5 \cdot 0.1^5 + B(10, 6) \cdot 0.9^6 \cdot 0.1^4 + B(10, 7) \cdot 0.9^7 \cdot 0.1^3$$

$$+ B(10, 8) \cdot 0.9^8 \cdot 0.1^2 + B(10, 9) \cdot 0.9^9 \cdot 0.1^1 + B(10, 10) \cdot 0.9^{10} \cdot 0.1^0.$$

With $B(10, 5) = 252$, $B(10, 6) = 210$, $B(10, 7) = 120$, $B(10, 8) = 45$, $B(10, 9) = 10$, and $B(10, 10) = 1$, $R = 0.0015 + 0.0112 + 0.0574 + 0.1937 + 0.3874 + 0.3487 = 0.9999$. This result means that the total probability of four and fewer gates functioning must be 0.0001 so that the complementary requirement, Equation (4.3-11a), is met: $F(10, 4, 0.9) + R(10, 5, 0.9) = 1$. The reader can convince himself by calculating the sum of probabilities from zero to four units functioning, which is $F(10, 4, 0.9) = 0.0001$. Thus the requirement of complementarity is met.

(b) For the two-crane system, a functioning system is a 1-out-of-2 system, that means a system in which at least one crane of a total of two is functioning. The reliability is

$$R = B(2, 1)\, p\, (1 - p) + B(2, 2)\, p^2 \;=\; 2 \cdot 0.9 \cdot 0.1 + 1 \cdot 0.9^2 \;=\; 0.99.$$

Here the complementary probability is $F(2, 0, 0.9) = 0.1^2 = 0.01$, which is the probability of none of the systems functioning. The two sums of probabilities encountered in (a) and (b) represent the probability of mutually exclusive events, which is the probability of one *or* another possible event to happen. The greater the number of possible events included in this sum, the greater is the probability that one of them will happen.

(5) Find p for a specified non-exceedance probability $F(k) = 0.95$, $n = 50$, and $k = 20$ from the implicit form

$$F(20) = \sum_{x=0}^{20} \{B(50, x)\, p^x (1 - p)^{50-x}\} \;=\; 0.95$$

where x is the current number of special events from $x = 0$ to $x = k = 20$. <u>Solution</u>: First, for an initial p, find an initial $P(n, x, p)$. Then calculate successive $P(n, x+1, p)$ using the recursive formula, Equation (4.3-9). Then cumulate $P(n, x, p)$ up to $x = k$ and check if $F(20) = 0.95$. If this is true, the assumed p is correct and the calculation is terminated. If no, make a new assumption for p, and recalculate P and F. After a few trials with different p's, the p is found that meets the requirement. Suppose (here with some prior knowledge) $p = 0.3$. Then, for $x = 0$,

$$P(50, 0, 0.3) = B(50, 0)\, 0.3^0\, 0.7^{50} = 1 \cdot 1 \cdot 1.8 \cdot 10^{-8} = 1.8 \cdot 10^{-8}.$$

This initial value is used in Equation (4.3-9) to calculate all others. Given $P(50, 0, 0.3)$, the subsequent probability for $x = 1$ is

$$P(50, 1, 0.3) = P(50, 0, 0.3)\, f_x.$$

The probability adjustment factor according to Equation (4.3-9a) is

$$f_x = (50 - 0)\, 0.3/[(0 + 1)(1 - 0.3)] = 21.4286,$$

so that for $x = 1$:

$$P(50, 1, 0.3) = P(50, 0, 0.3)\, 21.4286 = 3.857 \cdot 10^{-7}.$$

For $x = 2$: $P(50, 2, 0.3) = 3.857 \cdot 10^{-7} \cdot (50 - 1)/(1 + 1) \cdot 0.3/0.7 = 4.05 \cdot 10^{-6}.$

For $x = 3$: $P(50, 3, 0.3) = 4.05 \cdot 10^{-6} \cdot (50 - 2)/(2 + 1) \cdot 0.3/0.7 = 2.78 \cdot 10^{-5},$

and so on. The results of the recursive formula can be tested by calculating $P(50, x, 0.3)$ directly from the binomial formula. The calculation of P and F for the assumed $p = 0.3$ is summarized in Table 4-3. In row $x = k = 20$, the table shows $F(20) = 0.9522$, a good approximation of 0.95. Hence $p = 0.3$ is an acceptable estimate of p. The table also includes P and F for $p = 0.6$ as another trial value. The graphical display of P and F for $p = 0.3$ and $p = 0.6$ is given in Figure 4-6. It shows that the distribution for $p = 0.6$ is shifted to the right and $F \approx 0.946$ is reached for $k = 35$ instead of $k = 20$.

Table 4-3: Trial Calculation to Find p that Gives $F = 0.95$ for $k = 20$. Assumed Probabilities are $p = 0.3$ and $p = 0.6$

x	f_x	P(50, x, 0.3)	F(x)	f_x	P(50, x, 0.6)	F(x)
(1)	(2)	(3)	(4)	(5)	(6)	(7)
0	21.4286	0.00000002	0.0000	75.0000	0.0000	0.0000
1	10.5000	0.00000039	0.0000	36.7500	0.0000	0.0000
2	6.8571	0.00000405	0.0000	24.0000	0.0000	0.0000
3	5.0357	0.00002775	0.0000	17.6250	0.0000	0.0000
4	3.9429	0.00013973	0.0002	13.8000	0.0000	0.0000
5	3.2143	0.00055093	0.0007	11.2500	0.0000	0.0000
6	2.6939	0.0018	0.0025	9.4286	0.0000	0.0000
7	2.3036	0.0048	0.0073	8.0625	0.0000	0.0000
8	2.0000	0.0110	0.0183	7.0000	0.0000	0.0000
9	1.7571	0.0220	0.0402	6.1500	0.0000	0.0000
10	1.5584	0.0386	0.0789	5.4545	0.0000	0.0000
11	1.3929	0.0602	0.1390	4.8750	0.0000	0.0000
12	1.2527	0.0838	0.2229	4.3846	0.0000	0.0000
13	1.1327	0.1050	0.3279	3.9643	0.0000	0.0000
14	1.0286	0.1189	0.4468	3.6000	0.0000	0.0000
15	0.9375	0.1223	0.5692	3.2813	0.0000	0.0000
16	0.8571	0.1147	0.6839	3.0000	0.0000	0.0001
17	0.7857	0.0983	0.7822	2.7500	0.0001	0.0002
18	0.7218	0.0772	0.8594	2.5263	0.0003	0.0005
19	0.6643	0.0558	0.9152	2.3250	0.0009	0.0014
20	0.6122	0.0370	0.9522	2.1429	0.0020	0.0034
21	0.5649	0.0227	0.9749	1.9773	0.0043	0.0076
22	0.5217	0.0128	0.9877	1.8261	0.0084	0.0160
23	0.4821	0.0067	0.9944	1.6875	0.0154	0.0314
24	0.4457	0.0032	0.9976	1.5600	0.0259	0.0573
25	0.4121	0.0014	0.9991	1.4423	0.0405	0.0978

26	0.3810	0.0006	0.9997	1.3333	0.0584	0.1562
27	0.3520	0.0002	0.9999	1.2321	0.0778	0.2340
28	0.3251	0.0001	1.0000	1.1379	0.0959	0.3299
29	0.3000	0.0000	1.0000	1.0500	0.1091	0.4390
30	0.2765	0.0000	1.0000	0.9677	0.1146	0.5535
31	0.2545	0.0000	1.0000	0.8906	0.1109	0.6644
32	0.2338	0.0000	1.0000	0.8182	0.0987	0.7631
33	0.2143	0.0000	1.0000	0.7500	0.0808	0.8439
34	0.1959	0.0000	1.0000	0.6857	0.0606	0.9045
35	0.1786	0.0000	1.0000	0.6250	0.0415	0.9460
36	0.1622	0.0000	1.0000	0.5676	0.0260	0.9720
37	0.1466	0.0000	1.0000	0.5132	0.0147	0.9867
38	0.1319	0.0000	1.0000	0.4615	0.0076	0.9943
39	0.1179	0.0000	1.0000	0.4125	0.0035	0.9978
40	0.1045	0.0000	1.0000	0.3659	0.0014	0.9992
41	0.0918	0.0000	1.0000	0.3214	0.0005	0.9998
42	0.0797	0.0000	1.0000	0.2791	0.0002	0.9999
43	0.0682	0.0000	1.0000	0.2386	0.0000	1.0000
44	0.0571	0.0000	1.0000	0.2000	0.0000	1.0000
45	0.0466	0.0000	1.0000	0.1630	0.0000	1.0000
46	0.0365	0.0000	1.0000	0.1277	0.0000	1.0000
47	0.0268	0.0000	1.0000	0.0937	0.0000	1.0000
48	0.0175	0.0000	1.0000	0.0612	0.0000	1.0000
49	0.0086	0.0000	1.0000	0.0300	0.0000	1.0000
50	0.0000	0.0000	1.0000	0.0000	0.0000	1.0000

Notes: column 1: x starts with zero to obtain an initial ordinate for the integration in columns 4 and 7.
column 2: f_x for $n = 50$ and $p = 0.3$.
column 3: recursive calculation of $P(50, x, 0.3)$.
column 4: cumulation of column 4; for $x = k = 20$, $F(20) = 0.9522$; cumulation of the total column meets the requirement of $F(50) = 1$.
column 5, 6, and 7: repeat of columns 2, 3, and 4 for $p = 0.6$.

(6) Discuss the principal features that can be deduced from the binomial pdf's of Figure 4-6. <u>Solution</u>: We limit the discussion to $P(50, x, 0.3)$. The maximum of $P(50, x, 0.3)$ occurs at $x = 15$. The expectation of x for $n = 50$ and $p = 0.3$ is $E(x) = n p = 50 \cdot 0.3 = 15$. For $x = 15$, Table 4-3 gives $F(15) = 0.57 \approx 0.5$, which means that $x = 15$ is the expectation of x. Also, for $x = 15$, $E(x/n) = E(p_x) = E(15/50) = p = 0.3$. See binomial relations in Section 2.7.2.

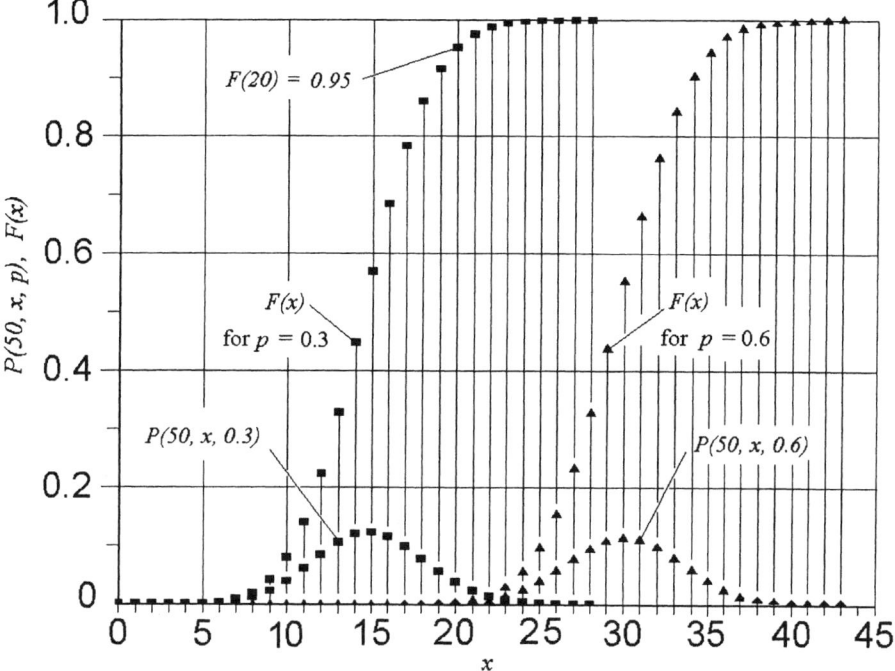

Figure 4-6: Binomial probability densities $P(n, x, p)$ for $n = 50$ and $p = 0.3$ and $p = 0.6$. The probabilities of special events x in n observations are shown as disconnected points because no values exist between the integers x for which the function can be calculated and thus no P's exist. The integral of $P(n, x, p)$ is $F(x)$, the non-exceedance probability of x. It is also represented by disconnected points as no values exist between the integers x. For $p = 0.3$, and for $x = k = 20$ (k being the specified upper integration limit), $F(20) = 0.9522$ (see Table 4-3). This means that the probability of k being equal to or less than 20 is about 95 %. For comparison, $P(50, x, 0.6)$ is also shown. In this case there is a probability of about 95 % that k is equal to or less than 35.

4.3.3 Reliability of Serial Systems

The probability that a set of dependent components $\{x_1, x_2, ..., x_n\}$ functions is the probability that component x_1 works, *and* that x_2 works, given x_1 works, *and* that

x_3 works, given x_1 *and* x_2 work, and so on. The probability of a component working is called an event. In the general case, events occurring jointly may depend in some way on each other. For example, after a first event occurs, the second or other events may be more or less likely to follow. The probability of the second event to happen, given the first event has happened, is a conditional probability. In a time sequence, follow-up events are conditioned on earlier events. In a space setting, events in one area may be conditioned on events in other areas. The probability of such joint, dependent events is also called *compound probability* (Goldberg, 1986, p.78; see Section 2.1.4, Equation (2.1-22)):

$$P(x_1 \cap x_2 \cap ... \cap x_n) = P(x_1)P(x_2|x_1)P(x_3|x_1 \cap x_2)... \qquad \text{(4.3-15)}$$

The compound probability of Equation (4.3-15) uses conditional probabilities to express the interrelationships among the joint events. Sometimes notation is used that omits the $\cap$ symbol. Then the compound probability of n events is expressed by

$$P(x_1, x_2, ..., x_n) = P(x_1)P(x_2|x_1)P(x_3|x_1, x_2)...$$

$$\text{(4.3-15a)}$$

$$P(x_n|x_1, x_2, ..., x_{n-1}).$$

If the occurrence of joint events does not depend on other events then these joint events are independent.

The difference between joint event probability, $P(x_2 \cap x_1)$, conditional probability $P(x_2|x_1)$, and independent event probability, $P(x_2) P(x_1)$, is best explained by their definitions (see also Section 2.1.4). If the individual events are independent of each other then the probabilities are simple event probabilities, $P(x_1) = n_1/N$, $P(x_2) = n_2/N$, and so on, with n_1, n_2, and so on, being the elements of the sets x_1, x_2 of the sample space that consists of a total of N elements. The joint probability of the two events x_1, x_2 is

$$P(x_2 \cap x_1) = n_{2,1}/N,$$

where $n_{2,1}$ is the number of the elements that are common to both sets, graphically the intersecting portion of sets x_1 and x_2 in a Venn diagram. The conditional probability of an event x_2 is

$$P(x_2|x_1) = n_{2,1}/n_1$$

where $n_{2,1}$ is referenced only to the elements n_1 of set x_1. Therefore $P(x_2|x_1)$ may be much larger than $P(x_2 \cap x_1)$. It is usually easier to deal with independent event probabilities but major differences in dependent and independent event probabilities exist and errors may result if independence does not hold. These differences are demonstrated by the relations for probabilities of dependent and independent events. For two dependent events, the probability of joint occurrence is

$$P(x_2 \cap x_1) = P(x_1)\, P(x_2|x_1) = n_{2,1}/N,$$

and for independent events, the probability of joint occurrence is

$$P(x_2 \cap x_1) = P(x_2)\, P(x_1) = n_2\, n_1/N^2,$$

The change in probability that occurs when the conditioning is dropped is demonstrated by

$$P(x_2|x_1) = n_{2,1}/\, n_1 \rightarrow P(x_2) = n_2/N.$$

This means that as long as dependence exists, $P(x_2|x_1)$ and $P(x_2)$ are different numbers.
The compound probability of *three* events is

$$P(x_3 \cap x_2 \cap x_1) = n_{3,2,1}/N,$$

where $n_{3,2,1}$ is the number of elements of the intersection of the three events x_3, x_2, and x_1. If x_3 is conditioned on the events x_1 and x_2 one obtains

$$P(x_3|x_1 \cap x_2) = \frac{P(x_3 \cap x_2 \cap x_1)}{P(x_2 \cap x_1)} = \frac{n_{3,2,1}}{n_{2,1}}$$

where $n_{3,2,1} \le n_{2,1}$. The probability of three joint events is

$$P(x_3 \cap x_1 \cap x_2) = P(x_2 \cap x_1) P(x_3|x_1 \cap x_2) = \frac{n_{2,1}}{N}\frac{n_{3,2,1}}{n_{2,1}} = \frac{n_{3,2,1}}{N}$$

When $P(x_3|x_1 \cap x_2)$ is simplified to $P(x_3)$ by the independence assumption the change in probability is expressed by

$$P(x_3|x_1 \cap x_2) = n_{3,2,1}/\, n_{2,1} \rightarrow P(x_3) = n_3/N.$$

These differences in dependent and independent event probabilities must be kept in mind when the independence assumption is made.

Suppose a bearing failure x_2 has the probability $P(x_2)$ and an oil pump failure x_1 has a probability $P(x_1)$, both being 0.01 for a selected period. Let the conditional probability of a bearing failure, given an oil pump failure, be $P(x_2|x_1) = 0.9$, which means that once the oil pump has failed, the bearing failure becomes very likely. The probability of a bearing failure as a consequence of an oil pump failure for dependent events is $P(x_2 \cap x_1) = P(x_1)\,P(x_2|x_1) = 0.01 \cdot 0.9 = 0.009$. The probability of the two events occurring independently is $P(x_2 \cap x_1) = P(x_2)P(x_1) = 0.0001$. This means the follow-up failure is 90 times more likely than the independent failures.

If independence can be assumed, the compound probability reduces to the product of simple probabilities:

$$P(x_1, x_2, x_3, \ldots, x_n) = P(x_1)P(x_2)P(x_3) \cdot \ldots \cdot P(x_n). \qquad (4.3\text{-}16)$$

The inevitable consequence of a component failure for the serial system is system failure. This result is the same regardless of how many components fail. However, the probability that failure occurs depends on the events being dependent or independent of each other, as was illustrated by the previous definitions.

The probability of a component not failing is the success probability. For clarity we introduce a success probability designation, $R(x_i)$, instead of the general event probability, $P(x_i)$, and a failure probability, $F(x_i)$, as the complementary (failure) probability of $R(x_i)$, so that $P(x_i) = R(x_i) = 1 - F(x_i)$. These notations are used in the following as component or event probabilities. The joint probability of success of Equation (4.3-16) can then be construed as the reliability function of the serial system of independent components,

$$R_s = \prod_{i=1}^{n} R(t_s, \lambda_i), \qquad (4.3\text{-}17)$$

where R_s is the reliability of an n-component serial system, t_s is the survival time of the system, λ_i is the failure rate of component i, and $R(t_s, \lambda_i)$ is the survival probability of component i at time t_s.

Special cases:

(1) All n components have the same success probability: $R(x_i) = p$. Then the probability of success of the serial system is

$$R(x_1, \ldots, x_n) = p^n. \qquad (4.3\text{-}18)$$

This result was also obtained as a special case of the binomial k-out-of-n system with $k = n$, or the probability of n successes in n trials (Section 4.3.2, Equation 4.3-12).

(2) If the success probabilities R_s and $R(t_s, \lambda_i)$ are substituted by the failure probabilities p and R are substituted by their complementary probabilities, $1 - F_{ns}$ and $1 - p_f$, respectively, then Equation (4.3-17) becomes

$$F_{ns} = 1 - (1 - p_f)^n .$$ (4.3-19)

where F_{ns} is the failure probability of the system and p_f is the failure probability of each of the n components.

The term $(1 - p_f)^n$ is the probability of n successes in a row and F_{ns} is the probability that this series of successes fails, in other words, that there is at least one failure. Solving for p_f gives

$$p_f = 1 - (1 - F_{ns})^{1/n}$$ (4.3-19a)

where p_f is the failure probability of a single event for a given F_{ns}. For example, if the allowable failure probability of annual events during successive 100 years of a project is one failure in hundred trials, then $p_f = 1/100 = 0.01$ and the probability of a failure to happen during successive hundred trials is $F_{ns} = 1 - (1 - 0.01)^{100} = 0.63$. Suppose this probability needs to be reduced to $F_{ns} = 0.01$, then the failure probability by an annual event must not exceed $p_f = 1 - (1 - 0.01)^{1/100} = 0.0001$.

(3) If the individual components or events have different failure probabilities, then Equation (4.3-19) becomes

$$F_{ns} = 1 - \prod_{i=1}^{n}(1 - p_{fi}) ,$$ (4.3-19b)

where p_{fi} is the failure probability of component i.

(4) The reliability is defined as an exceedance probability. Suppose the survival time of a system is exponentially distributed with the pdf (see Section 2.7.5)

$$f(t) = \lambda e^{-\lambda t} .$$ (4.3-20)

Then the exceedance probability of survival time t_s is the integral of $f(t)$ with t_s as lower limit:

$$R(t_s, \lambda) = P(T \geq t_s) = \int_{t_s}^{\infty} f(t)dt = \int_{t_s}^{\infty} e^{-\lambda t} d(\lambda t) = e^{-\lambda t_s} \qquad (4.3\text{-}21)$$

where t_s is survival time; λ is the failure rate, for example, the number of breakdowns per year on average; in this case, t would have the units of years, as the exponent must be a dimensionless number.

If the system is composed of n units, all of which are required to work in order for the system to work, then the probability of survival, given the failure events are independent of each other, is according to Equation (4.3-16)

$$R_s = e^{-\lambda t_s} e^{-\lambda t_s} \ldots e^{-\lambda t_s} = e^{-n \lambda t_s} . \qquad (4.3\text{-}21a)$$

The factor n in the final exponent means that the probability of survival of the system is much smaller than the probability of survival of the individual components. For small exponents, the exponential function can be approximated by

$$R_s = 1 - n \lambda t_s , \qquad (4.3\text{-}21b)$$

where t_s are n-component system survival times. The approximation holds for small exponents, $n \lambda t_s < 0.1$ if an error of about 5 % is acceptable. This approximate linearity between R_s and t_s is illustrated by the flat slope of the exponential pdf for small exponents in Figure 2-22 (see Section 2.7.5).

The probability of success of a serial system with n components may have a reliability requirement, R_s, that must be met by a sufficiently low failure rate of all n components. Solving Equation (4.3-21a) for λ gives

$$\lambda = -\frac{\ln R}{n t} , \qquad (4.3\text{-}21c)$$

where λ is assumed the same for all n components.

Examples: (1) A dam is to be designed for a 200 year life without failure. What is the implied probability of the failure event? Solution: Assuming the dam can fail after 200 successive years, Equation (4.3-19a) gives for the failure event probability $p_f = 1 - (1 - 1/200)^{1/200} = 0.000025 \approx 1/40,000$. To avoid failure, for example, associated with a flood or earthquake, the dam must be able to withstand a 1-in-40,000-year event (see also NRC, 1983, p. 53). It may be possible, albeit not realistic, to construct an event of such rarity. However, it conveys the message that design as well as O&M must keep very low to achieve a high long-term survival probability, here $R_s = 1 - F_{ns} = 1 - 1/200 = 0.995$, or 99.5 %.

(2) The reliability of a system, R_s, is the probability that failure will not occur before the end of some successful operation period, t_s, say a warranty time. The pdf of failure is given by Equation (4.3-20), $f(t) = \lambda e^{-\lambda t}$, where λ is a failure rate per year and t is the time to failure in years. For an estimated failure rate $\lambda = 0.01/$year and system reliability $R_s = 0.9$, calculate the warranty time. Solution: The time to failure t_s for $R(t_s, \lambda) = 0.9$ is obtained from Equation (4.3-21):

$$ t_s = -\frac{1}{\lambda}\ln[R(t_s, \lambda)], $$

which gives $t_s = -\ln 0.9/0.01 = 10.5$ years.

(3) A serial system has three components that all have the same reliability with an annual failure rate of $\lambda = 0.01$. Calculate the survival probability of the three-component system for two years and compare it with the approximation, Equation (4.3-21b). Solution: The n-component system reliability according to Equation (4.3-21a) for $n = 3$, $\lambda = 0.01$, and $t_s = 2$ is $R_s = e^{-3 \cdot 0.01 \cdot 2} = e^{-0.06} = 0.94$. The approximation formula, Equation (4.3-21b) gives $R_s = 1 - 0.06 = 0.94$.

4.3.4 Reliability of Parallel Systems

Parallel systems are side by side arrangements of subsystems or components to make the system more resistant to failure or more flexible for serving variable loads. The subsystem B in Figure 4.2c consists of three subsystems in side by side or parallel arrangement connected to serial components at the front end and at the back end. The subsystems in turn may be serial, parallel, or mixed systems. For B to partially function at least one of the subsystems must function. The probability of partial functioning is the probability of the union of subsystem success, as it is only required that one *or* the other subsystem functions. The probability of n parallel subsystems functioning is the joint probability of non-failure of all subsystems. This is expressed by (Moan, 1982, p. 4-46)

$$ R_p = 1 - \prod_{i=1}^{n}[1 - R(t_s, \lambda_i)] \tag{4.3-22} $$

where R_p refers to the reliability of the parallel system; $R(t_s, \lambda_i)$ is the probability that the time to failure exceeds t_s at a failure rate λ_i of component i. Equation (4.3-22) expresses the reliability of the parallel system by first calculating the joint *unreliability* of n components, $1 - R_1$, $1 - R_2$, and so on. The parallel system totally fails if component 1 fails and component 2 fails and component 3 fails, and so on. By subtracting this joint failure probability from 1 gives the complementary probability that this does not happen, which means that still one or the other component works. This is the parallel system's strength, its robust reliability or *survival probability*.

The probability that all components of the parallel system are working is given by Equation (4.3-17),

$$R_p = \prod_{i=1}^{n} R(t_s, \lambda_i) .$$

This probability is smaller that by Equation (4.3-22) because it is the joint success probability of all components. The difference between the two reliabilities is demonstrated by Example (1).

Examples: (1) A system has three parallel components with reliabilities R_1, R_2, and R_3. (a) Calculate the system reliability by Equation (4.3-22) and (4.3-17). (b) express the reliability if all components have the same reliability, R. (c) Compare the different results by using $R = 0.9$. Solution: (a) According to Equation (4.3-22), the system reliability of an n-component parallel system is according to Equation (4.3-22)

$$R_p = 1 - (1 - R_1)(1 - R_2)(1 - R_3)$$

$$= R_1 + R_2 + R_3 - (R_1 R_2 + R_1 R_3 + R_2 R_3) + R_1 R_2 R_3 .$$

This expression is the probability of the union of three ordinary events obtained by Equation (2.1-14). The joint success probability of three independent events is obtained from Equation (4.3-17) as

$$R_p = R_1 R_2 R_3 . \tag{4.3-17a}$$

(b) If all components have the same success probability R, then Equation (4.3-22) gives

$$R_p = 3 R - 3 R^2 + R^3 . \tag{4.3-22a}$$

For $0 \le R \le 1$, $3 R \ge 3 R^2$, so that $0 \le R_p \le 1$. This is true also for Equation (4.3-22) where each factor of the product is always less than or maximally equal to 1 so that $0 \le R_p \le 1$.

(c) For $R = 0.9$, one obtains by Equation (4.3-22a) $R_p = 0.999$. The parallel system has a sort of redundancy, which makes the probability that one *or* the other component works relatively high. In contrast, the probability of all components working, as expressed by Equation (4.3-17a), is $R_p = 0.9^3 = 0.729$.

(2) Suppose a parallel system's three components have reliability functions given by Equation (4.3-21): $R(t_s, \lambda) = e^{-\lambda t_s}$, with $\lambda = 0.01$ and $t_s = 1$. Write Equation (4.3-22a) for this system and check the numerical result. Solution:

$$R_p = +3\, e^{-\lambda t_s} - 3\, e^{-2\lambda t_s} + e^{-3\lambda t_s}$$

The numerical result is

$$R_p = 3 \cdot 0.990\ 049\ 8 - 3 \cdot 0.980\ 198\ 7 + 0.970\ 445\ 5$$

$$= 2.970\ 149\ 5 - 2.940\ 596\ 0 + 0.970\ 445\ 5 = 0.999\ 999 \le 1.$$

The seven digit accuracy is used to show that R_p is indeed within the limits $0 \le R_p \le 1$. In a general way, the limits of the exponential reliability R_p of this example are: for $\lambda t_s \to 0$, $e^{-\lambda t_s} \to 1$, and $R_p \to 1$, which means survival for very short periods is virtually assured; for $\lambda t_s \to \infty$, $e^{-\lambda t_s} \to 0$, and $R_p \to 0$; this means that over the long run failure is assured. Hence, R_p has the range $0 \le R_p \le 1$, as required.

4.4 Redundancy

4.4.1 System Redundancy

For a system that is critical for meeting an objective or for avoiding catastrophic failure, stepped-up surveillance and maintenance may not be sufficient to achieve the necessary level of reliability. In such cases, *redundancy* is a means of increasing reliability. There are two types of redundancy: doubling up components or *component redundancy*, and doubling up a system or *system redundancy*. The limit on redundancy is usually economic. The redundant component or system may work alongside the primary component or system, or be in waiting on hot *standby*, ready to immediately take over operation when called upon. A probabilistic aspect is introduced by the possibility that standby equipment expected to be operable when needed may be in any of two states: operable or not operable. Figure 4-7a shows an arrangement in which a serial system is backed up by a complete standby system which provides system redundancy.

The probability of a serial system backup is related to the probability of parallel serial systems. Assume a string of components 1, 2, 3, 4, ... is backed up by a string 1a, 2a, 3a, 4a,.... For the moment, assume a system has two serial components, x_1, x_2, backed up by x_{1a}, x_{2a}. Then the probability that either the system or its backup works is given by the serial system reliability

$$R_s = P[(x_1 \cap x_2) \cup (x_{1a} \cap x_{2a})]$$

(4.4-1)

$$= P(x_1 \cap x_2) + P(x_{1a} \cap x_{2a}) - P(x_1 \cap x_2)P(x_{1a} \cap x_{2a}).$$

Equation (4.4-1) gives the probability of one or the other system working for units with different reliabilities and which may be dependent on each other. If the success events of a unit working, x_i, are independent, with $P(x_1) = R_1$, and $P(x_2) = R_2$, and so on, then

$$R_{rs} = R_1 R_2 + R_{1a} R_{2a} - R_1 R_2 R_{1a} R_{2a}$$

(4.4-2)

where R_{rs} is the reliablity of the redundant serial system. If the success probabilities are all identical, then $R_1 R_2 = R^2$, and so on, and the reliability is

$$R_{rs} = 2R^2 - R^4 .$$

(4.4-2a)

If the system and its backup each have three components, then the probability that one of the two systems is working is

$$R_s = P[(x_1 \cap x_2 \cap x_3) \cup (x_{1a} \cap x_{2a} \cap x_{3a})]$$

$$= P(x_1 \cap x_2 \cap x_3) + P(x_{1a} \cap x_{2a} \cap x_{3a})$$

(4.4-3)

$$- P(x_1 \cap x_2 \cap x_3)P(x_{1a} \cap x_{2a} \cap x_{3a}).$$

If the units are all independent of each other, with $P(x_1) = R_1$, $P(x_2) = R_2$, and so on, the reliability is

$$R_{rs} = R_1 R_2 R_3 + R_{1a} R_{2a} R_{3a} - (R_1 R_2 R_3)(R_{1a} R_{2a} R_{3a}) .$$

(4.4-4)

If the redundant components have the same reliability as the principal components, $R_1 = R_{1a}$, $R_2 = R_{2a}$, and so on. Then Equation (4.4-4) becomes

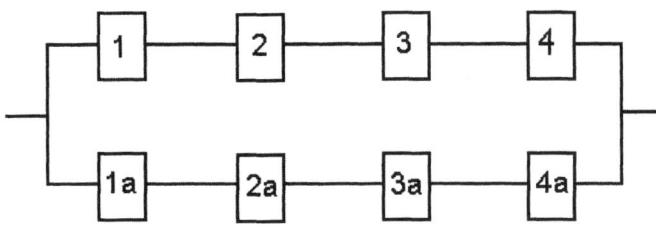

(a) serial system with system redundancy

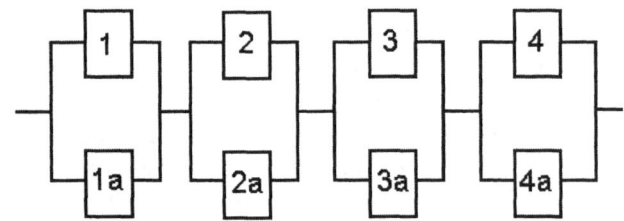

(b) serial system with component redundancy

Figure 4-7: Redundancy achieved (a) by system redundancy and (b) by component redundancy.

$$R_{rs} = 2\prod_{i=1}^{3} R_i - \prod_{i=1}^{3} R_i^2 .$$ (4.4-4a)

For identical components, all indices can be dropped and one obtains for the three-component system

$$R_{rs} = 2 R^3 - R^6.$$ (4.4-4b)

By comparing Equations (4.4-2) to (4.4-4), one recognizes a pattern in the reliability formulas. By induction one may expect a four-component system with identical components in principal and backup system to have the reliability

$$R_{rs} = 2 R^4 - R^8. \tag{4.4-5}$$

As the number of components increases, the backup cannot prevent the reliability from decreasing. For example, with $R = 0.9$, the two-component system with system backup has $R_{rs} = 0.96$, the three-component system has $R_{rs} = 0.93$, and the four-component system has $R_{rs} = 0.88$.

If corresponding components of the system and its backup have the same reliability, such that $R_1 = R_{1a}$, $R_2 = R_{2a}$, and so on, then Equation (4.4-4a) can be generalized for n components

$$R_{rs} = 2\prod_{i=1}^{n} R(x_i) - \prod_{i=1}^{n} [R(x_i)]^2 \tag{4.4-6}$$

where R_{rs} is the reliability of a system with a redundant system as backup, $R(x_i)$ is the reliability of component x_i and its backup x_{ia} , and n is the number of components of the principal system *not* counting the backup components.

A summary of reliability formulas for various component arrangements and backup configurations for the three unit types (the most general case: dependent different units, DDU; an intermediate case: independent different units, IDU; and a special case: independent identical units, IIU) follows.

(1) Serial system backed up by another serial system with DDU (dependent different units):

$$R_s = P(x_1 \cap x_2) + P(x_{1a} \cap x_{2a}) - P(x_1 \cap x_2)P(x_{1a} \cap x_{2a}). \tag{4.4-1}$$

(2) Serial system backed up by another serial system with IDU (independent different units):

$$R_{rs} = R_1 R_2 + R_{1a} R_{2a} - R_1 R_2 R_{1a} R_{2a} . \tag{4.4-2}$$

(2) Serial system backed up by another serial system with IIU (independent identical units):

$$R_{rs} = 2R^2 - R^4 . \tag{4.4-2a}$$

An overview of DDU and IIU formulas for different system configurations is given by Shooman (1968, pp. 130–131, Tables 3.3 and 3.4).

Examples: (1) Assume a serial system with three components has an identical redundant system. All components are independent and identical units (IIU). Calculate the system reliability and compare it to the system without backup. Assume a component reliability $R = 0.9$. Solution: The formula for the three-component system with system redundancy and IIU is according to Equation (4.4-4b)

$$R_{rs} = 2\,R^3 - R^6.$$

If $R = 0.9$, $R_{rs} = 2 \cdot 0.729 - 0.53 = 0.927$. If only one serial system with the same components were available, its reliability would be $R^3 = 0.729$. The redundancy raises the system reliability by $100 \cdot (0.927 - 0.729)/0.729 = 27$ %.

(2) A serial system of three components, x_1, x_2, x_3, is backed up by a system of one component, x_{1a}. Give the reliability formulas for the three unit types DDU, IDU, and IIU. Solution: For the case of dependent different units (DDU), the probability of one or the other system working is

$$R_s = P[(x_1 \cap x_2 \cap x_3) \cup x_{1a}] = P(x_1 \cap x_2 \cap x_3) + P(x_{1a})$$

$$- P(x_1 \cap x_2 \cap x_3)P(x_{1a}).$$

The reliability for independent but different units (IDU) is with $P(x_i) = R_i$,

$$R_s = R_1\,R_2\,R_3 + R_{1a} - R_1\,R_2\,R_3\,R_{1a}.$$

The reliability for independent identical units (IIU) is

$$R_s = R + R^3 - R^4.$$

4.4.2 Component Redundancy

Instead of backing up a system by a complete system, one can also back up each component or selected components by one or more components. Such an arrangement is called *component redundancy*. Figure 4-7b shows a serial system in which each component is backed up by an alternate component. The reliability of a system with component redundancy in which at least one component is functioning at a time can be derived from the joint probability of unions. The probability that one unit or its alternate works is the probability of a union. The probability that several of these unions in series work is the joint probability of unions. If the system has three components 1, 2, 3, and alternate components are $1a$, $2a$, and $3a$, then the system reliability can be stated as

$$R_{rc} = P[(x_1 \cup x_{1a}) \cap (x_2 \cup x_{2a}) \cap (x_3 \cup x_{3a})] \qquad (4.4\text{-}7)$$

where R_{rc} is the reliability for component redundancy. For independent but different components,

$$R_{rc} = P(x_1 \cup x_{1a}) \, P(x_2 \cup x_{2a}) \, P(x_3 \cup x_{3a}) \, .$$

The probabilities of the union $x_1 \cup x_2$ can be expanded into

$$P(x_1 \cup x_2) = P(x_1) + P(x_2) - P(x_1)P(x_2) \, .$$

Introducing the notation $R_i = P(x_i)$ for the success probability leads to

$$R_{rc} = (R_1 + R_{1a} - R_1 R_{1a})(R_2 + R_{2a} - R_2 R_{2a})(R_3 + R_{3a} - R_3 R_{3a}) \, . \qquad (4.4\text{-}7a)$$

This product is a sum of 27 terms of probability products ranging from 3 to 6 factors, as is seen by writing down the first term and the last term: $R_1 R_2 R_3$ and $R_1 R_{1a} R_2 R_{2a} R_3 R_{3a}$. If all probabilities are assumed to be the same, $P(x_i) = R$, then the 27 terms can be condensed to

$$R_{rc} = 8R^3 - 12R^4 + 6R^5 - R^6 \, . \qquad (4.4\text{-}8)$$

Note that the absolute values of the coefficients add to 27.

    If the backup units are given the same index as the principal units and have the same reliability, then Equation (4.4-7a) can be written as

$$R_{rc} = \prod_{i=1}^{n} \{2R(x_i) - [R(x_i)]^2\} \qquad (4.4\text{-}9)$$

where R_{rc} is the reliability of a system with component redundancy, and the $R(x_i)$ are the success probabilities of each set of principal and alternate components.

    The assumption of all component reliabilities being the same does not limit the use of the formulas. These simplifications are made to demonstrate principles and to keep the examples simple. In real-world applications, different probabilities for different components can be handled by using the more general models. For example, if Equation (4.4-7a) is used instead of Equation (4.4-8) different probabilities can be used for each set of component and its back-up.

Examples: (1) Use Equation (4.4-9) to derive the reliability for a system of two components with component redundancy for identical components. Solution: From Equation (4.4-9) one obtains

$$R_{rc} = (2 R - R^2)^2 = 4 R^2 - 4 R^3 + R^4.$$

Solutions like this for a number of different systems are given by Shooman (1968, pp. 130–131, Tables 3.3 and 3.4).

(2) Evaluate Equation (4.4-9) for $n = 3$, and compare the result with Equation (4.4-8). Calculate the reliability for component redundancy, R_{rc}, if all components have reliability $R = 0.9$. Solution: With the simplification $R(x_i) = R_i$, Equation (4.4-9) for three components and their backups is

$$R_{rc} = \prod_{i=1}^{n} \{ 2R(x_i) - [R(x_i)]^2 \}$$

$$= (2 R_1 - R_1^2) \cdot (2 R_2 - R_2^2) \cdot (2 R_3 - R_3^2) = 8 R^3 - 12 R^4 + 6 R^5 - R^6.$$

This is the same result as Equation (4.4-8). For $R = 0.9$, one obtains $R_{rc} = 0.969$. Without backup, it would be $R_{rc} = 0.729$. Providing redundancy increases system reliability. Installation cost and operation cost may make a large single unit apparently more economical than two or more smaller units (economy of scale). When life cycle costs are considered, which include probabilistic O&M costs, such as repair costs, loss of service costs, and so on, over the equipment life, a multiunit arrangement may turn out to be more economical and reliable than a single unit, as the former may also be able to meet redundancy requirements .

4.4.3 Comparison of System and Component Redundancy

Comparing system and component redundancy in Figures 4-7a and 4-7b, it can be recognized that the backup of a serial system of four components ($n = 4$) by an additional serial system creating two parallel serial systems ($m = 2$) is still rather sensitive to failure. All components in the redundant system must work in order to provide backup. In contrast, in a system with component redundancy, Figure 4-7b, one can visualize more functioning pathways through the system as long as only one component out of each pair ($m = 2$) in each of the four pairs ($n = 4$) or *component stacks* works. As long as only one chain of components reaches from beginning of the series to the end, the system will work. Hence the reliability with component backup is greater than with system backup.

Consider a comparison of the reliability of two types of serial systems: the first is a system that is backed up by another serial system, and the second is a system in which each component is backed up by a duplicate, as shown in the examples of Figure 4-7a and 4-7b. The comparison is expanded beyond Figure 4-7 to include systems with five serial components ($n = 1,..., 5$) and up to four parallel systems or backup components ($m = 1, ..., 4$). The reliability of these systems is obtained as the reliability of a k-out-of-n system. If the success probability of a component is p, then the system success probability for $k = n$ is p^n. The probability that the serial system fails is $1 - p^n$. Furthermore, the probability that all serial systems fail is $(1 - p^n)^m$. Hence, the probability that the system works is (Shooman, 1968, p. 284, Fig. 6.6)

$$R_s = 1 - (1 - p^n)^m \qquad\qquad\qquad (4.4\text{-}10)$$

where R_s is the reliability with *system redundancy*.

A similar derivation can be made for the reliability of a serial system with component redundancy. The probability that one component in the stack of m components ($m = 2$ in Figure 4-7b) fails is $1 - p$. The probability that all m components of one stack fail is $(1 - p)^m$. The probability that not all stack components fail is $1 - (1 - p)^m$. Hence, the probability that the series of n stacks does not fail is

$$R_c = [1 - (1 - p)^m]^n \qquad\qquad\qquad (4.4\text{-}11)$$

where R_c is the system reliability for *component redundancy*.

An illustration of Equations (4.4-10) and (4.4-11) for $p = 0.5$ is given in Figure 4-8. If the system has only one component $n = 1$, then the system reliability becomes the reliability of one stack,

$$R_s = R_c = 1 - (1 - p)^m \ . \qquad\qquad\qquad (4.4\text{-}11a)$$

These reliabilities lie on the ordinate $n = 1$ of Figure 4-8. If $m = 1$, the reliabilities are the same for all n,

$$R_s = R_c = p^n . \qquad\qquad\qquad (4.4\text{-}11b)$$

Equation (4.4-11b) is illustrated by the lower most curve of Figure 4-8. With the increase of the number of serial components n the reliability of both systems decreases.

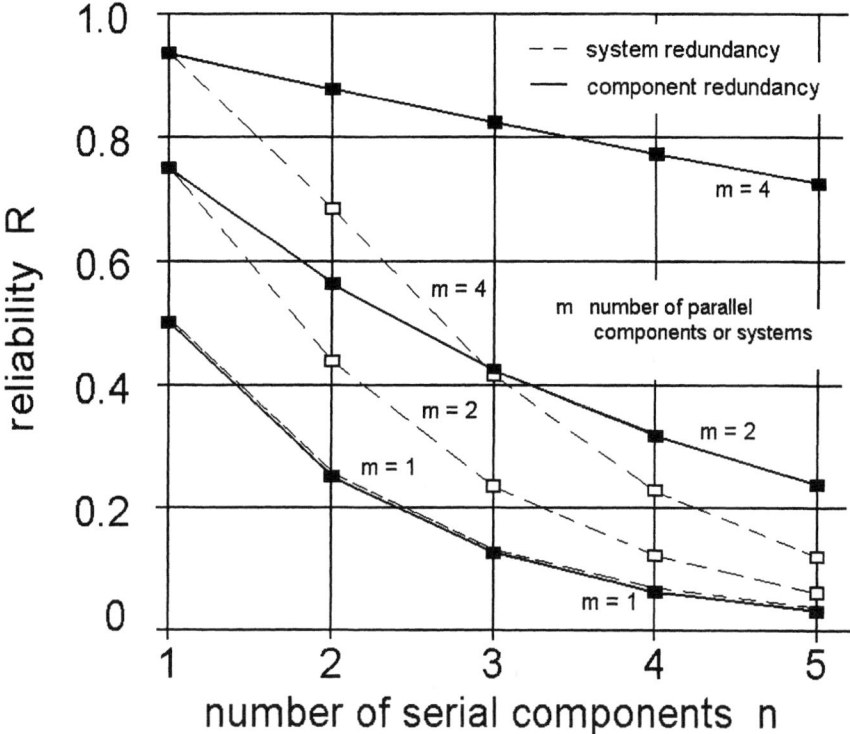

Figure 4-8: Comparison of system and component redundancy for two systems similar to those shown in Figure 4-7a and 4-7b. The number of serial components is *n* and the number of parallel components is *m*; $p = 0.5$. For *n* = 1, the reliability for system and component redundancy is the same for any number of backups *m*. The system reliability decreases with increasing *n* for both arrangements, but holds up better for component redundancy than for system redundancy (after Shooman, 1968, p. 284, Figure 6.6, which was adapted from Figures 7.10 and 7.11, "Reliability Engineering," ARINC Research Corp., Prentice Hall, Inc., Englewood Cliffs, N.J., 1964).

Component backup is seen to hold up system reliability better than system backup. The reason for this is the availability of multiple pathways through the system with component backup. For $m = 4$, $n = 5$, and $p = 0.5$, the reliability of a system with component backup is according to Equation (4.4-11) $R_c = 0.724$. For system backup, Equation (4.4-10) gives $R_s = 0.119$. Hence, in this case, reliability with component backup is about six times higher than with system backup. This case is also illustrated on the right-hand side of Figure 4-8.

4.5 Cut Sets and Tie Sets

In Sections 4.2.1 and 4.2.2, the problem was discussed of the number of functioning components necessary for the system to at least partially work, or the minimal number of failing components for the system to fail. The method of *cut sets and tie sets* can be used to analyze this condition (Shooman, 1968, p. 136). In the serial system, it was the failure of any one component that shuts down the system by cutting the connection between input and output. Therefore, in the serial system, the set of components required to cut the system, or the *cut set*, is $C = x_i$, where x_i can be any component, $i = 1$ to n. If the question is asked the other way around, how many units are required to maintain a tie between input and output, or at least the partial functioning of the system, then in an n-component serial system the number is $T = (x_1, x_2, ..., x_n)$, where T is the *tie set* of the serial system. Generally, systems have many cut sets and tie sets. They can be identified in advance. Then the reliability or the failure probability of the system can be determined as the success probability or failure probability of these sets. The sets are collections of components that are critical for the functioning or failure of the system.

The system in Figure 4-9 has numerous cut sets and tie sets. Examples of cut sets are

$$(x_1, x_2), (x_1, x_4), (x_3, x_7, x_4), (x_5, x_6), (x_3, x_9, x_6), (x_2, x_7, x_5, x_8), \text{ etc.}$$

Examples of tie sets are

$$(x_1, x_3, x_5), (x_2, x_4, x_6), (x_1, x_7, x_9, x_5), (x_2, x_4, x_9, x_5), (x_1, x_3, x_8, x_6), (x_1, x_7, x_6).$$

In reliability calculations, *minimal cut sets* and *minimal tie sets* are used. A minimal cut set is the smallest number of components that must be removed to cut the system. A minimal tie set is a path through the system that touches each node only once. In the preceding examples, all cut sets are minimal cut sets. If a cut set contains all elements of another cut set then it is not minimal, as the contained cut set obviously is smaller.

The system failure probability is the probability of at least one cut set failing. If the probability of cut set C_i working is $P(C_i)$, then the probability of cut set C_i failing is $P(C_{fi}) = 1 - P(C_i)$. The probability of at least one cut set failing is the probability of the union of all failing cut sets. A system may have n cut sets, but only the number of minimal cut sets, $j = 1, ..., k$, needs to be considered:

$$P_f = P(C_{f1} \cup C_{f2} \cup ... \cup C_{fk}). \tag{4.5-1}$$

The probability of a cut set failing is computed as the joint failure probability of all components in the cut set. According to definition of the cut set, all of its components must fail to disrupt the system. Using the first of the preceding cut sets and assuming independence among failing components, the failure probability of the cut set is

$$P(C_{f1}) = P(x_{f1} \cap x_{f2}) = P(x_{f1}) \, P(x_{f2}) \tag{4.5-2}$$

where $P(x_{fi})$ is the failure probability of component x_i.

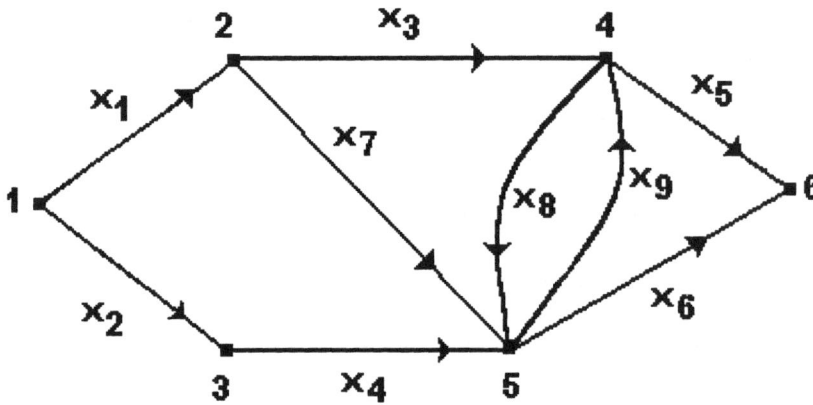

Figure 4-9: A system in the form of a directed graph indicates by an arrow on the branches the only possible direction of the process that transforms input at node 1 into output at node 6. A minimal tie set is a sequence of components x_i that connects node 1 to node 6 without visiting a node more than once, for example, $T = (x_2, x_4, x_6)$. A minimal cut set is the minimal number of components that disrupts the connection between input node 1 and output node 6, for example, $C = (x_1, x_2)$.

The probability of the system functioning is then

$$R = 1 - P_f \tag{4.5-3}$$

where P_f is the probability of failure of the union of all minimal cut sets, as expressed by Equation (4.5-1).

The probability of the system functioning can also be expressed as the probability of at least one tie set functioning, which is the success probability of the union of all minimal tie sets,

$$R = P(T_j \cup ... \cup T_k)$$ (4.5-4)

where j and k are indices of minimal tie sets. The probability of tie set T_1 functioning assuming independence among branches is

$$P(T_1) = P(x_1 \cap x_3 \cap x_5) = P(x_1)\, P(x_3)\, P(x_5)\ .$$ (4.5-5)

All components of a tie sets must function for the system to function, because each tie set represents a serial subsystem that connects input and output of the system..

The probability of the union of ordinary sets, as required by Equations (4.5-1) and (4.5-4), is difficult to evaluate if the number of sets exceeds a relatively small number, as Equation (2.1-14) Section 2.1.3 shows. An approach described by Shooman (1968, p. 139) is followed to circumvent this difficulty. Suppose there are four ordinary sets (meaning they are intersecting and not mutually exclusive). They are named here A, B, C, and D. The probability of their union is

$$P(A \cup B \cup C \cup D) = P(A) + P(B) + P(C) + P(D)$$

$$- P(A\,B) - P(A\,C) - P(A\,D) - P(B\,C) - P(B\,D) - P(C\,D)$$

$$+ P(A\,B\,C) + P(A\,B\,D) + P(A\,C\,D) + P(B\,C\,D) - P(A\,B\,C\,D)$$ (4.5-6)

where the $\cap$ symbols between the letters inside the parentheses have been omitted, for example, $P(A\,B)$ for $P(A \cap B)$. If the sets were mutually exclusive, the probability of their union would be just the sum of their probabilities:

$$P(A \cup B \cup C \cup D) = P(A) + P(B) + P(C) + P(D)\ .$$ (4.5-7)

Equation (4.5-7) represents an upper limit on the reliability calculated by Equation (4.5-6). For the minimal tie sets of Equation (4.5-4) this produces

$$R_u = P(T_1) + P(T_2) + ... + P(T_k) \geq R$$ (4.5-7a)

where R_u is the upper bound on system reliability R. Actually, for high component probabilities, p, Equation (4.5-7a) vastly exceeds 1. But for small probabilities, all the joint probabilities in Equation (4.5-6), if they exist, are very small, and Equation (4.5-7a) becomes a meaningful upper bound on R computed by Equation (4.5-6), as $R \leq R_u$. However, success probabilities of tie set components are usually high, because they represent system components connecting system input and output. Nevertheless, R_u is an adequate approximation of R in the range $0 \leq p \leq 0.3$, as shown in Figure 4-10.

If Equations (4.5-1) and (4.5-3) are combined, one obtains

$$R = 1 - P_f = 1 - P(C_{f1} \cup \ldots \cup C_{fk}) \ . \tag{4.5-8}$$

Here again one is confronted with the evaluation of the union of ordinary sets, which leads to an expansion similar to Equation (4.5-6), or worse, as there may be even more minimal cut sets than tie sets.

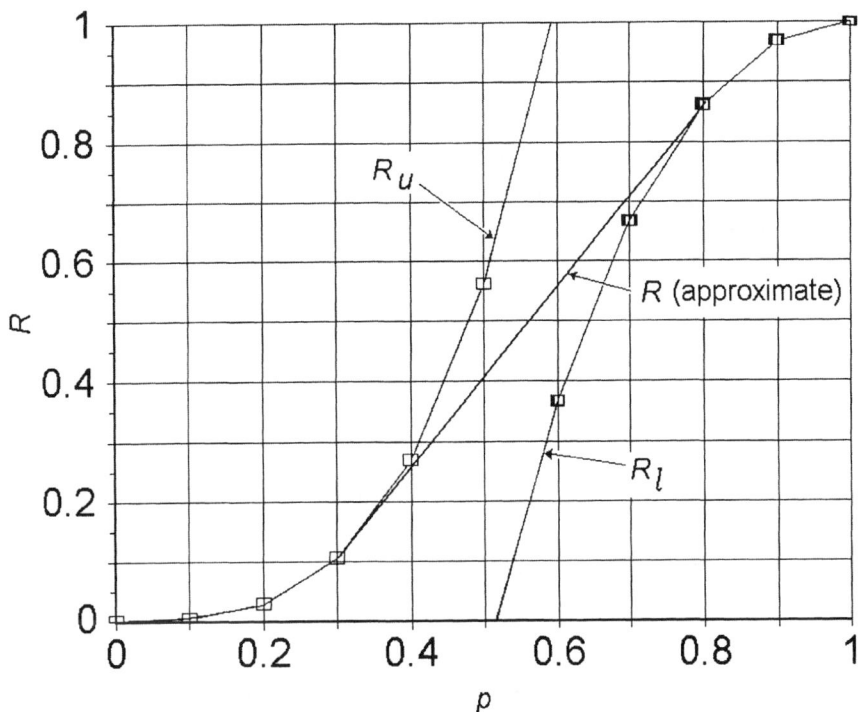

Figure 4-10: Calculation of system reliability R based on an upper bound, R_u, for low component success probabilities, p, and a lower bound, R_l, for low component failure probabilities, $1 - p$. Reliabilities for medium p are found by interpolation between both bounds (after Shooman, 1968, p. 140). The system reliability R is shown as an ascending function of p.

A similar boundary approach is taken that gives

$$R_l = 1 - P_f = 1 - [P(C_{f1}) + \ldots + P(C_{fk})] \leq R \tag{4.5-9}$$

where R_l is the lower bound on system reliability, and the index k designates the number of minimal cut sets. In Equation (4.5-9) the sum of probabilities is subtracted from 1, which produces a lower value than would be obtained by calculating the complementary probability of P_f using Equation (4.5-6). Hence, $R \geq R_l$. The failure probability must meet the requirement $P_f \leq 1$. This is the case if component reliability p is high, so that the component failure probability, $1 - p$, is low and accordingly P_f is low. The upper and lower bounds on system reliability R are illustrated in Figure 4-10. The upper bound is shown on the left for low component reliability, and the lower bound is shown on the right for high component reliability. Between these bounds, a connection in the form of a tangent on both bounds represents the reliabilities that would have to be calculated for medium probabilities in the range $0.3 \leq p \leq 0.8$ by Equation (4.5-6). Using this approach, the bounds can be evaluated as the sums of tie set probabilities using Equation (4.5-7a) in the range $0 \leq p \leq 0.3$, and as the sums of cut set probabilities using Equation (4.5-9) in the range of $0 \leq 1 - p \leq 0.2$. The reliabilities for medium p are then found by interpolation between the two bounding curves. This obviates the need for evaluating the complex formula for the probability of the union of intersecting sets.

Examples: (1) For the system shown in Figure 4-9, the following minimal cut sets and tie sets were identified:

Minimal cut sets: (x_1, x_2), (x_1, x_4), (x_3, x_7, x_4), (x_5, x_6), (x_3, x_9, x_6), (x_2, x_7, x_5, x_8).

Minimal tie sets: (x_1, x_3, x_5), (x_2, x_4, x_6), (x_1, x_7, x_9, x_5), (x_2, x_4, x_9, x_5), (x_1, x_3, x_8, x_6), (x_1, x_7, x_6).

Calculate the system reliability over the entire range $0 \leq R \leq 1$ for the entire range of component reliability $0 \leq p \leq 1$ assuming the same p for all components. Solution: Since p is the same for all components, the following abbreviations are introduced for cut set probabilities and tie set probabilities. For cut sets, $P(C_{f1}) = P(x_1 \cap x_2) = (1 - p)^2$, and so on; for tie sets, $P(T_1) = P(x_1 \cap x_3 \cap x_5) = p^3$, and so on. The upper bound R_u based on three tie sets with three elements and three tie sets with four elements is

$$R_u = P(T_1) + \ldots + P(T_k) = 3\,p^3 + 3\,p^4.$$

The lower bound R_l based on three cut sets with two elements, two cut sets with three elements and one cut set with one element is

$$R_l = 1 - P_f = 1 - [P(C_{f1}) + \ldots + P(C_{fk})]$$

$$= 1 - [3\,(1 - p)^2 + 2\,(1 - p)^3 + (1 - p)^4]$$

where k is the number of minimal cut sets or of minimal tie sets. The numerical evaluation of these formulas is given in Table 4-4.

Table 4-4: Lower and Upper Bounds and Interpolated System Reliability

p	0	0.1	0.2	0.3	0.4	0.5	0.6	0.7	0.8	0.9	1
R_u	0	0.003	0.029	0.11	0.27	0.56	1.04	1.75	2.76	4.16	6
R_l	− 5	− 3.5	− 2.4	− 1.4	− 0.6	− 0.06	0.37	0.67	0.86	0.97	1
Interpolated System Reliability:											
R	0	0.003	0.029	0.11	0.26	0.40	0.56	0.71	0.86	0.97	1

The upper and lower bounds, R_u and R_l, of Table 4-4 are plotted in Figure 4-10. Drawing a tangent from R_u to R_l gives approximate reliabilities R for the range $0.3 \leq p \leq 0.8$ for which the approximations by R_u and R_l are not valid. Values of R read from Figure 4-10 are given in the bottom line of Table 4-4. For low p, R starts with the R_u values and then shifts over to the R_l values for high p.

4.6 Standby

In the discussion of redundancy, it was assumed that the system consists of more than one identical or similar path between system input and output and that the system can randomly take any path that is available. This redundancy is characteristic of the parallel system. In contrast, *standby* means that the primary component is backed up by an identical or similar component or system but this backup is switched off and used only if the primary component fails, or if the system is willfully switched from its regular path to the standby path. The important element of the standby configuration is the *switch* from the primary path to the standby path.

As in the case of redundancy, there can be a *system standby* and *a component standby*. In the first case, the entire system can be switched by one switch to the standby in the event that the first system fails; in the second case, failing components are bypassed by switching to a standby component one at a time.

Not only the system but also an input to a system may require standby. Figure 4-11 shows a primary power source E that supplies a system. The standby is a central replacement A of E or a distributed replacement A_i of parts of the system. The normal state is "power on" with the system supplied by the primary power source. When the primary power source fails, the standby must function which includes the successful switch to the standby and the readiness of the standby for operation. The reliability of the standby is a joint probability of success of these two events. High reliability of the primary system must not be allowed to cause a deterioration of the reliability of the standby. The testing of switches and the test operation of the standby are measures to maintain high switching and operational success probability.

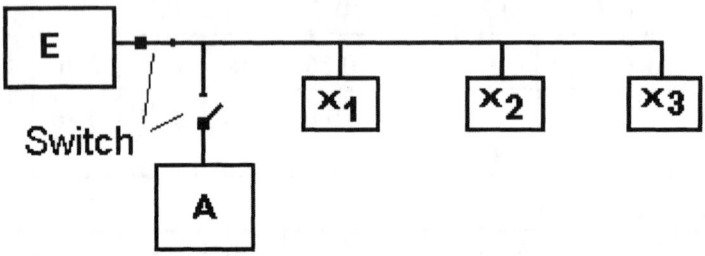

(a) centralized standby A for primary source E

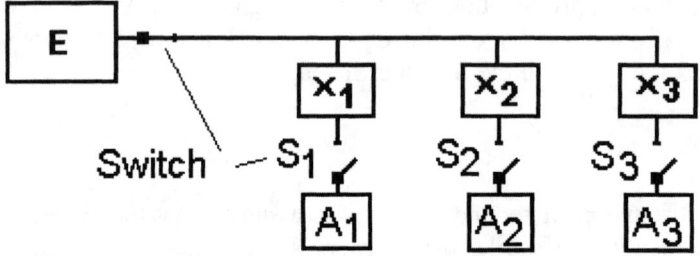

(b) distributed standbys for primary source

Figure 4-11: Standby configurations for a primary power source E. In the first system, the standby A is switched on as an alternative centralized power source for the components x_i when E fails. In the second system, failure of E triggers closure of switches S_i to distributed standbys A_i.

In Figure 4-11, E and A can be construed as two parallel, independent power sources. The system fails if both fail. Suppose E is the event of the primary power source being successful, A is the event of the standby being ready, and S is the event of a successful switch. If $P(E)$, $P(A)$, and $P(S)$ are the success probabilities, then the probability that the system is successful is

$$R = P[E \cup (A \cap S)], \tag{4.6-1}$$

which can be expanded into

$$R = P[E \cup (A \cap S)] = P(E) + P(A \cap S) - P(E) \, P(A \cap S).$$

If independence can be assumed, and the success probability of all events is p, then

$$R = p + p^2 - p^3. \tag{4.6-2}$$

When obtaining a result like Equation (4.6-2) it is useful to check if the requirements for R being a probability are met. To check $R \le 1$, we substitute p in Equation (4.6-2) by its complementary probability, $p = 1 - p_f$. This leads to

$$R = (1 - p_f) + (1 - p_f)^2 - (1 - p_f)^3 = 1 - 2 p_f^2 + p_f^3$$

Since $p_f \le 1$ and $2 p_f^2 > p_f^3$, the requirement $R \le 1$ is met.

As a standby, A can only fail after E has failed and A has been successfully switched on. Hence, the probability of A failing is conditioned on the failure of E and on the successful switching to A. This probability of primary source failure <u>and</u> standby failure is

$$P(E' \cap S \cap A') = P(E') \, P(S|E') \, P(A'|E' \cap S) \tag{4.6-3}$$

where E' and A' are the failure events of E and A. A failure of E and A working side by side in a parallel arrangement would have the probability $P(E' \cap A')$. If all events are independent, Equation (4.6-3) reduces to

$$P(E' \cap S \cap A') = P(E') \, P(S) \, P(A') \tag{4.6-3a}$$

and $P(E' \cap A') = P(E') \, P(A')$. If the success probabilities of all events are $p = 0.9$, then Equation (4.6-3a) gives $0.1 \cdot 0.9 \cdot 0.1 = 0.009$, and the parallel arrangement give $0.1 \cdot 0.1 = 0.01$. This shows that the parallel arrangement has a higher probability of failure than the standby arrangement (see also Shooman, 1968, p. 146).

Standby A is brought on line when the primary source E fails. According to the exponential survival time distribution, short periods have high survival probabilities. Hence, the probability of A failing right after E fails and A is switched on should be smaller than if A had been used jointly with E, as in a parallel system. After A takes over, unless there are multiple standbys or E can be quickly restored, A is operating without standby and the system is vulnerable to failure. This state may have associated with it a low reliability that may be unacceptable. In such a case, an orderly shutdown of the system may be required until the primary source E and standby A are restored to their respective functions.

Standbys occur in many forms and in practically all systems that require high reliability of operation. In a hydrosystem, power units are sometimes kept running drawing power from the electrical net, as *spinning reserve* so that they can be quickly accelerated to pick up load in periods of rapid load increases. If the less flexible thermal power sources must be used, thermal units are kept under steam pressure ready to begin operation when other units fail or loads increase, a condition aptly named *hot standby*. A familiar standby is a battery in the power supply of a computer. It has two functions. It bridges power fluctuations that would cause a computer failure even without complete power failure, and if the power completely fails, the battery allows an orderly computer shutdown.

****

5. Probabilistic Maintenance Concepts

5.1 Maintenance Options

Meeting a few basic maintenance needs can go a long way toward ensuring a system's continuous delivery of reliable service. For example, routinely maintaining water, oil, and air levels in a car not only increases reliability, but also extends service life. Similarly, maintaining a well-greased, well-cooled, and dust-free environment for turbine generators will keep these units running for years if not decades. Two popular, albeit seemingly contradictory, statements describe two fundamental maintenance policies:

"If it ain't broke, don't fix it. "
"An ounce of prevention is worth a pound of cure. "

Both approaches are valid for different conditions. Policy 1 lets failure occur and then repairs the damage. This is usually referred to as *corrective maintenance*. Policy 2 anticipates breakdown and makes the necessary corrections to avoid it. [1] This is usually referred to as *preventive maintenance*. The statement also makes a value judgment by ranking preventive maintenance over corrective maintenance. Both policies actually have a role to play in maintenance, as discussed in the following paragraphs.

Discussed first is *routine maintenance or operational maintenance* This is the simplest and most common form of maintenance that practically everybody applies or should apply in his or her own household. It consists of maintenance that can be done at any time, say once every day, whether there is an obvious need for it or not

---

[1] Ascribed to Henry de Bracton, twelfth century England, as cited by Jay Landers, Industrial Wastewater, May/June 2001, p. 4.

424 HYDRAULIC STRUCTURES

(Goldman and Slattery, 1964, p. 27). Routine maintenance is a basic strategy that accompanies operation. It keeps at least the easily serviceable parts of machinery and structures in good repair and maintains a clean and healthy operating environment. All moving parts are kept well greased and free of condensation, dust, and rust. Routine maintenance is performed without interrupting the operation of the system. From a probabilistic point of view, routine maintenance is a *hazard control* activity. It has the potential to discover minor irregularities and incipient flaws that, overlooked or unattended, might develop into major failure. For example, an oil or cooling water pump failure can lead to turbine or generator bearing failure that takes the production unit off-line, thus causing thousands or millions of dollars in damage and repair cost. Although there is nothing obviously probabilistic about it, routine maintenance plays a significant role in the generally probabilistic system operation environment by its contribution to *reliability* and *longevity* of systems and equipment.

Corrective maintenance is performed in response to failure. This type of maintenance may be justified when damage associated with failure is minor or if no serious consequences are associated with failure. The damage is corrected after it occurs without a possibly unjustified interference with the running system.

Preventive maintenance is performed in anticipation of failure or after deterioration of equipment has been detected. Preventive maintenance may be mandatory when system failure is associated with serious consequences for the continued operation of the system or because of serious or unacceptable consequences for the public, public and private property, and the environment. Thus, both corrective maintenance and preventive maintenance are legitimate approaches under proper circumstances. The criteria for selecting one or the other are essentially the costs and consequences of failure.

The most desirable approach is *maintenance on demand*. Suppose a system's normal performance signature is known and continuously monitored. If departures beyond a normal band of deviations occur, alarms are triggered that alert operators of the need for inspection or immediate maintenance action. But also this kind of maintenance has probabilistic aspects as it raises the problem of false or failed alarms, which may trigger premature maintenance or miss the opportunity for preventive maintenance. No type of maintenance can eliminate all probability of failure because of the probabilistic character of systems. Other maintenance-related activities include rehabilitation, replacement, retirement, and removal. Although these aspects will not be specifically discussed, they all include engineering work with probabilistic aspects that is addressed here under *maintenance*.

5.2 Maintenance Management

A probabilistic approach to maintenance considers the system to be maintained as an assembly of components that exhibit random behavior with respect to being operational or not. In developing a maintenance approach to such a system, the following steps need to be considered (LaPay, 1992):

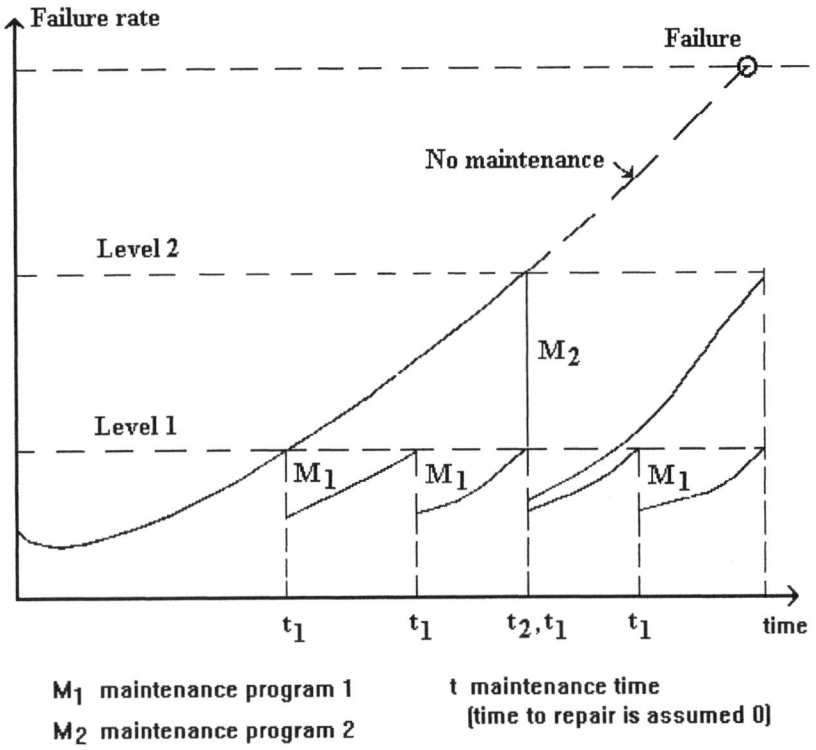

Figure 5-1: Maintenance applied periodically brings the failure rate down to or below an acceptable level. The indices identify two maintenance programs that differ by the time maintenance is applied and by the failure rate incurred (after LaPay, 1990).

(a) <u>Identify systems, structures, and components</u> : There should be a clear understanding of the scope of equipment and structures to be addressed by the maintenance program. Areas of commonality should be identified so that the scope and cost of the program can be minimized.

(b) <u>Understand system operational characteristics</u>: It is important to understand the total system operational behavior and the interrelationships between components; identify redundant paths that allow shutdown of equipment without loss of service.

(c) <u>Identify failure mode characteristics</u>: Failure modes typical to the system must be identified. Examples are seepage and erosion in embankment dams; overtopping of embankments; sliding of embankments, foundation seepage and pressurization of rock foundations; deformation of concrete dams by concrete swelling or foundation movements; tunnel, pipeline, and penstock leakage; valley slope creep and slides; settling of embankment dams and associated rupturing of the core; jamming of gates by concrete swelling, failure of gate operation device; failure of emergency gates; structural failure of gates; penstock corrosion; trashrack collapse; turbine runner failure, generator insulation and bearing failure; operation errors, external events, and so on.

(d) <u>Understand aging and deterioration behavior</u>: Generic and project specific data should be obtained.

(e) <u>Understand basis of design</u> : Strength inherent in the design should be identified and quantified.

(f) <u>Understand environment and loadings of system and components</u>: Data on the environment and loading conditions to which the equipment and structure are subjected should be collected. Historic records should be consulted and the exceedance probability for critical natural events (high winds, floods, snow, ice, earthquakes) should be prepared.

(g) <u>Understand existing maintenance procedures</u>: The current maintenance process should be reviewed and existing maintenance procedures should be incorporated into the evaluation process. The need for new procedures or the modification of existing procedures can then be identified.

The evaluation of these steps forms the basis for the selection of a maintenance program. The maintenance budget can be estimated and the extent and frequency of the most effective maintenance activities can be determined. An indicator of the need for maintenance is the increase of failure rates or other symptoms beyond an acceptable level. Figure 5-1 illustrates the increase of failure rate with time. If failure rates are not known or failures are acceptable, other measurable indicators must be defined and monitored instead. Such an indicator is the stress state in a component. When a safe level of stress is exceeded, maintenance, repair, or rehabilitation are explored as possible methods to correct the problem. Figure 5-1 also illustrates the use of two maintenance programs, one program with short maintenance intervals and another with longer intervals. The level of reduced stress state achieved by the two programs is not necessarily the same. A criterion for judging the two programs could be the expected remaining life of the component that results from the two maintenance programs.

5.3 Reliability-Centered Maintenance (RCM)

Reliability ranks high among maintenance objectives. This was brought out in the responses to survey questions (Section 1.6.2, Question 8). Other objectives are ensuring the safety of process operations, minimizing cost, completing the work on time, and meeting external requirements. Different emphasis on these objectives could lead to various "centered" maintenance approaches, such as *safety-centered maintenance, economy-centered maintenance*, or *reliability-centered* maintenance (RCM). Reliability-centered maintenance gives the highest priority to service reliability, which implicitly requires reliability of all contributing system components (e.g., structures and equipment), competent management, and expertise of O&R personnel. Prioritizing reliability could also mean compromising safety by skipping a preventive shutdown to avoid service disruption; or it could mean *overmaintaining*, thereby causing a higher than necessary cost as a consequence of unnecessary shutdowns, and premature replacement of operable components. It could also mean rushing repairs to minimize downtime at the expense of repair quality. If one objective is given the highest priority in a maintenance program, others must still be included as secondary objectives or constraints that put limits on trade-offs in favor of the *central* objective. Economy-centered maintenance could mean maintenance at minimum cost, or minimizing the cost of overall O&M. Also, safeguards would be required to prevent such an approach from becoming overly *centered* on economy, for example, by taking risks in stretching maintenance intervals, and by taking shortcuts on safety and reliability measures, to minimize maintenance cost and loss of service costs accruing from a shut down production process. *Safety-centered maintenance* could mean sacrificing service reliability and economy by frequent and extended downtime to avoid unexpected breakdown. For industries in which safety is of overriding concern, such as the nuclear industry, a safety focus should be mandatory. In the hydro industry, mandatory provisions require meeting minimum maintenance requirements that address the various objectives mentioned here. Such mandatory requirements are usually imposed on the industry as the consequence of damages and losses caused by the neglect of safety objectives or constraints in favor of other objectives (see Section 1.5.1).

LaPay (1992) states that "reliability-centered maintenance philosophy requires that resources be concentrated on those components which are critical to plant operation and safety. The prioritization process is based upon the impact of the equipment on plant availability, plant safety, economic, and other plant specific factors." It is a basic premise of multiobjective optimization that compromises among objectives are necessary in order to arrive at a feasible solution. This kind of optimality that restricts the optimum seeking procedure by side objectives and constraints of an administrative, political, environmental or other nature in favor of tradeoffs among objectives is known as *Pareto optimality*. To make a difference with respect to *general* maintenance, RCM must maximize reliability, while serving all other objectives at or within agreed upon limits.

A plant maintenance program under the RCM label must address the following aspects (LaPay, 1992):

(a) <u>Critical component identification</u>: A prioritized list of critical components and documentation of these components is prepared that must be included in the program. Critical components are those that must function for the process to function.

(b) <u>Requirements, definitions, and documentation for each critical component</u> For each identified component, the following information items are prepared and made part of a readily accessible database:

- related industry codes and standards
- regulatory requirements
- technical specifications
- vendor documentation
- vendor recommendations for inspection and maintenance
- vendor warranty (life expectancy)

(c) <u>Component performance definition</u>: The component's function, possible failure modes, root causes (trigger events), and impacts of subcomponent failure are documented. This information is used to identify key equipment problems to be addressed by maintenance.

(d) <u>Maintenance activity selection</u>: Equipment requirements and performance needs, including historical performance are evaluated. Any new techniques, such as diagnostic and monitoring techniques, that support more effective performance should be evaluated.

(e) <u>Activity breakdown into tasks</u>: Every activity is broken down into tasks, and tasks are broken down into steps. Related procedures may emerge.

(f) <u>Detailed task definition</u>: The major steps of each maintenance task must be described with detailed step-by-step instructions, including lists of spare parts, tools, and other necessary resources. These task descriptions form the basis for the plant maintenance procedures and training programs.

(g) <u>Procedure development</u>: The task descriptions are incorporated into preventive maintenance procedures, but are also applicable to corrective maintenance. Also, the human factor is addressed, which provides maintenance personnel with documentation for their activities by on-site (by desktop and notebook) computer displays.

The functional performance designed into equipment must be understood and combined with all external requirements imposed on the equipment to provide an effective and justifiable maintenance program.

5.4 Maintenance Times

5.4.1 Corrective Maintenance Time

Corrective maintenance, also called *breakdown maintenance*, is initiated by a breakdown event. The elapsed time from breakdown to when maintenance ends is a probabilistic quantity, because many events surrounding the maintenance activity contribute to the length of the outage. All maintenance policy can do is to reduce the uncertainty inherent in maintenance work without being able to completely eliminate uncertainty in all aspects. Hence, maintenance time is a stochastic variable and the occurrence of maintenance time over a period of time, such as the life of the component, has the characteristics of a stochastic process. A process that alternates between operation and shutdown periods is also called a *renewal process* or a *birth–death process*, with the startup of the operation being construed a birth or renewal, and the shutdown being a failure or death. The literature on such processes provides examples of the analytical treatment of such processes with applications to maintenance (Parzen, 1965; Ross, 1992).

The life span of a system or a component consists of periods during which it is operating and periods when it is shut down. The time from the start-up of an operation to a shutdown or unexpected breakdown, and on to a new start-up, is the *cycle time*. The cycle time consists of two periods, operation time and downtime. *Operation time* begins with start-up and terminates with failure. It is also called *time to failure TTF*. *Downtime* is the time the system or equipment is down for repair. It is also called *time to repair TTR*. Downtime begins with failure and ends with completed repair. Cycle time is also called *time between failures TBF*:

$$TBF = TTF + TTR \qquad (5.4\text{-}1)$$

where *TTF* and *TTR* are observed random realizations. The operation cycle is illustrated in Figure 5-2. Each cycle marks a period of random length, *TBF*, that is the sum of two random variables.

5.4.2 Preventive Maintenance Time

Preventive maintenance interferes with the occurrence of random failures by replacing the random variable *time to failure* by a scheduled *time to maintenance*, *TTM*, which is followed by the *preventive maintenance time*, *PMT*, the scheduled downtime. Thus, the duration of the preventive maintenance cycle, the *time between maintenance*, *TBM*, is

$$TBM = TTM + PMT. \hspace{3cm} (5.4\text{-}2)$$

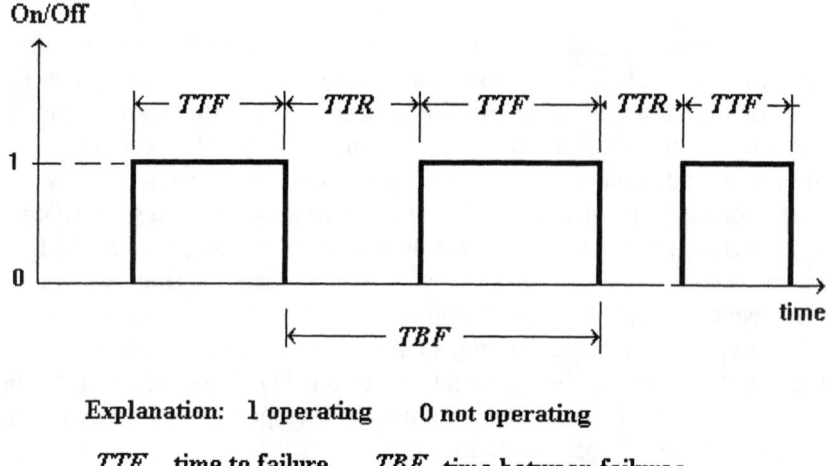

Explanation: 1 operating   0 not operating

TTF   time to failure   *TBF* time between failures
TTR   time to repair

Figure 5-2: Corrective maintenance is initiated by a random failure. The ensuing time to repair *TTR* is composed of many time contributions by work that is performed under uncertainty, which makes *TTR* a random variable. With the process start-up at the end of the repair period begins the next time to failure, *TTF*, that again ends with a random failure. The time between failures, or cycle time, *TBF*, is the sum of two random variables and therefore also a random variable.

Since failures cannot be eliminated completely, a series of scheduled maintenance cycles may be disrupted by an unexpected shutdown, a random failure, followed by a random *TTR*, as illustrated in Figure 5-3. A perfectly successful preventive maintenance program would eliminate all random *TBF*'s and replace them by an uninterrupted series of *TBM*'s. However, even if the safety issue is central, achieving a reduction of the probability of failure to zero may be prohibitively expensive or technically simply impossible. Some components of hydrosystems, such as large dams, qualify for safety-centered maintenance, and special maintenance approaches have been instituted for them (Section 1.3.4). The primary goal in these cases must be to minimize the probability of unexpected failure, subject to few, if any, secondary objectives and constraints.

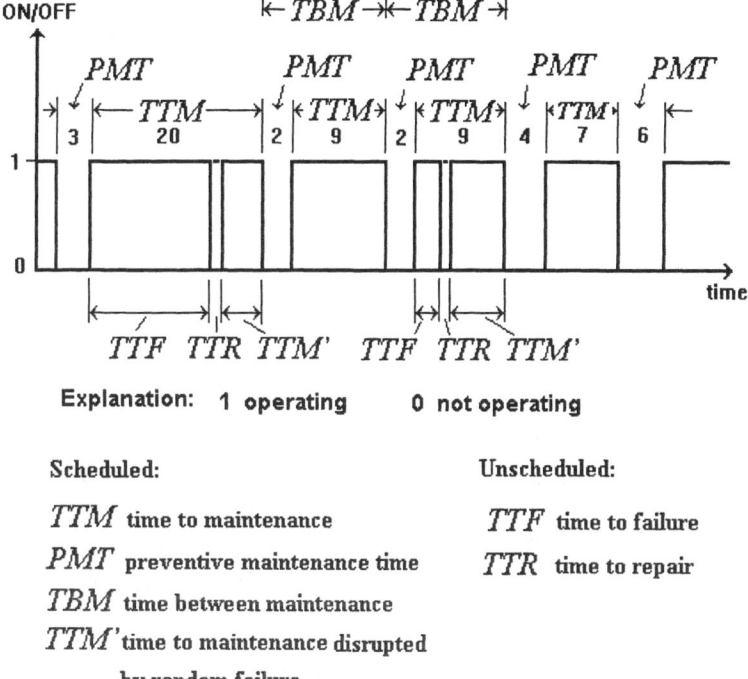

Explanation: 1 operating   0 not operating

Scheduled: Unscheduled:

TTM time to maintenance TTF time to failure

PMT preventive maintenance time TTR time to repair

TBM time between maintenance

TTM' time to maintenance disrupted
          by random failure

Figure 5-3 : For preventive maintenance, the time between two successive maintenance periods is the time between maintenance, *TBM*, or cycle time. It is the sum of the time to maintenance, *TTM*, and preventive maintenance time, *PMT*. Unexpected breakdowns during *TTM* are possible, but their duration should be short and their probability should be small.

In preventive maintenance, *TTM* is a scheduled time, whereas *PMT* is still subject to random fluctuations, as not all its time components can be accurately predicted. Generally, *PMT* is much smaller than *TTM* and the fluctuations of *PMT* can be absorbed in *TTM*, so that *TBM* can be part of a fixed maintenance schedule. From time to time, the scheduled *TBM*'s are disrupted by an unexpected shutdown. Then the *TBM* cycle has to be reset in some suitable way. All a preventive maintenance policy can accomplish is to reduce the frequency and size of failures, in other words, it can lengthen *TTF*, and reduce the size and variability of *PMT*. Thus, the frequency and size of the disruptions of the production process are reduced.

Example: Suppose the preventive safety-centered maintenance schedule of a structure is scheduled on an annual basis beginning with *PMT* = 10 days followed by *TTM* = 172 days. This cycle is followed by a second *PMT* = 10 days and a *TTM* = 173 days. Then the annual cycle starts all over again. If *PMT* has a standard deviation of 0.5 days, and is normally distributed around a mean of 10 days, there is a probability of 95 % that *PMT* will not exceed 11days (see Section 2.6.3). This day in exceess of the scheduled *PMT* is absorbed by the *TTM* period and the maintenance schedule remains fixed.

5.4.3 Downtime

The time between shutdown and start-up is the *time to repair*, *TTR*. This time is not the same as *repair time*. The repair time is just one component of the time to repair. Therefore, the total time from shutdown to start-up is also called *downtime* to avoid confusion. The various downtime components are to some extent project-dependent. Their names differ from source to source. Listings from two sources are given in Table 5-1. The activities associated with the time components are similar for corrective and preventive downtimes, but their length and variability can be significantly different.

The Navord time set is used in the time plan of Figure 5-4. The layout of a repair time plan shows analogy to serial and parallel systems (Chapter 4), as well as networks. A network consists of nodes connected by branches. In activity planning, a node represents an event, and a branch represents an activity. Passage through the event–activity network can only be in the direction of time, from a completed activity marked by an event or node to a subsequent activity marked by a branch emanating from the preceding node. Networks that simulate activity schedules are directed networks with restrictions on movements from a completed event to the next event. There is a critical sequence of activities with each depending on completion of a preceding activity and a resulting critical time which indicates the minimum time the completion of all sequential activities requires. Usually there are also parallel activities that can be carried out simultaneously, but they must all be completed before the next activity based on their completion can begin. Waiting times may result in parallel branches of the network. For example, obtainment of spare parts can begin as soon as the fault has been identified, but the spare parts must be available before replacement can begin, as illustrated in Figure 5-4. Since the minimum time excluding all waiting times is a stack of random variables representing the many individual activity times, it is likely to have a normal distribution (see Section 3.4.2).

Table 5-1: Terminology of Maintenance Time Components
(after Navord, p. 2.3, 1970; and Goldman and Slattery, p. 27, 1964)

Navord (1970)	Goldman and Slattery (1964, p. 27)
Fault identification time	Reaction time (delay time)
Team assembly time	Administrative time
Tool and equipment assembly time	Preparation time
Fault localization time	Fault location time
Gaining access time	Item procurement time
Spare part obtainment time	Supply time
Repair or replacement time	Fault correction time
Alignment and adjustment time	Adjustment and calibration time
Reassembling time	Final test time
Testing time	

The sum of fault location time and team assembly time is the *delay time*. In corrective maintenance, this time represents the *surprise effect* of the unexpected breakdown. In preventive maintenance, the delay time does not exist, because prior planning clears the way for an immediate start of the *active repair time*. This time consists of all time components that deal with repair in contrast to waiting for some activity to be completed. Preventive maintenance also may not require tool and equipment assembly time or spare part obtainment time after shutdown occurs. These activities can be completed in anticipation of the shutdown. Thus, preventive maintenance may shorten downtime considerably compared with corrective maintenance because of elimination of the surprise effect and by using preparatory parallel activities that would otherwise have to be serial activities commenced after shutdown. This means that time components may have smaller standard deviations than under corrective maintenance. This allows more precise prediction of downtime for preventive maintenance than for corrective maintenance. Successful preventive maintenance depends, however, on reliable diagnosis of faults for repair work preparation and good estimates of the repair time. Even the planned shutdown can hold surprises when the proverbial can of worms is opened and *Murphy's Law* comes into play (see Section 2.1.6).

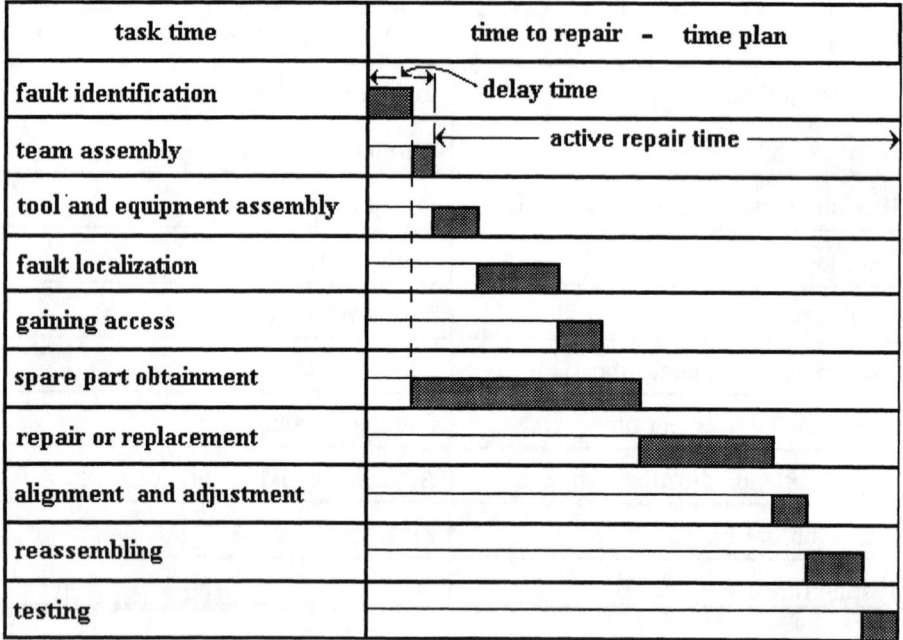

Figure 5-4: Downtime schedule for corrective maintenance (after Navord, p. 2-3, 1970). The delay time is due to the *surprise effect* of the unexpected breakdown. Time requirements up to and including spare part obtainment time can be significantly shortened or entirely eliminated from the downtime by preventive maintenance.

5.4.4 Downtime Distribution

For demonstration purposes, suppose the downtime is the sum of three random variables, delay time (TD), repair time (TR), and verification time (TV).

Assuming they are normally distributed, also their sum is normally distributed (Section 3.3.2). The total downtime is

$$TTR_0 = [E(TD) + z_1 \, \sigma(TD)] + [E(TR) + z_2 \, \sigma(TR)]$$

$$+ [E(TV) + z_3 \, \sigma(TV)] \tag{5.4-3}$$

where the three terms in brackets are the three random variables TD, TR, and TV; $E(TD)$, $E(TR)$, and $E(TV)$ are their expectations; $\sigma(TD)$, $\sigma(TR)$, and $\sigma(TV)$ are their standard deviations; and z_1, z_2, and z_3 are independent random variables that create the random variations with the range of the normal distribution; theoretically this range is $-\infty < z < +\infty$, but for practical purposes it can be limited to $-3 \leq z \leq +3$; 99.73 % of all realizations of a normal variable fall into this range (sometimes $z = \pm 4$ is used with a probability of containing the normal variable with 99.99 %; see also Table 2-3).

The normal distribution allows one to consolidate an expression of the type of Equation (5.4-3) by using the fact that sums of random variables tend to be normally distributed. This leads to

$$TTR_0 = E(TTR_0) + z \, \sigma(TTR_0) \tag{5.4-4}$$

with

$$E(TTR_0) = E(TD) + E(TR) + E(TV) \tag{5.4-4a}$$

and

$$\sigma(TTR_0) = [\sigma^2(TD) + \sigma^2(TR) + \sigma^2(TV)]^{0.5} \tag{5.4-4b}$$

where $E(TTR_0)$ is the expectation of the sum of the three variables TD, TR, and TV; $\sigma(TTR_0)$ is the standard deviation of the sum of the three variables; and z is the random variable that represents the sum of the random variables z_i. The pdf of this distribution is

$$f(z) = \frac{1}{\sqrt{2\pi}\,\sigma} e^{-0.5\frac{(z-\mu)^2}{\sigma^2}} \tag{5.4-5}$$

with $\quad z = \dfrac{x - \mu}{\sigma},$

$$\mu = E(TTR_0) \tag{5.4-5a}$$

and

$$\sigma^2 = \sigma^2(TTR_0) \tag{5.4-5b}$$

where x represents the sum of the random variables and only one z representing this sum needs to be generated as a random number; μ is the sum of the three expectations; and σ^2 is the sum of the three variances.

Since the downtime or its components cannot be negative and most elements of a normal distribution lie within a range of ± 3 σ, the normal distribution is applicable only if $\mu \geq 3$ σ, in other words, if the mean of the practical range lies at least 3 σ to the right of zero of the real variable x. This also means that the coefficient of variation should be $C_v = \sigma/\mu \leq 1/3$, which means that the data should not be excessively dispersed (see also Section 2.6.3).

Distributions other than the normal pdf that have been found adequate for repair time distributions are the exponential pdf, the Weibull pdf, and the log-normal pdf. A first attempt to fit the data might be to try the log-normal distribution, which is based on the assumption that the logarithms of the data are normally distributed. Since logarithms are dimensionless numbers and graphical representations use the decimal logarithm, it is recommended to use the decimal logarithm of x/μ. Using this ratio does not change the distribution of the original data x. The standard normalized variable is

$$z = \frac{\log(\frac{x}{\mu}) - \mu_{\log(\frac{x}{\mu})}}{\sigma_{\log(\frac{x}{\mu})}} = \frac{y_d - \mu_{yd}}{\sigma_{yd}} \tag{5.4-6}$$

where $y_d = \log(x/\mu)$ is the transformed variable; x are the original data; μ is the mean of the original data; μ_{yd} and σ_{yd} are the parameters calculated from y_d.

If the data analysis shows that the log-normal distribution is applicable, a probability can be attributed to $\log(x/\mu)$-values by the corresponding z. The relationship between x/μ and z according to Equation (5.4-6) is

$$\frac{x}{\mu} = 10^{\mu_{yd} + z\,\sigma_{yd}} \tag{5.4-7}$$

where x/μ is the nondimensional ratio of the real data, and μ_{yd} and σ_{yd} are the parameters of the logarithmic data, $\log(x/\mu)$.

Example: Suppose TTR/μ data are log-normally distributed with $\mu_{yd} = 0.1$ and $\sigma_{yd} = 0.025$, both parameters are based on the logarithms of TTR/μ, with $\mu = 40$ h. Find the TTR that has an exceedance probability of not more than 5 %. Solution: 95.45 % of all data of a normal pdf are contained in a band of $z = \pm 2$ around the mean (Table 2-3). The normal distribution is valid only for $\log(TTR/\mu)$. Using Equation (5.4-7), the ratio of TTR/μ for $z = 2$ is

$$TTR/\mu = 10^{0.1 + 2 \cdot 0.025} = 1.41$$

Hence, $TTR = 40 \cdot 1.41 = 56$ h has a exceedance probability of less than 5 %.

5.4.5 Maximum Downtime

If the distribution of the downtime is known, it can be used for calculating downtimes that will not be exceeded with a selected level of probability. A time to repair with a small probability of exceedance can be used as an estimated *maximum time to repair* TTR_{max}. The probability that this time is not exceeded is

$$P(TTR_{max} \leq t_u) = F(z_u), \tag{5.4-8}$$

or formulated as an exceedance probability,

$$P(TTR_{max} > t_u) = F(z_u) \tag{5.4-8a}$$

where t_u is the upper threshold on downtime not being exceeded; $F(z_u)$ is the non-exceedance probability of t_u, and $1 - F(z_u)$ is the complementary exceedance probability. The normalized variable

$$z_u = \frac{t_u - \mu}{\sigma} \tag{5.4-9}$$

ties t_u to a probability. Table 5-2 gives some guiding normal probabilities $F(z_u)$ and their complements for selected z_u. Cases (c) and (d) of Figure 2-14 apply, as high non-exceedance probabilities are of interest. For $z > 0$, $F(z) = 0.5 + F_0(z)$, where $F_0(z)$ is the table lookup for positive z (see also Table 2-3).

Table 5-2: Normal Non-exceedance and Exceedance Probabilities for Maximum Downtimes Estimates

Normalized Variable	Non-exceedance Probability	Exceedance Probability
z	$F(z)$	$1 - F(z)$
(1)	(2)	(3)
0.6745	0.750	0.250
1.00	0.841	0.159
2.00	0.977	0.023
3.00	0.999	0.001

Example: Suppose t_u, μ, and σ produce $z = 3$. What is the probability of this t_u not being exceeded? Solution: If t is normally distributed, $F(z_u) = 99.9\,\%$. The user may accept $TTR_{max} = t_u$ based on $z = 3$ as a "practical" maximum downtime estimate.

5.4.6 Expected Cycle Time

The cycle time in corrective maintenance is also the time between random failures, *TBF*, as given by Equation (5.4-1). The derivation of this parameter is given as a demonstration of how the probability of such a parameter can be derived once the pdf's of time to failure and time to repair have been identified by a data analysis. Exponential pdf's are used here because they occur frequently in nature and can be relatively easily handled by calculus (after Locks, 1973, p. 172). Let α be the rate of an exponential distribution of time to failure *TTF*, and let δ be the rate of the exponential distribution of the time to repair *TTR*. The probability of the occurrence of a failure at time x is

$$f(x)dx = \alpha\, e^{-\alpha x} \tag{5.4-10}$$

and the probability of completing a repair during an elapsed time $t - x$ is

$$f(t-x)dt = \delta\, e^{-\delta(t-x)} . \tag{5.4-11}$$

The probability of the equipment being repaired and available for operation at time t is the joint probability that the failure at time x has occurred and that the repair has been completed during time $t - x$:

$$f(t)dt = f(x)dx \, f(t - x)dt \, . \tag{5.4-12}$$

The total probability of all these joint, but mutually exclusive event pairs of failure and completed repair at time t is the sum of all these joint probabilities. Since the pdf's are given as integrable functions, the pdf of the total probability is obtained by the integral

$$f(t) = \int_0^t f(x) f(t - x) dx \, . \tag{5.4-13}$$

The total probability density, $f(t)$, is the sum of all contributing ordinates, $f(t)$, for a given argument t and is known as a convolution integral [see also Section 3.3.1, Equation (3.3-5)]. Numerically, the convolution integral is a sum of products. Substituting data-based functions for $f(x)$ and $f(t - x)$ gives

$$f(t) = \alpha \delta \int_0^t e^{-\delta(t-x)} e^{-\alpha x} dx \, . \tag{5.4-14}$$

The integration over x treats t as a constant parameter so that the integral can be simplified to

$$f(t) = \alpha \, \delta \, e^{-\delta t} \int_0^t e^{(\delta - \alpha)x} dx \tag{5.4-15}$$

where x is a time before and up to t, which is an upper boundary. Using the substitution $u = (\delta - \alpha) x$, and $du = (\delta - \alpha) dx$ leads to $\int \exp(u)du = \exp(u) + c$. For the range $(0, t)$:

$$f(t) = \frac{\alpha \delta}{\delta - \alpha} e^{-\delta t} e^{(\delta - \alpha)x} \Big|_0^t \, .$$

Substituting the integration boundaries leads to

$$f(t) = \frac{\alpha \delta}{\delta - \alpha} (e^{-\alpha t} - e^{-\delta t}) \tag{5.4-16}$$

where $f(t)$ is the pdf of the equipment being repaired and operable at time t, and $f(t)$ dt is the probability of the equipment being operable at time t.

The average time the equipment is operable is the probability-weighted sum of all t's, usually calculated as the first moment of t,

$$E(t) = \int_0^\infty t\, f(t)dt \tag{5.4-17}$$

After substituting Equation (5.4-16) into (5.4-17), one obtains

$$E(t) = \frac{\alpha\,\delta}{\delta - \alpha}\, (\int_0^\infty t\, e^{-\alpha t}dt - \int_0^\infty t\, e^{-\delta t}dt) . \tag{5.4-18}$$

The integration requires finding integrals of the kind $\int u^k \exp(u)du$. For $k = 1$, the general solution of each of the two integrals is $(u - 1)\exp(u)$. Expanding the t in front of the exponential function into $u = -\alpha\, t$ and applying the general solution leads to

$$\int_0^\infty t\, e^{-\alpha t}dt = \frac{1}{\alpha^2}(1 + \alpha\, t)e^{-\alpha t}\Big|_0^\infty = \frac{1}{\alpha^2} .$$

Evaluating the product of the two functions in t for $t \to \infty$ leads to the undetermined product $0 \cdot \infty$. Using the Bernoulli–de L'Hospital rule shows that the product is zero. For the lower limit, $t = 0$, the integral gives -1. The same values are obtained for the boundaries of the second integral. Thus the solution of Equation (5.4-18) becomes

$$E(t) = \frac{\alpha\delta}{\delta - \alpha}(\frac{1}{\alpha^2} - \frac{1}{\delta^2}) = \frac{1}{\alpha} + \frac{1}{\delta} . \tag{5.4-19}$$

where $E(t)$ is the average cycle time that includes an operation time, which ends with failure at time x, here called time to failure, TTF, and the time to repair from x to t, TTR. At time t the system is ready to function. The average cycle time is also the time between failures, TBF,

$$E(t) = E(TBF) = E(TTF) + E(TTR) \tag{5.4-20}$$

where $E(TTF) = 1/\alpha$, and $E(TTR) = 1/\delta$. The inverse of the failure rate is seen to be the average time to failure and the inverse of the repair rate is the average time to repair.

5.5 Availability

5.5.1 Availability Under Corrective Maintenance

A system that is available is ready to function. A system that is unavailable is out of service or shut down. By repair or replacement a system that is shut down can be made to function again. During an extended period, a planning period or a cyclic period, there are periods of availability alternating with periods of unavailability. Availability is the probability that the system functions. This does not mean the system actually is providing service when it is available. Availability is defined differently for corrective maintenance and preventive maintenance. Somewhat different definitions are used by different authors for availability parameters. The user must be sure to make proper use of the terms and associated numbers for his or her purposes (see Glossary; also see glossary by Goldman and Slattery, 1964, p. 27).

A process that relies on corrective maintenance runs its course from startup to failure with the time to failure *TTF* being a random variable. The operation of a lightbulb is an example. From the moment it lights up for the first time, it is available until it fails. At the time of failure a replacement is made (repair is not possible in this case), and the next operation period begins that ends with the next failure. The period from a failure to the next is a corrective maintenance cycle. Availability under corrective maintenance is therefore the ratio of two random variables, time to failure *TTF* and time between failures *TBF*,

$$A_c = \frac{TTF}{TBF} = \frac{TTF}{TTR + TTF} \tag{5.5-1}$$

where A_c is availability under corrective maintenance; *TTF* is the operational period of the system and *TTR* is the time to repair.

If corrective maintenance uses replacement without repair, then *TTF* is the operational life and *TTR* is the replacement time. The time between failures *TBF* is the sum of two random variables. If *TTR* is a fixed time, *TBF* is still a random variable because of the randomness of *TTF*. If time to replacement (or repair) is small compared to *TTF*, or $TTR \ll TTF$, then $A_c \to 1$. This means that availability is maximized by minimizing the time to replacement or time to repair.

5.5.2 Availability Under Preventive Maintenance

Under preventive maintenance, the time to maintenance, *TTM*, and the preventive maintenance time, *PMT*, are scheduled periods. Availability by definition is the probability of the system being ready to function. In analogy to corrective maintenance, availability under preventive maintenance, A_p, is

$$A_p = \frac{TTM}{PMT + TTM} = \frac{TTM}{TBM} \qquad (5.5\text{-}2)$$

where A_p is calculated for each maintenance cycle of length TBM. Figure 5-3 illustrates the preventive maintenance terminology. Equation (5.5-2) is also an expression for scheduled equipment use under preventive maintenance.

   If preventive maintenance is disrupted by random failure, then the actual time to maintenance, TTM_a, is shorter than the scheduled time to maintenance, TTM_s, and A_p is reduced. By excluding the scheduled preventive maintenance time, an *operational availability* under preventive maintenance can be defined as

$$A_{op} = \frac{TTM - \sum TTR}{TTM} \qquad (5.5\text{-}3)$$

where A_{op} is the operational availability, and $\sum TTR$ is the sum of all unscheduled shutdown periods during the scheduled operation period, TTM, as illustrated in Figure 5-3. One of the purposes of preventive maintenance is to keep $\sum TTR$ as small as possible so that $A_{op} \rightarrow 1$.

   A measure for *the effectiveness of preventive maintenance*, E_p, is the ratio of actual time to maintenance, TTM_a, to scheduled time to maintenance, TTM_s,

$$E_p = \frac{TTM_a}{TTM_s} \qquad (5.5\text{-}4)$$

where TTM_a is equal to the numerator of Equation (5.5-3). This ratio, which is a random number because of the random nature of TTM_a is computed for each maintenance cycle. If preventive maintenance is effective, $E_p \rightarrow 1$.

Example: Use the numbers given in the preventive maintenance illustration of Figure 5-3 to calculate the following: (a) Availability under preventive maintenance. (b) Operational availability under preventive maintenance. (c) Effectiveness of preventive maintenance. Solution: (a) The availability under preventive maintenance, A_p, for the four maintenance cycles is by Equation (5.5-2): 20/23 = 0.87; 9/11 = 0.82; 9/11 = 0.82; and 7/11 =0.64. (b) The operational availability accounts for two failures, one in the first cycle assumed to last 2 time units, and the second in the third cycle assumed to last 1 time unit. Equation (5.5-3) applied to each cycle gives: 18/23 = 0.78; 9/11 = 0.82; 8/11 = 0.73; and 7/11 = 0.64. (c) The effectiveness of preventive maintenance is by Equation (5.5-4); $18/20 = 0.9$; $9/9 = 1$; $8/9 = 0.89$; and $7/7 = 1$.

5.5.3 Transient and Steady-State Availability

The on–off random process illustrated in Figure 5-2 can be construed as an alternating renewal process (Ross, 1992, p. 43). The process is *on* for some period x_1, then *off* for some period y_1, then *on* again for some period x_2, and *off* again for some period y_2, and so on. If X and Y are independent random variables, then the probability of the process being *on* at time t, $P(\text{on at time } t) = P(t)$, as $t \to \infty$ (steady state), is

$$P(t \to \infty) = \frac{E(X)}{E(X) + E(Y)} \tag{5.5-5}$$

where X and Y represent collections of random realizations, $X = (x_1, x_2, ..., x_n)$, and $Y = (y_1, y_2, ..., y_n)$ with expectations $E(X)$ and $E(Y)$. If X stands for the random variable *TTF*, and Y for *TTR*, then $E(TBF) = E(TTF) + E(TTR)$, and $P(t \to \infty)$ becomes the *inherent availability*,

$$AI = \frac{E(TTF)}{E(TBF)} \tag{5.5-6}$$

It is shown in Section 5.4.6 that $E(TTF) = 1/\alpha$, and $E(TTR) = 1/\delta$, where α is the failure rate and δ is the repair rate. Substituting these rates into Equation (5.5-6) leads to

$$AI = \frac{\delta}{\alpha + \delta} \tag{5.5-7}$$

where AI is the availability theoretically at a long time after startup. At startup, starting with a working system, $A(t = 0) = 1$. Over time t, theoretically for $t \to \infty$, the availability decays to $\delta/(\alpha + \delta)$. It can be shown that this transient availability is described by the decay function (Shooman, 1968, p. 338)

$$A(t) = \frac{\delta}{\alpha + \delta} + \frac{\alpha}{\alpha + \delta} e^{-(\alpha + \delta)t} \; . \tag{5.5-8}$$

This function satisfies both boundaries of t. For $t = 0$, $A(0) = 1$. As t increases, the second term, which is a function of t, decays exponentially and disappears for $t \rightarrow \infty$, so that A becomes the steady state availability of Equation (5.5-7) that is independent of time.

5.5.4 Expected Availability

Availabilities for corrective as well as preventive maintenance include random elements. Availabilities are calculated for each maintenance cycle. Analysis of the observed availabilities includes calculation of the parameters of the data, such as expectations and probability distributions. The *expected availability* for the corrective maintenance cycle is

$$E(A_c) = \frac{1}{N} \sum_{i=1}^{N} \frac{TTF}{TBF} = E(\frac{TTF}{TBF}) \qquad (5.5-9)$$

where N is the number of observed data pairs (TTF, TTR); and $TBF = TTF + TTR$.

$$E(A_p) = \frac{1}{N} \sum_{i=1}^{N} \frac{TTM}{TBM} = E(\frac{TTM}{TBM}) \qquad (5.5-9a)$$

where (TTM, PMT); and $TBM = TTM + PMT$.

Example: For data of time to failure, time to repair, and time between failures calculate $E(A_c)$ and AI. Solution: The data sample is given in tabular form as follows.

No.	TTF	TTR	TBF	TTF/TBF
(1)	(2)	(3)	(4)	(5)
1	5	3	8	0.625
2	4	2	6	0.667
3	8	1	9	0.889
4	7	3	10	0.700
5	3	4	7	0.429

6	6	6	12	0.500
7	2	5	7	0.286
8	9	7	16	0.563
9	1	2	3	0.333
10	8	4	12	0.667
Average	5.3	3.7	9.0	0.566

The expected availability by Equation (5.5-9) is the average of column 5, $E(A_c) =$ 0.566. The inherent availability by Equation (5.5-6) is the ratio of columns 2 and 4, $E(TTF) = 5.3$, and $E(TBF) = 9.0$, respectively, or $AI = 0.589$.

5.5.5 Availability and Reliability

Availability and reliability have different meanings. Availability is the probability that the system is ready to function at a time t, and reliability is the success probability that the system has not failed during period t, or that some threshold value is exceeded, or some other defined success probability. The difference between the two parameters is demonstrated by an example.

Twenty availabilities were obtained from a corrective maintenance process (data after Locks, 1973, p. 178). These elements x of the sample space are arranged in ascending order in Table 5-3. A unit probability $f(x) = 0.05$ is given to each x (in contrast to a set or class probability, if a histogram had been chosen with one or more elements per set). The cumulation of the $f(x)$ leads to the non-exceedance probability $F(x)$. A data fit by assuming an exponential distribution (see Section 2.7.5)

$$F(x) = 1 - e^{-ax}$$

suggests a straight line fit by the logarithmic form

$$\log[1 - F(x)] = -a'x \qquad (5.5\text{-}10)$$

where $a' = 0.4343\ a$. This semilogarithmic plot is shown in Figure 5-5. If the reliability of x is defined as $R(x) = 1 - F(x)$, then the curve represents the reliability of the availability. The reliability varies inversely with availability. Low availabilities are exceeded with high probability and high availabilities are exceeded with low probability.

Table 5-3: Observed Availabilities and Their Reliabilities
(data after Locks, 1973, p. 178).

No.	x	$f(x)$	$F(x)$	$R(x)$ $1-F(x)$
1	0.615	0.05	0.05	0.95
2	0.625	0.05	0.10	0.90
3	0.734	0.05	0.15	0.85
4	0.734	0.05	0.20	0.80
5	0.829	0.05	0.25	0.75
6	0.870	0.05	0.30	0.70
7	0.889	0.05	0.35	0.65
8	0.936	0.05	0.40	0.60
9	0.967	0.05	0.45	0.55
10	0.971	0.05	0.50	0.50
11	0.978	0.05	0.55	0.45
12	0.982	0.05	0.60	0.40
13	0.983	0.05	0.65	0.35
14	0.985	0.05	0.70	0.30
15	0.985	0.05	0.75	0.25
16	0.986	0.05	0.80	0.20
17	0.991	0.05	0.85	0.15
18	0.992	0.05	0.90	0.10
19	0.996	0.05	0.95	0.05
20	0.996	0.05	1.00	0.00

Notes: column 1: number given to data item or element of sample space, $N = 20$.
column 2: observed availabilities x arranged in ascending order.
column 3: unit probability $f(x)$ allocated to data item, $1/N$.
column 4: non-exceedance probability $F(x)$ calculated by cumulating $f(x)$.
column 5: exceedance probability $1 - F(x)$.

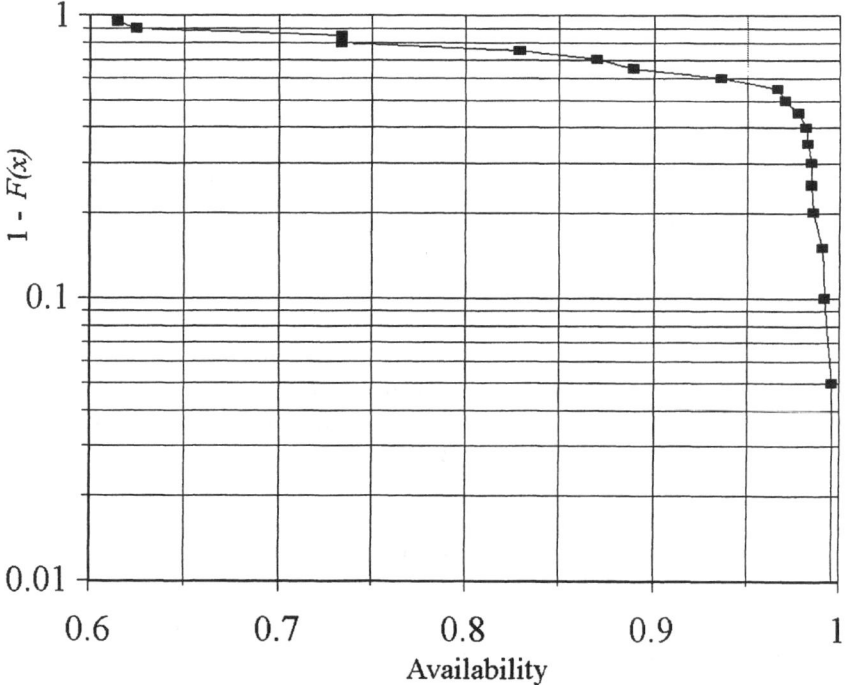

Figure 5-5: Exceedance probabilities of experimental availabilities. If reliability is defined as the exceedance probability of a given availability then the curve shows the reliability of availability (the data are from Locks, 1973, p.178).

5.6 Maintainability

5.6.1 Maintainability Measures

Maintainability is the ease with which maintenance can be performed (Bloch and Geitner, 1990, p. 17). Goldman and Slattery (1964, p. 28) define maintainability as "the probability that an item will conform to specified conditions within a given period when maintenance action is performed in accordance with prescribed procedures and resources." If a period is determined for a set of activities, then maintainability M can be expressed as the probability that the *actual* time to repair is within an admissible deviation from the *scheduled* time to repair:

$$M = P(TTR_a - TTR_s \leq \Delta TM_\alpha) \geq \alpha \tag{5.6-1}$$

where TTR_a is the actual time to repair, which in this context could be for a planned or for a corrective maintenance activity; TTR_s is the average time to repair or a standard time prescribed for a recurrent repair or replacement activity; $TTR_a - TTR_s$ is the difference between scheduled and actual time to repair, which may be positive or negative. The latter indicates maintainability beyond expectations and may indicate a higher maintainability than the one being tested; ΔTM_α is a prescribed positive difference between actual time to repair and scheduled time to repair that is associated with a measure α. If the ΔTM_α for $\alpha = 0.95$ is not exceeded, then maintainability is said to be 95 % (see Section 2.4.5). A larger ΔTM_α may be specified with a lower maintainability attached to it or, vice versa, a smaller ΔTM_α may be specified with a higher maintainability. If the actual maintenance time TTR_a consistently differs from TTR_s by significant under- or overestimation, an investigation of maintenance planning and application procedures is indicated. Maintainability is a quality control measure for maintenance performance that takes into account the probabilistic nature of maintenance work. It can also be a measure of design quality and operation quality.

It is often advantageous to use nondimensional measures. Equation (5.6-1) in nondimensional form is

$$M = P(\frac{TTR_a}{TTR_s} - 1 \leq \frac{\Delta TM_\alpha}{TTR_s}) \geq \alpha \tag{5.6-2}$$

for $TTR_a/TTR_s > 1$.

The maintainability index is defined as

$$MI = TTR_a / TTR_s \tag{5.6-3}$$

where MI can be greater or less than 1; $\Delta TM_\alpha/TTR_s$ is a tolerance for the index as long as it is greater than 1 for a specified α.

Example: Assume the tolerance for the maintenance index is $\Delta TM_\alpha/TTR_s = 0.05$, for $\alpha = 95$ %, and the observed data are in the range $TTR_a/TTR_s \leq 1.04$. Then the criterion of maintainability $M = P(TTR_a/TTR_s - 1 \leq 0.05) \geq 0.95$ is met.

5.6.2 Accessibility

The way a system is designed and components are arranged can either help or obstruct maintenance. Accessibility is the ease by which parts in need of surveillance and maintenance can be accessed. Important components of hydraulic structures are difficult to access. The lack of access is partly due to the nature of hydraulic structures but it may also be due to design judgments dictated by project economy and other reasons. Examples are lack of access to the interiors of dams, to water facing parts or water filled components of structures, and to possible seepage pathways. Hazards may develop in inaccessible areas and endanger structural integrity before any visual signs and measurable symptoms are detected. The hazard potential of hidden flaws is especially great at the start-up of operations, during the first loading of a dam by reservoir filling and other start-up operations. Foundation compression can cause interior arching and horizontal cracking in the core of an embankment dam with subsequent development of piping in the core that may be completely invisible from the outside (Sherard et al., 1963, p. 139). It is therefore important to have a comprehensive surveillance program in place at the start-up of a project. The high surveillance intensity at start-up can be subsequently reduced as initial problems are solved and O&M experience is gained. O&M may have to be increased again as project life advances to cope with an increasing age-related failure rate so that O&M intensity over the lifetime of a project may follow a bathtub curve (see Figures 3-15 and 6-3). Good accessibility aids maintainability and is desirable at all stations of a project's life.

The groundwork for various degrees of accessibility is laid in design and is sometimes a matter of judgment, culture, and economy. For example, European embankment dams include a control tunnel at the base of the dam, along its foundation across the valley (Sherard et al., 1963, pp. 74–76, and 108). It is considered an important accessibility and maintainability feature of major dams, because it provides access to areas of seepage, settlement, and deformation along the dam foundation across the valley; it allows the placement and control of measuring devices; and it provides access for remedial measures, such as foundation grouting (Blind, 1983, p. 20; Muckenthaler, 1989). Diverging expert opinions exist on the choice of hazard control methods in design and operation. Whereas some experts think that such galleries are essential for accessibility and maintainability, others think they may provide seepage paths and weaken the structure (ICOLD, 1991, p. 69). There are many dams in the world with and without inspection galleries in embankment dam foundations.

There are many components for which difficult and costly access cannot be avoided. Such components should be designed as low-maintenance components requiring little or no maintenance. Components that are difficult to maintain have most likely low maintainability, which means high probability for overruns of time to repair and maintenance cost. Low maintainability may be acceptable if it is paired with low investment cost and a relatively low hazard rating of the structure. Life-cycle cost analysis should provide the decision maker with information on whether low maintainability is an acceptable trade-off for a low-cost design with limited life expectancy, in contrast to a high-cost design with high maintainability and long expected service life (see Section 5.8.5).

Development and use of special technology is a means of coping with the intrinsic accessibility problems of hydraulic structures. For example, in high head power plants, including pumped storage plants, the main conduit provides the water to all generation units. If inspection and repair require dewatering the supply conduit on which all units depend, the entire plant needs to be shut down. There must be a good reason for initiating such a shutdown procedure. Dewatering of a conduit that uses the surrounding rock for support against inside water pressure, must be slowly depressurized to avoid possible cracking or crushing of the tunnel lining by outside pressure. Also refilling may take several days to allow all air trapped in the conduit to be safely evacuated. In the Mauvoisin power development, a 14.7 km tunnel had to be built with two culmination points to provide gravity water drainage during construction. Air that becomes trapped in the two culmination points on refilling the tunnel is drained by pipes, 92 mm and 80 mm in diameter, that run from the culmination point, through pipelines in the tunnel floor, to exhaust points. It takes four days to fill the conduit from the moment air becomes trapped during refilling, and six days until all air is evacuated (Electro-Watt, 1959). For remote tunnel lining inspections, unmanned submarines and robots have been developed that navigate such conduits under full pressure and detect lining problems (Heffron, 1990). This allows costly and hazardous dewatering procedures to be reserved for cases in which there is almost certain knowledge of damage and need for repair.

5.7 Maintenance Strategy

5.7.1 Maintenance as Interference with a Random Process

If a system is observed at a sequence of points in time, it may be found in various states: the normal operational state, the shutdown state, or other states of various degrees of deterioration in between. The movement of the system between these states, if it were completely random, would represent a stochastic or random process. Since each new state depends on the state before it, this movement from one state to another over time can be construed as a Markov process. If a system cannot heal or repair itself, then it can only progress from the normal operation state to defective states and finally to the failure state. A random failure means that the system has moved into the failure (or shutdown) state from which it can only be removed by repair. A random failure is an event beyond the control of the operator, whereas a planned shutdown is a deliberate action by the operator in control of the system. Corrective maintenance is a directed Markov process with its random movements limited to pathways from good to deteriorated states. After failure has occurred corrective action is required to set the process on a new path.

For important systems, the operator usually prefers to be in control of the system states, even if complete control is usually elusive. A Markov process that is under partial control of the decision maker is a *Markov decision process*. The control strategy in the form of a preventive maintenance program attempts to identify and control hazards that arise as consequence of the design, construction, operation, and aging of a project. The preventive maintenance strategy anticipates the development of new hazards, recognizes their early symptoms, and mitigates and/or eliminates them to the extent possible or economically feasible. Even if fixed pathways cannot be prescribed, the probability of failure can be reduced. Such a strategy is illustrated in Figure 5-1. The continued increase of failures rates with time is stopped and reverted to a lower rate and random failure is avoided. The effect is a deliberate *interference* in the random process that would otherwise lead to failure.

5.7.2 Maintenance Actions

The actions that interfere with the random deterioration process are the maintenance actions. They are illustrated in some detail in Figure 5-6. At the start of the observations, the process is in the state of normal operation, as shown in the lower left-hand corner of Figure 5-6. Ideally the process should remain in this state. The high probability of normal operation is expressed by the transition probability ($p = 0.99$) on the loop that indicates the process remaining in this state in successive periods or maintenance cycles.

There is usually at least a small probability of an excursion out of the normal state. If such an excursion is detected and communicated to experts who recognize its importance and take proper action, maintenance returns the process to the normal state. If detection fails or if even one wrong decision is made following detection, the process moves on to the next deteriorated state and toward failure. Thus, maintenance action aims at changing the random Markov process of deterioration into a Markov decision process by influencing the probabilities of transitions from a less deteriorated state to a more deteriorated state so that the transition to the failure state becomes less likely than if the random process is allowed to run its course. However, complete control of the transitions is usually not achieved so that the deterioration process remains a probabilistic process.

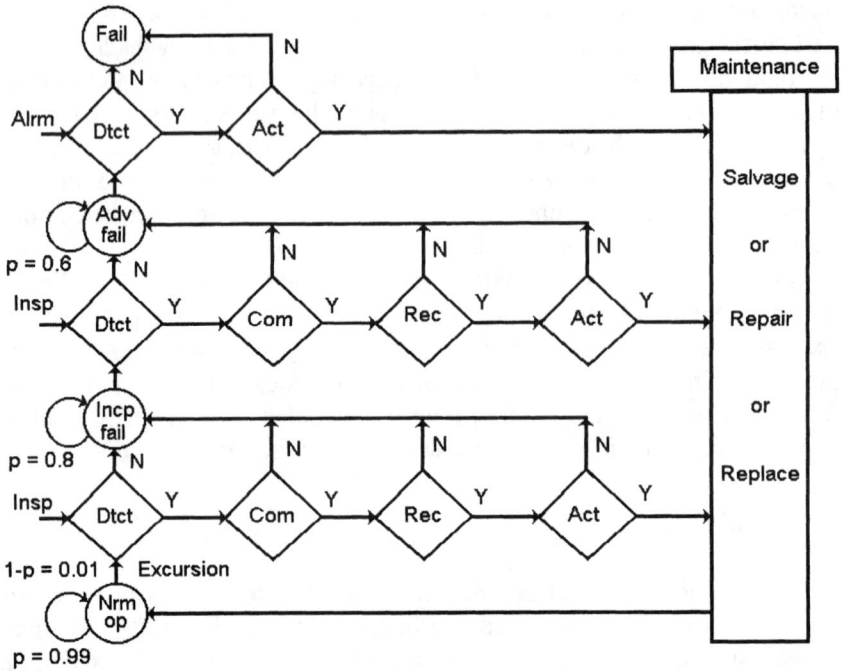

Figure 5-6: Random deterioration process with interference by maintenance. On the left, the upward directed arrows mark the path that leads from the normal operation state via several increasingly degraded states to failure. The horizontal paths are triggered by inspection followed by detection. For detection to be successful the findings must be communicated, their importance must be recognized, and action must be taken. Any failure along this path returns the process to the path to failure. If the interference is successful, maintenance returns the process to normal operation. Abbreviations: Nrm op = normal operation with high probability p of the system being found there; Insp = inspection or monitoring; Dtct = detection of defect; Com = communication of defect; Rec = recognition of importance of the defect; Act = decision on corrective action to be taken; Incp fail = incipient failure state with reduced probability of the system being found there; Adv fail = advanced failure state with further reduced probability of the system being found there; Alrm = alarm to alert to override normal procedures; Fail = failure.

There is usually at least a small probability of an excursion out of the normal state. If such an excursion is detected and communicated to experts who recognize its importance and take proper action, maintenance returns the process to the normal state. If detection fails or if even one wrong decision is made following detection, the process moves on to the next deteriorated state and toward failure. Thus, maintenance action aims at changing the random Markov process of deterioration into a Markov decision process by influencing the probabilities of transitions from a less deteriorated state to a more deteriorated state so that the transition to the failure state becomes less likely than if the random process is allowed to run its course. However, complete control of the transitions is usually not achieved so that the deterioration process remains a probabilistic process.

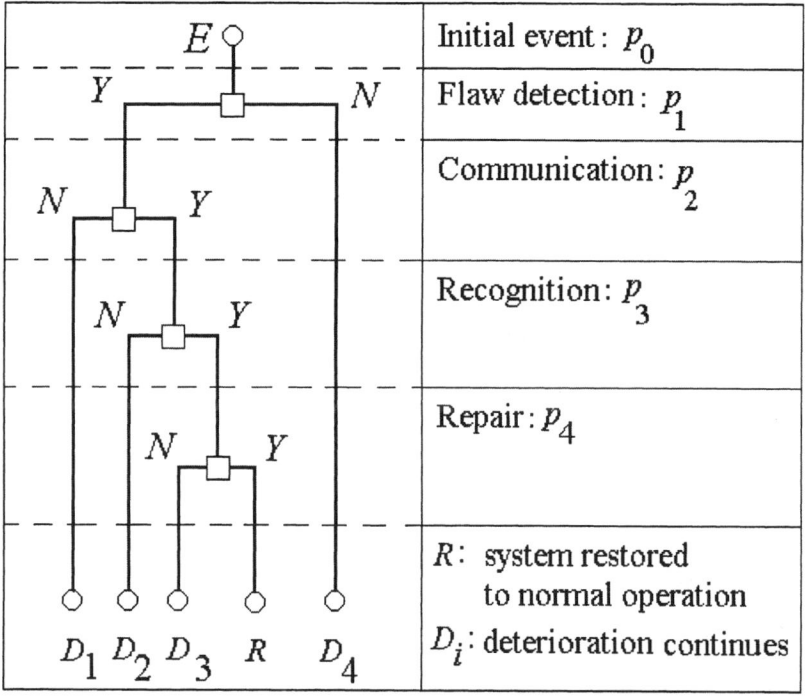

E	Initial event: p_0
Y ⬚ N	Flaw detection: p_1
N ⬚ Y	Communication: p_2
N ⬚ Y	Recognition: p_3
N ⬚ Y	Repair: p_4
D_1 D_2 D_3 R D_4	R: system restored to normal operation D_i: deterioration continues

Figure 5-7: Event tree for detection and repair of potentially fatal system flaw (after Rissler, 1988, p. 200). E is the initial event that diverts the system from normal operation; Y marks the successful branch that leads to restoration R; N marks the branches that lead to further deterioration. D_i are outcomes indicating further deterioration.

Redundancy in safety measures may be necessary to detect a hazard that has evaded detection. In the scheme in Figure 5-6, three barriers prevent the system from failing. They are activated by three detection levels, which give the system supervisor three opportunities to prevent failure. Often the decision to initiate maintenance must be made with incomplete information. Such decisions in the face of uncertainty are risky decisions, because they may turn out to be wrong. Risky decision making will be discussed in more detail in Chapter 6.

5.7.3 Event Tree as Decision Aid

Analytically, the probabilistic outcome of the deterioration process can be represented by an *event tree*. An example is shown in Figure 5-7. It represents the same events that lead either to further deterioration or to restoration after a departure from normal operation has occurred. At each branch point there is a probability of the system either continuing on its path to deterioration or to restoration to normal operation. At each branch, the path to success has probability p_i. If the flaw is detected, communicated to the experts, recognized as what it means and if the requered repair action is taken, the system is restored to normal operation, otherwise deterioration continues and may reach the failure state.

Example: If all branches in Figure 5-7 have a probability of 0.5, what are the probabilities of the outcomes? Solution: The probabilities of the outcomes in the lower right portion of Figure 5-7 are as follows:

$$NOP = p_0\, p_1 \qquad\qquad\qquad = 0.500\, p_0$$
$$D = p_0\,(1 - p_1)\, p_2\, p_3\, p_4\, p_5 \qquad = 0.031\, p_0$$
$$F_1 = p_0\,(1 - p_1)\,(1 - p_2) \qquad = 0.250\, p_0$$
$$F_2 = p_0\,(1 - p_1)\, p_2\,(1 - p_3) \qquad = 0.125\, p_0$$
$$F_3 = p_0\,(1 - p_1)\, p_2 p_3\,(1 - p_4) \qquad = 0.063\, p_0$$
$$F_4 = p_0\,(1 - p_1)\, p_2\, p_3\, p_4\,(1 - p_5) \qquad = 0.031\, p_0$$

_____

Sum of probabilities $= 1.000\, p_0$

The events at each branch point are mutually exclusive and exhaustive, if the decision tree shows all possible branches. In this case, the probabilities at each branch point add to 1: $p_i + (1 - p_i) = 1$. The final outcomes, NOP through F_4, are also mutually exclusive and exhaustive so that their probabilities add to 1. The total failure probability is $\sum F_i = 0.469\, p_0$. This result is the joint probability that the initial event E occurs and that one or the other of the alternative pathways to F_1 or F_2 or F_3 or F_4 is followed. If E has occurred, then $p_0 = 1$. The probability of failure can be stated as $P(F) = P[E \cap (F_1 \cup F_2 \cup F_3 \cup F_4)]$.

5.7.4 Examples of Successful and Failed Maintenance

Usually only personnel closely associated with operation and/or maintenance become aware of instances in which maintenance successfully interferes with the deterioration process and prevents it from reaching the failure state. The case histories become more widely known only when failures occur that are spectacular enough to be newsworthy. If the schematics of Figure 5-6 or 5-7 are overlaid on the activities described in such reports, the ultimate failure or its prevention can be associated with one or more wrong branches taken in the event sequence. A relatively inconspicuous *cause event* can produce a sequence of *follow-up events* that turn what would have been a manageable repair job into a full-blown system failure. Some examples follow.

A *maintenance action at Grand Coulee Dam*, on March 14, 1952, almost caused an unscheduled powerplant shutdown and a regional power blackout. On that day, the Grand Coulee powerplant carried 40 % of the Northwest's power load. There are three levels of tube outlets through the dam, at 290 m, 320 m, and 350 m elevations, with ten pairs of outlet tubes at each level. Each tube is equipped with two gates in series near the upstream face of the dam. At the 320 m level, all downstream gates had been checked the day before as closed and all upstream gates as open, except for two where paint experiments were conducted for repainting the tubes. The strategy decided on by the senior operator was to go from one gate chamber to the next, closing all the upstream gates, then come back and open all the downstream gates. The three-man crew started the procedure, but without the plan in hand that showed all the gate positions. In block 55, the space between the gates had been used for a paint experiment and the manhole between the gates had remained unbolted. The senior operator, not being fully informed of the status of the gates, tried to close the upstream gate in block 55 that was already closed. When it didn't work, apparently thinking he had pushed the wrong button, he tried the other button and water began discharging against the closed downstream gate. The water rapidly filled the space between the gates and rose through the manhole shaft. Even when the water began spilling on the gate chamber floor, the man was so stunned that he didn't manage to stop the gate and activate the closing mechanism. A jet of cold water shot out of the well against the chamber ceiling, transformed the gate chamber into a water tank and spilled out into galleries and stair wells which became rushing water flumes and cascades. Finally the water found its way into the turbine pits of the power house. The power plant operator made the decision to continue running the turbines with water in the bearings, until the bearings reached critical temperature, before he shut them down. He did this despite advice to shut down the units as soon as water reached the bearings. His concern was to avoid a regional power blackout. This delay provided warning time for alerting power consumers to bring down the power load. Finally, an engineer who was intimately familiar with the dam approached the gate location from the upper gallery at the 350 m level. He met a painter crew who knew of the unlocked manhole cover and directed him to block 55. In anticipation of the need for some tool, he took a wrench with him from one of the painters. It turned out he needed it to get into the control box of the downstream gate. Fortunately, the gate mechanism still worked, the gate opened and released the water through the tube. By that time, of 18 generators (115 MW each), six had been shut down, the rest were under severe overload. One engineer's comment was "The important thing about the flood was the realization that it was possible"(Morgan, 1954).

An 8 MW private hydroplant next to the Corps of Engineers' Dam 2 on the Mississippi River in Minneapolis had been in operation for more than 90 years. On November 9, 1987, at 18:00, the Corps of Engineers' head lock operator at the adjoining navigation dam noticed a drop in the level of the navigation pool. The dam's gates were closed and after the power company had shut down its plant, the pool kept dropping. A large flow was noticed exiting the power plant. An hour after the first sign of water level drop, the 4 m deep, relatively small pool was empty. Three barges, one loaded with grain, lay stranded on the river bed. The next day, an inspection revealed that some piers of the arched substructure of the powerhouse had been lost by foundation washout. In the afternoon of the same day, a portion of the powerhouse next to the left river bank collapsed. Eight loaded barges, six empties, and a motor vessel were blocked upstream in the Port of Minneapolis. A crash program of dike construction was started to wall off the intake bay from the river. The river was flowing at a low rate of 85 m³/s baring most of its bottom, which facilitated quick repair. After a week, the 6 m high earth dam was ready, the pool was refilled and navigation resumed (USACE, 1987). The powerplant was not rebuilt.

The Alaska oil pipeline showed corrosion damage penetrating 10 % to 20 % of the 12 mm thick pipe wall after only 13 years of operation, a fraction of the expected service life of 30 to 40 years. This damage was detected along buried sections of the 1,500 km pipeline. The repair costs were estimated to be on the order of hundreds of millions of dollars. Keeping pipeline downtime at a minimum was important, because it provides 25 % of the U.S. oil supply (Bloomberg, 1990). Interference by maintenance was successful: it led to incipient flaw detection and repair and returned the process to the normal state of operation. It probably saved the Alaska pipeline company enormous embarrassment and most likely enormous costs had a spill occurred.

The Chicago flood of April 13, 1992, was caused by the rupture of an underground transport tunnel system in downtown Chicago. A tunnel about 2 m high crossing under the Chicago Canal had existed as part of an 80 km underground transportation system dating from around 1900. During all this time, there was a possibility of the system being flooded, because a rupture of the tunnel by natural causes or human activity could not be ruled out with certainty. A company driving piles for a river pier in the summer of 1991 had no inkling that they were about to puncture the tunnel because their work map did not show the tunnel, which was either not known to those who made the map or was intentionally not shown to the pile drivers because of its sensitive nature, providing underground access to building basements (McManamy, 1992). The pile drivers rammed piles into the riverbed without knowledge of what might be down there. They may have actually tried to ram a pile right through the tunnel roof but abandoned the effort just before succeeding, assuming it was an unsuitable (rocky) spot. Instead of causing an immediate disaster, the job created a hidden hazard for the tunnel system, because it substantially increased the probability of a later rupture of the shattered tunnel. A cable company that wanted to use the tunnel for stringing cables under the river discovered the damaged area in January 1992. The cable company could not find a city office that would accept responsibility for the tunnel for six weeks. Previous reorganizations had confused competencies. Finally, in March, the city inspected the tunnel and estimated the repair cost at $10,000 (the "ounce of prevention"). This assessment most likely was too low because the true extent of the damage done to the tunnel was not recognized, and neither was the hazard of the damaged area being under the canal with a practically unlimited water supply under a head of more than 10 m on top of it. Evidently there had been no major leakage during the past six months. On April 2, the grave danger of flooding of the tunnel system and the potential for a catastrophe was finally recognized. But still no emergency action was taken. The city administration indulged in seeking bids that matched their $10,000 estimate and began to reexamine their own estimate when the incoming estimates were all in the $55,000 to $70,000 range. The tunnel failed on April 13, 1992. It submerged the basements of many downtown skyscrapers more than 10 m deep. The flooding forced the shutdown of electrical systems located in or near the flooded areas, which caused power outages in some 120 buildings in downtown Chicago for several days, in 11 buildings for more than five days. Among others, the 110-story Sears Tower and the 80-story Amoco Building (now the Aon Center) had to be evacuated. Altogether, some 300 buildings were affected. The activities of the Chicago Board of Trade, the Chicago Board Options Exchange, and the Chicago Mercantile Exchange came to a halt. Financial repercussions were felt as far away as the New York Stock Exchange, where the Chicago shutdown caused the second slowest trading day of the year. The city transportation commissioner and four engineers had to surrender their jobs. The damage was estimated to be on the order of $1 billion. The inner city was declared a state and national disaster area.

A *Folsom Dam tainter gate* failed in July 1995. The right-hand side radial struts that support the gate's steel skin buckled while it was opened under nearly full water pressure. The cause for the failure was not given but probably vibration induced by the flow under the fully loaded gate combined with rust-weakened gate struts ripped the gate out. The discharge of 1,133 m³/s through the fully open gate bay (about 16 m wide by 13 m high) drained $500 \cdot 10^6$ m³ of stored water in about a week, or 40 % of the total Folsom Reservoir content. There was no way to control the flow until the water level had receded to the spillway crest. The cost for the washed out gate, and the stilling basin damage was estimated at about $1 million. There was probably also a cost due to loss of service in connection with the forcibly spilled water. As a follow-up of the unexpected failure the Bureau of Reclamation issued a contract to inspect and repair as necessary the four remaining spillway gates and the stoplog guide frame structure (www.usbr.gov/mp).

5.8 Economy of Maintenance

5.8.1 Benefit-Cost Analysis

Benefit-cost analysis that encompasses the entire project life of water projects with federal government participation can be traced back to the Flood Control Act of June 22, 1936 (Grant and Ireson, 1970, p. 135). Since the benefit and cost streams of projects stretch over years or decades, two major considerations in a lifetime benefit-cost are the *time value of money* and the *uncertainty* of costs and benefits. A component that adds uncertainty to costs and benefits and that has become very apparent since the 1930s are project *externalities* that are often difficult to quantify or are not even recognized at the time of the analysis. In a lifetime benefit-cost analysis *all* benefit and cost streams over the project's life should be considered. Because of the time value of money, the values of benefits and costs that accrue at different times during a project's life are not immediately comparable. They must be converted to an equivalent money value, which is *present worth*. This is the value of all benefits and costs at the time a decision for or against a project is to be made. This conversion to present worth is known as *discounting*. Discounting makes all costs and benefits over the project life commensurate with the money value of the initial investment. This method has come into use in the 1930s with federal water projects. A similar approach with emphasis on costs has become known as *life cycle costing*, or LCC (Dell'Isola and Kirk, 1981).

A major incentive for using benefit-cost analysis was created by the Flood Control Act of 1936, which called on the federal government to improve or participate in improvement of navigable waters, their tributaries, and watersheds if the benefits were in excess of the estimated costs (Grant and Ireson, 1970, p. 135). This requirement was limited to federal water projects. Over the years, benefit-cost analysis has become a standard procedure for all major projects, both public and private.

The *time value of money* manifests itself by the interest that a borrower must pay to the lender for using his money. As interest accrues, the lender obtains a benefit for loaning his money over the payback period; this benefit is not immediate, however, but occurs as a benefit stream over a long time, such as 30 to 50 years. *Inflation* is a devaluation of money over time. For one reason or another, the government increases the amount of money in circulation, thereby decreasing its value. According to the Office of Management and Budget Circular A-94 (OMB, 1992), costs and benefits can be measured in *real* dollars or *nominal* dollars. Real dollars are *constant* dollars that do not include inflation. Nominal dollars represent the future buying power of the dollar and include inflation. Both can be used in benefit-cost analysis. Market interest rates are nominal interest rates. The analysis that uses real dollars must also use a *real discount rate* that excludes the effect of expected inflation. A real discount rate can be approximated by subtracting the expected inflation rate from a nominal market interest rate. The lender who wants to earn money by investing must charge a positive *net* interest, an interest that exceeds the inflation rate by an amount that covers the cost of lending and profits. In 1992, the OMB proposed "a real discount rate of 7 percent. This rate at that time approximated the marginal pretax rate of return on an average investment in the private sector in recent years." In the 1990s this rate fluctuated at the 5 % level. In 2000 it was at 6 %.[2] Project planning involves long-term capital investments not short-term loans. The interest rate used, which is also called the *discount rate*, real or nominal, should not be confused with the Federal Reserve's discount rate. The discount rate that is used to discount future project costs and benefits of multi-year planning periods is the *opportunity cost of money* or the interest rate that can be obtained by alternative use of the capital invested in the project. The three main alternative uses of money are current consumption, capital investment, and financial investment. Each of these alternative uses may have a different interest rate. The

[2]The Federal Reserve Bank *discount rate* is the interest rate charged by Federal Reserve Banks to private banks on short-term loans. It has nothing to do with the discount rate used to calculate present worth, which has more the nature of a long term mortgage rate. The *federal funds rate* is the interest rate that banks charge one another for short-term loans. The federal funds rate is sensitive to changes in the demand for and supply of reserves in the banking system and thus provides an indication of tightness or ease in monetary conditions. On October 13, 2002, the federal funds rate was 1.73 %, the three-month Treasury bill rate was 1.58 %, and the discount rate was 1.25 %. The federal funds rate usually envelops the Federal Reserve's discount rate. The interest rates used in assessing the economic viability of long term investments dealt with in this chapter resemble more the rates of seasoned corporate bonds (Moody's AAA) at 6.25 %, and 30-year home mortgage rates, which also are at about this level.
www.kc.frb.org/fed101html/Monetary/basics.htm#top;
www.newyorkfed.org/maghome/dirchrts/1-page18.pdf.

highest rate should be used as the discount rate (Jaffe, 1989, p. 114). An alternative financial investment rate is the seasoned corporate bonds rate (Moody's AAA) of about 6.25 %, which is not too far from the OMB's   *real discount rate* of 7 % proposed in 1992.

A summary of compounding and discounting methods that are used for obtaining commensurate present worth values of project net cash flow (benefit minus cost) streams under deterministic and probabilistic scenarios follows.

(1) The *single payment-compound amount factor* (following the terminology of Grant and Ireson, 1970, pp. 34 and 594) is

$$F/P = (1+i)^n \tag{5.8-1}$$

where F is a sum of money that results from compounding a single payment P starting at the beginning of the first interest period over n consecutive periods; i is the interest rate per period, usually a year. If a monthly compounding period is used at an annual interest rate i, Equation (5.8-1) becomes

$$F/P = (1+\frac{i}{12})^{12\,n} \tag{5.8-1a}$$

where n is the number of years and i is the annual interest. If $i = 0.1$ (10 %), $n = 10$ years, annual compounding gives $(F/P)_a = 2.594$, whereas monthly compounding gives $(F/P)_m = 2.707$. The total gain by monthly compounding over annual compounding over 10 years is $[(F/P)_m - (F/P)_a]/(F/P)_a = 0.044$, which is 4.4 % over 10 years, or about 0.43 % per year.

For probabilistic compounding it is necessary to go back to the derivation of formulas such as Equation (5.8-1). If the interest rate changes over the n periods, the closed formula, Equation (5.8-1), is not applicable. If an amount of money P is put into a fund at the beginning of a period of n years, then at the end of the 1. year, the fund is $F = P + i_1 P = P(1 + i_1)$. Supose the interest rate changes from year to year, then at the end of the second year, the fund is $F = P (1 + i_1) + P (1 + i_1) i_2 = P (1 + i_1) (1 + i_2)$. At the end of the third year, the fund is

$$F = P (1 + i_1) (1 + i_2) + [P (1 + i_1) (1 + i_2)] i_3$$

$$= P (1 + i_1) (1 + i_2) (1 + i_3).$$

At the end of n years, one obtains the compound sum

$$F = P \prod_{j=1}^{n} (1 + i_j) \tag{5.8-1b}$$

where the i_j are random interest rates occurring during the n periods. If i = constant, Equation (5.8-1b) reverts to Equation (5.8-1).

(2) The *single payment-present worth factor* is the reciprocal of the single payment compound amount factor

$$P/F = \frac{1}{(1+i)^n} \tag{5.8-1c}$$

where *P/F* is the *discount factor df* for period n and discount rate i. If F is obtained by Equation (5.8-1b) by random interest rates over n years, discounting in a planning study has to be done with the discount rate that is used at planning time.

(3) The *uniform series-compound amount factor* is obtained by compounding the same amount A at the <u>end</u> of a number of successive intervals. But before proceeding with uniform series compounding we deal with the more general case of compounding a variable amount Aj, which could be a random net annual cash flow that is earned by an investment in year j. For simplicity, a constant interest rate is assumed. If A is paid at the end of the first year, then the fund at the end of the first year is $F_1 = A_1$. At the end of the second year, the first payment has accumulated interest $i\,A_1$, which is added to A_1, and a second payment A_2 is made so that the fund is

$$F_2 = A_1 (1 + i) + A_2$$

At the end of the third year, the fund consists of the starting value plus interest plus a new payment,

$$F_3 = + A_1 (1 + i) + A_2 + [A_1 (1 + i) + A_2]\, i + A_3 = A_1 (1 + i)^2 + A_2 (1 + i) + A_3.$$

The pattern of fund accumulation becomes apparent, and the series can be expanded to n years,

$$F_n = A_1 (1 + i)^{n-1} + A_2 (1 + i)^{n-2} + ... + A_{n-1} (1 + i)^1 + A_n . \tag{5.8-2}$$

For uniform compounding, the same amount A is added every year. If also the interest remains the same, Equation (5.8-2) becomes

$$F_n = A\left[(1 + i)^{n-1} + (1 + i)^{n-2} + \ldots + (1 + i)^2 + (1 + i)^1 + 1\right] \quad (5.8\text{-}2a)$$

where F_n is a uniform compound amount series for n periods. The expression in the brackets is a geometric series. For a variable $q = 1 + i$, its sum is

$$S_{n-1} = \frac{1 - q^n}{1 - q} = \frac{(1 + i)^n - 1}{i}.$$

For equal A and i, and $F = F_n$, Equation (5.8-2) can be written as

$$F/A = \frac{(1 + i)^n - 1}{i} \quad\quad\quad (5.8\text{-}2b)$$

where A is an end-of-period investment for n subsequent periods. F/A is the uniform series- compound amount factor.

(4) The *sinking fund factor* is the reciprocal of Equation (5.8-2b),

$$A/F = \frac{i}{(1 + i)^n - 1} \quad\quad\quad (5.8\text{-}2c)$$

where A is an end-of-year annuity or mortgage payment until an amount (initial investment) F has been paid off over n years at interest rate i. The remaining factors can be derived from the preceding factors.

(5) The *capital recovery factor*, $A/P = (A/F)(F/P)$, or

$$A/P = \frac{i(1 + i)^n}{(1 + i)^n - 1}. \quad\quad\quad (5.8\text{-}3)$$

A/P is also known by the abbreviation *crf*. It converges to i for large n.

(6) The *uniform series-present worth factor* is $P/A = (P/F)(F/A)$, or

$$P / A = \frac{(1+i)^n - 1}{i(1+i)^n} .$$
(5.8-3a)

For large n, an approximation is $P/A \approx 1/i$, which is always greater than P/A by Equation (5.8-3a). P/A is the *capitalization factor*, because P/A multiplied by A gives the capital P that is needed to pay the annuity A. A \$1,000 annuity at 6 % interest on the capital, by Equation (5.8-3a) requires $P = 1000 \cdot 15.762 = \$15,762$. The approximation requires slightly more, \$1000/0.06 = \$16,667.

In a real world situation, a net cash flow may consist of random amounts of A_j over the n years of the planning period, such as given by Equation (5.8-2). Discounting such a random cash flow is accomplished by dividing all terms by $(1+i)^n$, which gives its present worth as

$$P = \frac{F_n}{(1+i)^n}$$

$$= A_1 (1+i)^{-1} + A_2 (1+i)^{-2} + ... + A_{n-1} (1+i)^{-(n-1)} + A_n (1+i)^{-n}.$$
(5.8-3b)

Equation (5.8-3b) shows that the first item A_1 is discounted for one period, the second item A_2 for two periods, and so on. No further simplification of Equation (5.8-3b) is possible, unless the A_j are all equal to A. This allows taking A to the left side to create P/A. Multiplication with $F/P = (1+i)^n$ produces F/A according to Equation (5.8-2a), whose closed form solution is given by Equation (5.8-2b). Multiplying both sides of Equation (5.8-2b) by $P/F = (1+i)^{-n}$ gives P/A, the present worth factor, Equation (5.8-3a).

(7) *Internal rate of return.* A criterion of economic project performance is the *rate of return*. Suppose there is an interest rate for which the present worth of all the cash flows of a project sums to zero. This rate, applied to all cash flows like a discount rate, is the rate of return, also called *internal rate of return*. The difference between the annual benefits and costs are the net annual (positive) cash flows. Other cash flows include the initial investment cost, construction costs, repair and rehabilitation costs, restoration and cleanup costs, which are negative cash flows, and salvage values at the end of the planning period or project life, which is a positive cash flow. Since the annual cash flows usually vary from year to year, and all other cash flows have a probabilistic nature, Equation (5.8-2) is a representation of compounded annual net cash flows, and Equation (5.8-3b) represents the discounted compounded net cash flows. The latter equation is augmented to represent all project cash flows. Furthermore, instead of using an externally determined discount rate it uses an *a priori* unknown rate of return, r. The rate of return is determined by trial and error.

The computation starts with a guess and obtained rate is modified until a slightly positive and a slightly negative value close to zero is found, whereupon linear interpolation between the two values is used to find the approximate rate of return (Grant and Ireson, 1970, pp. 111- 112).

The equation for the calculation of the rate of return is based on Equation (5.8-3b):

$$\sum_{j=1}^{n} \frac{R-D}{(1+r)^{j}} + \frac{S-C}{(1+r)^{n}} - I = 0 \qquad (5.8\text{-}4)$$

where R are receipts; D are disbursements, S is the salvage value after n periods, C is a disposal or cleanup cost at the end of the project life or planning period; I is the front end investment; r is the rate of return, n is the number of years of the planning period. The values of all items are those at the time of their occurrence with inflation not included. They are discounted to present worth by r. The idea behind Equation (5.8-4) is that the rate that makes the sum all discounted net cash flows (net annual benefits stream plus final costs) equal to the initial investment cost provides the dividing value between profitable and non-profitable investment returns, regardless of the external discount rate. If the numerators of the two left-hand terms in Equation (5.8-4) are large, then r may be large to match I, and vice versa if they are small, then r has to be small. In the first case, the investment is a good one, in the second it is not. According to the OMB (1992), the rate of return provides useful information, particularly when budgets are constrained or when there is *uncertainty* about the appropriate discount rate. Since practically all items in Equation (5.8-4) are probabilistic quantities, the rate of return should be evaluated as a probabilistic variable with a range and a pdf in order to make an informed project decision.

Real or constant dollars are used in benefit-cost analyses without adjustment for *inflation*. The probabilistic approach can include predicted inflation and interest rate changes, as it was outlined in the previous paragraph. Since the economic mechanisms that produce inflation are poorly understood, it would introduce an additional source of uncertainty as the prediction would have to be based on past records. One of the basic problems with predictions of economic phenomena is process stationarity. In other words, will the process repeat itself based on parameters derived from past records? Each benefit-cost calculation can be evaluated with different stochastic series representing possible financial scenarios. This provides a sensitivity analysis of the performance of long-term investments under random interest and inflation rates. OMB (1992) recommends that for projects "that extend beyond the six-year budget horizon, the inflation assumption can be extended by using the inflation rate for the sixth year of the budget forecast. The Administration's economic forecast is updated twice annually, at the time the budget is published in January or February, and at the time of the Mid-Session Review of the Budget in July. Alternative inflation estimates, based on credible private sector forecasts, may be used for sensitivity analysis."

The Flood Control Act of 1936 required that project benefits exceed cost in order to warrant federal participation. This led to the use of benefit-cost analysis for all federal projects (Grant and Ireson, 1970, p. 135). According to (OMB 1992), "the standard criterion for deciding whether a government program can be justified on economic principles is *net present value*—the discounted monetized value of expected net benefits (i. e., benefits minus costs)."

Some simplified calculations are used to demonstrate the effect of including or excluding benefits and costs from the analysis. If I is the initial investment, B is the total benefit over the planning period to the general public, D_b are disbenefits to the general public, and $O + M$ are O&M costs, the *net benefit* is

$$N_B = B - D_b - (I + O + M). \tag{5.8-5}$$

For simplicity, only the most important benefit and cost items are used here for demonstration purposes and all items are assumed to be in commensurate present worth money values. If a benefit-cost analysis compares two alternatives 1 and 2, and if both provide exactly the same net benefit, $B - D_b$, then the benefits cancel out and the criterion becomes

$$C_D = I_2 + O_2 + M_2 - (I_1 + O_1 + M_1). \tag{5.8-6}$$

where C_D is a cost difference. If the O&M costs of the two alternatives are also the same, then the cost difference is further reduced to the difference in initial costs,

$$C_D = I_2 - I_1. \tag{5.8-6a}$$

This shows that only those costs and benefits that differ from one another in the alternatives make a contribution to the *criterion* that judges the merit of the alternative. Such an analysis based only on costs is a *cost effectiveness analysis*. If costs are the same and only benefits vary, then the difference in benefits becomes the criterion. Equations (5.8-6) and (5.8-6a) also make it clear that neglecting benefits or costs is tantamount to setting them to zero. This simplification may falsify the criterion by which project alternatives are judged. Economic failure, which would otherwise be indicated by the criterion, may go unnoticed, and a project with a nonnegligible failure probability may be implemented.

A criterion that has been traditionally used to compare projects is the benefit-cost ratio

$$B_C = \frac{B}{I + O + M + D_b} \qquad\qquad (5.8\text{-}7)$$

where B_C is a dimensionless benefit-cost ratio, B is the total benefit and $I + O + M + D_b$ is the total cost. The *net benefit-cost ratio* compares net benefits with costs,

$$NB_C = \frac{B - D_b}{I + O + M} \qquad\qquad (5.8\text{-}8)$$

The B_C and the NB_C are dimensionless numbers. They are indices of cost effectiveness, because they give the benefit dollars per dollar of expenditure. Both ratios must exceed 1 for the project to be economically meaningful. They can be used to compare projects or alternatives of different sizes and with very different sizes of benefits and costs. It should be clearly specified what is meant with net benefits: the discounted net social benefits, which remain after social disbenefits have been subtracted. The OMB Circular A-94 of 1992 does not specify the benefit-cost ratio as a decision criterion.

A problem with project benefits is that they are sometimes ambiguous (Grant and Ireson, 1970, p. 141). What may be a benefit to some, may be a cost or a loss to others. The criterion for a benefit of a public project is its contribution to public welfare. If it enhances public welfare, it is a socially desirable benefit. If it does not, it is subtracted from the benefits as a disbenefit. A benefit can be tangible, if it is a marketable product, or intangible if it is non-marketable, accessible to all, or a contribution to general quality of life. The same holds for costs. Both types of benefits and costs should be evaluated. There may be public benefits that result in disbenefits to special interests. These disbenefits are not deductible from the public benefits. For example, if a regulated reservoir water level eliminates malaria along the river, then the health benefits to the public are project benefits, whereas the losses to doctors, hospitals, and undertakers due to the reduction in morbidity are suppressed.

There are *direct and indirect benefits* and costs. An investment that increases reliability may increase structural reliability and service reliability. Increased structural reliability may reduce the risk of accidents to operators and to the public at large, it may increase service reliability that translates into operation cost savings by a reduction of forced shutdowns, unscheduled loss of service, repair costs, and unreliability costs to the public in the form of expenditures for private back-up systems. The problem of quantifying benefits, difficult as it may be, cannot be ignored just by omitting benefits from project analysis. Methods of including subjective measures in benefit-cost analysis are addressed in Chapter 6.

A major difficulty of using benefit-cost analysis besides recognizing and quantifying all benefits and costs is the lack of knowledge of what will happen in future, as the end of the planning period may be 50 years away. These difficulties have been recognized from the beginning of benefit-cost analysis use. The OMB Circular A-94 of 1992 (OMB, 1992, Sec. 9) states on the subject of treatment of *uncertainty*: "Estimates of benefits and costs are typically uncertain because of imprecision in both underlying data and modeling assumptions. Because such uncertainty is basic to many analyses, its effects should be analyzed and reported. Useful information ... would include the key sources of uncertainty; expected value estimates of outcomes; the sensitivity of results to important sources of uncertainty; and where possible, the probability distributions of benefits, costs, and net benefits... It should be recognized that many phenomena that are treated as deterministic or certain are, in fact, uncertain. In analyzing uncertain data, objective estimates of probabilities should be used whenever possible. Market data, such as private insurance payments or interest rate differentials, may be useful in identifying and estimating relevant risks. Stochastic simulation methods can be useful for analyzing such phenomena and developing insights into the relevant probability distributions.... Any limitations of the analysis because of uncertainty or biases surrounding data or assumptions should be discussed."

The probabilistic approach is a sort of extensive sensitivity analysis. Parameters that have uncertainty associated with them are represented as variables with ranges and frequency distributions from which selections are made. Numerous benefit-cost calculations are executed for selected parameter combinations, and the outcomes are statistically analyzed. Typical parameters represented as probabilistic variables are maintenance frequency, maintenance cost, external input to the system (water flow rates for a hydroplant), system use frequency and duration, and others (Dell'Isola and Kirk, 1981, p. 25). OMB (1992, Sec. 9) states: "In general, sensitivity analysis should be considered for estimates of: (i) benefits and costs; (ii) the discount rate; (iii) the general inflation rate; ..."

Examples: (1) Over a 50-year period, an initial capital of $P = \$1$ is compounded by a 6 % annual interest rate to $F = \$18.42$. This is the time value of $1 in 50 years. Discounting a benefit of $F = \$18.42$ accrued over 50 years to present worth at a constant discount rate of 6 % gives $P = \$1$. Interest rates fluctuate over extended periods. From 1950 to 2001, the short-term lending rate rose from about 1 % to approximately 16 % around 1981, then dropped back to about 2 % by the end of 2001. The long-term contractual rates, like bond and mortgage rates do not fluctuate to this extent. The 30-year mortgage rate was 6.45 % early in November 2001, the lowest rate since 1971; by the third week of December 2001, it was 7.17 %, where it had been the year before. If a high rate contract matures, a lower rate contract may take its place, and vice versa.

(2) A comparison is given of the life-cycle costs of a baseline project *A* and its three alternatives *B*, *C*, and *D*. The comparison includes construction cost, indirect costs, operation cost, facility use, and gross income. The baseline and alternatives produce a range of returns on equity investment. Equity is the amount of total project cost not covered by the mortgage loan. The mortgage loan is calculated as 75 % of capitalized net income. The discount rate is 10 %. All numbers are assumed to represent present costs (after an example by Dell'Isola and Kirk, 1983, p. 86).

Table 5-4: Example of a Life Cycle Cost Sensitivity Analysis

		Alternatives			
No.	Items (in $1,000)	*A*	*B*	*C*	*D*
1	Total construction cost	35,000	*38,500*	*31,500*	35,000
2	Indirect costs	9,000	*9,900*	*8,100*	9,000
3	Land cost	4,500	4,500	4,500	4,500
4	Total project cost	48,500	52,900	44,100	48,500
5	Mortgage loan	41,007	41,007	38,816	34,715
6	Equity investment required	7,493	11,893	5,284	13,785
7	Gross income	8,900	8,900	8,900	*8,010*
8	Operation costs	3,100	3,100	*3,410*	3,100
9	Net income	5,800	5,800	5,490	4,910
10	Mortgage payment (debt service)	4,350	4,350	4,118	3,682
11	Before tax stabilized cash flow	1,450	1,450	1,372	1,228
12	Return on equity investment %	19.4	12.2	26.0	8.9

Notes: Numbers in *italics* are fixed percentages higher or lower than the corresponding amount for A.

row 1: the total construction cost of alternative *B* is 10 % higher than *A*; alternative *C* construction cost is 10 % lower than A.

row 2: alternative B indirect cost is 10 % higher than *A*; alternative C indirect cost is 10 % lower than *A*.

row 3: the land cost is the same for all alternatives.

row 4: total project cost is sum of total construction cost, indirect cost, and land cost.
row 5: mortgage loan is 75 % of capitalized net income (in row 9) over 30 years at 10 %: $P/A = 9.427$; row 5 = row 9 multiplied by $(9.427 \cdot 0.75)$.
row 6: equity is total project cost minus mortgage loan.
row 7: alternative D gross income is 10 % lower than A.
row 8: alternative C operation cost is 10 % higher than A.
row 9: net income is gross income minus operation cost.
row 10: 75 % of net income is assumed to go to mortgage debt service.
row 11: before tax stabilized cash flow is 25 % of net income.
row 12: return on equity is cash flow divided by equity, in percent.

The type and number of alternatives in the example are selected by modifying cost and benefit parameters of the baseline. A probabilistic approach would use a similar evaluation of alternatives but more extensive random sampling of the cost items, which would be assumed as random inputs to the analysis to obtain an experimental pdf of the return on equity and other parameters of interest.

(3) It has long been known that the energy consumption of the conventional incandescent lamps is large compared to that of fluorescent lamps, and that 95 % of the energy consumed by incandescent lamps is converted into heat and only 5 % produces light. Fluorescent lamps have a much higher efficiency of turning power into light than incandescent lamps but they had the disadvantage of a disagreeable light color. In recent years, the quality of their light has improved. On the one hand, fluorescent lamps contain small amounts of mercury, lead, and cadmium, and create a potential disposal hazard in landfills, similar to batteries. On the other hand, they save energy and thereby reduce mercury, lead, nitrogen oxide, and sulfur dioxide output into the atmosphere by power plants. Before advocating widespread use of fluorescent lamps, an analysis of all benefits and costs would be necessary. An outline of a cost effectiveness analysis is given in Table 5-5.

A probabilistic aspect of the problem is the lamp life or the number of lamps needed per analysis period; one or the other can be introduced as a random variable. Only the mean life length was used here in a deterministic pilot analysis. If lamp life distributions for both lamp types can be deduced from data, the preceding calculation can be expanded into a probabilistic analysis that finds the pdf of costs and of the cost difference between the two light sources. A statement can then be made about the range of the cost difference and about the exceedance probability of a cost threshold.

Table 5-5: Cost Comparison of Lighting Using Incandescent Lamps and Fluorescent Lamps for the Period of One Year (VDI, 1995/11)

Item	Incandescent Lamps	Fluorescent Lamps	Cost Reduction[1]
Average lamp life, h	800	8,000	
Number of lamps needed	11	1.1	
Unit cost per lamp	$1	$15	
Unit energy cost	$0.063/kWh	$0.063/kWh	
Power of lamp[2], W	100	20	
Initial cost of lamps (investment)	$11	$16.5	+$5.5
Annual energy cost[3]	$50.2	$10	−$40.2
Total present cost	$61.2	$26.5	−$34.7

Notes: 1) The cost reduction uses the incandescent lamp cost as reference: + is a cost increase; − is a cost reduction or benefit by using fluorescent lamps.
2) The power required to achieve the same luminous flux.
3) The energy cost is assumed to be accumulated to the end of the year assuming 8,760 h (hours) of use and discounted to present: $0.063 kW/h · 0.1 kW · 8760 h · 0.909 = $50.2.

(4) A company won a job by a bid of $45 million to build a flood control project consisting of a 34 m high and 222 m long gravity dam flanked by 34 m high earth embankments of altogether about 6 km in length. As excavation proceeded, foundation conditions deteriorated so fast that instead of opening up 760 m segments only 7 m to 15 m segments could be exposed. The embankment foundation had to be excavated deeper than the specified 30 m, the foundation surfaces had to be cleaned with water jets, and the concrete dam foundation required blasting. Also, there were unforeseen problems with the quality of the material for the embankment filter blanket. The unexpected geological conditions and the insufficient specifications caused the construction costs to more than double and completion to be delayed by more than a year. The choice for the owner was to accept the contractor's claim of $57 million through alternative dispute resolution or incur an estimated eight weeks of court hearings and a possible three-year wait for legal resolution compounded by possible interest costs and legal fees (ENR, 1990). In this case, the geologic information was incomplete and made the project cost a random variable with a range that none of the project participants had anticipated. The low bid may have been near

the lower bound of the range with a high exceedance probability of perhaps 95 % (the "if all went well" case); a cost of $80 million may have had exceedance probability of 50 %; and a cost of $100 million may have an exceedance probability of about 10 %. Had the owner approached the problem by a probabilistic benefit-cost analysis he or she might not have been surprised by the outcome, namely that a massive overrun of the accepted bid had a high probability.

5.8.2 Probabilistic Benefit and Cost Streams

In Section 5.8.1 it was shown that compounding and discounting formulas become more complex when periodic benefits and costs are no more constant but vary from period to period in a random way. Also, there may be sufficient uncertainty about the discount rate so that it also may have to be considered a random variable. The resulting variable benefit and cost streams can still be compounded and discounted, although in a less convenient way than by the analytical formulas for constant additions and subtractions and constant interest rates. A cost or benefit stream with arbitrary *end of period additions* is given by Equation (5.8- 2). The end of period additions A_i are compounded for $n - i$ years to the end of the planning period n. For example, A_n accrues at the end of the nth year and is not compounded at all. The compound amount for *beginning of period additions* is obtained by multiplying Equation (5.8-2) with $1 + i$, which gives

$$F = A_1(1+i)^n + A_2(1+i)^{n-1} +...+ A_n(1+i) \ . \tag{5.8-9}$$

.

For the beginning of period compounded sum, the present worth is obtained by dividing Equation (5.8-9) by $(1 + i)^n$,

$$P = A_1 + \frac{A_2}{(1+i)} +...+ \frac{A_n}{(1+i)^{n-1}} \ . \tag{5.8-10}$$

For an end of period compounded sum, the present worth is given by Equation (5.8-3b).

The properties of random variables allow one to make two assumptions on their characteristics that can be used with advantage in a benefit-cost analysis:

1. The A_i for each period i being either benefits or costs, or differences of benefits and costs (net cash flows) consist of many individual contributions and therefore may have a normal distribution so that the A_i can be considered normal variables.

2. The sums of A_i over the planning horizon may exhibit normal variable characteristics for two reasons. First they may be normal according to point 1, and second, sums of random variables with any original distribution tend toward being normally distributed.

The advantage of knowing about a distribution of variables is significant. If numbers A_i that represent benefits and costs are developed and are considered to have equal probability of occurring, then by calculating their mean and standard deviation one immediately knows the entire distribution of benefits and costs and also the distribution of their difference, the net benefit. This allows to draw conclusions on economical viability and risks associated with the project.

Example: A new machine costs \$500,000. It is expected to produce annual cash flows that pay off its cost with a profit over the machine's life. The machine's life is a random variable with a uniform pdf over a range $6 \leq t \leq 15$ years. Each year's cash flow is also a random quantity and may be \$100,000, \$300,000, or \$500,000 with equal probability. (a) Determine the probability that the project breaks even; (b) Determine the probability that the sample mean represents the true mean of the net benefit; (c) Determine the probability that the net benefit exceeds \$1.5 million. (The example is developed after D. Samson, 1988, Example 9-2, pp. 320–322.) Solution: Two random numbers are generated, one for the length of the compounding period, which is also the machine life, and one for the cash flow of each year of machine life. The compound sum of machine benefits discounted to present worth is

$$B_i = \sum_{j=1}^{t} \frac{CF_j}{(1+k)^j}$$

where B_i is the sum of random cash flows for each year of machine life discounted to present worth; $j = 1, ..., t$ is the count of the years of machine life, where t is a random variable that can vary from 6 to 15; CF_j is the annual random cash flow that can assume a value of either \$100,000, \$300,000, or \$500,000 and accrues at the end of the year j; $k = 0.15$ is a constant discount rate. The machine cost is a front end investment with a present worth of $C = \$500,000$. The net benefit is $NB_i = B_i - C$, a random number because of the randomness of B_i. A sample of 100 random NB_i, $i = 1, ..., 100$, was generated by sampling the two random variables. The smallest possible net benefit is a cash flow of six times \$100,000 discounted to present worth

$$B_{min} = 100,000 \frac{1.15^6 - 1}{0.15 \cdot 1.15^6} = 378,448$$

which results in a net benefit $NB = \$378,448 - \$500,000 = - \$121,552$, a loss. The maximum net benefit is the highest possible cash flow compounded over the longest machine life and discounted to present worth minus the machine cost, which gives $3,773,500. A random sample of one hundred NB_i is shown in Table 5-6. Both estimated limiting values are not part of the sample range $\$52,300 \leq NB \leq \$1,630,300$. The highest cash flow has a probability of 1/3 and the longest machine life has a probability of 1/10, the joint probability of both events is $(1/3)^{15} \cdot (1/10) = 7 \cdot 10^{-9}$. The lowest cash flow also has a probability of 1/3 and the shortest machine life also has a probability of 1/10, but the requirement of fewer minimum cash flows in a row gives the lowest possible cash flow a probability of $(1/3)^6 \cdot (1/10) = 1.4 \cdot 10^{-4}$.

The data set of 100 random NB values is tested to see how closely it approaches a normal distribution. The plot of z_F versus z in Figure 5-8 shows that the normalized variable z and the inverse z_F of the empirical $F(x)$ are very close. The regression gives

$$z_F = 0.0309 + 0.991 \, z$$

with a coefficient of determination $R^2 = 0.965$. Based on these numbers it is concluded that the generated net benefits are approximately normally distributed, but major departures occur in the tails of the sample.
(a) From the previous estimate of the limiting values of net benefits and their probabilities it is known that negative NB values exist but they do not appear in this sample. A larger sample is required to obtain an estimate of the probability $P(NB \leq 0)$. Based on the sample on hand all one can say is that this probability is very small.

Table 5-6: Random Sample of Sums of Net Benefits and Test of Their Normal Distribution (the NB-values are taken from D. Samson, 1988, Example 9-2, Table 9-7, pp. 320–322)

No	NB	F(x)	z	z_F					
(1)	(2)	(3)	(4)	(5)	11	648.7	0.11	-1.112	-1.227
1	52.3	0.01	-2.973	-2.327	12	650.5	0.12	-1.106	-1.175
2	145.6	0.02	-2.682	-2.054	13	655.0	0.13	-1.092	-1.126
3	227.9	0.03	-2.425	-1.881	14	687.1	0.14	-0.992	-1.080
4	352.6	0.04	-2.036	-1.751	15	698.4	0.15	-0.957	-1.036
5	414.1	0.05	-1.844	-1.645	16	710.7	0.16	-0.918	-0.994
6	415.0	0.06	-1.841	-1.555	17	726.6	0.17	-0.869	-0.954
7	487.9	0.07	-1.614	-1.476	18	726.7	0.18	-0.868	-0.915
8	491.8	0.08	-1.601	-1.405	19	734.7	0.19	-0.843	-0.878
9	506.4	0.09	-1.556	-1.341	20	761.6	0.20	-0.759	-0.841
10	544.3	0.10	-1.438	-1.282	21	784.5	0.21	-0.688	-0.806
					22	794.4	0.22	-0.657	-0.772

23	816.0	0.23	-0.590	-0.739	63	1134.7	0.63	0.405	0.331
24	823.4	0.24	-0.567	-0.706	64	1136.0	0.64	0.409	0.358
25	825.5	0.25	-0.560	-0.674	65	1155.7	0.65	0.470	0.385
26	827.1	0.26	-0.555	-0.643	66	1159.2	0.66	0.481	0.412
27	829.0	0.27	-0.549	-0.612	67	1175.7	0.67	0.533	0.439
28	837.1	0.28	-0.524	-0.582	68	1189.4	0.68	0.576	0.467
29	846.4	0.29	-0.495	-0.553	69	1191.7	0.69	0.583	0.495
30	848.9	0.30	-0.487	-0.524	70	1192.4	0.70	0.585	0.524
31	855.5	0.31	-0.466	-0.495	71	1193.9	0.71	0.590	0.553
32	875.9	0.32	-0.403	-0.467	72	1199.1	0.72	0.606	0.582
33	876.1	0.33	-0.402	-0.439	73	1204.3	0.73	0.622	0.612
34	883.0	0.34	-0.381	-0.412	74	1216.4	0.74	0.660	0.643
35	902.4	0.35	-0.320	-0.385	75	1222.0	0.75	0.677	0.674
36	913.4	0.36	-0.286	-0.358	76	1240.3	0.76	0.734	0.706
37	915.5	0.37	-0.279	-0.331	77	1251.9	0.77	0.771	0.739
38	916.2	0.38	-0.277	-0.305	78	1269.5	0.78	0.826	0.772
39	916.8	0.39	-0.275	-0.279	79	1270.1	0.79	0.827	0.806
40	932.3	0.40	-0.227	-0.253	80	1273.6	0.80	0.838	0.841
41	932.9	0.41	-0.225	-0.227	81	1291.1	0.81	0.893	0.878
42	945.0	0.42	-0.187	-0.202	82	1301.3	0.82	0.925	0.915
43	947.5	0.43	-0.179	-0.176	83	1303.0	0.83	0.930	0.954
44	949.9	0.44	-0.172	-0.151	84	1316.2	0.84	0.971	0.994
45	966.0	0.45	-0.122	-0.125	85	1328.2	0.85	1.009	1.036
46	981.5	0.46	-0.073	-0.100	86	1357.9	0.86	1.101	1.080
47	996.4	0.47	-0.027	-0.075	87	1367.0	0.87	1.130	1.126
48	1000.9	0.48	-0.013	-0.050	88	1371.6	0.88	1.144	1.175
49	1006.4	0.49	0.005	-0.025	89	1372.3	0.89	1.146	1.227
50	1032.6	0.50	0.086	0.000	90	1374.5	0.90	1.153	1.282
51	1048.7	0.51	0.137	0.025	91	1374.7	0.91	1.154	1.341
52	1049.0	0.52	0.137	0.050	92	1399.6	0.92	1.232	1.405
53	1061.6	0.53	0.177	0.075	93	1402.9	0.93	1.242	1.476
54	1064.1	0.54	0.185	0.100	94	1444.6	0.94	1.372	1.555
55	1065.7	0.55	0.190	0.125	95	1455.3	0.95	1.405	1.645
56	1066.1	0.56	0.191	0.151	96	1494.2	0.96	1.527	1.751
57	1067.5	0.57	0.195	0.176	97	1523.4	0.97	1.618	1.881
58	1075.9	0.58	0.221	0.202	98	1574.1	0.98	1.776	2.054
59	1088.7	0.59	0.261	0.227	99	1625.8	0.99	1.937	2.327
60	1090.8	0.60	0.268	0.253	100	1630.3	0.999	1.952	3.091
61	1094.5	0.61	0.279	0.279					
62	1122.0	0.62	0.365	0.305					

Notes: column 1: current number of element.

column 2: random generated elements NB; arranged in ascending order.

column 3: the unit probability 1/100 given to each element cumulated into an empirical $F(x)$, where $x = NB$. The value $F(x) = 1$ is replaced in the tabular calculations by 0.999 to avoid dividing by zero in the calculation of z_F. This substitution is also justified by the knowledge that benefits higher than shown by the sample are possible.

column 4: $(z - \mu)/\sigma$ calculated for elements NB and for $\mu = 1004.95$ and $\sigma = 320.44$ which here are sample parameters.

column 5: z_F obtained by inverting $F(x)$ of column 3 (see Section 2.6.4).

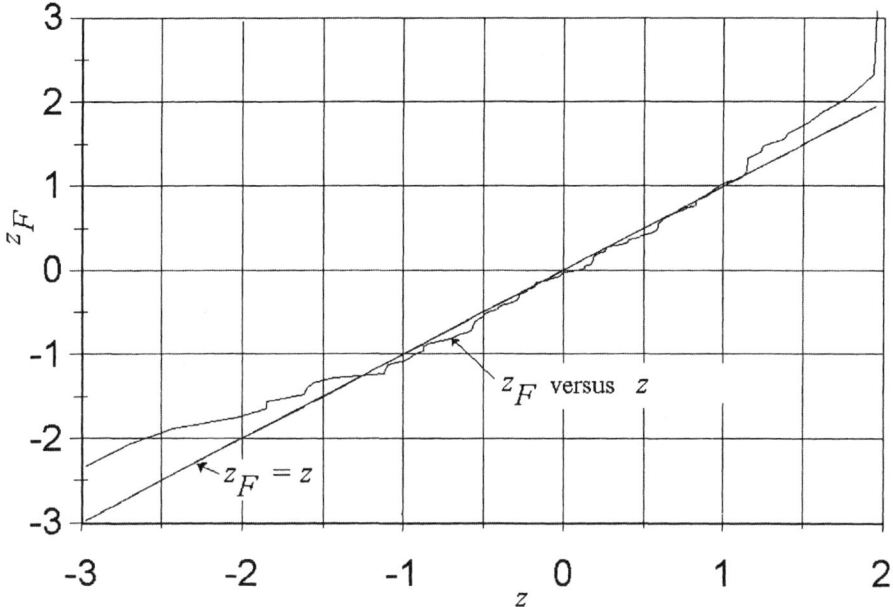

Figure 5-8 : Comparison of a random generated net benefit distribution represented by z_F with the normal distribution represented by z. The diagonal $z_F = z$ would represent the perfect normal fit. The empirical fit is $z_F = 0.991\ z + 0.0309$ with $R^2 = 0.965$.

(b) The mean of the random variables X_i is $\overline{X} = \$1.005$ million with a standard deviation $\sigma = \$0.320$ million. This standard deviation based on the sample X_i is used as an approximate population standard deviation. The standard deviation of the sample mean is then $\sigma_{\overline{X}} = \sigma / \sqrt{n} = 0.320/10 = 0.032$. Hence the mean can be bracketed using Equation (3.4-4) of Section 3.4.1 by

$$P(\overline{X} - z\sigma_{\overline{X}} \le \mu \le \overline{X} + z\sigma_{\overline{X}}) = F(z) - F(-z)$$

The sample mean $\overline{X}$ is based on sums of random variables and is assumed to be normally distributed. The interval that contains μ with a probability of 95 %, or $F(z) - F(-z) = 0.95$ is obtained by using $z = 1.96$ (Table 2-3). Then,

$$P(1.005 - 1.96 \cdot 0.032 \leq \mu \leq 1.005 + 1.96 \cdot 0.032) = 0.95$$

or $P(0.942 \leq \mu \leq 1.068) = 0.95$. The remaining uncertainty of the mean at the 95 % confidence level is $1,068,000 - \$942,000 = \$126,000$. If a higher confidence level is sought, then a larger range must be accepted; if a lower confidence level is acceptable, then the range can be smaller. Selected ranges and probabilities are given in Table 2-3.

(c) The net benefit of \$1.5 million is exceeded with a probability of about 3 %.

5.8.3 Corrective Maintenance Economy

Both corrective and preventive maintenance policies can be policies of choice as well as optimal policies under suitable circumstances. For corrective maintenance to be optimal, little or no incentive must exist for sacrificing remaining service life of a component by *premature replacement* or repair, before breakdown occurs. The longer the service life, the lower the average *lifetime cost*, A. The average annual cost of a component is

$$A = \frac{I}{T} \tag{5.8-11}$$

where A is the average annual cost, \$/year (sometimes "a" is used as unit symbol from Latin *annum*, but this is sometimes confusing); I is the initial cost, \$; and T is the lifetime, or time to failure or replacement, in years. If T is a variable t, then A declines hyperbolically with t, as illustrated in Figure 5-9 by the curve marked "residual cost." The value loss with time is expressed by the gradient

$$\frac{dA}{dt} = -\frac{I}{t^2} \tag{5.8-12}$$

where dA/dt is the decrement of value per time increment; t is the time elapsed from $t = 0$ to the instant where the derivative is taken. For finite time increments, which are small compared to t, the differential quotient can be replaced by a finite difference, $\Delta A/\Delta t$. The minimum of A occurs where t reaches its maximum, which is at the boundary of t, here the end of the lifetime, $t = T$. This is an important difference compared to the time of occurrence of the minimum cost of preventive maintenance, as will be seen in Section 5.8.4.

The annual cost A as expressed by Equation (5.8-11) is also known as *straight line depreciation* cost. If T is a random variable "lifetime" with units of years, then A is a random annual cost. The straight line depreciation can be shown to be a special case of the capital recovery factor

$$A/P = \frac{i(1+i)^n}{(1+i)^n - 1} \qquad (5.8\text{-}13)$$

where A is an annual cost over n periods and P is the present worth of the compounded costs A. For $i \to 0$, $A/P \to 0/0$, which is undefined. By applying the Bernoulli-L'Hospital rule, which requires separately taking derivatives versus i of the numerator and the denominator and testing for $i = 0$ gives

$$\lim_{i \to 0} \frac{i(1+i)^n}{(1+i)^n - 1} = \lim_{i \to 0} \frac{(1+i)^n + in(1+i)^{n-1}}{n(1+i)^{n-1}} = \frac{1}{n}. \qquad (5.8\text{-}14)$$

Hence, Equation (5.8-13) becomes

$$A/P = \frac{1}{n}. \qquad (5.8\text{-}15)$$

Equation (5.8-15) is the same as Equation (5.8-11), if $P = I$, and if one reconciles the unit problem. In the compound interest calculus, n is a pure number, the number of compounding periods, usually years. This gives A and P money units, but A is understand as an annual quantity.

A comparison of capital recovery factors for various interest rates and straight line depreciation is shown in Table 5-7. The capital recovery factors increase with the interest rate, because they have to recover the interest on investment. Equation (5.8-15) is derived for $i = 0$ and therefore has the smallest capital recovery factors, like straight line depreciation, Equation (5.8-11). As i approaches zero, the capital recovery factor approaches straight line depreciation.(compare columns 2 and 6 of Table 5-7).

Table 5-7: Capital Recovery Factor A/P as a Function of Interest Rate Compared to Straight Line Depreciation A/I

	A/P				A/I
n	$i = 1\%$	$i = 5\%$	$i = 10\%$	$i = 15\%$	$1/n$
(1)	(2)	(3)	(4)	(5)	(6)
1	1.010	1.050	1.100	1.150	1.000
2	0.508	0.538	0.576	0.615	0.500
5	0.206	0.231	0.264	0.298	0.200
10	0.106	0.129	0.163	0.199	0.100
20	0.055	0.080	0.117	0.160	0.050
50	0.026	0.055	0.101	0.150	0.020
100	0.016	0.050	0.100	0.150	0.010

Table 5-7 also shows that the straight line depreciation factors and the capital recovery factors for low interest rates continue to decline with increasing n. However for large i, the capital recovery factors tend toward a plateau that is equal to the interest rate. For $i = 0.15$, A/P is almost equal to i after about 20 years. From then on, the annual depreciation rate stalls out at the limiting value $A/P = i$. This plateau is reached the sooner the higher the interest rate. For any greater n, the down payment stays at that rate in perpetuity.

The conclusion for maintenance is: Low interest rates may produce less growth over time for invested capital than high rates. But low discount (interest) rates also produce higher present worth for net cash flows than high interest rates. At low interest rates, debts are paid off at declining capital recovery factors. At high interest rates, capital recovery factors tend to stall out after about twenty years at a level that is equal to the interest rate. For these reasons, long term investments are more attractive at low discount rates.

Examples: (1) An accountant wants to minimize system cost so that the product cost can be kept at low. He wants to know the probability that a critical component will last at least 10 years so that he can use this period as the payback period. What is the probability that the component will survive 10 years, given the records indicate that such components have an exponential survival time pdf with a failure rate $\lambda = 0.1/a$? Solution: Based on Section 2.7.5, the reliability or probability that the component will survive a period t_f is

$$R(t_f) = P(T > t_f) = 1 - F(t_f) = e^{-\lambda t} \ .$$

For $t_f = 10$ a, $R(10) = e^{-0.1 \cdot 10} = 0.368$. Given this relatively low reliability of about 37 %, the accountant wants to shorten the payback period, so that $R(t_f)$ is at least doubled. Setting $e^{-\lambda t_f} > 0.75$ gives $t_f < -\ln(0.75)/\lambda = 2.9$ a. For a more acceptable reliability of about 75 %, the write-off period must be reduced to about three years.

(2) Compare the uniform series compound amounts and the uniform series discounted compound amounts for annual cash flows over 20 years at 5 % and 10 % interest. Solution: The results are summarized in tabular form as follows.

Interest	Uniform Series		Single Payment
	Compound Amount F/A	Present Worth P/A	Present Worth P/F
5 %	33.066	12.462	0.3769
10 %	57.275	8.514	0.1486

In Section 8.5.1 it was shown that $P/A = (P/F)(F/A)$. This means that the uniform series compound amount is discounted by the same discount factor as the single payment compound amount, $(P/F) = (1 + i)^{-n}$. For 5 %, $P/A = 0.3769 \cdot 33.066 = 12.462$. The factors F/A and P/A show that at 10 %, the uniform series compound amount reaches a higher compounded value but is discounted to a lower present worth than at 5 %. In other words, the cash flow from an investment over a 20-year period at 5 % is \$12.5 for each dollar, whereas at 10 % it is only \$8.5. This reduced present worth of a multi-year cash flow at high interest rates has a negative effect on the outcome of a benefit-cost analysis.

5.8.4 Preventive Maintenance Economy

A policy of preventive maintenance can be justified if it is either mandatory or economical. It is mandatory if breakdown is unacceptable. It is economical if benefits result from repair before serious damage has occurred or replacement before the failure of the component. Preventive maintenance causes a cost in the form of the benefit forgone from utilizing the remaining life of equipment that is prematurely repaired or replaced. Conversely, preventive maintenance can eliminate the cost of declining efficiency and possible failure. The total cost of preventive maintenance, C_T, is the sum of residual value forgone, C_R, plus the cost of declining efficiency plus the cost of failure, C_F,

$$C_T = C_R + C_F. \tag{5.8-16}$$

C_R is a cost that decreases with time as the residual work life of the component runs out. C_F increases with time as inefficiency and other prefailure costs increase. Thus, the two costs have gradients with opposite signs, dC_R/dt has a negative sign and dC_F/dt has a positive sign. There may be a point in time at which the gradients though opposite in sign are equal in numbers. This is the time instant of interest, as at this point the total cost gradient is zero:

$$\frac{dC_T}{dt} = \frac{dC_R}{dt} + \frac{dC_F}{dt} = 0. \tag{5.8-17}$$

Equation (5.8-17) is the necessary condition for an extremum of total cost, either a minimum or maximum. Here without further check, a cost minimum is assumed. Equation (5.8-17) can also be written as

$$-\frac{dC_R}{dt} = \frac{dC_F}{dt}. \tag{5.8-18}$$

This balance of incremental costs states that the total cost is a minimum when the incremental decrease of the cost of the equipment in place, $-dC_R/dt$, is equaled by the incremental increase of the cost of inefficiency and possible failure, dC_F/dt. The two cost curves with positive and negative gradients with time, and the resulting total cost are shown in Figure 5-9. The total cost curve passes through a minimum at t_m. At any time before t_m, the cost reduction from continued use is greater than the cost increase from inefficiency and possible failure so that total cost decreases. At any time after t_m, the cost increase from inefficiency and possible failure exceeds the cost reduction by continued use of the equipment and the total cost increases. In a probabilistic evaluation of C_R and C_F, many pairs of curves may be obtained with many time instants t_m so that t_m is a random variable.

The optimal time for preventive maintenance to take place is t_m. If maintenance is scheduled before that time, increased costs are incurred due to *over-maintenance*. If maintenance is delayed after that time, losses by inefficiency and the cost of exposure to *corrective or breakdown maintenance* in the form of insurance or self insurance costs exceed the savings from making use of the residual life of equipment.

The fact that t_m has a probabilistic nature may entice risk-prone decision makers to *gamble* with whether they should perform maintenance now or wait for a later time. The range of t_m is caused by the lack of knowledge and information that goes into the construction of the cost curves C_F and C_R. If the incremental costs of incipient failure are underestimated, then C_F will rise more slowly. The resulting change in incremental cost moves the optimal maintenance time to the right. In other words, a risk-free or too low assessment of C_F may lead the decision maker to delay maintenance beyond what would otherwise be the optimal time. The gamble of delaying maintenance may pay off if no breakdown occurs, but it may also result in substantial loss if a breakdown occurs. This shows that including all possible costs of failure is a conservative approach to finding the optimal maintenance time.

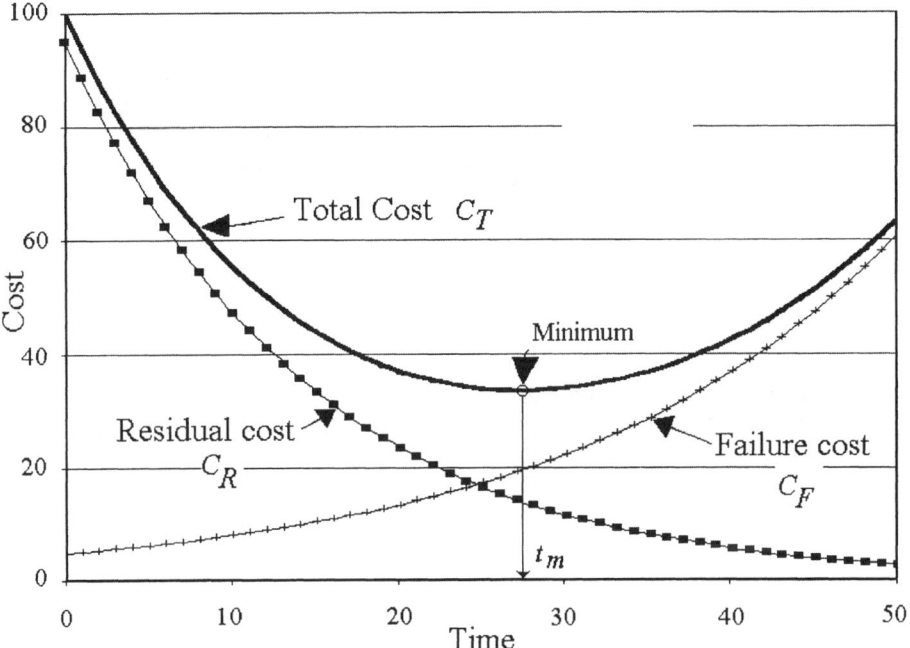

Figure 5-9: Maintenance policies and their economic justifications: *If it ain't broke, don't fix it* is demonstrated by the residual cost curve C_R that exhibits diminishing residual cost or remaining value with time. *An ounce of prevention is worth a pound of cure* is illustrated by the total cost curve C_T that pinpoints the most favorable time for maintenance when cost is a minimum. The total cost is the sum of residual value forgone C_R and the cost of inefficiency and possible failure C_F.

The criteria for whether corrective maintenance or preventive maintenance is indicated are as follows:

- If breakdown or failure is unacceptable, preventive maintenance *must* be used.
- If there is no cost associated with failure or outage, or if these costs are negligibly small compared to repair or replacement costs, then repair should occur after failure has taken place because this policy maximizes equipment use and minimizes the investment cost prorated over the lifetime.
- If the cost decline by continued use is outstripped by an increasing cost of inefficiency and possible failure, the optimal repair or replacement time is found as the time at which the minimum of the total cost occurs.

From the preceding it follows that corrective maintenance time is just a special case of a cost-oriented maintenance approach in which the costs of rising inefficiency and failure are ignored or do not exist. The *general principle of minimum-cost maintenance* is to apply maintenance when the incremental benefit (or declining cost) of continued use is balanced by the incremental cost of inefficiency and failure.

Example: Assume the costs in Figure 5-9 can be represented by differentiable functions of time. The decreasing cost of equipment in place is $C_R = Ae^{-at}$, and the increasing cost due to inefficiency and possible breakdown is $C_F = Be^{bt}$, where t is the time from the start of equipment use or from the beginning of a maintenance cycle; A, B, a, and b are empirical coefficients obtained by fitting cost data. Then the total cost is

$$C_T = Ae^{-at} + Be^{bt}.$$ (5.8-19)

The necessary extremum condition, Equation (5.8-17), applied to Equation (5.8-19) produces

$$-aAe^{-at} + bBe^{bt} = 0.$$

Solving for t produces the optimal time of maintenance

$$t_m = \frac{1}{a+b} \ln(\frac{aA}{bB}),$$ (5.8-20)

where t_m is the time at which C_T is a minimum. For the curves in Figure 5-9: $A = 95$, $B = 5$, both in monetary units; $a = 0.07$, a cost decay rate, and $b = 0.05$, a cost growth rate. These numbers produce

$$t_m = \frac{1}{0.07 + 0.05} \ln(\frac{0.07 \cdot 95}{0.05 \cdot 5}) = 27.34 \quad \text{(time units)}.$$

Substituting t_m into C_T yields

$$C_{T\min} = 95e^{-0.07 \cdot 27.34} + 5e^{0.05 \cdot 27.34} = 33.63 \text{ (monetary units)}.$$

The calculation is illustrated in Figure 5-9. In a probabilistic analysis, this calculation is carried out numerous times for random selections of cost curve parameters (a, A) and (b, B), which are obtained by maintenance cost analysis for a range of possible maintenance scenarios. Any functional forms or tables of numerical data can be handled to produce the probabilistic coordinates of the optimal maintenance point, (t_m, C_{Tmin}).

5.8.5 The Economy of Premature Demise

A system that has its life cut short by some unexpected external event or failure ends in *premature demise*. The cost of premature demise is the investment whose benefits are never realized. Some spectacular examples of premature demise are hydroprojects that failed before they ever saw operation, for example, dams that collapsed before the project started generating or failed early in their lives. The costs of projects that are never producing benefits are also called *stranded costs*. This term became known in connection with the many finished or partly finished nuclear plants that never saw operation, or operated only a fraction of their expected lives when they were shut down for repair, modifications, or unacceptably high risk of failure. Many of these plants never reentered the operational state. A dam is put to test by the first filling of the reservoir. Hidden hazards due to errors of design, construction, and operation (lack of experience with a new project) increase the probability of failure, and in some cases failure becomes a reality. Examples in the hydro sector are Malpasset Dam in France, Teton Dam in Idaho, and St. Francis Dam in California.

In context with maintenance, premature demise is the end of the line when maintenance becomes excessive or impossible due to massive damage or material failure. Most maintenance practitioners shy away from the use of untried and unproven technology because of the perceived high probability of cost overruns and failure by premature demise. However, this has not prevented the use of new methods and new materials throughout the ages. Premature demise costs are often a necessary cost of progress as many methods after initial failures proved to be not only more economical but also safer and more effective. Innumerable examples can be cited of innovations that have advanced engineering even if they were not successful from the first moment. In many cases, premature demise was traced to flaw in design, construction and operation. In these instances premature demise served as a learning experience to eliminate flawed construction methods and materials. An example in this category is the hydraulic fill method in embankment dam construction. Embankment slides already in the construction stage indicated a basic flaw of this method of oversaturating and undercompacting fine earth materials. The recognition of these flaws and the hazards they created especially in earthquake prone areas led to the abandonment of this method and the replacement of the structures by layered and compacted embankments. Inflatable rubber dams were introduced as low-cost water barriers and in the beginning failed long before their anticipated replacement time. Material improvements now have considerably improved the life expectancy of such dams (Post and Stussman, 1989; Daus, 2001; Jansen, 1988).

The availability of new techniques and materials offers trade-offs between a high-cost initial one-time investment with low probability of failure and multiple low-cost investments staged over time with a medium to high probability of premature demise. The consideration of the possibility of premature demise thus has been integrated into the planning and investment strategy.

Examples: (1) Single lifetime installation and less expensive but repetitive installations may be competitive. An example are steel weirs (gates) versus rubber weirs for small installations. Rubber weirs are made of sealed rubber-coated nylon fabric tubes with about 12 mm skin thickness. They developed from an originally troublesome novelty into a rather reliable structure for impounding water (Post and Stussman, 1989). Earlier material and operational problems caused failure of rubber weirs long before the end of their expected 20-year life. Rubber weir advantages include reduced initial cost (80 % of the cost of steel gates), less expensive support structures, capability of opening the full cross section during floods, quick and relatively easy installation, and low maintenance costs. These advantages helped rubber weirs overcome their initial problems. Now, a 30-year life is guaranteed. The rubber material has improved resistance against puncturing, vandalism, and environmental impacts, such as solar radiation; abrasion caused by sand, gravel, and rocks going over the weir in the deflated position; it is less sensitive to temperature changes; and the weirs have better operating mechanisms. About 1,100 rubber weirs had been installed by 1989 in Japan alone. Material reliability is still a problem for material manufacturers. Typical applications include replacement of flashboards on fixed gravity weirs of small hydro plants, tidal barriers, and diversion weirs. Rubber weirs can be raised and lowered much faster than flashboards. Automatic deflation of air-inflated weirs within 30 minutes is especially fast, which means that it cannot hold a major reservoir, but more a run-off-the-river type impoundment. Automatic water level adaptation provided "exceptionally fast" payoff for small hydro installations (Takasaki, 1989). Water or air can be used for inflating the weir. Air inflation is advantageous because inflation and deflation are relatively quick, requires smaller pipes, is not sensitive to freezing, and does not depend on water quality (interior fouling). However, there is the ever-present danger of vandalism, especially from firearms. Air pumps must have reserve capacity to keep the dam from collapsing too rapidly in case of a leak. Vibrations of the dam body caused by the overflowing water intermittently clinging to the round shape of the dam and separating from it have been effectively eliminated by a fin running along the dam body downstream from the crest, which throws the nappe clear of the downstream dam face. Rubber dams have also been used in tidal channels where they are exposed to water pressure from both directions, the impounded river and the tide. Maintenance is considered minimal because no rust protection is required. For low head applications, when safety is not a dominant concern, a short life expectancy design may be able to economically compete with a long-lived design, if low initial requirements compensate for the costs of possible multiple renewals.

(2) Two alternatives are considered for a project: (a) A one-time initial construction cost C that is paid off over n years at an interest rate i. The project incurs an annual maintenance cost M. (b) The alternative to (a) is to use emerging technology, but the construction is expected to require replacement after $n/2$ years. The first cost and the replacement cost are both $C/2$, and the required maintenance cost is $3M$ during the first period, and $2M$ during the second. The benefits are the same for both alternatives and remain unconsidered. Compare the total costs of the two alternatives over the project's life of n years. Solution: The calculations are given in deterministic form, but a corresponding probabilistic analysis could proceed according to the formulations of Sections 5.8.1 and 5.8.2 with time to replacement and maintenance costs being random variables.

(a) Single investment: A money balance is written for the end of the planning period that equates the outlay, C, made at the beginning of the period of n years, as if it were an interest bearing investment, plus the annual maintenance cost stream over n years, M, to the uniform series compound amount of annual costs, A_s, that match the compounded C and M:

$$C(1+i)^n + M\frac{(1+i)^n - 1}{i} = A_s\frac{(1+i)^n - 1}{i}. \tag{5.8-21}$$

The cost C is present worth, as it is an outlay at the beginning of the n-year period; the interest on C accrues from the beginning of the first year on; M and A_s are due at the end of each year. Solving Equation (5.8-21) for A_s produces

$$A_s = C\frac{i(1+i)^n}{(1+i)^n - 1} + M \tag{5.8-22}$$

where A_s is the annual capital recovery of investment C plus the annual maintenance cost M.

(b) Multiple investments: The first investment is $C_1 = C/2$. The second investment is $C_2 = C/2$ and is made n_1 years from the present. The annual maintenance costs for period n_1 are $M_1 = 3\ M$, and for the following n_2 years they are $M_2 = 2\ M$. The annualized cost streams are

$$[C_1(1+i)^{n_1} + M_1\frac{(1+i)^{n_1} - 1}{i}](1+i)^{n_2} + C_2(1+i)^{n_2} + M_2\frac{(1+i)^{n_2} - 1}{i}$$

$$= A_m\frac{(1+i)^n - 1}{i}. \tag{5.8-23}$$

In this case, the first investment, C_1, and the associated maintenance cost, M_1, are compounded over the first period, n_1; after that, both costs are compounded as a lump sum over the second period n_2. The second investment, C_2, and the associated maintenance cost, M_2, are compounded over the second period n_2. The costs of both periods are matched at the end of the period $n = n_1 + n_2$ by an average annual cost stream, A_m, over the total length of the planning period n. Solving for A_m leads to

$$A_m = [C_1\frac{i(1+i)^{n_1}}{(1+i)^n - 1} + M_1\frac{(1+i)^{n_1} - 1}{(1+i)^n - 1}](1+i)^{n_2}$$

$$+ C_2\frac{i(1+i)^{n_2}}{(1+i)^n - 1} + M_2\frac{(1+i)^{n_2} - 1}{(1+i)^n - 1}. \tag{5.8-24}$$

The formulation of Equation (5.8-24) expresses the commitment of financial resources to the project for the entire planning period, n. This example should be considered a deterministic *pilot scheme* for a probabilistic analysis. Whereas the calculations use deterministic formulas, the variables and parameters used in the formulas, such as C, C_1, C_2, M, M_1, M_2, n_1, n_2, and possibly i, are candidates for probabilistic variables.

(3) Evaluate the formulas of Example (2) for the following numerical values: $C = 100$, $C_1 = 50$, $C_2 = 50$, $i = 0.08$, $n = 60$, $n_1 = n_2 = 30$, $M = 1$. Solution: (a) Single investment by Equation (5.8-22):

$$A_s = C \cdot 0.0808 + M = 8.08 + 1 = 9.08$$

(b) Multiple investments by Equation (5.8-24):

$$A_m = [C_1 \frac{0.08 \cdot 1.08^{30}}{1.08^{60} - 1} + M_1 \frac{1.08^{30} - 1}{1.08^{60} - 1}] \cdot 1.08^{30} +$$

$$+ C_2 \frac{0.08 \cdot 1.08^{30}}{1.08^{60} - 1} + M_2 \frac{1.08^{30} - 1}{1.08^{60} - 1}$$

$$= 50 \cdot 0.00803 \cdot 10.063 + 3 \cdot 0.0904 \cdot 10.063$$

$$+ 50 \cdot 0.00803 + 2 \cdot 0.0904 = 7.35$$

The difference is $D = A_s - A_m = 9.08 - 7.35 = 1.73$. This means that in this case the single investment is more expensive than the multiple investment.

(4) Assume M, M_1, and M_2 are normally distributed random variables with means equal to the values used in Example (3), and with the following ranges: $0.4 \leq M \leq 1.6$, $1.95 \leq M_1 \leq 4.05$, and $1.1 \leq M_2 \leq 2.9$. Assuming that the ranges correspond to six times their standard deviations (Section 2.6.5), the following standard deviations are obtained: $\sigma = 0.2$, $\sigma_1 = 0.35$, and $\sigma_2 = 0.3$, all in the same units as the maintenance costs. Then, with $E(M) = 1$, $E(M_1) = 3$, and $E(M_2) = 2$, the probabilistic maintenance costs are

$$M = 1 + 0.2 z \tag{5.8-25}$$

$$M_1 = 3 + 0.35 z_1 \tag{5.8-26}$$

$$M_2 = 2 + 0.3 z_2 \tag{5.8-27}$$

where z, z_1, and z_2 are random numbers in the usually assumed practical range of normalized variables, $-3 \leq (z, z_i) \leq +3$. All other variables are assumed single valued variables. Then the cost recovery payments are

(a) $\qquad A_s = C \cdot 0.0808 + M = 9.08 + 0.2 z \tag{5.8-28}$

where A_s is a random variable with mean of 9.08 and an added or subtracted random component $0.2 z$ depending on the sign of z.

(b) $A_m = C_1 \cdot 0.0808 + M_1 \cdot 0.9097 + C_2 \cdot 0.00803 + M_2 \cdot 0.0904$

$= 4.04 + 0.9097 \, (3 + 0.35 \, z_1) + 0.40 \ + 0.0904 \, (2 + 0.3 \, z_2)$

Multiplying out gives

$$A_m = 7.35 + 0.318 \, z_1 + 0.027 \, z_2 \qquad\qquad\qquad (5.8\text{-}29)$$

where A_m is a function of two independent random variables z_1 and z_2.

A further simplification can be made by recognizing that A_m by Equation (5.8-29) is the sum of a constant plus two random variables. The sum of the two normal random variables is also a normal random variable with the mean being the sum of the means and the standard deviation being the root of the sum of variances (see Section 2.4.3). Disregarding the constant for the moment, the two random variables have zero means so that the sum of their means also is zero. The standard deviation of the sum is

$$\sigma = [(0.91\sigma_1)^2 + (0.09\sigma_2)^2]^{0.5} = [(0.91 \cdot 0.35)^2 + (0.09 \cdot 0.3)^2]^{0.5} = 0.32.$$

This gives A_m as function of one random variable,

$$A_m = 7.35 + 0.32 \, z \qquad\qquad\qquad (5.8\text{-}29a)$$

with mean 7.35 and standard deviation 0.32.

The difference between the single investment and the double investment can be obtained as the difference of two random variables. Since A_s and A_m are normal variables, their difference, D, is again a normal variable:

$$D = A_s - A_m = 9.08 - 7.35 + z \, (0.2^2 + 0.32^2)^{0.5}$$

so that D becomes the new random variable

$$D = 1.73 + 0.377 \, z \, . \qquad\qquad\qquad (5.8\text{-}30)$$

This result not only gives the average difference between the alternatives, $\mu = 1.73$, but also its range. Values of special interest are the smallest value and the biggest value of D. Given the practical range of a normal variable is $z = \pm 3$, one can state that D is within the range $D_l = 1.73 - 3 \cdot 0.377 = 0.60$ and $D_u = 1.73 + 3 \cdot 0.377 = 2.86$ with a probability $F(3) - F(-3) = 0.9973$ (Table 2-3). One can also say that the upper bound of the range, D_u , has an exceedance probability of $1 - F(3) = 1 -$

0.9986 = 0.0014 and the lower bound D_l has an exceedance probability of $1 - F(-3)$ = 0.9986. Hence, for all practical purposes, $D > 0$, and $A_m < A_s$. The average difference, $\mu = 1.73$, has a probability of 50 % of either being exceeded or not exceeded and no statement can be made on the probability of the difference becoming negative. Therefore the average by itself is not a good criterion for economic success or failure.

5.9 Maintenance Scheduling Methods

5.9.1 Critical Path Method (CPM)

A major project consists of numerous activities that need to be completed, some in parallel, some in series, in order to complete the entire project. A time plan for a project can be construed as a network with the branches being the activities and the nodes being points in time where activities reach completion and become starting points for the new activities. Sometimes a new activity cannot be started until all or at least some previous activities have been completed. If one activity is completed before others because it needs less time than others, then this activity has a waiting time at the end of the activity or a delay time by which its start can be delayed and still end on time without delaying a subsequent activity. The delay time that precedes or the waiting time that follows an activity is the *slack time*. An activity with slack time is noncritical as a time overrun does not immediately cause an exceedance of total project time.

The *critical path* of a project plan is the path with minimum or no slack time. It is the longest path through the activity network. The sum of activity times along the critical path is the minimum time needed to complete the project. A failure to successfully complete any of the activities in the critical path on time, in the absence of slack time, leads to a time overrun for the entire project. There usually is a trade-off between the time spent to carry out an activity and the cost. If the activity must be completed in a very short time, the cost usually is high. If a longer completion time is acceptable, the cost is lower. Each activity may have a cost versus time function. At least two points of this function can be established, a *normal time* and a *crash time* and their associated costs. A straight line through the two points gives a linear cost–completion time function. The critical path method (CPM) uses these linear cost-time functions to determine which time-cost point should be selected for each activity so that the total project cost is minimized. Such a time plan optimization by activity cost minimization is accomplished by a linear program (Hillier and Lieberman, 1974, p. 237). A sensitivity analysis of the total cost can be carried out by varying the total project time from the time minimum, when all activities are completed by crash time, to the time when all activities are completed by normal time. If the calculations show that one or several activities are carried out at a boundary of their limited cost-time ranges, these constraints can be relaxed by extending the cost-time range until the cost-time points selected by the program for these activities move away from the boundaries of their ranges. Such activities may indicate bottlenecks in the project time plan that may require particular attention, such as a change in resource allocation, to improve the reliability of a desired project completion time.

5.9.2 Project Evaluation and Review Technique (PERT)

An adaptation of network techniques to project control is PERT (Project Evaluation and Review Technique). PERT can also include probabilistic aspects, such as the probability of meeting deadlines and the probability of total project time and cost. A method of doing this is the *PERT Three-Estimate Approach* (Hillier and Lieberman, 1974, pp. 230 and 234–235). For each activity, three time estimates are made, the optimistic, a; the pessimistic, b; and the most likely, m. Another assumption is that the activity times are made up of several smaller activities and their completion times, so that each activity's completion time is a sum of random times. Such sums are likely to be normally distributed, or at least can be approximated by a normal distribution. The parameters of the normal distribution can be estimated from a, b, and m (see Section 2.6.5). First, an estimate of the standard deviation is

$$\sigma = \frac{b-a}{6}.$$ (5.8-31)

In a normal distribution, 99.73 % of all data fall within the range $b - a = 6\,\sigma$ (Table 2-3). The expected activity time t_e is approximated by a weighted average of two thirds of the mean plus one third of the mid-range,

$$t_e = \frac{2}{3}m + \frac{1}{3}\frac{a+b}{2} = \frac{a+4m+b}{6},$$ (5.8-32)

as the normal distribution requires a symmetric pdf over the range. The total project time can be calculated by making three additional assumptions:

- Activity times are statistically independent.
- The critical path time is the longest total expected time.
- The total project time has a normal distribution.

The third assumption is justified by the central limit theorem, which states that whatever the probability distributions of the elements of a sum, the sum tends toward the normal distribution if the number of elements of the sum is sufficiently large and if the individual element contributions to the sum are all small and approximately equal size (Section 3.4.2). If the individual activity times already have an approximately normal distribution, then their sum is normally distributed without any additional conditions (see Section 3.3.2). The mean of the total project time is the sum of all means of activity times, t_e, in the critical path:

$$\mu_T = \sum_{i=1}^{n} t_e . \tag{5.8-33}$$

The standard deviation of the total project time is

$$\sigma_T = \sqrt{\sum_{i=1}^{n} \sigma_i^2} \tag{5.8-34}$$

where the σ_i^2 are the variances of the individual activities in the critical path (see Section 2.4.3). The normalized variable of the total project time t is

$$z_t = \frac{t - \mu_T}{\sigma_T} . \tag{5.8-35}$$

The probability of a project time t being met is

$$P(T \leq t) = F(z_t) \tag{5.8-36}$$

where T is the random variable representing all possible project times, $F(z_t)$ is the non-exceedance probability of z_t, and t is the total project time being tested for its probability. The probability of t being exceeded is

$$P(T > t) = 1 - F(z_t) . \tag{5.8-37}$$

For a calculated t that is normally distributed, and the corresponding z_t, $F(z_t)$ is found by an empirical polynomial (Section 2.6.4) or by table lookup (Table 2-3). Complete tabulations of $F(z_t)$ for $0 \leq z_t \leq 5$ with $0.5 \leq F(z_t) \leq 0.999\ 999\ 713\ 3$ are given by Abramowitz and Stegun (1970, pp. 966–972).

Example: Calculate the total project time t that has an exceedance probability of 5 % if t has been determined to be normally distributed as a consequence of the central limit theorem. Solution: The total project time t that has an exceedance probability of 5 % has the complementary non-exceedance probability of $F(z_t) = 0.95$. The corresponding normalized standard variable is $z_t = 1.645$. Thus, the time that has 5 % exceedance probability follows from Equation (5.8-35) as $t = \mu_T + 1.645\,\sigma_T$, where μ_T and σ_T are mean and standard deviation, respectively, of the normally distributed random variable of total project time T.

****

6. Risk in Maintenance

6.1 Nature of Risk

6.1.1 Definition of Risk

Practically everybody, sooner or later in his or her lifetime, will be confronted with some manifestation of risk. Some people seek it for the thrill of it or for perceived gains, but many try to avoid it or transfer it to professional risk takers or insurers. The possibility of an unexpected major loss may ruin a business or a private life. Many people seek protection from such losses by buying insurance. This demand has spawned an entire industry that offers insurance as a monetary compensation for the loss of life, health, home, car, accidental damage done to others, and so on. In return, the insured pays an insurance premium whether damage actually occurs or not.

There is no single precise definition of the word *risk*. The word *risk* is derived from the Italian *risco* (modern Italian *rischio*). The Lombard merchants and moneylenders of the Middle Ages faced many adversities in carrying on their trades. Ship fleets and wagon convoys were used to protect transports of goods from robbers and pirates. Today's dictionaries give many meanings and synonyms for *risk*. Engineers have not helped to promote a clear and meaningful definition of what they mean when they talk about risk. The following definition from the revised Water Resources Council Principles and Standards is an example: "situations of risk are conventionally defined as those in which the potential outcomes can be described in reasonably well-known probability distributions. For example, if it is known that the river will flood to a specified level on the average of once in twenty years, a situation of risk, rather than uncertainty exists" (U.S. WRC, 1980). This definition completely misses the meaning of risk. Calling a probability of an event a risk is a widespread misuse of the word among engineers. The probability of a river level or other potentially harmful event is an element of risk, but not a complete representation of risk. Uncertainty, a varying degree of lack of knowledge, is a major ingredient of risk. None of the previously mentioned insurances would be sellable were it not for the uncertainty of the harmful event. The probability of floods of rivers in the wilderness may be of interest to some hydrologists but otherwise poses no danger to life or property and their occurrence is not an element of risk. In contrast, a high stage of the Red River at Grand Forks, North Dakota, may represent a potentially harmful event for the lives and properties of many people and thus poses a high risk.

The United Nations Department of Humanitarian Affairs defines risk as "expected losses (of lives, persons injured, property damaged, and economic activity disrupted) due to a particular hazard for a given area and reference period. Based on mathematical calculations, risk is the product of hazard and vulnerability" (UN, 1992). This is another poor definition of risk that does not address the nature of risk and uses other poorly defined terms to further obscure the meaning by multiplying them with each other. A mathematical expectation of loss is not risk, but, as will be seen later in this chapter, it is some measure of risk.

For a definition of risk one should look to the insurance industry, which has been professionally dealing with risk for centuries, According to Greene (1977, p. 2), there is *objective risk* and *subjective risk*. Objective risk refers to the variation that occurs when actual losses differ from expected losses, and subjective risk refers to the mental state of an individual who experiences uncertainty or doubt or worry as to the outcome of a dangerous event. These definitions show that the concept of risk comprises factual and emotional aspects. The concept of "variation from the expected" is an important ingredient of risk. The more variation there is about the expectation, the more risk there is, as the variation quantifies the spread of the individual outcomes that make up an expectation. The following definition of risk is proposed here:

> **Risk is** (possible) **exposure to** (possible) **harm**.

Exposure to harm means that personal values, such as life and/or health, or material values, such as property, are vulnerable to damage or loss if an event occurs that has the potential of causing harm. For risk to exist there must be uncertainty about at least some aspects of *exposure* and *harmful event*. Here the word *possible* means that there is a probability of some degree of exposure to a probable harmful event. The quantification of uncertainty is preferably based on objective measures (observations) but often also involves subjective measures (judgement). The uncertainty about the exposure may refer to extent of the exposure and the uncertainty about the event may refer to timing among others. A fire is a harmful event if there is exposure of personal and/or material values to it. The time when a fire occurs is unknown. For a given period it is therefore possible that a fire occurs or does not occur. Uncertainty also may be associated with the event-exposure relationship that determines the extent of damages. Whenever the word *risk* is used, it should be possible to substitute *exposure to harm* for it. Webster's (1993) definition of the verb *to risk something* is "to expose something to danger," which supports the definition of risk used here. In summary, risk involves the following elements:

1. An event and associated probability.
2. An exposure to harm or loss and associated probability.
3. A relationship between exposure and harm.
4. Objective and subjective measures of the severity of the exposure.

An event that has the potential of causing harm under one set of circumstances may not be harmful at all under other circumstances. Only events that have a damaging effect on people, property, or other societal values are called harmful. For example, floods and earthquakes are natural events. If humans enter into their path, these events become potentially "harmful events" or "disasters." Persons or values that they may damage or destroy are said to be *at risk*.

Event and exposure combinations that cause risk are given in Table 6-1. A *certain* event with a *certain* exposure has no risk associated with it. This case has some practical importance. If a harmful event is imminent and the exposure that is associated with this event will cause certain damage, then there is no risk but only certainty. If either event or exposure is missing, or if both are missing, there also is no risk. The most typical risk situation is that of a *possible event* producing *possible harm*. A good test for risk being present or not is to determine if the event is insurable (see Section 6.2.3). In the certainty case, the insurer would ask for a premium equal to the full impending damage, which would make insurance an absurdity. In a no-risk case, the insurer would see no honest reason to get involved.

Table 6-1: Event-Exposure Combinations With and Without Risk

Exposure	Event		
	Certain	Probable	Does not exist
Certain	Certainty - No Risk	**Risk**	No Risk
Probable	**Risk**	**Risk**	No Risk
Does not exist	No Risk	No Risk	No Risk

The relationship between the harmful event and the exposure to damage may be well defined or it may be vague and poorly understood. For example, a stage-damage curve can be constructed for a flood site based on inundation areas for a given stage and existing properties and values within this area. From a known level on, inundations of farmland, roads, bridges, houses, industrial complexes and so on occurs. A reasonably accurate damage estimate can be associated with each stage producing a deterministic stage-damage curve. Similarly, earthquakes of magnitude M5 on the Richter scale or below are rather insignificant and exposure to them is inconsequential. But from M5.5 on, damage to structures and fatalities may occur. The severity of damage and loss of life depend on the distance of structures from the epicenter, the resistance of the structures to shaking, the characteristics of the quake, the geologic properties of the terrain, and the ground movements. The time of the day has an effect on the vulnerability of the populations because there are different exposures of people in their homes, during night and day, during transit, or at work. In this case, the event-exposure relationship is very complex with many uncertainties.

In maintenance, event-exposure relationships are used to justify investments for upgrading dams and equipment. This includes the upgrading of generating equipment, the repair or renewal of water control equipment, such as gates and valves, the earthquake proofing of dams, the expansion of spillway capacity, the repair or rebuilding of dams with foundation weakness and concrete deterioration, in short the elimination of defects that are considered to have serious consequences, if they should become the root cause of accidents. These interventions are expected to reduce exposure to harm of downstream populations and properties and improve the reliability of systems that serve power supply, flood control, water supply, and other functions.

6.1.2 Types of Risk

There are various types of risk that are distinguished by what they expose and which parts of society they affect (individuals, industry, economy, etc.). A number of examples according to these criteria are given in Table 6-2. The table is not exhaustive and other exposures exist (Greene, 1977, p. 53). The exposure of a person's life and health to harm is also called *personal risk*. An especially difficult problem is the assignment of a monetary value to a human life. This is usually considered impossible, immoral, or simply taboo. But the facts tell a different story. The value of life for the person who owns it is usually very high, if not infinite, but it is finite for almost everybody else. Examples from the insurance industry, the transportation industry, and the justice system, to name only a few sectors that frequently must deal with personal loss, demonstrate that the value of life can range from relatively modest amounts of less than a hundred thousand dollars to hundreds of millions of dollars, usually assessed by others, and not the owner of the life himself. A recent attempt of putting a monetary value on the loss of life is the compensation of victims' relatives of the World Trade

Center destruction. Wrongful death suits against well-endowed defendants known as *deep pockets* have resulted in awards of hundreds of millions of dollars to the victim's survivors.

Table 6-2: Types of Risk

No.	Exposure	Type of Risk	Description
1	Individual	Personal risk	Loss of life and health through exposure to natural events or unintentional technology failures, such as structures collapse, fires, explosions, crashes, etc.
2	Real estate	Property risk or material risk	Loss of property through exposure to natural events or unintentional technology failures
3	Society	Liability	Loss of property and assets through legal penalties imposed for damage caused to workers, the public, the environment and others through negligence, accidents, professional misconduct, etc.
4	Economy	Market risk	Loss of property, investments, and income through change in demand for a product
5	Politics	Political risk	Loss of facilities or markets due to instability; change of government, adverse political and governmental action, such as changes of laws, treaties, taxation, etc.
6	Industry	Production risk and **maintenance risk**	Failure of process, machinery, and facility, health and safety considerations, labor problems, exhaustion of raw materials, unintentional human errors and oversights
7	Professional	Insurance risk	Default of insurance company by fraudulent business conduct, invalid coverage

8	Environment	Environmental risk	Exposure of employees or the public to process externalities causing danger to life and health, damage to cultural values, wildlife, climate, etc.
9	Nature	Pure risk	Natural events due to meteorological and geological processes, such as floods, earthquakes, landslides, volcanic eruptions, tsunamis, etc.
10	Investment	Speculative risk	Investment with uncertain benefits; gains or losses are possible
11	Recreation	Recreational risk	Voluntary exposure to hazardous conditions for personal satisfaction
12	Transport	Transportation risk	Exposure to accidents of land, air, and water modes of transportation
13	Human actions	Deliberate damage risk	Exposure to vandalism, war, terrorism, and others

Notes: column 2: area of exposure; column 3: name of risk; column 4: types of possible exposures of life and property to events.

When risk analysis puts a value on life, this value serves as a weight that directs the decision maker's choice to a risk-averse solution. Nowhere in a risk analysis is money actually exchanged for a life. For example, an analysis that puts no value on possible loss of life may find a thin arch dam more economical than a gravity dam, or a rockfill dam. If an appropriate value is put on possible loss of life, a safer but more expensive gravity dam may become the economically preferred solution over a riskier thin arch dam. The symbolic input of a monetary value for a life into the analysis may result in the implementation of a risk-averse alternative and thus actually may save lives. Therefore, there is no rationale for opposing the use of monetary values for human lives in engineering studies.

Since risk is the exposure to harm, there is no reason why a person should accept risk unless there is compensation for possible harm or loss. Many risks indeed offer rewards in monetary and nonmonetary forms. *Pure risk*, however, is a risk without reward. It results from exposure to natural and human-caused events, such as floods, earthquakes, fires, foundation failures, bearing failure, structural and equipment failures, accidents, acts of war, and so on. Those exposed to pure risk may try to reduce it, or transfer it to a professional risk taker through insurance (see Section 6.2).

A freely accepted risk is a *voluntary risk*. If people are forced to accept exposures, such as a polluted air and water environment, they are placed under *involuntary risk*. The federal government through health and safety regulations seeks to reduce involuntary risks imposed on the population by entrepreneurs who attempt to maximize profits by using so-called open-ended processes by which waste is released into the environment without proper cleanup. The boundary between voluntary exposure and involuntary exposure may become blurred when there are no realistic alternatives for avoiding involuntary exposure.

Risky trade-offs are choices a decision maker may have between accepting one risk for another. Popularly this is known as *making a choice between a rock and a hard place*. Many people, especially workers, have no realistic choice when it comes to exposure to injury in the workplace on the one hand and the possibility of finding a safer work place elsewhere on the other hand. An example of a choice between risky alternatives is to either preventively shut down equipment that is thought to be on the verge of breakdown, incur all the costs that a shutdown entails and possibly find nothing wrong with it versus not shutting it down and continuing operation under the threat of failure that would cause even greater costs.

Risk under uncertainty exists if the probabilities of incidents are vague or unknown. Such risk is harder to deal with than risk that is based on a record of past occurrences. An example is home fires versus dam breaks. There are hundreds of thousands of homes in a region and there are numerous fires every year but there are relatively few dams which break. Thus probabilities can be derived for home fires that provide guidance to the insurer but it is more difficult to provide probabilities for dam breaks, especially for kinds of dams and for organizations whose dams have never failed. It does not change, however, the basic concept of risk as exposure to possible harm. As long as an object has any possibility of failing, thereby creating death and destruction, it is a source of risk.

An example of the difference between statistical probability and risk is the coin toss experiment (Section 1.2). The long-range average probability, $p = 0.5$, provides some guidance for the decision maker of what to expect, but he or she still faces the full consequences of one or the other *real* outcomes, a gain or a loss, and not of some more benign average of the two. The increased risk under uncertainty is a possible bias of the probability toward loss that may exist and that the decision maker would not know if p were not known. For example, if the decision maker knew the probability of losing in the toss were $p = 0.9$ instead of 0.5, he or she may refuse to make a toss. If p is not known, this option does not exist. Therefore, in general, efforts must be made to obtain all information about the probabilities of process outcomes.

Speculative risk results from exposures that may yield a gain or a loss (Greene, 1977, p. 10). This risk usually accompanies gambling and stock market investing, but many engineering ventures expose the decision makers to speculative risk. Sometimes these exposures are extremely serious. If the company promises to finish the job by a fixed date and faces penalties of thousands of dollars for every day this date is exceeded then a serious exposure to economic failure is incurred because there is usually uncertainty about foundations, construction difficulties, weather, labor force, accidents, and so on. Historically, economic risk was often transferred to workers' risk. By saving cost on workplace safety and aiming at timely work completion, workers's lives and health were put at risk and lost, but the project was completed on or before deadline resulting in bonuses for the company and management. Exposure to economic risk by inadequately assessing project costs has caused serious cost overruns, project shutdowns, and bankruptcies. Another exposure to speculative risk is the decision to install less costly, but possibly more maintenance-intensive or failure-prone structures and equipment in order to save on first investment and hope it will function satisfactorily. The calculation of the probability of economic success is a typical application of probabilistic methods. By assessing the range of possible outcomes, failures and successes, and associated probabilities, for several possible alternatives, the probabilistic benefit-cost analysis (Chapter 5) can identify acceptable and unacceptable economic exposures.

Unavoidable risk arises from events beyond human control, or at least beyond the project owner's or operator's control. If an originally rural area downstream from a dam is turned into an urban development, the project becomes a high-hazard project because of the increased life and property values that are exposed to possible harm in the path of a potential project failure. *Avoidable risk* is caused by an unnecessary exposure that is accepted for some reason, usually for some perceived tangible or intangible gain. The commonsense reaction to risk is to avoid it. If this is not possible, then compensation should be provided by those who impose risk. Whenever those on whom risk is imposed fail to perceive a reward, resistance arises against those who impose risk or against the object that represents risk.

Maintenance risk is associated with industrial risk that is incurred with the operation of a process and its production facilities. The exposure occurs in the form of structural failure; equipment failure or malfunctioning; hidden flaws; lack of or insufficient monitoring and surveillance; defective instrumentation; lack of timely data analysis; failure to recognize the meaning of symptoms; lack of follow-up action to correct identified defects; dangerous work environment; accident potential due to insufficient health and safety measures; operator error; and external events. Maintenance is aimed at reducing these exposures by recognizing and anticipating potential problems before they manifest themselves, and being ready for corrective action when needed. The objective is to keep the process and its facilities protected from exposures to internal and external disruptions. Aside from the mostly involuntary exposures that exist in the maintenance sector, there may be voluntary exposures motivated by perceived profits or cost savings. Decisions based on such motives may include delaying maintenance when it is due in the hope that the system will not fail until some future time when maintenance is thought to be less disruptive or less costly. Like in a coin, the decision maker must be ready to accept the favorable and the unfavorable outcome whatever their probabilities may be.

6.1.3 Hazard and Risk

The word *hazard* is of Arabic origin and is associated with the game of dice. Like risk, it has assumed many meanings. Sometimes it is used synonymously with risk. Here, hazard is defined as a condition or activity that increases risk. Government regulatory agencies use hazard ratings to judge the exposures created by dams. Table 1-1 (Section 1.3.5) gives the *hazard ratings* used by FERC. A hazard can increase risk by increasing the probability of the harmful event, the magnitude of the event itself, and the magnitude of the exposure to harm. The hazard definition used in Table 1-1 is compatible with the definition used here.

In medicine, hazards are called *risk factors*. They are conditions that, if present, increase the likelihood of onset of a serious disease and its consequences. For example, atherosclerosis is a blood vessel disease (arterial lesions), and its complication is coronary heart disease, a major cause of death. The risk factors that increase the likelihood of onset of atherosclerosis are hypertension, elevated serum lipids (from fatty foods), cigarette smoking, diabetes mellitus, obesity, and others (Merck, 1982, p. 387). If a hazard, such as hypertension, is recognized and proper action is taken, the probability of occurrence of the disease and its complications is reduced.

Hazard analysis deals with the identification of conditions in various project phases: planning, design, construction, operation, maintenance, and rehabilitation of systems, that enhance the chance of malfunction or failure. *Hazard control* is aimed at reducing or eliminating hazardous conditions through maintenance, repair and rehabilitation, changes in operation, and by increasing human awareness about the source and existence of hazards. Hence, hazard control is part of risk control.

6.1.4 Attitude Toward Risk

A person's or an organization's attitude toward risk can take three forms: *risk-neutral*, *risk-averse*, and *risk-prone*. The motivations for these attitudes are diverse and shed light on the character of the person, on organizational culture, and on the nature of risk.

(a) A *risk-neutral* attitude does not recognize that an exposure to possible harm exists or is insensitive to it. Such an attitude may be intentional or unintentional. If it is intentional it may express denial that risk exists or that it exists but nothing can be done about it. If it is unintentional, the risk taker may be ignorant of existing risk and may react risk-averse when informed that risk exists.

(b) A *risk-averse* attitude avoids or attempts to reduce, circumvent, or eliminate potentially harmful exposures. Risk averseness may be the result of the following factors:

- Noncompromising rejection of involuntary (imposed) risk.
- The risk is perceived as unacceptably high.
- No reward is recognized that would justify a potentially harmful exposure.
- Reaction to deceptive risk-related practices (lack of credibility of those who impose risk).

(c) A *risk-prone* or *risk-accepting* attitude in return for a reward in the form of a reduced cost or an increased profit. This attitude may be the result of the following factors:

- The perceived reward is commensurate with the perceived exposure.
- The situation offers no alternative but accepting the risk.
- The exposure is unknown to those called on to accept it.

Sometimes people who voluntarily assume risk are strongly averse to imposed risk. These differences in attitude toward voluntary and imposed risk are an indication of the complexity of risk that consists of several components, a quantitative or objective component and a subjective or emotional component. The reaction of the population to imposed risks in the form of dangerous industrial installations became known as the *NIMBY syndrom*e (Not-In-My-Backyard). It was an expression of a severe reaction of the public to incomplete information about nuclear risks, which led to a severe loss of credibility of private industry and government regulatory decision makers, and in the government's effectiveness in and willingness of protecting public health.

6.1.5 Risk Perception

Risk assessment by experts and *risk perception* by people on whom risk is imposed are frequently at odds with each other. The discrepancies are due to judgmental biases. Slovic et al. (1981) give the following reasons for this bias:

- *Recall and frequency of an event*: Laypeople overestimate events that are spectacular and have happened rather recently, such as tornadoes, floods, fires, dam breaks, industrial accidents.

- *Overconfidence*: This bias also affects expert judgment. It may lead to overestimation or underestimation of the extent of exposure and event probability. Failure of structures and systems has resulted from complacency, poor judgment, and lack of timely response. Many human-caused disasters can at least partially be blamed on the overconfidence of experts.

● *Desire for certainty*: The lack of certainty produces risk. The desire for certainty drives the *insurance industry*. People are willing to make periodic payments of agreed upon sums of money for having the burden of possible large losses removed from them.

The exposure to possible harm is difficult to measure in quantitative terms. Risk analysts may claim that their results are based on facts and thus are objective measures of risk with the implication that their results are "real" or "true." The population that is being exposed to possible harm may view the exposure more by commonsense rather than by quantitative measuring methods which are not at their disposal. Thus risk is rarely measured by the same measuring stick on the part of those who impose risk and by those on whom risk is imposed, who are the ultimate risk owners and who have to live with it. Risk assessment professionals may find the risk perceived by laypeople exaggerated and unsupported by facts. The laypeople who are put in harm's way will tend to be risk-sensitive and risk-averse especially to involuntary risk. In order to capture the true nature of risk, risk evaluation must reflect the *risk imposer's* assessment as well as the *risk owner's* perception. Sasser (1996) describes the formula for satisfactory risk assessment by *perception equals reality*. Both perception and reality must be satisfied in a compromise solution that is acceptable to both risk imposers and risk owners.

In many cases, risk can be reduced by controlling, reducing, or eliminating hazards that may increase the occurrence of (human-caused) harmful events and the extent of associated exposures. Complete elimination of risk usually is not possible, but risk perception often becomes more flexible through efforts to diminish risk on the part of the risk imposer. This may help to approach, if not reach, the *perception-reality balance*. In many cases, a lengthy education effort may be necessary to lower or eliminate barriers to risk acceptance. Working toward a compromise solution includes offering rewards to the risk owners. Risk perception also is modifiable by providing information, building credibility, using persuasion, and expressing empathy and concern (Sasser, 1996). It has been found extremely difficult to change risk perception and risk acceptance. These obstacles have forced abandonment of projects and of entire technologies (Slovic et al., 1981). The nuclear industry is an example of an industry almost driven to extinction after insurmountable risk acceptance barriers had arisen by the industry withholding information from those who would ultimately own the risk. Also hydropower's decline began in the 1970s with growing public dissatisfaction over a perceived lack of commitment to environmental quality by the hydropower industry.

6.2 Risk Transfer

6.2.1 The Principle of Insurance

A method of mitigating risk is spreading the possible damage that may be incurred by a few people over a large number of people who are exposed to the same or similar damage, but are not likely to suffer loss all at the same time. An huge industry makes a thriving business by collecting contributions from millions of people who are willing to pay for the promise of compensation when damage should occur to them. The risk-sharing concept is as simple as it is effective. It has its origin in medieval trade. If n ships go out and k ships don't return, then each of the owners of lost ships incurs a loss C per ship. It includes the value of the ship, its freight, and the harm to or loss of its crew. If the k losses are balanced by n contributions, then

$$k\,C = n\,c \tag{6.2-1}$$

where $k\,C$ represents the total loss incurred and $n\,c$ represents the n payments c per ship paid by the owners. The probability of loss is defined as

$$p = \frac{k}{n}. \tag{6.2-2}$$

The contribution each participant has to make to cover the total loss is

$$c = p\,C. \tag{6.2-3}$$

The insurance industry has made a business out of collecting the *insurance premiums,* c, and paying for the *claims*, C. The insurer is a speculative risk taker who takes on the risk of others for profit. He has to deal with two uncertainties: the number of losses k and the premium c.

If the insurer makes correct predictions of the number of losses and good estimates of the premiums, the money he collects will cover and exceed the losses. Some of the surplus must be transferred into an asset fund to build a hedge against future losses. This money is invested in stocks, bonds, money market shares, or other relatively safe interest-bearing instruments. In this way, additional income is produced for meeting claims and for increasing profits. Competition may put a cap on premiums the insurer can charge in order to prevent a shrinking of his client base n. The larger the number of participants, the more predictable becomes the loss probability p and the profits. If the probability of failure is $p = 0.5$ and the insurer has two premium paying clients, the probability that both suffer loss in one contract period is $0.5^2 = 0.25$. If he collects premiums from one hundred clients, the probability that they all suffer loss in the same period is $0.5^{100} = 0.8 \cdot 10^{-30}$. The complementary probability that none of the clients suffers a loss is practically 1. Between these two extremes, the loss probability can be estimated by the formula for the k-out-of-n system (Section 4.1.4),

$$P(p_k = k/n) = B(n,k) p^k (1-p)^{n-k} \tag{6.2-4}$$

where $P(p_k = k/n)$ is the probability of k losses out of n loss opportunities, given the overall system failure probability is p. The premium, c, approaches the insured asset's value C as k approaches n. Hence, for insurance to be affordable, k must be much smaller than n. The expected loss accruing in a clientele of n premium payers can be calculated as the expected loss of the k-out-of-n system (see Section 2.7.2, Equation (2.7-4))

$$E(C_T) = \sum_{k=0}^{k=n} p_k \, k \, C = E(k) \, C = n \, p \, C \tag{6.2-4a}$$

where $k\,C$ is the number of losses that occurs with probability p_k; $E(k)$ is the expectation of k and C is a constant asset value. The expected loss, $E(C_T)$, divided by the number of participants n is the insurance premium, $c = p\,C$, a result that also was obtained by Equation (6.2-3).

The probabilistic cost, c, is sometimes called "risk." It is more appropriately called *risk premium*, or *risk cost*. It is the cost that must be paid to compensate an insurer for taking over a risk. The risk transfer is agreed upon for a limited period, usually a year with c being the annual insurance premium. If C represents numerous exposures, then c represents as many premiums. Insurance premiums are often individually crafted. Inclusions and exclusions are used in a contract to define the scope of the risk transfer. If damage or loss occurs during the contract period, the insurer pays the agreed upon sum which may be a multiple of the insurance premium. If no damage occurs, the insurer keeps the premium as reward for carrying the risk during the contract period. An insurer feels no loyalty for a long term client. Each contract period is contracted separately.

Example: (1) The flood record in Table 3-14 is used as an event set for a flood site. It covers 38 years, from 1911 through 1948, with 120 flood events above a threshold level of 800 m³/s. For the purpose of this example, it is assumed that the events are independent. It is further assumed that above a specified level, $Q > 1{,}200$ m³/s, flood damage occurs. All flows above this level are assumed to cause damage as if they were unrelated events. Determine an annual insurance premium for a given flow-damage relationship. Solution: The 120 events are concentrated in an annually recurring 80-day period, as shown in Figure 3-21. The flow rates range from 800 m³/s to 2,800 m³/s. This range is subdivided into classes with a width of 400 m³/s to construct a histogram. The histogram represents all 120 events of the annual flood period. No further distinction is made when the events occur during the year. The histogram is tabulated in Table 6-3. The number of events, k, in each class, j, is divided by $N = 120$, which produces an experimental frequency $p_j = k_j/N$, in column 3 of Table 6-3.

Table 6-3: Probabilistic Flood Damage and Insurance Premium Calculation for a Project

No. j	Event Class	Events per Class k_j	Class Frequency $p_j = k_j/N$	Real Damage per Event $M	Probabilistic Damage per Event Class $M	Annual Insurance Premium $M/Year
	(1)	(2)	(3)	(4)	(5)	(6)
1	800	—	—	—	—	—
2	1200	68	0.567	0	0.00	0.00
3	1600	27	0.225	1	0.23	0.73
4	2000	18	0.150	5	0.75	2.37
5	2400	5	0.042	25	1.05	3.32
6	2800	2	0.016	50	0.80	2.52
7	Total	120	1.000	—	2.83	8.94

Notes: column 1: upper class limit of flow events in units of m^3/s.
column 2: absolute frequency or number of events in class j; e.g., class 800—1,200 has 68 events.
column 3: relative frequency: column 2 divided by total number of events, $N = 120$.
column 4: incremental damage for event class j, D_j, according to flood level-damage curve; M is a monetary unit which could be a million dollars.
column 5: probabilistic incremental damage for event class j, $p_j D_j$.
column 6: annual probabilistic damage by multiplying column 5 by (120/38) where 120 is the total number of events in 38 years.

It is assumed that flood damage occurs for $Q > 1,200$ m^3/s. The no-damage class has probability $P(Q \leq 1,200) = p_2 = 0.57$. The damages in column 4 are taken from a flow-damage function. Multiplied by the associated event probability they become the probabilistic damages (or insurance premiums), $p_j D_j$, for the class of events per event period. For the annual damage calculation it does not matter when during the year the events occur, as long as one assumes that damage is immediately repaired so that full damage can reoccur within the same year. It is assumed in the ranking of events that each event occurs in any of 120 periods of the 38-year record, and the histogram is based on such an event period. The damage in column 5 is the damage for an event period, with a length of $(38/120) = 0.32$ years. This damage per event period is converted to annual damage in column 6 by multiplying column 5 with $(120/38) = 3.158$. The probabilities of the classes stay the same as they are defined on the sample space of 120 elements. For example, 18 of 120 events are in the flow class 1,600—2,000, which gives this flow class the probability $p_4 = 0.15$. The associated damage for the time slot of 0.32 years is $p_4 D_4 = 0.15 \cdot 5 = 0.75$ M. The *annual* probabilistic damage is then about three times that much, namely, $0.75 \cdot 120/38 = 2.37$ M, which is shown in column 6. The column entries represent premiums for all mutually exclusive events that can occur during a year, and each of them has to be covered by an annual insurance premium, if all people are affected by all floods. The sum of all probabilistic damages in column 6 is \$8.94 M. It is the total annual insurance premium for this flood site. This amount must be collected through individual insurance premiums each tailored to the value of the exposed asset.

(2) At the flood site of Example (1), all those who are exposed to damage for a given flow rate have to come up with their share of the total premium shown in the bottom line of column 6, if they want to be insured. The dwellers low down in the floodplain should contribute more to the total premium, because they are more frequently affected by damaging floods than those higher up. Those at the 1,600 m^3/s flood level face a flood probability of $1 - 0.567 = 0.43$, or 43 %, and participate in all flood damages, whereas those at the highest level face a flood probability of 1.6 % and are affected only by the biggest flood. However, average premiums usually are levied from all insurance participants. It is obvious that those low down have the greatest incentive to buy insurance. If they are uninsured, they pay the damages out of their pockets, seek handouts from charities, or demand public assistance through taxpayer subsidies. In publicized major disasters, such as floods and earthquakes, it is often surprising how small the liabilities of insurance companies are in comparison to total damage, because most victims are uninsured (see Section 6.2.3).

(3) Insurance contracts are renewed or are cancelled at the end of a contract period. A home insurance premium of \$600 per year for a \$125,000 home over 40 years fully invested at 6 % according to Equation (5.8-2) returns about \$93,000, much less than the value of the insured asset. Insurance only works by many participants not making claims and not by the amount of the premiums. For this reason an insurer feels little or no loyalty for a client, especially one that makes claims. He may be glad to see him go or terminate him.

(4) Suppose the probability p of a dambreak is the same each year. Then the probability that the dam will *not* break in n years is $(1 - p)^n$ or approximately $1 - np$, if p is very

small. This means that the probability that the dam breaks during n years is approximately $n\,p$ or n times the one-year probability. Therefore, in first approximation, the insurance premium for an n-year contract is $c = n\,p\,C$.

6.2.2 Types of Insurance

The various types of risk, some of which are shown in Table 6-2, have kindled insurance activities in various areas of exposure. This has led to many types of insurance, some of which are summarized in Table 6-4. The list is not exhaustive, as lines of insurance appear and disappear according to needs and sellability. An important aspect of all insurance contracts is limitations of exposure demanded by the insurer known as exclusions (see also Greene, 1977, p. 105).

Table 6-4: Types of Insurance

No.	Type of Insurance	Exposure Addressed
1	Accident and health or Personal	Addresses two distinctly different exposures, but may be considered one exposure by insurer; also in the form of group accident and health insurance; various limiting clauses to prevent abuse and high payout rate, which in turn would require high premiums; exclusions include willful exposure to harm, damage caused by influence of drugs, recklessness, violation of the law; preexisting and chronic health conditions, etc.
2	Property or General	Many specific lines, such as fire, marine, liability, casualty, etc.
3	Casualty	Unexpected calamity; property title protection; protection of businesses from disloyal employees; protection against bad debts, burglary, bank robbery, industrial accidents (steam boiler explosion, flywheel disintegration), car accidents, elevator accidents, sports accidents, etc.; also liability
4	Life	Livelihood of widows, children, and dependents
5	Liability	Losses by injury and/or damage done to a person or property; also part of casualty insurance

No.	Type of Insurance	Exposure Addressed
6	Fire	Homeowners; expanded into many special lines, such as tornadoes and windstorms, automobile fires, rain, frost damage on crops, tourist's baggage, personal effects, etc.
7	Flood	Water damage
8	Industrial	Workers accidental injury and sickness; alternative to or competitive with workmen's compensation
9	Marine	Ships and their cargo
10	Environmental	Contamination of air, water, and soil; brownfields
11	War	Losses by hostile actions; was offered by government during World War I

Insured losses in the U.S. from 1992 to 2001 are summarized in Table 6-5. They range from slightly less than $3 billion to $29 billion (all in 2001 dollars) and seem to be completely random in terms of damage amount. However, the causes frequently are summer and winter storms. Insurers have long data records and can fit probability density functions to such records to get an idea of what lies ahead in terms of annual expenses. Drawing a histogram of the short record in Table 6-5 would indicate a bi-modal distribution with peaks between $(5 and 10) billion, and $(25 and 30) billion. The column of adjusted dollars shows that the losses of 1992 with Hurricane Andrew and other damaging events exceeded the year 2001 with the World Trade Center loss.

An insurance activity that dates back to antiquity is fire insurance. It may have existed in some form already in Roman times. In modern times, it began as an adjunct to marine insurance. In 1667, the year after the Great Fire of London, the losses suffered by or demonstrated to many merchants allowed fire insurance to emerge as an independent insurance of buildings against fire. By 1720, fire insurance existed in all major cities in Great Britain. In 1752, the first fire insurance was established in Philadelphia as a *mutual organization,* which means that insured clients are also members of the organization who maintain a common fund for each other's assistance. Insurance companies in other cities followed, among them Aetna in Hartford in 1819. Insurance companies were also organized as *stock companies*, which raise capital for insurance by issuing shares. A third form are the *Lloyds* which are organized similar to Lloyds with individual members being liable for losses.

Table 6-5: Insured Losses for U.S. Catastrophes 1992—2001 (The Insurance Information Institute Fact Book 2003, p. 84; I.I.I, 2003)

Year	$Billion at Time of Occurrence	$Billion in 2001 Dollars[1]	Major event
1992	23.0	29.0	August: Hurricane Andrew
1993	5.7	7.0	March: 20-state winter storm
1994	17.0	20.3	January: Northridge earthquake
1995	8.3	9.7	October: Hurricane Opal
1996	7.3	8.3	September: Hurricane Fran
1997[2]	2.6	2.9	
1998	10.0	10.9	September: Hurricane Georges
1999	8.3	8.8	September: Hurricane Floyd
2000[2]	4.3	4.4	
2001	27.8	27.8	September: terrorist attacks

Notes: (1) adjustment amounts to about 2.6 % annual increase. $1 billion = $1,000 million. (2) no extraordinary disasters occurred but there was the usual assortment of plane crashes, fires, explosions, marine disasters, mining and construction accidents, ground transportation accidents, and natural disasters.

A substantial amount of insurance in the United States is provided by foreign companies. Property and casualty insurance sales by foreign companies have grown from $38.9 billion in 1991 to $63.3 billion in 1999. In 2001, the top three property and casualty insurers in the world by revenue were Allianz, Germany, with $85.9 billion, American International Group, USA, with $62.4 billion, and State Farm Insurance, USA, with $46.7 billion (I.I.I., 2003, p. 2). The Munich-based Allianz in 1997 had 68,000 employees, 238 subsidiaries worldwide, $200 billion in assets, $49 billion in annual premiums, and $41 billion in readily available capital (Fireman's Fund, 1997).

At times, there is much capital in search of profits in the insurance sector. Then the insurers must be very competitive in their underwriting rates. A simple but risky approach is to sign up as many clients as possible accepting all kinds of risks on the assumption that the bigger the premium pool, the more likely it is that the various kinds of losses can be covered with a profit. Unless an insurance company is managed in a conservative manner, failure is possible when a series of losses is encountered. The number of insurance companies that have failed over the years are legion. After the Chicago fire of 1871, from 1870 to 1880, forty-six companies failed in New York State, causing a loss of $35 million to policy holders. A similar folding of insurers followed the widespread damage caused by Hurricane Andrew in 1992.

The history of Lloyds of London in the late 1980s and early 1990s provides an example of what can happen to a professional risk taker. During a period of continued business expansion that lasted until 1988, Lloyds had acquired 32,433 members (lenders), some $17 billion capacity, and 376 syndicates. Then a string of disasters over five consecutive years almost brought down the house. By 1996, Lloyds had shrunk to 12,900 members, $15.7 billion capacity, and 167 syndicates. In July 1988, the Piper Alpha oil-drilling platform explosion in the North Sea killed 167 workers (Britannica, 1989, p. 153); in March 1989, there was the Exxon Valdez spill in Alaska; in August 1989, Hurricane Hugo devastated the eastern United States; in October 1989, the San Francisco earthquake struck; in the same month, the Philips Pasadena (Texas) refinery explosion killed 22 people; in August 1992, Hurricane Andrew struck Florida and Louisiana causing over $20 billion damage. From 1988 to 1992, both years included, Lloyds lost $12.5 billion (Lloyds, 1996). In 1993, for the first time in five years, Lloyds made an overall profit of $1.67 billion. By 1997, Lloyds had made a comeback as a competitor in the global insurance market (Britannica, 1998, p. 165). Insurance companies hedge against random losses by creating a relatively stable income stream through investment of income surplus.

The stabilizing role of investment income in the probabilistic income-loss streams of the U.S. property and casualty insurance industry from 1990 to 2001 is shown in Table 6-6. The investments of property and casualty insurers in 2001 were about 66 % in bonds, 21 % in stocks, the rest in real estate and mortgages, preferred stock, cash and short term investments (I.I.I., 2003, p. 22). Column 3 shows the net underwriting gain and loss which only in 1997 came close to being balanced. Investment income made up for underwriting losses in all years except 1992, when they almost matched each other, and in 2001, when underwriting losses exceeded investment income by $12.5 billion. This loss could not be compensated by the relatively small capital gains of $6.6 billion so that a negative after tax income of $7 billion was produced, the first net after tax loss ever. The 2001 losses led to a considerable hardening of the insurance market which means that insurers became risk-averse. It took twenty successive years of underwriting losses for this to happen. Maintaining a comfortable pool of assets is vital for staying afloat in the insurance business. For this reason, U.S. insurance companies maintain a guarantee fund that covers claims against insolvent insurers. Contributions by solvent insurers to this fund are assessed annually. In 2000, insurers were assessed a total $329 million to pay claims against insolvent insurers. Since 1969, this fund has collected $7 billion for the benefit of clients (I.I.I., 2003, p. 31).

Table 6-6: U.S. Property and Casualty Insurance Industry Results for 1990 to 2001 (The Insurance Information Institute Fact Book 2003, p. 17; I.I.I, 2003)

Year	Net Written Premiums	Net Under-writing Gain/Loss	Net Invest-ment Income	Net Realized Capital Gain/Loss	Policy Holder Dividends	Taxes Paid	Net After-Tax Income
			Billion (10^9) Dollars.				
(1)	(2)	(3)	(4)	(5)	(6)	(7)	(8)
1990	218.1	– 18.6	32.9	2.9	2.6	3.3	10.8
1991	223.3	– 17.1	34.2	4.8	2.8	4.4	14.2
1992	227.5	– 33.3	33.7	9.9	2.6	1.5	5.8
1993	241.6	– 15.1	32.6	9.8	2.7	5.1	19.3
1994	250.6	– 19.0	33.7	1.7	3.2	2.4	10.9
1995	259.7	– 14.2	36.8	6.0	3.4	4.9	20.6
1996	268.6	– 13.8	38.0	9.2	2.9	5.6	24.4
1997	276.4	– 1.1	41.5	10.8	4.7	9.5	36.8
1998	281.5	– 12.0	39.9	18.0	4.7	10.6	30.8
1999	286.9	–19.7	38.9	13.0	3.3	5.6	21.9
2000	299.6	– 27.3	40.7	16.2	3.9	5.5	20.6
2001	323.4	– 50.2	37.7	6.6	2.4	–0.2	–7.0

Notes: column 2: total property/casualty premiums written.

column 3: negative numbers are losses; an underwriting loss occurs when insurance premiums are insufficient to cover claims and expenses.

column 4: col. 4 – col. 3 is approximately the pretax income (not shown); it was $0.4 billion in 1992, and –$12.5 billion in 2001.

column 8: col. 8 = col. 3 + col. 4 + col. 5 – col. 6 – col. 7; does not exactly equal the sum in col. 8 due to omission of miscellaneous income.

If the income and outlay streams in Table 6-6 are random variables, or if only the outlays are random realizations, their sums and differences are again random variables.

Sums and differences of random variables have greater variances than the original variables (see Section 2.4.3). If one visualizes the average after-tax income positioned on the *x*-axis by some positive amount to the right of zero, then it is easy to see that the after-tax income pdf may reach into the negative range to the left of zero. The area under the pdf to the left of zero represents the probability of a negative or zero after-tax income as it actually did in 2001. This probability is a measure of the risk for clients that their insurance company may fail. Apprehension over such a trend increased during the 1990s. From 1983 through 1993, the number of major insurance companies liquidated per year ranged from 20/year (1983) to 130/year (1992), with an average of 74/year for the 11-year period (KNS, 1995/12). These failures count companies that could not meet their obligations to redeem claims for catastrophe, accident, health, and life insurance. In an insurance review of February 2001, an insurance official is quoted as saying: "It has never been more important for an A&E (architect and engineer) firm to have a relationship with a strong insurance company—an insurer that will be healthy 3–5 years down the road" (Hirsh, 2001, p. I-17). Insurance companies buy insurance for themselves from reinsurers. Only about 7 % of all premiums written in 2001 were reinsurance premiums (I.I.I., 2003, p. 24).

Another source of risk for both the insured and the insurer is insurance fraud and insurance error. Insurance companies themselves may be involved in fraudulent practices. Two of the most widely known insurance companies, Prudential Insurance Company of America, and John Hancock Mutual Life Insurance, were involved in court proceedings in 1997 and were facing fines in the billions and hundreds of millions of dollars, respectively, for deceptive sales practices (Britannica, 1998, p. 165). Insurance company fraud includes misrepresentation of risk to the customer by its agents, and collecting premiums without intending to pay claims. Other risky business practices include underestimation of risk, miscalculation of premiums, and violation of insurance principles, such as setting the stage for a great number of simultaneous claims. Improper business conduct is sometimes motivated by tough competition. It may lead to inability of the insurer to provide coverage when claims arise. Typically insurance fraud is not discovered until claims are presented to the insurer. Insurer fraud exposes clients to involuntary risk in the form of worthless policies. But also clients indulge in fraudulent practices by misrepresenting or faking damages. Client fraud causes inaccurate risk assessment by the insurer and increased premiums for the clients. In 1997, insurers spent $650 million on client fraud detection and prevention. Nationally, the insurance industry estimates that insurance fraud committed annually amounts to $80 billion (KNS, 1995/2, and KNS 2001/10).

The insurer must also guard himself against overexposure by avoiding two *underwriting errors* (Hammond, 1980, p. 160):

• *Prognosis error* caused by wrong prediction of the probability of loss resulting from an insurance contract, and

• *Diagnosis error* caused by wrong estimation of the extent of loss involved in a failure event.

If there is plenty of capital available for speculative risk taking, the insurers becomes very competitive and tend to issue policies based on poorly researched exposures at low cost, just to obtain a premium for their pool. When disaster strikes, the losses caused by underinsured exposures may inflict severe, if not crippling loss on the insurer and also cause loss of protection for other clients.

6.2.3 Insurable and Uninsurable Risk

An insurer distinguishes between *insurable risk* and *uninsurable risk*. Categories 1 through 3 of Table 6-2 are insurable risks. These are risks that can be pooled for a great number of people, are individual random events, and have a probability that can be reasonably well established by observation. For example, the probability of a residential house fire would be the number of fires per year divided by the total number of houses in the area. Random occurrence means that the events are isolated and not part of a great number of similar events. Categories 4 through 6 are uninsurable risks because of the unpredictability of events or because of the excessive scope of the exposure. If a risk premium cannot be determined or if it is too high then the risk is uninsurable. According to Equation (6.2-3), as $p \to 1$, $c \to C$. If the insurance premium approaches the asset value insurance becomes meaningless. This is why insurance cannot insure preexisting conditions, as it would be against the insurance principle expressed by Equation (6.2-3). According to Table 6-1, if the event is certain and the exposure is certain, there is no risk, hence there can be no insurance.

Uninsurability also characterizes events that produce many simultaneous losses compared to the total number of insurance policies. This kind of exposure is called *risk concentration*. It is produced by floods, earthquakes, and hurricanes. If an entire city can be damaged or destroyed at the same time by the same event, it would be very risky for one insurer to issue a large number of policies, all in the same area, with premiums that are but a small fraction of the total value of each insured risk. When Hurricane Andrew leveled entire neighborhoods, several insurance companies that had violated this important insurance principle declared bankruptcy. To properly insure such a situation would require insurance premiums close to the full value of each client's assets, which, of course, defeats the insurance principle of paying relatively small premiums far below asset values. Instead of paying such a high premium, the client might as well practice self-insurance, in other words, have practically no insurance at all, and, of course, pay for the damage whenever it occurs.

The risk concentration by exposure to floods was made insurable in 1968 by the federal National Flood Insurance Program (NFIP) (Greene, 1977, p. 569). Floodplain dwellers were offered *federally subsidized insurance* (Simmons, 1981, pp. 42–48). Still the people were unwilling to buy insurance at rates of only about one third what private insurance could offer. After a 5-year trial period of voluntary insurance, the insurance was made mandatory. The federal government was temporarily the sole administrative and financial agent of the NFIP. As insurer, it assumed financial responsibility for flood losses and premium payments. In effect, by subsidizing people who were willing to accept flood risk, taxes from general tax revenues were transferred to this special group to cover the damages they might suffer. In 1977, $612 million were expended by the program, $211 million on flood losses, the balance on administering the program, such as flood studies and program maintenance.

Premium subsidies had the effect of encouraging settling in floodplains and other hazardous areas, or at least staying there thus continually increasing exposure for the federal insurer. It was realized that risk control measures by those who requested insurance had to become part of the program before they would be allowed to participate. In 1993, over 60,000 flood insurance claims were paid in the aftermath of the Great Mississippi–Missouri River Flood for a total insured loss of $1 billion. By 1997, approximately 18,000 communities across the United States had obtained certification to participate in the NFIP. The total insured property value exceeded $250 billion. Risk control was recognized as a necessary element for the insurability of concentrated risk. If risk control cannot be obtained, the insurer may find this sector unattractive and cease offering insurance policies.

Holding a policy for an insurable risk is not a fail-safe protection for the client. The policy defines the transfer of the exposure from the client to the insurer by exclusions stated in the policy, the so-called fine print. The clients must meet the conditions stated in the fine print or else they may offer the insurer an opportunity to back off from the policy. If a claim arises, the insurer checks all conditions and exclusions, as any violations of the contract may allow rejection or reduction of the claim.

To cope with occasional very large damages, insurers share risks with other insurers. Insurance taken out by insurers from other insurers is called *reinsurance*. The insurer contracts with the reinsurer for coverage of excess loss at a premium. If the insurer's loss remains below the contract level, the reinsurer earns the premium; otherwise the reinsurer assumes the excess damage. Total disaster damages of the year 1995 were estimated by the Munich Reinsurance at $180 billion. The previous record had been $65 billion in 1994. The Kobe Earthquake of January 1995 (7.2 on the Richter scale) caused estimated damages of $100 billion. Some 6,400 people died, 300,000 people were left homeless, some 200,000 buildings were destroyed, and the city infrastructure was badly damaged. Only 30,000 people had insurance, and the cost to insurance companies was a mere $3 billion. Legal limits on insurance coverage were in effect. The Japanese government acted as self-insurer, paid the damages to private households out of pocket, and assumed the cost of the infrastructure. Risk management and risk control in that earthquake-prone zone seemed to be lacking. Reconstruction was estimated to cost some $120 billion (Britannica, 1996, p. 47). Cost to the insurance industry was $17 billion in 1994, and $14 billion in 1995 (KNS, 1995/12), which amounts to 26 % and less than 8 % of total estimated damages, respectively. The examples show that only a fraction of total damages is insured. A great part of the exposures is either assumed by those exposed or covered by self-insurance of industry and government.

Some insurance trends observed from 2000 to early 2001 are summarized (from Hirsh, 2001). Insurance rates and the willingness of insurers to accept risk changes with the economic climate and with the risk climate. The period 2000–2001 was one of a *hardening insurance market*, which marks a trend when insurers take a harder look at the profitability of risk taking. When there is a lot of capital around seeking profits, insurance premiums tend to be low and underwriters are eager to take on risk, sometimes without proper investigation of the associated exposure. Competition and availability of risk-seeking capital tends to keep premiums down.

Premiums may rise for various reasons, such as an underestimation of the risk; an unforeseen increase in claims; a reduction in the number of risk takers and the amount of capital they are willing to expose at existing rates of return. Firms that depend heavily on insurance, like construction firms, may face higher insurance premiums, added exclusions, and increased deductibles at policy renewal time. Those with poor claims records may face major premium increases or denial of policy renewal. In early 2001, premium changes ranged from no raise for the best clients, perhaps a 10 % raise for normal clients, to several hundred percent for clients with poor records. Firms with risk control in place can keep premium increases down. The response of the insurer to a hardening market depends on both its financial strength and its standing with its reinsurer. The entire string from the insured via the insurer to the reinsurer reverberates when a severe accident occurs. Disasters and fatalities tend to cause reinsurers to withdraw from clients, to increase rates, or to clamp down on underwriters who incur risky exposures. In such situations, insurers who depend on reinsurance may have to comply with more restrictive coverage. Reinsurers who cover catastrophic damages pay more attention to risk concentration, such as earthquake and hurricane exposures. Smaller insurers may be forced to leave the market if they cannot get reinsurance. Insurers may not offer or renew multiyear policies. The annual policy gives them room for policy adjustment, premium increase, or cancellation of specific coverage. Various business sectors may face varying problems with the insurance industry at any given time. The construction industry was booming in 2000, but there was a tendency toward rising severity of claims due to shortness of staff and lack of staff training and experience, which in turn led to breaches of the standard of care. The deterioration of risk control may set in motion a cycle of increases in claims, a loss of returns for the insurance industry, a substantial hardening of the market, and a steep rise in premiums for the construction industry.

The destruction of the World Trade Center in September 2001 has rendered the insurance market "rock-hard" (ENR, 2001/10/22). Many insurers have exclusions in place for "an act of declared or undeclared war," but insurance claims are nevertheless expected to reach double-digit billion dollars. Lloyds of London estimated its liability at $1.9 billion. A Swiss reinsurer and a German reinsurer estimated their involvement at $730 million and $903 million, respectively (KNS, 2001/9/13 and KNS 2001/9/27). An American insurer estimated its loss at $700 million. The latest estimate of claims for the World Trade Center is $50 billion (ENR, 2001/10/22). In comparison, the Northridge Earthquake of January 1994 produced claims of $14.5 (12.5) billion, and Hurricane Andrew of August 1992 produced claims of $19 (15.5) billion (the numbers in parentheses being the dollar values at the time of the disaster). When reinsurers fail they may pull down with them primary insurers. Insurance against terrorism, at first "virtually unobtainable" from private insurers, became available by the end of 2002. Insurance against acts of terror is comparable to flood insurance because of the severe *risk concentration*. A reinsurer with practically unlimited capital (government) is required. After severe losses by major disasters, the insurance industry is vulnerable to repeat events that may strike during the following two- to three-year recovery period (ENR, 2001/10/22).

Earthquake insurance, like flood insurance, is not attractive to private insurers because of risk concentration, the area-covering, simultaneous occurrence of damages. State laws may force insurers to take on these risks, even if this does not make sense from an insurance point of view. In California a state law required that every homeowner who was offered home insurance had to be offered earthquake insurance as well. The 1994 Northridge Earthquake (M = 6.8 Richter, 61 deaths) produced $8.5 billion in damage claims for private insurers. Private insurance wanted to get out of home insurance altogether. The state then established a state earthquake insurance. A debate ensued over rate differentiation according to exposure. Those exposed to active faults should pay more than those in more quiescent areas. The law allowed only minor differences between the San Francisco area and the Los Angeles area (area rates were $4.60/$1000 coverage, and $3.30/$1000 coverage, respectively; Sieh and LeVay, 1998, p. 286).

6.2.4 Habitual Risk and Self-Insurance

Small and frequent exposures and repetitive events create *habitual risk*. The resulting damages are usually assumed by the person under risk (risk owner) and paid *out of pocket* as part of normal operation expenses. Accepting an exposure and its possible consequences is called *risk assumption*. A person or business with sufficient wealth can afford to assume some degree of exposure and deal with the consequences as they may occur. This is called *self-insurance*. A deductible in an insurance contract is self-insurance. It has a upper limit that can be set at what the risk owner can afford should the damage become a reality. In an attempt to minimize operation costs, an individual or an industry may want to reduce insurance costs by insuring only large and crippling damages, which exceed the risk owner's *affordability of risk*. For example, collision damage on a car is limited by the car's value and is often self-insured whereas liability that could mean an enormous loss needs to be transferred to a risk taker (insurer) at the highest affordable level, as it would vastly exceed most car owners' financial means.

Sometimes it is proposed to build up *individual savings accounts* as a method of coping with occasional major loss. This concept is contrary to the idea of insurance and is bound to fail for at least two reasons. First, during the years the savings accumulate, no fund is available to deal with loss, and second, if a failure occurs, no second accumulation is usually possible because of lack of time and lack of capability. Individual savings accounts only can meet relatively small needs, such as temporary loss of income or some minor unexpected expenses. They cannot provide coverage for major exposures, such as loss of home, loss of health, or loss of life, unless the self-insurer can set aside from the start a large fund commensurate with the exposure. Therefore, self-insurance is risk assumption equivalent noninsurance (Greene, 1977, pp. 11-12, and p. 408). An example for the difference of what insurance can do in lieu of a savings account is insurance for premature death. If a young person has made a commitment to a family and dies young, the life income is cut off at a time when the self-insurance fund is small to nonexistent, so that the dependents would be left without support. Only the sharing of similar exposures by many through insurance can prevent economic disaster for the individual.

An insurer accumulates information on his client base, or even nationally through census data, that allows him to estimate premiums on a reasonably well founded loss probability, given the loss process is stationary, which means that past history can be expected to repeat itself. The loss probability is based on the *law of large numbers*, which says that as n (number of trials or exposures) becomes very large, the ratio of the number of successes (or failures) k to the total number of trials, k/n, converges toward the average success (or failure) probability p. An *individual risk*, in contrast, in each insurance period is "loss or no loss." The individual does not know if his loss probability will be $p = 1$ or $p = 0$ in the next period. Only averages over time, for a large group, produce probabilities between these two extreme limits. For example, a home fire insurance may use a probability $p = 0.006$ to calculate a client's annual premium of, say, $600 on a $100,000 home. The individual for whom there is no middle ground, would have to lay aside $100,000 assuming $p = 1$, to be equally well insured. Through insurance, the generally unaffordable risk of catastrophic loss for the average individual is transformed into a deterministic and affordable loss in the form of the insurance premium.

Examples: (1) Suppose the head of a family of four pays $600 per month into a savings account for health self-insurance at the going annual interest rate of 1.5 %. After 10 years he gets sick and incurs doctor's and hospital bills of $100,000. By monthly compounding, his account has accumulated $77,628. He is unable to pay $23,000 and charges it to his credit card at 20 %. He is lucky that he can still put aside $600 a month but he has to pay off the debt as fast as possible so that he can rebuild his self-insurance account. Solving the uniform series present worth factor, Equation (5.8-3a), for n he finds that he needs five years to pay off the debt. After that he can start accumulating his self-insurance fund from scratch.

(2) Suppose a company has 1,000 employees; each pays the same health insurance premium of $600 per month. There is a 5 % probability that one of the participants gets sick each year and incurs a $100,000 bill but is fortunate enough to be able to continue paying his monthly insurance premium of $600. The expected annual cost to the company is according to Equation (6.2-4a)

$$E(C_T) = npC = 1,000 \cdot 0.05 \cdot 100,000 = \$5,000,000.$$

The individual annual insurance premium would have to be

$$c = E(C_T)/n = pC$$

or c = $5,000 per participant, which is covered by the monthly $600 premium. The success of the insurance hinges on the estimates of p and C, which in turn determine the premium c. The mutual insurance of many guarantees the support of those who may need it at any time.

6.2.5 Effect of Insurance on Financial Position

The value of insurance to a business can be estimated by calculating the difference in financial positions with and without insurance at the end of a business (insurance) period after D. Houston (as cited by Greene, 1977, pp. 77–80). The equations developed here are inspired by but not necessarily the same as those in the reference.

(a) Financial position *with insurance*:

$$F_i = (1+r)(NW - P) \tag{6.2-5}$$

where F_i is the financial position or net worth at the end of the year with insurance; r is the internal rate of return of the business; NW is the net worth at the start of the year;

P is the insurance premium paid in full at the start of the period. The end-of- year financial position is assumed to be the result of compounding the difference of net worth reduced by the insurance premium at the internal rate of return of the enterprise.

(b) Financial position *without insurance*:

$$F_{ni} = (1+r)(NW - F) + (1+i)(F - L) \tag{6.2-6}$$

where F_{ni} is the financial position or net worth at the end of the year; F is the reserve fund set aside for contingencies in lieu of insurance at the beginning of the year; L is a possible loss that may occur during the year and that is paid out of the reserve fund; and i is the interest earned by the liquid assets account in which the fund reduced by any loss is held. It is assumed that L does not exceed F. The financial position without insurance at the end of the year is the result of investing the initial net worth reduced by the reserve fund at the internal rate of return, and investing the reserve fund minus losses at the liquid asset interest rate i, which may be substantially less than the internal rate of return r.

(c) The *value of insurance* is the difference between the financial positions with and without insurance

$$V_i = F_i - F_{ni} = (1+r)(F - P) - (1+i)(F - L) \tag{6.2-7}$$

where V_i is the difference between the financial positions with and without insurance. According to definition, if $V_i > 0$, the financial position with insurance is superior to the financial position without insurance. Two cases are readily apparent:

(α) $L = F$; since $F \gg P$, $V_i > 0$, which makes the financial position with insurance the preferred position.

(β) $L = 0$; $V_i = F(r - i) - P(1 + r)$. The first term is the loss of revenue by investing the fund in the liquid asset account instead of at the internal rate of return. If the insurance premium compounded at the rate of return exceeds this amount, then $V_i < 0$, and the financial position without insurance is superior to the financial position with insurance.

(d) Financial position of *insurance with a deductible*. The deductible plays the role of a small reserve fund for small risks by which the enterprise wants to reduce the insurance premium. For such cases the financial position at the end of the year with a deductible is

$$F_{di} = (1+r)(NW - P' - D) + (1+i)(D - L) \tag{6.2-8}$$

where P' is the reduced premium, $P' = P - R$; P is the regular premium; R is the reduction of the premium as a reward for the deductible D. The deductible serves as a reserve fund and is kept in the liquid assets account. Suppose D is either consumed by loss or remains intact in case of no loss. Then the financial position at the end of the period is

$$F_{di} = (1+r)(NW - P' - D) + (1+i)\, qD \tag{6.2-9}$$

where q is the fraction of the deductible that earns interest.

(e) The *value of insurance with a deductible* is computed as the difference between the financial positions of insurance without and with deductible at the end of the period

$$V_{di} = F_i - F_{di} = (1+r)(D - R) - (1+i)(D - L). \tag{6.2-10}$$

For L expressed as a fraction of D, $L = (1 - q)\, D$,

$$V_{di} = (1+r)(D - R) - (1+i)qD. \tag{6.2-10a}$$

Two cases are readily apparent:

(α) $q = 0$; then $L = D$; $V_{di} = (1+r)(D - R)$; since $D \gg R$, $V_{di} > 0$, and the financial position with insurance is superior.

(β) $q = 1$; then $L = 0$; $V_{di} = (r - i)\, D - (1+r)\, R$; if $r - i$ is small, the interest gain by the premium reduction R at the internal rate of return may exceed the interest lost by keeping D in the liquid assets account and $V_{di} < 0$, which means the financial position with deductible is financially superior.

The deterministic relationships of this section should be considered as pilot calculations that should be expanded into a more comprehensive probabilistic analysis by treating L as a random variable with a range and a distribution. The time when loss occurs is another random variable that should be expanded to a multi-year analysis to examine various insurance premiums, deductibles, and reserve funds for decision making on their respective sizes. The differences in financial positions, V_i and V_{di}, become random variables with ranges and distributions. This generated data base provides a broader foundation for decision making than the analysis outlined here.

Examples: (1) A self-insurance reserve fund $F = \$100,000$ is proposed instead of an annual insurance premium, $P = \$3,000$. The expected loss per year is $L = \$6,000$. The internal rate of return of the project is $r = 0.1$, and the market rate is $i = 0.04$. Should the reserve fund proposal be accepted or insurance be taken? Solution: Equation (6.2-7) gives the difference in financial position with and without insurance for a one-year period

$$V_i = F_i - F_{ni} = (1+r)(F-P) - (1+i)(F-L)$$

$$= 1.1\,(100,000 - 3,000) - 1.04\,(100,000 - 6,000) = \$8,940 > 0.$$

$V_i > 0$ means that taking out insurance results in the preferred financial position.

(2) What would the upper limit on the premium be that makes the financial position with insurance superior for the conditions of Example (1)? Solution: As long a V_i exceeds zero, the financial position with insurance is preferred. From Equation (6.2-7) one obtains

$$(1+r)(F-P) - (1+i)(F-L) > 0$$

Taking $(1+r)\,P$ to the right side and rearranging gives

$$P < (r-i)F + \frac{1+i}{1+r}L$$

$$= (0.1 - 0.04)\,100,000 + (1.04/1.1)\,6,000 = \$11,672.$$

Up to $P = \$11,127$, insurance gives a financial position that is superior or at least equal to the financial position without insurance.

(3) Suppose the annual loss is \$5,000, the deductible is \$10,000, the premium reduction is \$1,500 and all other conditions are the same as in Example (1). What is the financial position for insurance without deductible compared to insurance with deductible? Solution: The difference between the financial positions of insurance without and with deductible is according to Equation (6.2-10)

$$V_{di} = (1+r)(D-R) - (1+i)(D-L)$$

$$= 1.1\,(10,000 - 1,500) - 1.04\,(10,000 - 5,000) = \$4,150 > 0$$

The financial position without deductible is superior to the financial position with deductible, because the partial loss of the deductible exceeds the premium reduction.

6.2.6 Liability

Liability is responsibility for damages inflicted on other persons or property. Here the emphasis is on the probabilistic aspects of liability, not on the legal aspects. The fact that probabilistic aspects exist is demonstrated by the existence of liability insurance (see Section 6.2.7). *Negligence* is the basis for establishing liability. Negligence can occur in many ways. For a legal case, negligence must be established in context with the damage. Establishing negligence also depends on the relationship between the party that is held liable and the party that levels the charge, for example, the owner of the dam and the person(s) dispossessed or killed by a dam accident. In dam incidents, most state legislatures have made dam owners responsible for remedial action. A legal injury or *tort* is an act that wrongfully invades a person's rights. Negligence is a tort arising out of a failure to exercise a customary degree of care required by law.

Negligent acts can occur in the following ways (Greene, 1977, p. 289):

- *A positive or negative act*: The positive act is doing something, the negative act fails to do something both resulting in a breach of duty to another person.
- *A voluntary or involuntary act*: Both voluntary acts and involuntary acts may result in the charge of negligence even if they are not meant to do harm.
- *An imputed act*: A person or company may be liable for acts of others, such as relatives, employees, contracted persons; a seller of a product may be liable for acts of persons who bought and consumed the product.
- *A proximate cause*: An act is the sure or most likely cause (proximate cause) of the damage if an unbroken chain of events leads from the act to the damage. Since one damage can lead to another damage in a sort of chain reaction, the court must draw the line where the sequence ends.

Since negligence, if established, can have severe legal consequences, there are defenses against the charge of negligence (Greene, 1977, p. 293):

- *Contributory negligence*: If the defendant is 90 % at fault and the plaintiff is 10 % at fault, then both are guilty of contributory negligence and cannot collect damages. One party must have *clean hands* for the negligence charge to stick.
- *Assumed risk*: The plaintiff participated in a risky activity and incurred loss despite reasonable attempts made by the defendant to reduce the risk

to the participant. The participant should have known that a residual risk
of damage existed and assumed it by his participation.
● *Expanded liability*: The producer of a product, the dealer selling it, and
the consultant recommending it may all be held liable for damages.

In employer-worker relationships, establishing the existence of negligence or
degrees of it by litigation proved unsatisfactory especially for the worker. It has been
replaced by the *workers' compensation* insurance, a sort of no-fault insurance by which
workers are compensated for on-the-job injuries regardless of who is at fault. All states
have workers' compensation insurance but each state has its own laws. In Tennessee,
all businesses with five or more employees are required by law to workers'
compensation insurance. The compensation rates to be paid by employers vary
significantly from state to state and are a factor in the competition of the states for jobs
(for details see website on workers' compensation).
        There is a trend toward a weakening of the defense of contributory
negligence. Instead of no liability at all in the absence of*clean hands*, the defendant
may be liable according to the percentage of his contribution to the damage by the
doctrine of *comparative negligence*. If the percentage of contributory negligence by
the plaintiff is 10 % he may receive a 90 % compensation from the defendant.
Another weakening of the defense of contributory negligence is the *last clear chance
rule*. If a plaintiff got in harm's way, negligence on the part of the defendant may still
exist if harm could have been avoided by an effort of last resort by the defendant
(Green, 1977, p. 293).

6.2.7 Professional Liability Insurance

Professional liability is the liability of individuals who render professional
services and the liability of contractors who deliver completed projects. *Professional
liability insurance* protects professionals from errors and omissions arising out of their
practice. It is also known as *malpractice insurance*. This insurance is special in some
respects. It may not allow out-of-court settlements without permission of the insured,
as the professional may believe his reputation is damaged without court-proven
innocence. The insurance does not cover damages arising out of dishonest or criminal
acts. Professional liability covers deliberate acts with unintended consequences. If the
consultant suggests a method of repair that fails and causes damage, he is covered. But
the policy excludes claims arising out of any guarantee that the service rendered will
accomplish a certain result. If the professional guarantees that a particular method will
secure the dam against seepage and it does not work, the resulting damages are
excluded (Greene, 1977, p. 329). This exclusion, a risk control measure taken by the
insurer, is well founded in experience. Insurers may be selective in accepting
applicants for professional liability insurance in high-risk areas (Reynolds, 1976, p.

165).

Examples: (1) Coolidge Dam on the Gila River in Arizona was completed in 1928, a quite unusual structure of three cupolas (domes) supported by arched buttresses at 30 m distances between centers, 75 m high. It was judged "highly probable" to fail in 1989 by dam experts of the Bureau of Reclamation. It was also a high-hazard dam with as many as 500 people exposed in the path of a dam break flood over some 130 km downstream. It had the dubious reputation of being "one of the five most dangerous dams in the United States" according to documents quoted in the *Tucson Citizen*. Rehabilitation was estimated to cost $50 million. An official of the Bureau of Indian Affairs (BIA), owner and operator of the dam, challenged the warning of the BUREC experts by claiming "Coolidge Dam is not going to fail today, tomorrow or any time in the near future." Rehabilitation of the dam was requested, however, by the BIA in 1990, and repair work was undertaken to prepare the dam for passing the probable maximum flood (PMF) and stabilizing the buttresses and the spillway floors against erosion of deteriorated concrete (KNS,1989); www.usbr.gov/cdams/dams/coolidge.html.

(2) The 1963 Vaiont landslide disaster killed over 2,000 people. In the trial, the prosecution inquired into whether officials and consultants had acted negligently by ignoring or belittling the imminent threat of a landslide, especially after some movement on the mountain slope had been observed in 1960, before the project went into operation. Eleven engineers were criminally charged, including manslaughter charges. Two engineers had died by the time of the trial. Two defendants, management employees of the power company, were charged with too rapid test fillings of the reservoir, given the slide proneness of the slope. The others were charged with being accomplices. They included the director-general of the power company, public works officials, a local civil engineer, and a hydraulics professor who had served as a consultant. The hydraulics professor, who was well known in the profession and served as consultant to the owner-operator, was charged with having refused to consider other consultants' geologic and seismic studies and having guaranteed the stability of the site "even in the most catastrophic foreseeable event of a landslide." He was right insofar as the dam was concerned, which withstood a 100 m high overtopping wave caused by the slide. All were accused of knowing about the 1960 slide and should have been able to figure out that impoundment would increase the instability (Reynolds, 1976, pp. 157–165). Of eight persons tried, five were acquitted, among them the hydraulics professor. Three were found guilty of simple manslaughter and received six-year sentences of which they were to serve two years. Among them was a member of the company that operated the dam, an official of the ministry of public works that was in charge of testing Vaiont, and the civil engineer consultant. The mitigating circumstance for all accused was the unpredictability of the

extent and the velocity of the slide (ENR, 1970/1).

(3) A $500 million lignite thermal power plant fell behind several months in its schedule by a combination of design delays, material delivery, early commissioning problems, and lack of skilled labor. The skilled labor problem became so severe that the contractor wanted to invoke *force majeur*. The engineer-procure-construct contract had a guaranteed price and incentives for timely completion. It specified that if power would not flow by the new date, about five months beyond the original deadline, the developer would face follow-up damages caused by replacement costs for power not being delivered and fuel not being bought (ENR2001/2/12). In a hardening insurance market, contractors cannot afford the risk of completion guarantees because of the unacceptably high probability of time and cost overruns and the associated failure penalties.

6.2.8 Surety

In the construction industry, owners (and developers) need to entrust projects they cannot handle themselves to contractors who claim they can do the work according to the owner's specifications and come up with timely and quality-wise satisfactory results. A project that may represent a sizable monetary investment exposes the owner to the possibility of being presented with an unsatisfactory product. A way of dealing with a specific contract risk is to require the contractor to bring in a third party that assures the owner of satisfactory implementation of the contract, should the contractor fail, in other words, taking on the liability of the contractor. This third party is the surety (from Latin *securitas*).

The three participants in contract risk mitigation are the principal (*obligor*), the person protected (*obligee*), and the insurer (*surety*). A contractor or obligor promises the owner or obligee to construct a building according to agreed upon specifications. The owner wants assurances that the job will be done right and asks the contractor to post a *surety bond*. The contractor purchases this bond from a surety. This bond is the assurance the obligee wants or needs to entrust the contractor with the job. The surety does not expect to pay out any money. The surety must "make good" only in the event that the obligor (contractor) fails. The reputation of the contractor determines whether he can obtain a surety bond and at what cost. Thus, the ability to obtain a bond qualifies or eliminates a contractor from seeking to do work for the obligee. If several contractors compete for the job, the obligee must require that they are all bonded so that all contractors are competing on an equal footing and make serious bids to do the work at the stated price, and not just submit low bids to get the foot in the door. These surety instruments are the *construction bond* and, for the special purpose of the bid, the *bid bond* (Greene, 1977, p. 376).

The cost of a bond varies from contractor to contractor depending on the risk perceived by the surety. In a strong business climate with sureties awash in money, sureties may be easily obtainable by almost everyone and at a low price. As the business climate deteriorates, the cost gap between good and not-so-good contractors widens, and it may become more difficult for some contractors to get a bond at all. The surety also relies on reinsurance to cover its risks. As the number of claims for failed bonds increases and reinsurance losses increase, the surety itself will be charged higher rates for reinsurance. Also, as losses increase, reinsurance terms may become more restrictive in the form of limits on exposure. Sureties then in turn must apply risk control measures by carefully analyzing their exposures (Hirsh, 2001).

6.2.9 Liability Risks in the Deteriorating Infrastructure

Liability risk has become one of the most dreaded exposures in the construction sector as lawsuits seeking enormous compensations have proliferated in recent years. The basis for these lawsuits is usually establishing negligence on the part of the liable individual or firm and teaching them a lesson by imposing large punitive damages. Given this environment, it is surprising that so many dams that are considered unsafe on inspections still exist in high-hazard locations. Of all hydraulic structures, a dam can produce the largest exposures of people and property, given that it impounds a large mass of water with an inherent large amount of energy. A sudden water release by a dam failure, be it by piping, an overtopping and breach in a flood, or a misoperation of its release facilities can quickly produce a threat to life and property downstream. To avoid liability by negligence, a dam owner must exercise strict risk control through a high standard of operational safety and maintenance. Dams have been terminated by breaching them because owners could not afford the associated liability or the cost of rehabilitation that would reduce the risk of liability.

The dam inspections authorized by the Dam Safety Act of 1972 (PL92-367) were not undertaken until several years later, after several more dam breaks, because of lack of funding. Some progress has been made since then. Thousands of dams have been inspected, but the lack of funding to improve the overall state of repair and safety of nonfederal dams has continued to be a problem. The ASCE infrastructure report card of March 2001 provided evidence that a defective water infrastructure persists (ENR, 2000/10/9, p. 18; ASCE, 2001; see also Chapter 1, Sections 1.4.3 and 1.4.4).

Examples: (1) In the Buffalo Creek Dam disaster of March 1972, a mine waste retention dam failed resulting in a flood that killed 118 people. The Corps of Engineers blamed the U.S. Bureau of Mines, the state of West Virginia, and the Buffalo Mining Co., who were legally responsible for the inspection and safety of the structure. In the court proceedings, survivors claimed that the mining company had warned them several times before, when rains threatened a dam break. When it finally happened, people were tired of false alarms and did not respond (Reynolds, 1976, p. 159).

(2) The inquiry into the causes of the St. Francis Dam collapse was concluded on April 13, 1928, one month after the accident that toppled the dam of a reservoir filled to the rim and drowned some 500 people. The investigators attributed the responsibility for the dam's collapse to the Los Angeles Water Works and Supply, its chief, William Mulholland, and the employers of the chief, the Department of Water and Power Commissioners; the legislative bodies of the city and state; and the public at large (Davis, 1993, p. 237). The cause of the disaster was identified as foundation failure. Of the various theories, the one by E. L. Grunsky (Jansen, 1988, p. 33), seems to best put the pieces of the puzzle together: the dam came under massive uplift, tipped over to the downstream side, and broke apart right and left of the center block more or less simultaneously, whereupon the center block tipped back into place and was left standing. Photographs of this block show that its toe was sheared off, probably when it tipped (Davis, 1993, photos following p. 146). The dam had a rather small top angle that made it vulnerable to instability under uplift. A conspiracy theory claiming that an explosion had caused the dam to fail was considered much less likely than foundation failure. The jury foreman summed up the liability situation: "The construction of a municipal dam should never be left to the sole judgement of one man, no matter how eminent" (Davis, 1993, p. 238). The state of California enacted the first dam safety law in 1929 requiring that "all dams must be reviewed by a board of eminent engineers and geologists retained by the state engineer before and during construction. This commission became the California Division of Safety of Dams (DSOD)" (Davis, 1993, p. 243).

(3) In the area of environmental exposure, the U.S. Environmental Protection Agency (EPA) and the U.S. Justice Department are moving aggressively against environmental polluters. Negligent pollutant discharge into a river may constitute a civil offense as well as a criminal offense. Prime targets are unlawful activities that cause exposure of environmental and human health to damage. "Federal prosecutors are committed to indicting as many responsible individuals as possible. Lower level employees and line workers often receive immunity in return for testifying against supervisors who ordered them to violate environmental statutes. The general rule is to indict as high as possible up the corporate chain to ensure that corporate management takes notice" (Marzulla, 1990, p. 12).

6.3 Risk Management

6.3.1 Elements of Risk Management

Risk management is the handling of risk from its origins to its reduction or elimination. Risk management includes *risk assessment* and *risk control* . Risk assessment begins with the identification of potentially harmful events and hazards that may trigger such events or increase their probability of occurrence. Then, for a given event or a group of mutually exclusive events, realistic quantifications of the related exposures are established, usually in monetary terms. Then, methods of risk control are selected and their effectiveness and costs are evaluated. Risk control also includes protection of company assets by risk transfer. The cost of risk transfer (insurance premium) is usually directly related to the extent of risk control in place. Finally, the decision maker reviews the decision criteria for the various alternatives of risk control and selects the preferred alternative for implementation.

The phases of risk management are illustrated in Figure 6-1. A hydraulic structure is used as a backdrop, but the approach is general. The sources of risk are identified first. For an identified critical event (flood) the search begins at the left of the figure with uncovering the underlying causes or hazards (design flood, underdesign of spillway, spillway malfunction) that may lead to the critical event (Gruetter and Schnitter, 1982). The result is a tree structure that leads from identified causes to the critical event. This part is called *cause-event analysis*. Then the analysis moves forward, from the critical event toward the identification of potential consequences in the form of *failure modes*, or system responses, such as overtopping and breaching of the dam, outlet failures, and so on, and associated costs. The consequences may be controllable by repair, reconstruction, or operational measures, unless they are rapidly progressing fatal flaws. In the latter case, the remaining options may be warning, evacuation, and rescue measures. The part of the analysis that deals with events and failure modes is the *event-consequence analysis*. It is also called *fault tree analysis*. A fault tree is a graphic depiction of a chain of events that leads from the detection of a defect by sensors to failure of the system. Several sources of defects and associated chains of events, in series and/or parallel may lead to the same outcome (Henley and Kumamoto, 1985, p. 66 and p. 151). In Figure 6-1, a fault tree would cover the event-failure mode section.

The complete analysis, from the identified causes to the quantified consequences, is the *cause-consequence analysis*. It covers the cause-event and event-failure mode sections in the left half of Figure 6-1. It provides the input to the *decision analysis,* which determines the preferred risk control measure that should be implemented. The decision analysis is illustrated in the right half of Figure 6-1. A flow chart for a *cause-consequence analysis* for a dam is given in Figure 6-2 (after Gruetter and Schnitter, 1982; and after Boggs et al., 1988, p. 514). From left to right the analyst follows a path through a possible sequence of causes, events, and activities. Probabilities are finally tagged on all uncontrolled causes, events, and consequences. Thus, the outcome can be associated with a resulting probability, usually a product of probabilities as a possible path through the system represents a series of dependent or independent events. For the universe of all possible events including that of the "no failure" event (which is normally the one with the highest probability), the sum of the probabilities of all outcomes must equal 1. The outcomes are mutually exclusive as only one path through the cause-consequence scheme can be realized at a time. Since it is usually impossible to cover the total number of possible paths, a finite number of paths is selected that represents the universe of all paths, similar to a histogram that lumps numerous similar elements into a set (bar of the histogram).

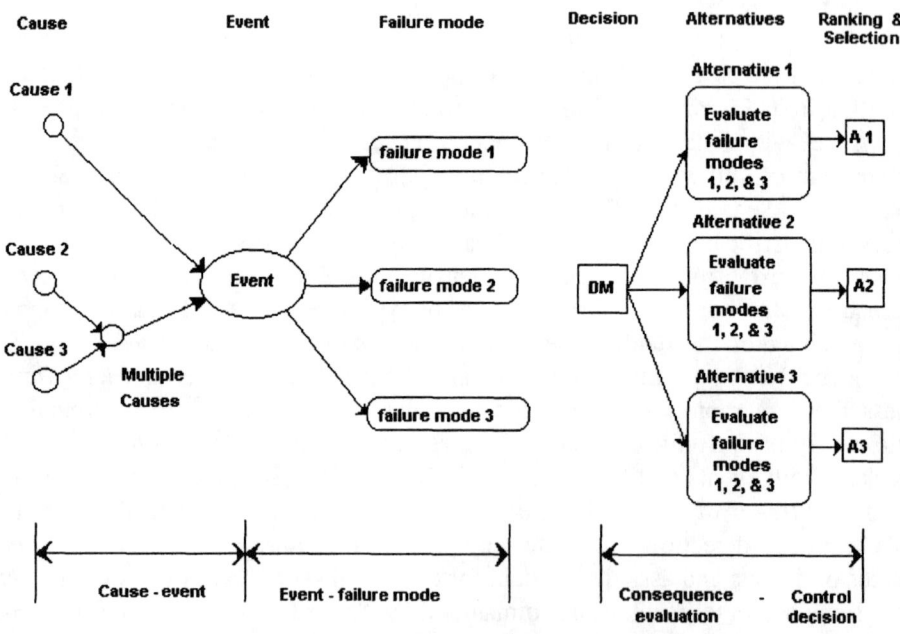

Figure 6-1: Phases of risk management: (1) Cause-consequence analysis: identifies cause-event relationships, and event-consequence (failure mode) relationships. (2) Decision analysis: identifies and evaluates control measures (alternatives) with respect to how they deal with event consequences; compares and selects the preferred alternative for implementation. DM = decision maker.

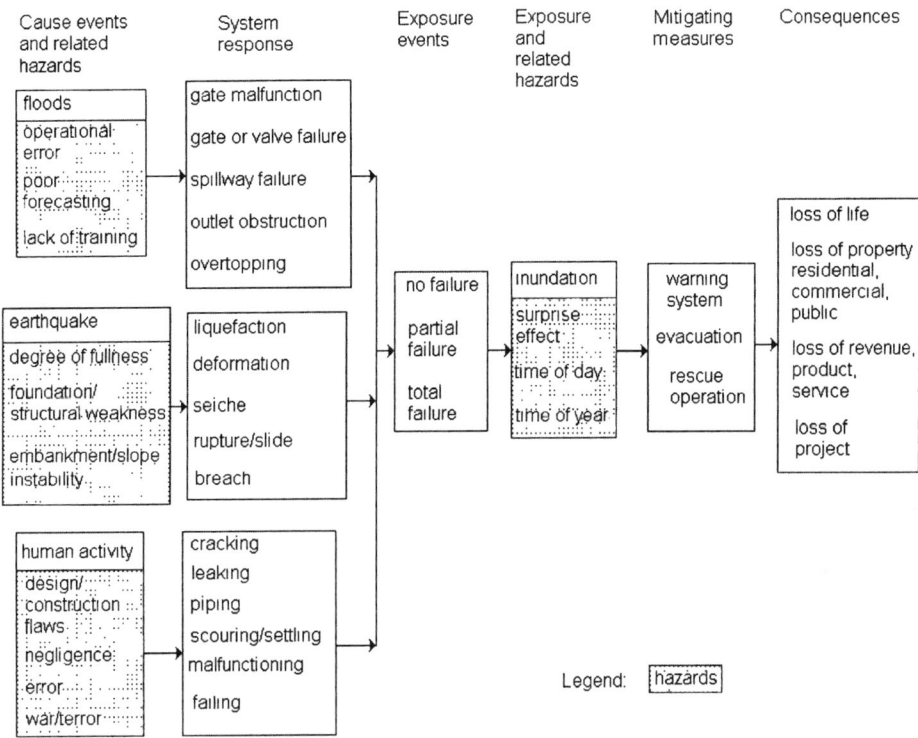

Figure 6-2: Cause-consequence diagram for a dam analysis. The analysis starts with assumed cause events, such as floods, earthquakes, human activities, and so on, all of which are often aggravated by known or hidden hazards. They lead to a system response in the form of symptoms, such as cracks, blockage, malfunctioning, and so on, which can either be resisted by the system or controlled by system management to avoid or reduce system failure. The impending consequences may be amenable to mitigation measures. The final consequences of the exposure are measured in terms of loss of life, personal property damages, and material damages. The sequence consisting of cause event and consequences can be construed as a serial The sum of the probabilities of all outcomes including that of the null event (nothing happens) must add to 1.

The various pathways from left to right in Figure 6-2 illustrate more risky or less risky scenarios for the project. They represent a selection of possible pathways consisting of events and possible consequences that unfold through several stages:

• *Cause events and related hazards*: A cause event or trigger event can be of two types: a natural event or a human-caused event. Natural events are external events; they are independent of the structure's presence or its state. Examples are floods and earthquakes. Human-caused events are structural failures due to hidden material or design flaws (structural hazards), foundations failures, equipment failures, operator error, and decision errors. Hazards include all conditions, natural or human-caused, that increase the frequency of events and make consequences more severe. A flood can be aggravated by a high lake level; by poorly maintained equipment (outlet works); by general lack of operational readiness; or by poor forecasting.

• *System response*: Between performance as expected and failure is a range of possible failure modes. Some are more likely with specific (cause or trigger) events than others. For example, failures due to gate malfunction and overtopping are more likely with floods, whereas failures due to cracking (leading to joint pressurization and piping), toppling, and sliding are more likely with foundation failure and earthquakes.

• *Exposure event*: The exposure event or critical event is the event that causes the exposure to harm, for example, the overtopping of the dam that results in breach formation and the release of a large amount of water within a short period. The Teton Dam breach was initiated by piping. During six hours, which included the development of the breach from the initial small trickle at the downstream toe of the dam to a major gap through the dam, $300 \cdot 10^6$ m^3 were released, which amounts to an average six-hour flow rate of 14,000 m^3/s, with a peak outflow rate of about twice that much.

• *Exposure and related hazards*: The critical event creates the exposure of life and property to harm. Scenarios of such exposures are flooding of settlements, disruption of transport and communication lines, destruction of infrastructure (highways, railroads, and bridges), inundation and destruction of farmland, drowning of livestock, damage and destruction of industrial and navigation facilities along the river, and so on. Where people get in the way, casualties may result. The exposure may be aggravated by inopportune occurrence time of events (nighttime, winter) and by the surprise effect (out of the blue). The way in which an event evolves over time, such as instantaneous collapse, progressive collapse, gradual collapse, or partial collapse, and so on, may aggravate or mitigate the exposure.

• *Countermeasures*: Possible countermeasures depend on available warning time, repair possibilities, response time (readiness), rescue possibilities, protection measures, and so on (see Section 6.3.4, Table 6-7).

• *Consequences*: Loss of life, loss of property, loss of commercial and industrial facilities, damage to hazardous (industrial) facilities, damage to infrastructure (roads, railroads, bridges, navigation facilities, dams, dikes, and so on), possibly permanent loss of the failing project, loss of jobs, and loss of income.

The strategy of controlling consequences by avoiding or mitigating them is *risk control*. Alternatives are identified that reduce the probability and extent of exposure and consequences. The final step of risk management is the *decision process* that selects a preferred alternative including its risk control measures for implementation (Section 6.6.5).

6.3.2 Event-Consequence Relationships

The exposure of values to a harmful event creates the potential for loss. Sometimes losses are just one alternative outcome of accepting risk, the other being a gain. To engineers it may sound frivolous that serious exposures would be acceptable for gains but it actually happens all the time. Accepting a deteriorating infrastructure year after year while allocating money to other purposes is a decision process that accepts the possibility of severe losses from a failing infrastructure while going for gains in some other domain. There is much controversy about measuring the values of intangibles for including them in the analysis of risk-protective measures. In August 2003, the Libyan government offered $10 million for every passenger killed in the 1989 PanAm crash. Probably nobody would want to trade his life for any amount of money. But monetary awards tagged on persons' lives exposed to risk can serve as inducements for steering problem solutions away from potentially high loss alternatives toward more conservative alternatives without any money changing hands.

An exposure may have various levels of consequences. The event-exposure relationship may consist of different probabilistic pathways emanating from the same event that lead to different mutually exclusive outcomes. The consequences may be predictable (deterministic; e.g., flood damage) or unpredictable (probabilistic; e.g., earthquake damage). Often they are a mix of predictable effects and unpredictable effects. For example, in the dam failure-exposure example, the breach can take place in many different ways that produce different time distributions of the water release and thus different levels of exposure. The resulting water levels downstream are functions of many conditions that require assumptions, such as channel roughness, terrain roughness, known and unknown obstacles (debris jams), water retention in the flood plain and side valleys, and so on. Examination of the numerous conditions produces a range of possible outcomes in the form of consequential damages.

Earthquake-damage relationships are more uncertain than flood-damage relationships for at least three reasons. First, the epicenter, the point on the earth surface at the shortest distance from the focus, is the point of the highest intensity but it cannot be pinpointed before the quake. Second, spreading occurs by body waves traveling directly from the focus through the interior of the earth, and by surface waves radiating from the epicenter, so that all surface points around the epicenter receive impacts by both body waves and surface waves. Third, the effect on structures is a function of the resistance of the structure to shock, shear, horizontal and vertical acceleration, and of the geology of the foundations. Building codes have been designed for making structures earthquake resistant, but much uncertainty remains about their actual performance. Also, in zones with very infrequent but large possible earthquakes, earthquake-proofing of structures by rehabilitation is still more the exception than the rule. These uncertainties make exposure-consequence relationships dependent on many variables with ranges and probability distributions. Their evaluation leads to a large number of possible paths producing a large number of structural responses and associated costs.

A probabilistic event-consequence relationship is a conditional probability because a consequence will occur with some probability, given the causative event occurs with some probability. Figure 2-7 illustrates the probability of such follow-up events. If the consequence is triggered by one or more cause events, then the probability of the consequence is (see also Section 2.1.5)

$$P(C) = P(C|E_1)P(E_1) + P(C|E_2)P(E_2) + \dots \tag{6.3-1}$$

where $P(C)$ is the total probability of the consequence of several causes, $P(C|E_i)$ is the conditional probability of the consequence, given the cause-event E_i occurs; and $P(E_i)$ is the probability of the cause-event E_i. Equation (6.3-1) states that one or the other event can trigger consequence C. It is possible that one or more $P(C|E_i)$ are close to 1 so that $P(C)$ becomes the sum of the event probabilities $P(E_i)$. If also the $P(E_i)$ are large then there is obviously overlap among the event sets E_i which must be eliminated to obtain the mutual exclusive probability $P(C) \leq 1$. If only one event triggers the consequence, then

$$P(C) = P(C|E)P(E) \ . \tag{6.3-1a}$$

If a series of consequences results from a cause-event, then the probability of the event-consequence sequence is (see also Section 2.1.4)

$$P(E \cap C_1 \cap C_2 \cap C_3 \cap \dots)$$

$$\tag{6.3-2}$$

$$= P(E)\,P(C_1|E)\,P(C_2|E \cap C_1)\,P(C_3|E \cap C_1 \cap C_2) \ \dots$$

where E is the cause-event and the C_i are costs of consequences that are caused by E

and previous consequences. If the cost dependencies do not exist, then the probability of the event-consequence is a function of the product of terms $P(C_i|E)$ only. In the simplest case of only one consequence, Equation (6.3-2) reduces to

$$P(E \cap C) = P(E) P(C|E) \cdot \tag{6.3-2a}$$

Risk control attempts to reduce costs and exposures and their probabilities to the extent that this is possible. Probability reduction can be achieved by reducing the joint probability $P(E \cap C)$ in Equation (6.3-2a). This can be accomplished in two ways: by reducing the event probability and by reducing the consequence probability or both. The consequence probability can be reduced by weakening or disrupting the event-consequence relation. If flood damage in a sensitive area can be reduced by safer arrangement of equipment, or by completely moving it out of the reach of flooding, then in the case of a pump failure the consequential damage is reduced or eliminated. If surveillance detects an incipient hazard, which could trigger a cause-event, and this hazard is eliminated by preventive maintenance, then the probability of the cause-event is reduced or eliminated, which also reduces the probability of consequential damage. For human-caused events, usually both event probability and consequence probability can be reduced through risk control measures. The probability of *natural events*, such as a flood or an earthquake cannot be modified. In these cases, risk control is limited to reducing the consequence probability $P(C|E)$ and the consequence costs.

Event-consequence relationships often are difficult to establish. Good diagnostic skill may be required to identify consequences of events or to conclude from observed consequences backward on what the causative event may have been. This backtracking from known consequences is especially difficult when there are long delay times between event, exposure, and the resulting consequences. In an earthquake that occurs during the dry season, a dam may crack without visible damage. When the reservoir is impounded during the following flood season, the cracks may develop piping and destroy the dam. Industrial exposures to dust, radiation, asbestos, beryllium, and so on, may result in health damages many years later. Ten years after the Chernobyl accident, the scope of the damage was still only vaguely known (VDI, 1996/4). The methyl isocyanate spill in Bhopal, India, in 1984, killed some 4,000 people outright. Some 15,000 people are estimated to have died from accident-related illnesses over the following 14 years (Cohen, 1998). The event-consequence relationships and especially their intensity are usually contested because a clear cause-effect relationship may be hard to establish. Such exposures with long-range consequences are some of the most dreaded risks for companies and their insurers.

Examples: (1) Suppose a dam has an inadequate spillway, it is exposed to overtopping. The consequence of being overtopped may be erosion of the downstream embankment, and possibly breaching of the dam, or no damage at all, if the dam has been modified to sustain overtopping. This means that whereas the event probability $P(E)$ is not under the control of risk management, the consequence probability $P(C|E)$ can possibly be reduced, perhaps even to zero, by modifying the dam to the extent that overtopping can occur without consequence. Hence, even if $P(E)$ is relatively high and uncontrollable, the probability of a harmful consequence, $P(C)$, can be made arbitrarily small by reducing $P(C|E)$ through risk control measures.

(2) In a risk analysis application to the Delta Works of the Netherlands, Vrijling and de Graaf (1987, pp. 16 and 36) conclude that the permissible probability of harm is represented by (they use a somewhat different form)

$$P(D_i) = P(D_i|E_i) P(E_i) = \beta \cdot 10^{-4} \qquad (6.3\text{-}3)$$

where D_i is a specific damage; E_i is a specific event; β is a discretion factor that varies with the degree of voluntariness of the activity undertaken by the participant, ranging from 10 for voluntary activity (recreation) to 0.1 for involuntary activity (factory work); and i is an index of the activity. For a number of mutually exclusive activities or events E_i, the sum of the $P(D_i)$ over all i gives the total permissible probability, $P(D)$.

(3) The experience with construction and operation of nuclear power plants has advanced safety analysis, safety consciousness, and risk analysis in all fields of engineering. In nuclear plant risk analyses, the so-called Farmer's curve relates the release of ^{131}I (iodine) to a return period measured in reactor years (Brown, 1994; Farmer, 1967). The empirical formula of the curve is

$$\log Q_R = 1.024 + 0.6945 \log T \qquad (6.3\text{-}4)$$

where Q_R is the amount of ^{131}I release, in curies (Ci), and T is the return period in reactor (operation) years. Equation (6.3-4) is similar to Equation (3.7-10) except for its use of $\log Q_R$ instead of Q_R. The transformation $\log Q_R$ can also be used in Equations (3.7-11) and (3.7-12) for the calculation of T and $F(Q_R)$, respectively. The validity of Equation (6.3-4) is limited to the range $700 \le T \le 30,000,000$. For the range $100 \le T \le 700$, a similar formula holds but with different empirical coefficients. For thirty million reactor years, Equation (6.3-4) gives $Q_R = 1.6 \cdot 10^6$

Ci. The Chernobyl accident of April 1986 released an estimated $50 \cdot 10^6$ Ci.[1]

6.3.3 Risk Management in Maintenance

The risk management approach has great similarity to the probabilistic approach to preventive maintenance. Both approaches address process safety and service reliability. Pursuing these goals minimizes event likelihood and the extent of exposure. An outline of a risk management approach that can also serve as an outline of a probabilistic preventive maintenance approach follows:

I. Probabilistic cause-consequence assessment for the status quo and for possible alternatives to improve existing conditions (risk assessment):

1. Identification of possible causes.
2. Identification of critical events.
3. Identification of possible consequences (damages and losses) associated with the critical events and possible failure modes (existing hazards).

II. Measures to eliminate or reduce exposures, their causes, and possibly their consequences (risk control):

4. Identification of alternatives for mitigation or elimination of event causes and hazards.
5. Establishment of cause-event relationships, hazard-event relationships, and event-consequence relationships to support potential loss evaluation for the existing conditions and for proposed changes; establishing risk quantification measures (utility functions).
6. Calculation of decision criteria for choosing among (risk control) alternatives. Criteria include net benefit (the difference between total benefits and total costs ("economic safety margin"), benefit-cost ratio ("economic safety factor"), the

---

[1]The curie compares the radioactivity of isotopes with that of 1 g radium. The SI unit for radioactivity is the becquerel: 1 Bq = 1 nucleus disintegration (transition) per second. Radioactive decay is a random process. On the average, 1 g radium has an activity of 1 Ci $= 3.7 \cdot 10^{10}$ Bq $= 3.7 \cdot 10^{10}$. The curie is a very large quantity; therefore, permissible exposure to radioactive material over some period is measured in microcurie.

variance of probabilistic outcomes of a decision alternative (by testing alternatives on sets of chance events), the no-damage probability, and the probability of exceedance of economic safety margin, economic safety factor, and possibly others (see Section 3.5.2, Table 3-8; and the risk evaluation example in Section 6.6).

III. Decision making and selection of implementation alternative.

7. Ranking of alternative(s) and selection and justification of the preferred alternative.
8. Implementation of the preferred alternative(s).

In the *public health* sector, where risk management plays an important role, risk assessment (control measures not included) consists of similar steps (NRC, 1984, p. 29):

• Hazard identification (identify linkage of chemical compound to health effects)
• Dose-response assessment (probabilistic relationship between magnitude of exposure and health effects)
• Risk characterization (nature and magnitude of consequences on humans).

In engineering work, risk taking by exposures to harmful events of various kinds is an unwanted but often unavoidable circumstance. The benefits derived from engineering activities do not result primarily from risk taking, as is the case in the banking and investment sectors, but from planning, building, and operating production processes, usually one of a kind, or by plowing new ground. Risk management is part of the activities associated with a project from cradle to grave. Concern for risk starts with planning and design and continues through construction and operation. In the operation phase, risk management is incorporated into maintenance that includes inspection, surveillance, monitoring, repair, and rehabilitation.

6.3.4 Risk Control

Risk control is the implementation phase of risk management. It includes the following activities:

- Hazard surveillance and detection
- Maintenance action aimed at elimination of identified hazards
- Maintaining readiness of response to natural and human-caused events.

The insurance industry has a high stake in risk control, because it reduces losses and increases profits. The client also profits from risk control because loss reduction should be reflected in low premiums. Risk control can make otherwise uninsurable risks insurable (e.g., flood insurance). Insurance policies usually include clauses that limit coverage. This is part of the insurer's risk control. For example, if a driver is killed without wearing a seat belt, no death benefit will be paid that would otherwise have been due. If a worker is insured or killed in a work accident whose severity is increased by not using safe work procedures or protective equipment, supervisory personnel may face negligence charges and withdrawal of insurance coverage. Also, an insurer may refuse coverage outright in cases that have a high probability of loss.

Insurers exercise risk control by avoiding *adverse selection*. If an insurer has too many high-risk clients in his portfolio, he exposes himself to heavy, possibly simultaneous losses. Clients with a high probability of loss are commonly known as "bad risks." As damage probability approaches 1, risk rises to a very high level only to plummet to zero at probability 1. Risk does not exist under certainty. If the event is imminent or in progress, it represents a "preexisting" or "existing condition" and as such is uninsurable.

In the construction industry, risk control through *safety programs* pays off for several reasons. Indirect (uncovered or secondary) losses always far exceed anything a contractor can hope to recover in the form of direct losses. Indirect losses include disruptions that come with accidents, such as time loss, work disruption, litigation, and so on. It is thought that safety programs in construction, operation, and maintenance can make a difference in the overall costs of these programs that exceeds savings by minimizing the cost of labor, material, and equipment (ENR, 1997, p. I-20).

The importance and effectiveness of emergency risk control when a disaster (cause-event) is imminent is demonstrated by the effect of *warning time* on fatalities. For a number of flash floods and dam failures reported by Viessman et al. (1989, p. 366), the ratio of loss of life (LOL) to population at risk (PAR) was found to be sensitive to the length of warning time. Table 6-7 summarizes these data for warning times from zero to 1.5 hours and for warning times exceeding 1.5 hours.

Table 6-7: Effect of Warning Time on Exposure (from data by Viessman et al., 1989, Table 16.4, p. 366)

(a) Warning time from zero to 1.5 hours

PAR	LOL	LOL/PAR	Average LOL/PAR
4 to 17,000	1 to 421	0.0006 to 1*	0.13 (16 ratios)

Notes: PAR is the number of individuals exposed; LOL is the number of fatalities. *This 100 % fatality event happened with no warning time and included 4 persons (PAR = 4). If this case is excluded because of the small PAR, the max LOL/PAR is 0.3, and the average LOL/PAR is 0.074.

(b) Warning time exceeding 1.5 hours

PAR	LOL	LOL/PAR	Average LOL/PAR
100 to 58,000	0 to 11	0 to 0.0020	0.0004 (8 ratios)

For warning times from zero to 1.5 h, LOL/PAR is considerably higher than for warning times exceeding 1.5 h. Large involved populations indicate high-hazard situations. A warning time of as little as 1.5 h reduces the maximum fatality rate by a factor of (0.3/0.002 =) 150 and reduces the average fatality rate by a factor of (0.074/0.0004 =) 185.

Warning time should be a design consideration for any structure that represents a major exposure. The beneficial effect of even a short warning time was demonstrated by the World Trade Center collapse. The resilient outer tower shells withstood the airplane impact and thus provided about an hour's evacuation time, which allowed thousands of people below the impact levels to escape (Brouwer, 2002).

The possibility of a sudden failure in the absence of any general failure conditions creates a severe exposure, as is evident from Table 6-7. Earthquakes, explosions, or structural collapse by foundation subsidence occur at the worst possible time (in the middle of night), and there may be not enough time to contact people and to warn or organize a rescue. In the hydrosector, dam failures are the most dreaded of these events. Malpasset Dam and St. Francis Dam both collapsed instantly during the night due to foundation failure. Earthquakes have been the cause of sudden failures and near failures. A pre-warning of the release in such events does not exist, and warning time is limited to water slug travel time. The St. Francis flood traveled at about 20 km/h to 30 km/h. About 100 km of downstream valley were devastated leaving some 450 dead. The size of the exposed population is not known.

The fatalities occurred within a short time, within minutes to perhaps five hours after the collapse when the flood emptied into the ocean (Davis, 1993, p. 172). For Malpasset Dam (upstream from the city of Frejus), the preceding data give LOL/PAR = 412/6000 = 0.07. Teton Dam was breached gradually, but within about 5 hours a water mass of some 310 $\cdot$ 10^6 m^3, about the volume of the full reservoir, had passed through the breach and was traveling down the Teton River, 10 m and more deep in the downstream vicinity of the dam. Warning time for the exposed downstream population of about 23,000 was about 2 hours and more. There were 11 fatalities, 8 connected directly to drowning on the day of the flood (June 5, 1976). LOL/PAR = 8/23,000 = 0.0004 (Jansen, 1988, p. 28; Davis, 1993, p. 264; Viessman et al., 1989, p. 361 and 366; World Almanac 1999, p. 233; Teton Dam website: http://www.geol.ucsb.edu/faculty/sylvester/Teton%20Dam /narrative.html).

In many cases, complete risk control is not possible, the main reason being cost. In the rehabilitation of dams, an approach that the Corps of Engineers suggested was to expand the spillways of all dams so that they can withstand the most severe events, the probable maximum flood (PMF) and the maximum credible earthquake (MCE). The Committee on Safety Criteria for Dams was authorized by the Department of the Interior and organized by the National Research Council to investigate approaches that were more affordable. It presented its results in 1984 (ENR,1985/9). One of the findings was that dam breaks in the course of very large floods produce only incremental damage, as there is already a lot of water around. Hence the dam break during the PMF is most likely not the critical event in terms of exposure attributable to a dam failure.

The Committee on Safety Criteria for Dams goes on to suggest three approaches in estimating the flood the dam should withstand: first, the deterministic approach described as using a possible maximum precipitation, and calculate the hydrograph at the dam site; second, what is called a "probabilistic approach" that uses historical records and selects a flood, probably by extrapolating the record, similar to the procedure in Section 3.7, to obtain a flood peak with a long enough return period; and third, a risk assessment or risk analysis that "takes into account the risk to people and things downstream from different events and balance that against the cost of various additions to the dam"(ENR, 1985/9, p. 33). The third approach is what one could call a *probabilistic approach*.

An in-house study by the Hartford Steam Boiler Inspection and Insurance Company evaluated small hydroelectric power plants as to their frequency of loss and severity of loss over a 10-year period from 1980 to 1989. The evolution of loss frequency and loss severity for a small sample of hydroplants is given in Table 6-8 (after Lau and Sohre, 1992, p. 173). The data show an increase in loss frequency and loss severity over time. One would expect the opposite trend of loss frequency and severity reductions that was established during the first three years of the insurance to continue (frequency dropped from 100 to 50), but the early decline was followed by 8- and 10-fold increases in the fourth and fifth years. There also is a trend to higher loss frequencies in later years. Furthermore, a steep increase in the severity of losses occurred. Losses below deductibles are not included. The majority of losses was attributed to delays in repair and procurement of replacement parts. Keeping parts inventories is expensive, but chances are taken with and without inventories. The specific part may not be in the inventory after all when it is needed and the part that is in the inventory may have to be adapted for the installation for which it is needed. Inventory-related problems significantly increased downtime and power replacement costs. Delaying maintenance in favor of continued production only to be followed by major breakdowns was another cause for extended outages. Some hazards were caused by poor design and poor quality parts. These problems triggered early breakdown and demise of parts, such as turbine shafts, turbine blades, gears, bearings, and trash racks. Inexperienced operating personnel and inappropriate use of equipment possibly contributed to high stress and early failure of equipment. Inadequate maintenance due to inaccessibility of a turbine bearing led to failure and outage. The insurance experts recommend risk control measures that include conservative plant and equipment design, adherence to a preventive maintenance schedule, and an adequate replacement part or spare part inventory.

Table 6-8: Small Hydro Insurance Statistics (Lau and Sohre, 1992, p. 173)

Year	1980	1981	1982	1983	1984	1985	1986	1987	1988	1989
FI	100	50	50	400	500	100	550	500	500	300
SI	100	104	81	1095	1302	52	549	1162	2188	1153

Notes: FI is the frequency index; SI is the severity index; indexes were set to 100 for 1980.

Experience shows that for insurance to be of mutual benefit, risk control is required by both client and insurer. The protection of clients by insurance can lead to sloppy risk control by clients who rely on collection of claims for breakdowns and losses. This attitude leads to unexpected losses for insurers and ultimately to higher premiums and loss of coverage for the clients. A few quotes underline the importance of this client-insurer relationship: "... controlling losses is the key to controlling rates ... risk management programs (must be) integrated into the firm's overall management processes ... loss reduction can bring as many dollars to the firm's bottom line as delivery of the actual product or service" (ENR, 1997, p. I-24).

There is a similarity in probabilistic performance between a human being and technical systems. The way systems are operated and maintained and the overall operating environment suppresses or favors the development of exposures and hazards that increase the likelihood of system malfunction and failure. Figure 6-3 represents the number of fatal vehicle crashes as function of (system) age for three different parameters: distance, driver population, and general population (I.I.H.S., 1992). These parameters can be construed as measures of exposures and hazards which influence the failure rate. The failure rate versus age displays the typical shape of a bathtub curve (see also Section 3.6.3, Figure 3-15).

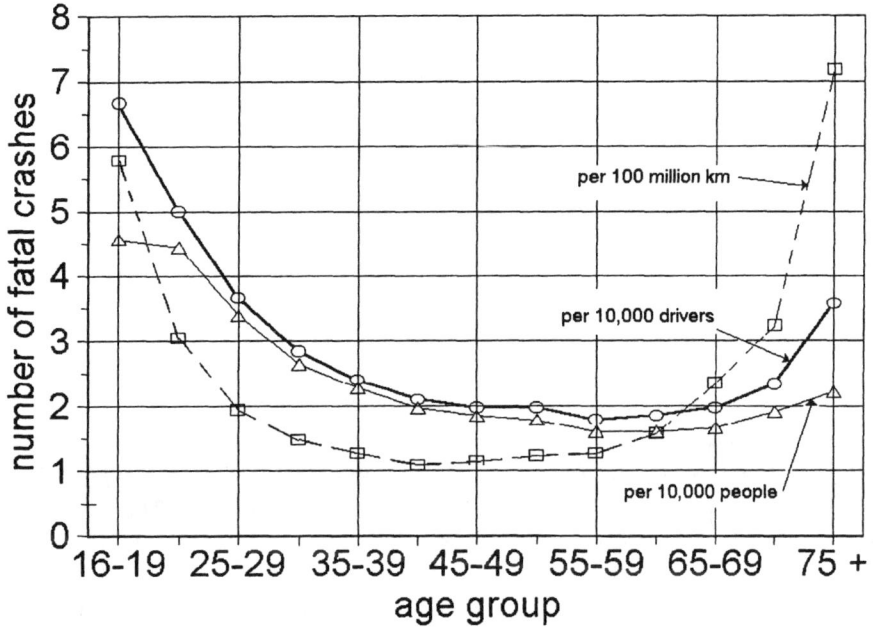

Figure 6-3: Passenger vehicle driver involvement in fatal crashes - 1990. The numbers of fatal crashes for five-year age groups (20 to 24 years, 25 to 29 years, etc., except for those younger than 20 years and older than 75 years) are given as rates for distance traveled, driver population, and general population. Rates are high in all cases for the risk-prone young age groups, and low for the risk-averse middle age groups, but vary according to reference base for old age groups. The failure rate versus age distribution shows sensitivity to exposure, special hazards, risk attitude, skill, experience, impairment, and other risk-related parameters (I.I.H.S. Status Report, Vol. 27, No. 11, September 5, 1992, p. 3).

Young drivers between the ages of 16 and 25 years have a relatively high accident rate, which may be due to risk-prone behavior, lack of experience, and increased exposure by extensive participation in traffic. At about 35 years, the rate reaches a minimum and stays there until about 60 years. Then it starts rising again, at first slowly, and then steeply, as age-related failures become dominant. Car insurers respond to this rate distribution by classifying drivers according to risk factors, such as age and driving record, and charge substantially higher premiums for drivers in high risk groups, such as drivers younger than 25 years. Any indication of more than average risk, such as a speeding citation or an accident, may be a justification for the insurer to increase the insurance premium or deny coverage.

The bathtub curve of failure rates is a time distribution of failures. It is not a pdf (see also Section 3.7.4). It is found to characterize the life of many systems. According to an ICOLD report (IWPDC, 1995), 70 % of dam failures occur in the first 10 years of a dam's life. The most common failure causes are foundation problems for gravity dams and overtopping for earth, rockfill, and masonry dams. Laying proper dam foundations and dimensioning spillways are the most error-prone design and construction activities. Dams built before 1950 have a failure rate of 2.2 %, whereas those built since 1950 have a failure rate of less than 0.5 %. Newer dams, as a result of improved design and construction procedures (improved understanding of rock and soil mechanics, of concrete quality and behavior, and of floods and earthquakes), and better hydrologic analysis and flow forecasting techniques, may have failure rates that are close to the bottom of the bathtub curve. But the bathtub curve also predicts a change of failure rate with age. Ultimately, what went up will come down. There is hardly a structure left from antiquity. Two factors seem to play a major role in longevity: a continued need for the services provided by the structure and maintenance. Continued need and maintenance motivated by it has allowed some structures to survive millennia under changing administrative systems.

According to the bathtub curve, over the life of a system, maintenance must watch for changing types of hazards that can put the system at risk (see also Stamatis, 1995, p. 113):

- *Break-in period*: During this period, the failure rate is declining with time; maintenance has to cope with early failure of parts by hidden design, construction, and installation errors, operator error caused by unfamiliarity with the system, and so on.
- *Useful life period*: After the failure rate of the break-in period has bottomed out, the failure rate may remain about constant over a number of years, say, 50 years of the useful life of the structure; failure arises mostly from external events that are not necessarily aggravated by structural flaws; but occasional

damages or failures may continue, caused by hidden design and construction defects and early wear-out failures; generally smooth operation can be maintained by a well-adapted operation and maintenance team.

● *Wear-out period*: Depending on the kind of use, environmental exposure, material deterioration, obsolescence, and so on, age-related failures increase slowly, steeply, or abruptly, as wear-out incidents occur sequentially or jointly, independently or dependent on each other; the rising failure rate requires an intensification of surveillance and inspection, and more frequent forced and preventive maintenance shutdowns to keep the level of the failure rate below an acceptable threshold (see Figure 5-1); renewal or rehabilitation may become a consideration when frequent shutdowns interfere with system reliability and product quality, and a decisive reduction in failure rate becomes economically or otherwise the preferred maintenance option.

Example: In a survey of 700 persons in The Netherlands, the respondents distinguished two main dimensions of risk: extent of accident and degree of organized protection (Vrijling and de Graaf, 1987, p. 14; Vrijling, 1993). With vast expanses of land threatened by floods from rivers and from the sea (the flood of 1995 forced the evacuation of 250,000 people), and thousands of kilometers of dikes threatened by millions of "piping" muskrats (a North American import from 1906; in 1996 alone, close to 327,000 muskrats were killed by trappers), the perception of extensive exposure to events beyond a person's control exists, and there is desire for exposure reduction by organized protection through the government (KNS, 1997/8).

6.3.5 Risky Decisions

Decision making means selecting among alternative actions. A decision is preceded by an analysis of the alternatives and a ranking of the alternatives according to adopted *decision criteria*. The decision criteria include total benefits, total costs, net benefits (benefits minus costs), total benefit-cost ratios, and net benefit-cost ratios, and so on, for each alternative (see Section 5.8.1). The comparison of the decision criteria then identifies the alternative that is preferred for implementation. So-called *risk-based decision analysis* has been applied to safety and rehabilitation analysis of dams (NRC, 1983; ACER, 1986). The approach typically includes a probabilistic hazard assessment in the form of an event tree that determines loading conditions (various flood and/or earthquake events), system response analysis (spillway performance) under these loadings, assessment of exposure (expected damage) by downstream flooding, structural costs, and benefits forgone by curtailing other uses (see Section 6.7). In comparative studies, project changes (upgrades) are compared to the project as is. The analysis may consider only costs; benefits can be construed as reductions of costs and damages in comparison to a baseline alternative, which could be the "as is" or "no action" case. For example, if a proposed alternative reduces or suppresses damages caused by overtopping or breaching of the "as is" alternative, then this damage reduction can be construed as a benefit of the proposed alternative. According to federal guidelines, the economic analysis can be carried out as a *cost effectiveness analysis* if benefits are essentially the same in the baseline case and the considered alternatives (see Section 5.8.1).

In a probabilistic analysis, inputs, outputs, and also parameters in input-output relationships are variables with ranges. This makes also the decision criteria, which are sums, differences or ratios of these variables, variables with ranges. The decision maker (DM) faces ranges of outcomes, which are not under his control. For example, the performance of a proposed structural alteration (spillway enlargement), say spillways alternative *A*, depends on how it copes with the events it will encounter. Each event causes costs and benefits for alternative *A* that have associated with them the probability of the event under which they accrue. The probability-weighted average of all outcomes is the expected value of alternative *A*, but this value never actually occurs. It is a guiding value that can be construed as the *insurance premium* of alternative *A*. A DM who insures his project, by selecting the preferred alternative selects the associated insurance premium. This premium, if accepted by an insurer, relieves the DM of the risk of facing real outcomes that the selected alternative may produce. The DM has three options: (a) Find an insurer to whom he can transfer the risk of incurring one of the mutually exclusive outcomes of the selected alternative; (b) Set aside a reserve fund and practice self-insurance; and (c) Ignore the risk and the analysis altogether and wait to see what will happen when a real outcome occurs. Government projects are usually self-insured and expect that any incurred real damages will be paid or at least alleviated by payments out of the national treasury. A *risky decision* is a decision made when confronted with the range of possible real outcomes. This is why the spread of outcomes plays an important role in risky decisions (see also Section 6.4.1).

The expected value analysis (EVA) that determines the insurance premium has recognized shortcomings all rooted in lack of information. Despite its shortcomings, EVA is commonly used nevertheless, because there is no known alternative. Arguments that the necessary information for such analysis does not exist, or that the dearth of information available is not good enough for "objective" analysis (NRC, 1983) fails to recognize the real meaning of EVA as an insurance premium. The opposing arguments are wrong on two counts. First, complete and precise information is practically never available and second, one must make the best of the available information, even if it is imprecise and incomplete. The use of judgmental probabilities in lieu of lacking observed ones is not as far-fetched as one might think. Judgment is used in the traditional *informal decision process* that leads to choices based on what are called *seat-of-the-pants decisions*, or *back-of-the-head decision*, or decisions made in a smoke-filled room. These processes use experience, intuition, and judgment in a subjective way. They are inferior to EVA because they lack *repeatability* and *verifiability*. Also, an intuitive decision cannot be subjected to sensitivity analysis, which would show the effect of an incremental change in variables and parameters on the outcomes of alternatives and thus on the decision. Such sensitivity analysis can be especially revealing when applied to the most uncertain input parameters. Second, the analysis of risk must somehow account for the differences in risk perception of those who are under risk (risk owners) and those who impose risk. These are usually different groups with different interests. Therefore, an "objective" risk analysis may be a contradiction in itself or at least incomplete if the subjective component of risk is omitted.

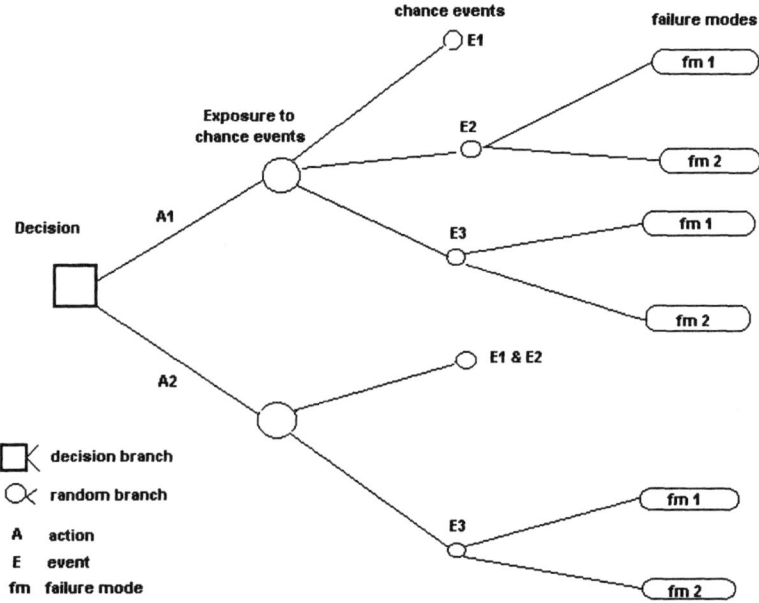

Figure 6-4: Decision tree for a risky decision. The alternatives A_1 and A_2 can be selected by the decision maker (DM). The alternatives evaluate exposures to random events (E_i) which are *not* under control of the DM. They may cause no damage or trigger failure modes (*fm*).

Using the EVA approach explained in the previous paragraph is a *structured decision* approach in contrast to guessing which is an informal decision approach. The structured decision approach is based on the evaluation of a *decision tree*, as shown in Figure 6-4. The structure consists of the evaluations of all branches that emanate from an alternative similar to the cause-consequence analysis of Section 6.3.1. The approach sets out to include *all* information not just firm information that is strictly based on observations or hard data. This approach is recommended here.

Raiffa (1970, p. 155) suggests that "when there is a paucity of objective evidence at hand, we require a methodology that brings information, however vague and imprecise, into the analysis, rather than a methodology that suppresses information in the name of scientific objectivity." On the same issue, Cornell (1972) states: "It is important to engineering applications that we avoid the tendency to model only those probabilistic aspects that we think we know how to analyze. It is far better to have an approximate model of the whole problem than an exact model of only a portion of it."

The decision tree in Figure 6-4 has two kinds of branches and two kinds of nodes:

- *Decision branches* which represent alternatives selected for analysis and available for choice by the DM. They emanate from a decision node shown as a square box, and
- *Chance (or random) branches*, which point to mutually exclusive random events that the alternative must cope with, and which produce the probabilistic outcomes of the alternative. They emanate from chance nodes shown as circles. The outcomes of chance branches are not under the control of the DM.

A maintenance or rehabilitation alternative can influence the extent and type of exposure, and probability of failure modes, but it cannot influence the probability and magnitude of external events. In Figure 6-4 this means the probabilities on the branches leading to the E_i cannot be influenced but the probabilities on the branches to the *fm* can be influenced by the alternative. The DM by choosing an alternative commits the future of the project to the part of the decision tree that unfolds from that decision branch.

Example: A decision concerning spillway rehabilitation has two alternatives: (1) expand the spillway from the 100-year flood to the 500-year flood, or (2) leave as is at 100-year flood capacity. Each of the two alternatives faces a universe of external probabilistic events: (a) floods up to the 100-year flood, (b) floods up to the 500-year flood, and (c) floods beyond the 500-year flood. For each day, year, or other period selected for analysis, the associated exposure to harm is incurred. If the first alternative is chosen, it means exposure to harm by floods in excess of the 500-year flood, which should have a reduced probability. If the second alternative is chosen, it means exposure to floods greater than the 100-year flood. This may mean little or no investment but also no increased protection. Selecting one of the two alternatives is a deliberate act of the DM. After that the random occurrences of nature take over beyond the control of the DM.

6.4 Objective Risk Evaluation

6.4.1 Variance and Standard Deviation as Risk Measures

Objective risk evaluation uses measures that do not depend on subjective judgment, at least not with the purpose of satisfying personal risk perception. Only what one usually calls *hard facts*, observed data, enter the decision analysis. Suppose Alternative 1 has two possible outcomes with net benefits NB_{11} and NB_{12}, with probabilities p_{11} and p_{12}, respectively; Alternative 2 has net benefits NB_{21} and NB_{22} with probabilities p_{21} and p_{22}, respectively. For each alternative, the probabilities of the two mutually exclusive outcomes satisfy

$$p_{11} + p_{12} = 1 \text{ and } p_{21} + p_{22} = 1,$$

respectively. The expected values of the two alternatives then are

$$EV_1 = p_{11} NB_{11} + p_{12} NB_{12} \tag{6.4-1}$$

and

$$EV_2 = p_{21} NB_{21} + p_{22} NB_{22} \tag{6.4-2}$$

It is important to realize that the expected value of the decision may not have a chance of occurring, as is the case here. Only one of the possible outcomes can occur, and the expected value is not associated with a possible outcome. This usually is the case in practical problems that have only a few discrete outcomes. The mean is a possible outcome only if a very large number of outcomes exists, as may be the case in statistical problems, or if the outcomes can be described by a continuous probability distribution.

Figure 6-5 illustrates a decision with two alternatives each having two chance outcomes: Alternative 1: 8 (gain) and -4 (loss), each with a probability of $p = 0.5$; and Alternative 2: 4 (gain) and 0, each with a probability of $p = 0.5$. The expected outcomes of the alternatives are

$$EV_1 = 0.5 \cdot 8 + 0.5 \cdot (-4) = 2$$

$$EV_2 = 0.5 \cdot 4 + 0.5 \cdot 0 = 2$$

According to the expected value, both alternatives are the same. The exposure to loss is, however, quite different. The worst outcome of Alternative 1 is a loss of -4, with 50 % probability, whereas the worst outcome of Alternative 2 is 0, with 50 % probability, no gain, but also no loss. A *risk-prone* DM may prefer Alternative 1 in the hope of gaining 8 instead of only 4 by Alternative 2. But going for the big gain means exposure to a loss of -4. A *risk-averse* DM may prefer Alternative 2 in the hope of winning 4, but at the same time avoiding a loss. Obviously, the   *EV* of these alternatives is not a criterion to judge the exposure to loss or the risk, namely facing one of the real outcomes that are associated with these two alternatives.

The variance measures the spread among outcomes. The larger the spread, the more uncertainty is associated with the outcomes, as very different results are possible.

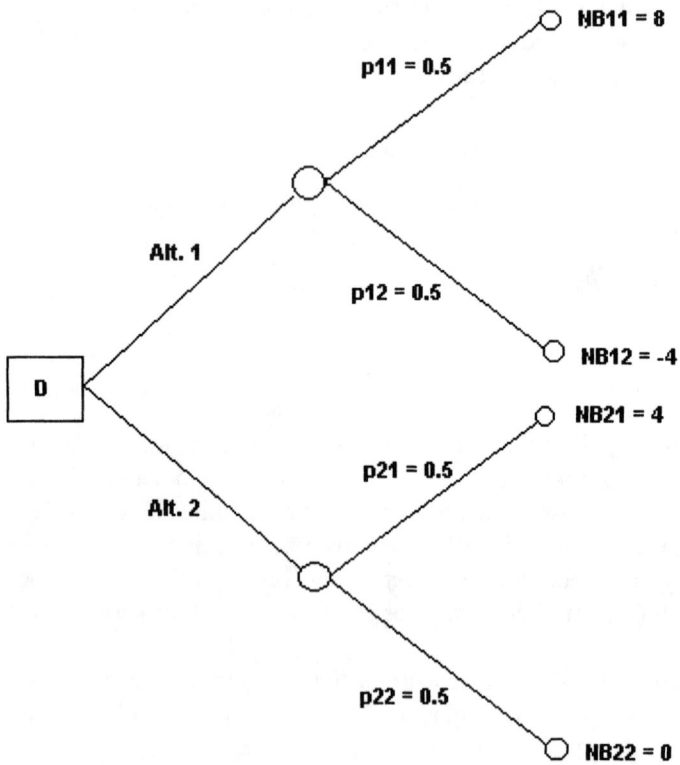

Figure 6-5: Decision between Alternative 1 and Alternative 2 that have the same expected value, $EV_1 = EV_2 = 2$, but different variances: $Var_1 = 36$, and $Var_2 = 4$. Alternative 1 has a much larger variance, a larger gain, but also a larger loss. It is the riskier of the two alternatives.

For the preceding example, the variances of the two alternatives are

$$Var_1 = p_{11} (NB_{11} - EV_1)^2 + p_{12} (NB_{12} - EV_1)^2 \tag{6.4-3}$$

$$= 0.5 \cdot (8 - 2)^2 + 0.5 \cdot (-4 - 2)^2 = 36$$

$$Var_2 = p_{21} (NB_{21} - EV_2)^2 + p_{22} (NB_{22} - EV_2)^2 \tag{6.4-4}$$

$$= 0.5 \cdot (4 - 2)^2 + 0.5 \cdot (0 - 2)^2 = 4$$

Since $Var_1 > Var_2$, Alternative 1 is clearly riskier than Alternative 2. This intuitively makes sense because of the large loss associated with Alternative 1. Therefore, the variance is an objective risk measure. The disadvantage of using the variance is that it is not in units commensurate with the quantity measured, here cost. An alternative measure is the standard deviation, $\sigma = Var^{1/2}$, which has the units of a cost.

Example: A decision is to be analyzed of whether an object that could be destroyed by fire is to be insured (after an example by Lapin, 1978, p. 742). An insurance offers to insure the replacement value of \$120,000 by a \$750 annual premium with a \$200 deductible. The insurance bases its premium on an approximately 0.006 probability that the object will have to be replaced during the insurance period of one year. The insurer also requires a deductible of \$200. If the owner does not insure and carries the risk himself, he pays nothing per year, but if a fire occurs, he faces the replacement value of \$120,000. The decision is illustrated in Figure 6-6. In risk analysis, the real outcomes count, not the expectations. This is borne out by the insurance example. The interesting fact is that the possible outcomes of "buy insurance" alternative come close together, ranging from \$750 to \$950, respectively, whereas for the "no insurance" alternative they are very dispersed, ranging from 0 to \$120,000. The "no insurance" alternative has the smaller expected value and could be falsely interpreted as the optimal alternative were it not for its high risk content which is clearly expressed by the very large standard deviation compared to that for the "buy insurance" alternative. A risk-averse decision maker will select the "buy insurance" alternative for the transfer of the risk of losing \$118,000.

6.4.2 Decision by Hypothesis Testing

A traditional method used in statistics is deciding whether a parameter obtained from a test is representative of a desired property. The procedure requires a test statistic and the definition of two regions, the *acceptance region* and the *critical region* or *rejection region* of the test statistic. Usually two hypotheses are formulated, one that proposes no change of the process represented by the test statistic and the other that proposes a change. The *null hypothesis* is usually used to test for no change, whereas the *alternative hypothesis* tests for change. As long as the test statistic is in the acceptance region, the null hypothesis cannot be rejected. The acceptance and rejection regions are separated by the *critical value* or *demarcation point* as illustrated in the sketch. This point can be set arbitrarily or it

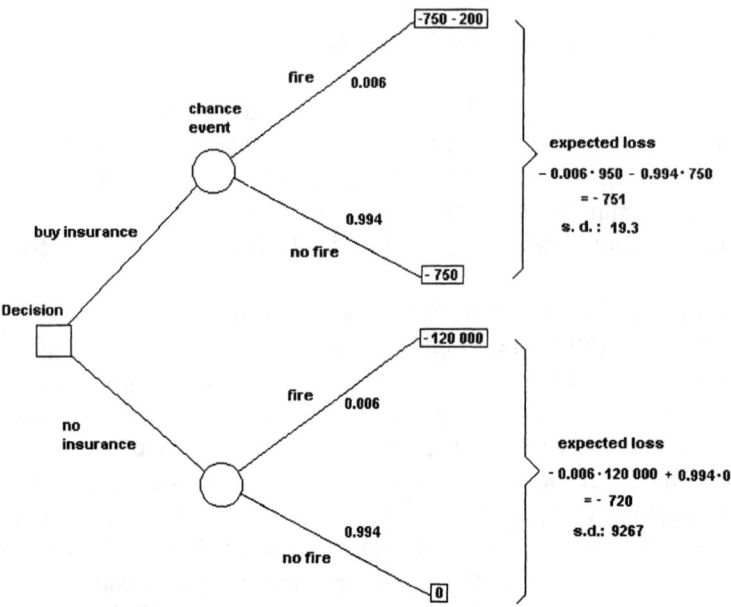

Figure 6-6: Decision between buying insurance and not buying insurance. Buying insurance at a premium of $750 and a deductible of $200 results in annual cost of $750, as long as there is no fire. In the case of a fire, the premium plus the deductible are lost. The maximum loss with insurance is $950. In case of not buying insurance, there is no annual expense. But in case of fire, $120,000 are lost. The difference in risk between the two decisions is expressed by the different standard deviations of the possible outcomes of each alternative. A high standard deviation indicates a high dispersion among outcomes and a high risk.

can be calculated. If the test statistic is a sample mean, $\overline{X}$, as is often the case, then, because of the asymptotically normal distribution of sample means (Section 3.4.2), the normal pdf can be used to calculate the critical point so as to make it compatible with another test parameter, the significance level, α. There may be advantages for both ways: setting the critical point and use the normal pdf to calculate the significance level, or setting the significance level and compute the critical point (Lapin, 1978, p. 280).

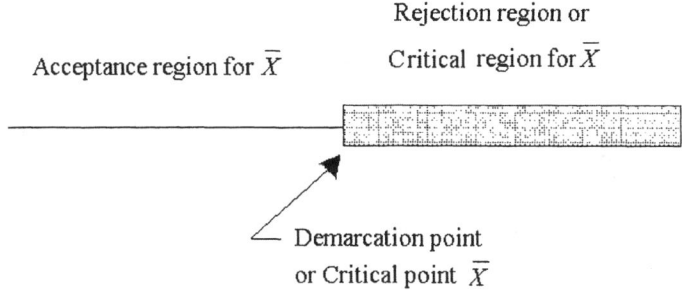

The demarcation point may also be to the left of the acceptance region if the test statistic is tested for exceeding the critical region. Also, a critical region may be on both ends of the acceptance region if the test statistic is tested for being within a double-sided bounded acceptance region.

Suppose the demarcation point is chosen such that 90 % of the test results are expected to fall into the acceptance region and the rest fall into the critical region. Then, if the test result falls into the acceptance region, the null hypothesis cannot be rejected. If the test result falls into the critical region, then the null hypothesis is said to be *rejected at the 10 % significance level*. Statistical tests generally are not powerful enough to determine the true "acceptable" distribution. They are only powerful enough to reject clearly wrong distributions. Therefore, *accepted* means the hypothesis "cannot be rejected." Walpole and Myers (1989, p. 288) state that "the acceptance of a hypothesis merely implies that the data do not give sufficient evidence to refute it. On the other hand, rejection implies that the sample evidence refutes it."

Hypothesis testing is not entirely objective; it includes objective and subjective elements. The sampling distribution and the sampling parameter are determined by objective data obtained from tests. The demarcation point and the significance level are subjective choices, but only one or the other can be independently set. Testing the null hypothesis has four possible outcomes, two correct outcomes and two erroneous outcomes, as summarized in Table 6-9.

Table 6-9: Decision Table for Testing the Null Hypothesis, H_0 (after Lapin, 1978, p. 279)

Alternative Actions	Possible States of Null Hypothesis H_0	
	H_0 is true	H_0 is false**
Accept* H_0	**Correct decision** Probability $1 - \alpha$	Type II error Probability β
Reject H_0	Type I error Probability α	**Correct decision** Probability $1 - \beta$

Notes: *Some statisticians advocate that "cannot be rejected" should be used instead of "accepting" the null hypothesis. After some mulling over the problem, the decision was made here to use "accepting" whenever the double negation of the statistical idiom "not rejecting" becomes too cumbersome. But the reader should know that problems exist with how to express oneself statistically correctly. **Some statisticians frown on calling the null hypothesis "true" or "false" with some probability. It should be said that the hypothesis was rejected at a specified significance level (Benjamin and Cornell, 1970, p. 408 and p. 622).

Suppose null hypothesis H_0 stands for no change and the alternate hypothesis H_1 stands for change. A type I error occurs if H_0 is rejected at significance level α when in fact H_0 is true and H_1 is false. A type II error occurs by not rejecting H_0 when H_0 is false and H_1 is true. The rejection of H_1 occurs at a significance level β. Both errors can be costly. The choice of the significance level directly affects the type I error and through it also affects the type II error. A reduction of the type I error leads to an increase in the type II error. The normal distribution can be used as a sampling distribution because it is the limiting distribution for sums of random variables, such as means (see Section 3.4.2).

The general procedure of hypothesis testing is summarized as follows:

1. Specify the null hypothesis H_0 and the alternate hypothesis H_1.
2. Specify the test statistic and its distribution parameters.
3. Specify the significance level α.
4. Determine the critical value of the test statistic that divides its range into an acceptance region and a rejection region for H_0, based on the assumption that H_0 holds with a rejection probability α.
5. Carry out the test by comparing the test statistic with the defined regions: If the test statistic falls into the acceptance region, the null hypothesis cannot be rejected. If the test statistic falls into the rejection region, the null hypothesis is rejected and the test result is

"significant at level α." In other words, α is the probability that H_0 is rejected while it may be true.

6. Carry out type II error analysis as needed using the alternative hypothesis H_1.

In the following, the hypothesis testing procedure is illustrated by using the example of an upper-tailed test, as illustrated in Figure 6-7:

1. Suppose the null hypothesis states that an output in the form of a population mean concentration, μ_0, has not changed and will continue to produce the present value or a smaller value. Hence, the null hypothesis is H_0: $\mu \leq \mu_0$, and the alternative hypothesis is H_1: $\mu > \mu_0$, where μ is a mean different from the present mean μ_0.

2. As long as μ is smaller than μ_0, it is in the acceptance region; if it exceeds μ_0, it is in the critical region. The test statistic is the sample mean $\overline{X}$ and its distribution, which is $N(\mu_0, \sigma_{\overline{X}}^2)$, with

$$\sigma_{\overline{X}} = \frac{\sigma}{\sqrt{n}} \approx \frac{s}{\sqrt{n}} \tag{6.4-5}$$

where $\sigma_{\overline{X}}$ is the standard deviation of the sample mean $\overline{X}$; σ is the population standard deviation of the unchanged process μ_0; if σ is not known, it can be approximated by the sample standard deviation, s; n is the sample size used in the calculation of $\overline{X}$ and s. From Equation (6.4-5), it follows that $\sigma_{\overline{X}}$ is usually much smaller than σ or s, so that $\overline{X}$ is increasingly tightly distributed around μ_0 with increasing n.

3. A demarcation point or critical value is selected such that a small exceedance of μ_0 will not immediately force a rejection of the hypothesis, but it will be rejected when some significance level α is exceeded. The sampling distribution of the test statistic, $N(\mu_0, \sigma_{\overline{X}}^2)$, implies an unchanged μ_0 for the location of the $\overline{X}$ distribution. The normalized variable of the $\overline{X}$ distribution is

$$z = \frac{\overline{X} - \mu_0}{\sigma_{\overline{X}}}. \tag{6.4-6}$$

The critical value of the test statistic follows from Equation (6.4-6) by solving for $\overline{X}$ with z being computed by inverting the cdf $F(z_\alpha) = 1 - \alpha$. Typically, α is chosen as 0.1, 0.05, and 0.01. Values of z for frequently used one-sided and double-sided critical regions, z_α and $z_{\alpha/2}$, are as follows:

Significance Level	Normalized Variable	
α	$z_{\alpha/2}$	z_α
0.10	1.64	1.28
0.05	1.96	1.64
0.01	2.57	2.33

4. The critical value of the test statistic that divides its range into an acceptance region and a rejection region for H_0 follows from Equation (6.4-6) as

$$\overline{X}_u = \mu_0 + z_\alpha \, \sigma_{\overline{X}} \, . \tag{6.4-7}$$

5. The test of H_0 requires one to check whether the sample statistic $\overline{X}$ falls into the acceptance region. For the upper-tailed test, the probability statement for this to happen is

$$P[\overline{X} \leq \mu_0 + z_\alpha \sigma_{\overline{X}} | H_0] = 1 - \alpha \quad , \tag{6.4-8}$$

which says that the probability that the test statistic is in the acceptance region, given H_0 is true, is $1 - \alpha$. The probability of the sample statistic to fall into the rejection region is

$$P[\overline{X} > \mu_0 + z_\alpha \sigma_{\overline{X}} | H_0] = \alpha \, . \tag{6.4-9}$$

If Equation (6.4-9) is true, the result is said to be statistically significant at level α.

6. It is possible that $\overline{X}$ belongs to a population with another mean; this means that μ_0 has changed to μ_1; the test statistic is distributed as $N(\mu_1, \sigma_{\overline{X}}^2)$; and the alternative hypothesis is true: $H_1: \mu > \mu_0$. This case is illustrated in Figure 6-7. The test statistic is now distributed around μ_1. The part of the sampling distribution that reaches to the left into the acceptance region of the null hypothesis produces the type II error of size β. The probabilities of making a decision error are expressed as

$$P(\text{type I error}) = P[\text{reject } H_0 | H_0 \text{ is true}] = \alpha \tag{6.4-10}$$

and

$$P(\text{type II error}) = P[\text{accept } H_0| H_0 \text{ is false}] = \beta \qquad (6.4\text{-}11)$$

where $P[.]$ are conditional probabilities that rejection or acceptance occurs when the state of H_0 demands the opposite action; α is a preset significance level; and β is computed by evaluating the alternative hypothesis. In the case illustrated in Figure 6-7, according to the rules of calculating the non-exceedance probability $F(z_u)$, which represents β, one obtains

$$\beta = 0.5 - F_0(z_u), \qquad (6.4\text{-}12)$$

with

$$z_u = \frac{\overline{X}_u - \mu_1}{\sigma_{\overline{X}}} \qquad (6.4\text{-}13)$$

where $F_0(z_u)$ is the part of the cdf on the positive branch of the z-axis usually tabulated in lookup tables. For the case illustrated in Figure 6-7, z_u is negative because $\overline{X}_u - \mu_1 < 0$. For negative z, $\beta = F(z) = 0.5 - F_0(z) < 0.5$. Suppose μ_1 is located to the left of $\overline{X}_u$, inside the acceptance region. This would make $\overline{X}_u - \mu_1 > 0$, and $\beta = F(z) = 0.5 + F_0(z) > 0.5$. In Figure 6-7 the null hypothesis cannot be rejected as long as $\overline{X}$ falls into the acceptance region despite a high type II error. A summary of null hypothesis formulations for the three possible configurations of acceptance and rejection regions is given in Table 6-10. The three cases are illustrated in Figures 6-7, 6-8, and 6-9.

Table 6-10: Summary of Null Hypotheses and Test Statistics for Hypothesis Testing

Null Hypothesis	Accept If		Reject If	
Upper-tailed test:				
$H_0 : \mu \le \mu_0$	$\overline{X} \le \overline{X}_u$	$z \le z_\alpha$	$\overline{X} > \overline{X}_u$	$z > z_\alpha$
Lower-tailed test:				
$H_0 : \mu \ge \mu_0$	$\overline{X} \ge \overline{X}_l$	$z \ge -z_\alpha$	$\overline{X} < \overline{X}_l$	$z < -z_\alpha$
Two-sided test:				
$H_0 : \mu = \mu_0$	$\overline{X}_l \le \overline{X} \le \overline{X}_u$	$-z_{\alpha/2} \le z \le z_{\alpha/2}$	$\overline{X}_l > \overline{X} > \overline{X}_u$	$-z_{\alpha/2} > z > z_{\alpha/2}$

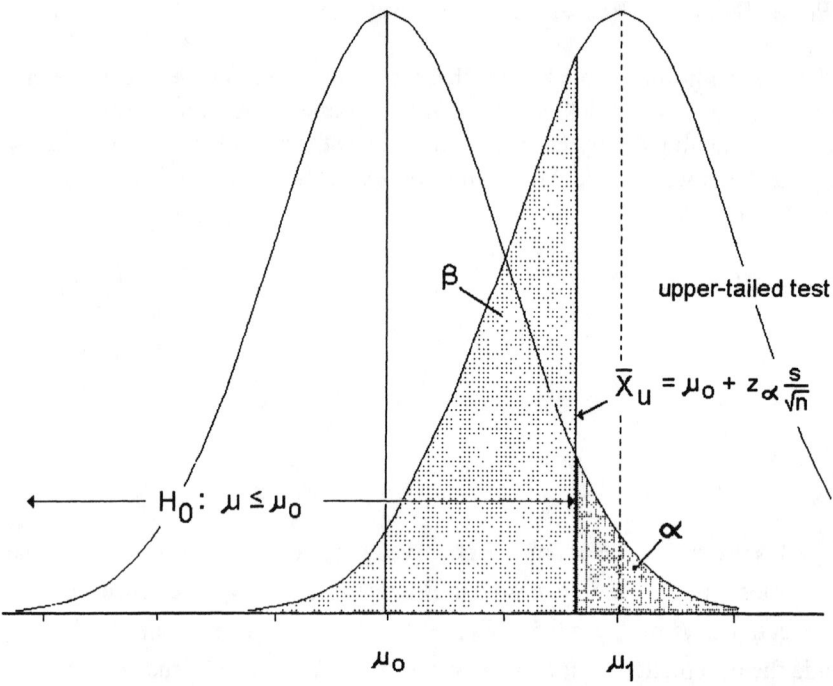

Figure 6-7: Upper-tailed test. The null hypothesis H_0 is not rejected if the test result $\overline{X}$ is less than μ_0. To avoid rejection of H_0 when it is true, thereby committing a type I error, an allowance is made for the test result to reach the critical value $\overline{X}_u$ before H_0 is rejected with probability α. It is said that H_0 cannot be rejected at the significance level α. The null hypothesis is based on the expectation of the sample mean being μ_0. If the sample mean is actually μ_1 and H_0 is not rejected, then a type II error is incurred. If α is made small, it is more likely that a changed μ is included in the null hypothesis thus increasing the probability β of the type II error.

Both errors cannot be reduced at the same time. Therefore, the null hypothesis in its formulation should emphasize keeping the type I error small and making the type II error less serious than the type I error. For example, if a new batch of concrete should meet a minimal strength to minimize the probability of structural failure, then an upper-tailed null hypothesis would be: $H_0: \mu \le \mu_0$, i.e., the concrete is too weak. As long as the test statistic falls into the acceptance region, the null hypothesis cannot be rejected, meaning the concrete is judged as not meeting the required strength. If the test statistic belongs to an alternative distribution with a mean greater than required strength, and the test statistic still falls into the acceptance region (concrete rejection), then the contractor is falsely blamed for low-quality concrete, but the structure is not endangered. Hence, the type II error of not rejecting the null hypothesis when it is false, i.e., erroneously rejecting good quality concrete, is less serious than the type I error of rejecting the null hypothesis when it is true, i.e., erroneously accepting poor quality concrete (Goldberg, 1986, p. 275).

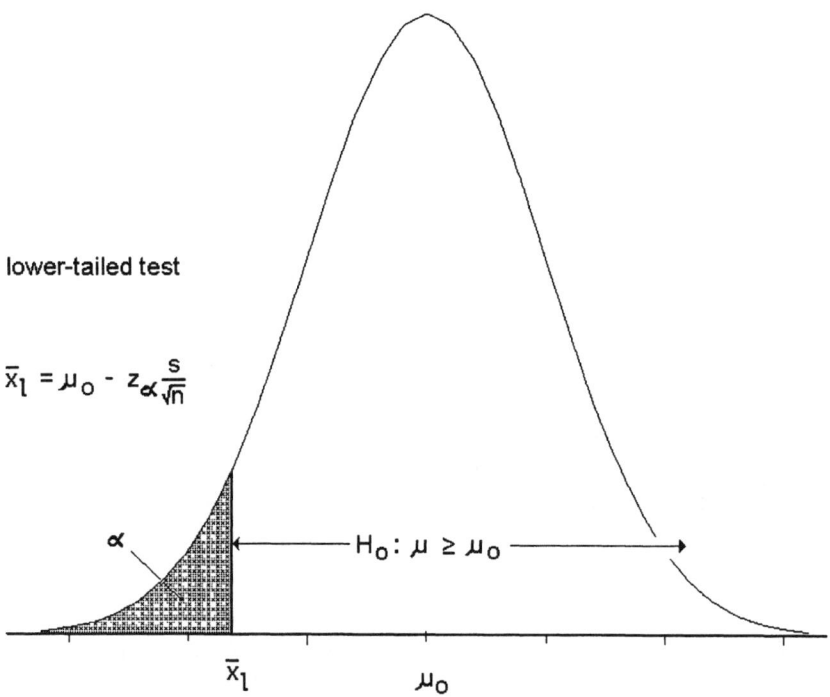

Figure 6-8: Lower-tailed test. The null hypothesis H_0 cannot be rejected as long as the test statistic falls into the acceptance region to the right of the critical value $\overline{X}_l$. If the test statistic falls into the critical range to the left of $\overline{X}_l$, H_0 is rejected and the test result is said to be statistically significant at level α.

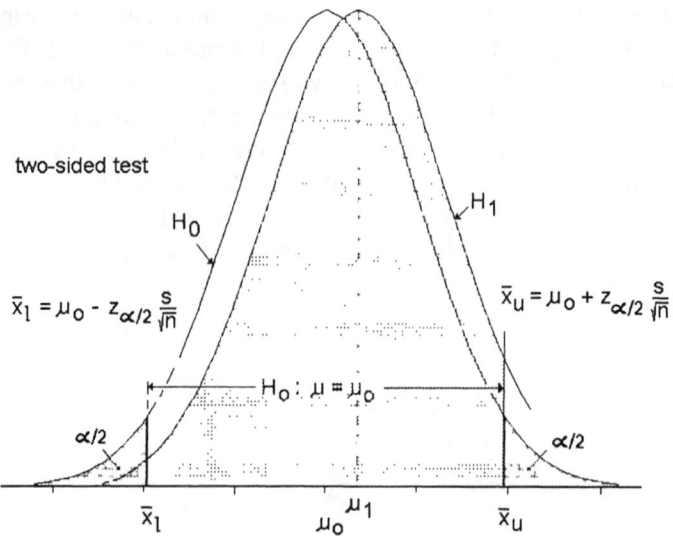

Figure 6-9: Two-sided hypothesis test. The null hypothesis H_0: $\mu = \mu_0$ cannot be rejected if the test statistic $\overline{X}$ falls into the acceptance region $\overline{X}_l \leq \overline{X} \leq \overline{X}_u$. If it falls into one of the critical regions on either side of the acceptance region the null hypothesis is rejected and the test result is said to be statistically significant at the significance level α. A type II error is committed if $\overline{X}$ falls into the acceptance region but is actually distributed according to the alternative hypothesis H_1: $\mu = \mu_1$. The probability of the type II error β is the shaded area over the range $\overline{X}_l \leq \overline{X} \leq \overline{X}_u$.

Often samples are small, $n < 30$, where n is the number of elements in the sample. Then z_α must be replaced by t_α of the Student's t distribution (Section 2.7.9). If the population also is small, then the *finite population correction factor*, a number smaller than 1, must be used to determine the band width around the mean, which has the effect of reducing the width of the confidence interval around the mean. In the case of the *two-sided test* based on a small sample from a small population, the test criterion may look like the following:

$$\mu_0 - z_{\alpha/2}\frac{s}{\sqrt{n}}\sqrt{\frac{N-n}{N-1}} \leq \overline{X} \leq \mu_0 + z_{\alpha/2}\frac{s}{\sqrt{n}}\sqrt{\frac{N-n}{N-1}} \qquad (6.4\text{-}14)$$

where $z_{\alpha/2}$ is the normal standard variable for a predetermined $\alpha/2$, or $t_{\alpha/2}$ of the Student's t, if $n < 30$. The finite population correction factor, $[(N - n)/(N - 1)]^{0.5}$, approaches 1 as N becomes large and n remains small compared to N. The *lower-tailed test* would use the left-hand side, and the *upper-tailed test* would use the right-hand side of Equation (6.4-14).

6.5 Subjective Risk Evaluation

6.5.1 Personal Probability

The mathematical definition of probability is the ratio of occurrences of a special event in a total number of trials when the number of trials approaches infinity. This probability is called *statistical probability* or objective probability and is based on observation data. Savage (1972, p. 3) calls this view of probability the objectivistic view which accepts only probabilities based on hard data. This probability measures the repetitions of a special event in a process over time or space. Another kind of probability "measures the confidence ... in the truth of a particular proposition." This view is called the personalistic view (Savage, 1972, p. 3 and pp. 28–30). This view of probability is contested because it is not based on the observation of repetitive events. Probability based on judgment, called *personal probability*, is the type of probability of interest here because it is indispensable for the quantification of risk. Personal probability is a subjective degree of possibility. If something can happen it is possible and if it can not happen it is impossible. If possibility approaches certainty, then probability approaches 1, and if possibility approaches impossibility, then probability approaches 0. Personal probability is based on experience, judgment, and on the attitude of a person toward risk. It is not based on hard data. It may be different for different persons, even if it refers to the same event. Therefore it is also called *subjective probability* in contrast to *objective probability*. If two persons are called upon to give an estimate of repair time, one may come up with 10 h, the other with 15 h. If these numbers are each person's best guess, one can assume they are their 50 % probability estimates (mean). It is also possible that instead of estimating the probability of a specific number, better agreement among experts can be obtained on a range of numbers, say from 5 h to 20 h, all having the same probability. This would mean that a subjective estimate of a repair time pdf is being suggested. An engineer trained in probabilistic thinking may well be able to cast his reasoning into the quantitative form of a subjective probability distribution. The solicitation of personal probabilities by interrogating a person through a more elaborate procedure that leads to a measure of personal attitude toward risk was proposed by Von Neumann and Morgenstern (1953) and will be described in the next section.

Judgmental establishment of a Gaussian pdf in Section 2.6.5 shows how a pdf can be constructed with nothing more than a variable range. Any other type of a personal pdf can be developed by eliciting probabilities through interview with an expert (Benjamin and Cornell, 1970, p. 542; Lapin, 1978, p. 773). The questions establish the beginning point and ending point of the range and some probability estimates for characteristic values of the range: mid point, quarter points, one eighth points, and perhaps one sixteenth points. Beyond 1/16 of the range no meaningful probability assignment may be possible. This splitting process can be done also by developing a probability tree. The first split of the range represents the root branches. The next split grafts secondary branches onto each root branch, and so on. Each new set of branches that splits from one previous branch forms a mutually exclusive and exhaustive set whose probabilities add to 1. The joint probabilities along each path through the tree produce the probabilities of the final range partitions (histogram bars). The same result can also be accomplished by developing a successively more refined histogram starting with two or more primary subdivisions of the range and subsequently splitting the subranges and giving them probabilities.

Instead of developing a pdf by a histogram one can develop a cdf. A cdf is a monotonically rising function, often with a typical S shape, and is therefore relatively easy to construct. The derivative of such a personal cdf then becomes a personal pdf. Depending on how much is know about such curves, such pdfs may assume symmetric, asymmetric, or even double-peaked shapes (Lapin, 1978, p. 777). The development of a personal cdf and pdf may evolve iteratively until a shape is found that best reflects what is known about the problem. If little or nothing is known about the probabilities over a range, it is always possible to give all values of the range the same probability, which amounts to specifying a uniform pdf.

It is not only the probability that may have to be given a personal touch for quantitatively approaching a real world probabilistic problem. In traditional decision theory, when dealing with uncertainty the expected value is used as decision criterion. It is known, however, and it will be shown in Section 6.5.2 that expected values are not risk-sensitive; in other words, they do not reflect the possible range of outcomes of which may actually happen. The expected value in the form of a risk premium, however, is a risk measure, given it represents a self-insurance premium or an insurer's premium. This chapter's main thrust is aimed at modifying the expected value in such a way that it is also a risk measure of the personal attitude toward risk. This is achieved through the use of personal utility functions, which allow a quantification of the risk attitude of either the insured or the insurer.

6.5.2 Subjective Risk Measures

The subjective nature of risk has been recognized for a long time, and methods have been proposed for its evaluation (Von Neumann and Morgenstern, 1953; Benjamin and Cornell, 1970, pp. 578–581; Raiffa, 1970, p. 289). In practical engineering, quantitative methods for calculating risk have yet to catch on. The following quantitative risk measures are discussed:

- Indifference probability and utility function
- Expected utility and certainty equivalent
- Risk premium and insurance premium.

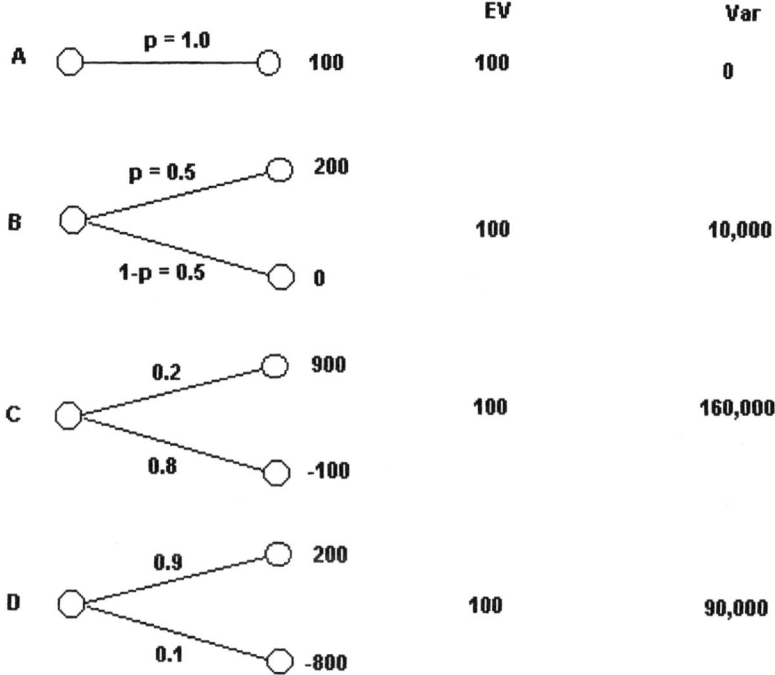

Figure 6-10 : Examples of chance branches which, in contrast to deliberate decisions, are not under the control of the decision maker and are referred to as lotteries. They happen to have the same expected values (*EV*) but different variances (*Var*), demonstrating that *EV* by itself is not a decision criterion.

Von Neumann and Morgenstern (1953) proposed a method for measuring a DM's perception of risk by *indifference probability*. A DM is presented with a chance branch, called a *lottery*, that has two chance outcomes, a favorable outcome and an unfavorable outcome. He is then asked the following question:

"What probability would the favorable outcome need to have, in order for you (the DM) to be *indifferent* between accepting the lottery with the possibility of one of the two outcomes, or a known value for certain?"

When thinking about answering the question the emphasis is on "being indifferent" in contrast to "preferring." Several examples of lotteries are shown in Figure 6-10. Case A is a trivial case that has only one outcome for certain ($p = 1$). A lottery does not work under deterministic conditions. Cases B through D have two chance outcomes: B has outcomes 200 and 0, each with $p = 0.5$, where p is the probability of the favorable branch and $1 - p$ is the probability of the unfavorable branch; C has outcomes 900 and -100, with $p = 0.2$, and $1 - p = 0.8$, respectively; and D has outcomes 200 and -800, with $p = 0.9$, and $1 - p = 0.1$, respectively.

The expected values for lotteries B through D are all the same, $EV = 100$. The variances, however, are different: A has no variance. $Var(B) = 10,000$; $Var(C) = 160,000$; and $Var(D) = 90,000$. These differences indicate that the exposures presented by these lotteries are not the same. C and D are the most risky lotteries because they include high losses or high probability of loss. Thus, the variance is sensitive to the spread of the real outcomes, which in the last two cases reach into the loss domain (negative outcomes).

Lottery B has a favorable outcome of 200 and an unfavorable outcome of 0. The DM, whose attitude toward risk is to be determined, is presented with the choice of this lottery and a value-for-certain of 100. Ignoring the probabilities for the moment, the question is: "what would p_s, the probability on the favorable branch, have to be so that you (the DM) are indifferent between accepting lottery B, or 100 for certain?" In figuring out an answer to the question, one has to consider that lottery B can produce 200 or 0 in lieu of a fixed return of 100. The emphasis of the question is on finding the p_s that makes the decision maker *indifferent*, not the p_s that makes him prefer one or the other. The answer can be found be raising p_s step by step from a low value to an ever higher value. The reader may ask himself if he would be indifferent for $p_s = 0.5$, a 50-50 chance of getting 200 or nothing, in lieu of 100 for certain? Probably not! A *risk-averse* person would perhaps settle for $p_s = 0.9$. With this probability, the chances of getting 0 may be small enough for that person, and there is a good probability that the favorable outcome will happen, but there is still an exposure to getting 0. Assume the DM settles for 0.9, which is then the *indifference probability*, p_s, for the value-for-certain of 100.

Finding an indifference probability for two more values-for-certain further illustrates the approach. If 0 were offered for certain, then indifference with lottery B would occur at $p_s = 0$. Preference for the lottery cannot be denied in this case, because the lottery's unfavorable outcome is 0, but it still offers a chance of winning

200. If 200 were offered for certain, then any $p_s < 1$ would make the lottery clearly inferior to a return for certain of 200, hence, $p_s = 1$. In summary, the elicitation using lottery B has now produced three pairs of values (x, p_s), where x is the value-for-certain, and p_s is the indifference probability: $(0,0)$, $(100, 0.9)$, and $(200, 1)$. Any number of other values can be found by repeating the procedure for other x-values, but usually it can be done only for a few values, as it is difficult to consistently assign a p_s to narrowly spaced values x.

The indifference probabilities for a lottery are most likely not the same for different people. The choice of $p_s = 0.9$ for lottery B may characterize a risk-averse person. A risk-prone person who is eager to get a gain out of a risky situation may become indifferent for $p_s = 0.4$. Hence, the indifference probability is a measure of the attitude toward risk of the person from whom it is elicited. This person can be an individual or a "corporate" person. An organization would act through an appointed DM (or a sort of oracle), a so-called "surrogate decision maker." The representative indifference probability is not always easy to find. The DM must be *truly indifferent* between the value for certain and the outcomes of the lottery; in other words, he must not prefer one over the other even slightly, or the indifference probability is biased.

Indifference probabilities usually are solicited by setting up a so-called *standard lottery* (SL) that spans the range of all values-for-certain, x_i. In Figure 6-11, a SL is attached to each of the values x_1 through x_4. S_1 and S_2 are the values of the x_i that bracket all outcomes of the standard lottery. The x_i are considered as the values-for-certain, and the p_{si} elicited from the DM become the indifference probabilities associated with the x_i. An expectation of these p_{si} is formed, which becomes the indifference probability of the so-called *equivalent standard lottery* (ESL):

$$p_e = \sum_{i=1}^{n} p_i \, p_{si} \qquad\qquad (6.5\text{-}1)$$

where p_e is the total probability of the S_1 for each x_i, $i = 1, ... , n$ or the indifference probability of the ESL; $1 - p_e$ is the complementary probability of the unfavorable outcomes S_2 for all x_i of the ESL. The summation of Equation (6.5-1) includes the universe of possible, mutually exclusive outcomes x_i of a lottery, whose probabilities p_i sum to 1: $\Sigma\, p_i = 1$.

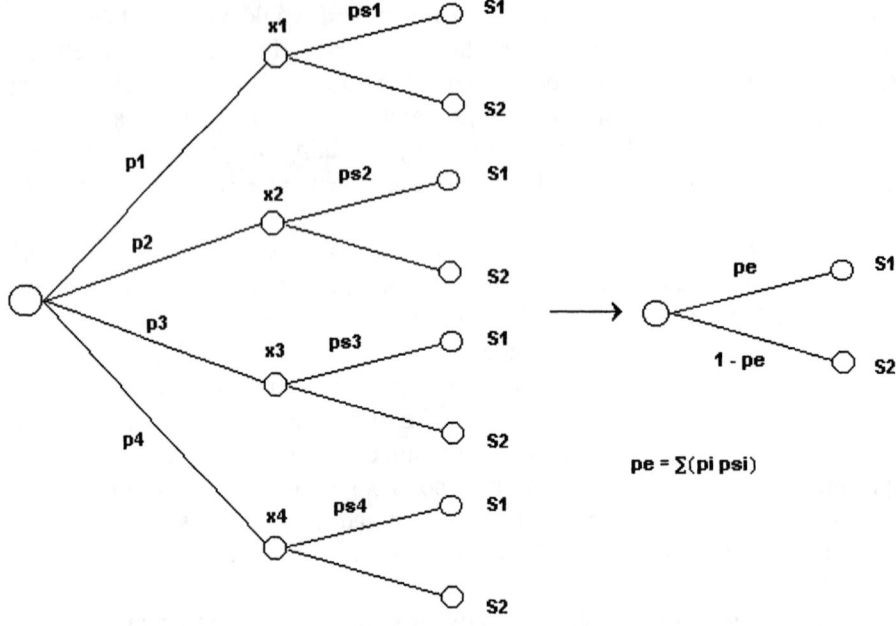

Figure 6-11: A lottery with several possible outcomes x_i is reduced to a two-pronged lottery for comparison with others. For this purpose, a standard lottery (SL) is attached to each outcome x_i, and a p_{si} is determined for each x_i. The total probability of the favorable outcome S_1 is p_e, which becomes the indifference probability of the equivalent standard lottery (ESL). The ESL replaces the multipronged lottery which may represent the probabilistic costs of a decision alternative.

(2) An alternative that leads to four possible outcomes is shown in Figure 6-11. Each outcome x_i has a probability p_i of occurring. Suppose x_1 and x_4 span the outcomes which become S_1 and S_2, respectively, of the SL. This SL is attached to each outcome x_i and a p_{si} is solicited. The multipronged lottery is then reduced to a two-pronged ESL. The indifference probability of the ESL is according to Equation (6.5-1) $p_e = p_1 p_{s1} + p_2 p_{s2} + p_3 p_{s3} + p_4 p_{s4}$. Any other lottery with any number of chance outcomes can be reduced to a two-pronged ESL so that they can be compared with each other. The p_e of the ESL becomes the "utility" of the lottery or alternative.

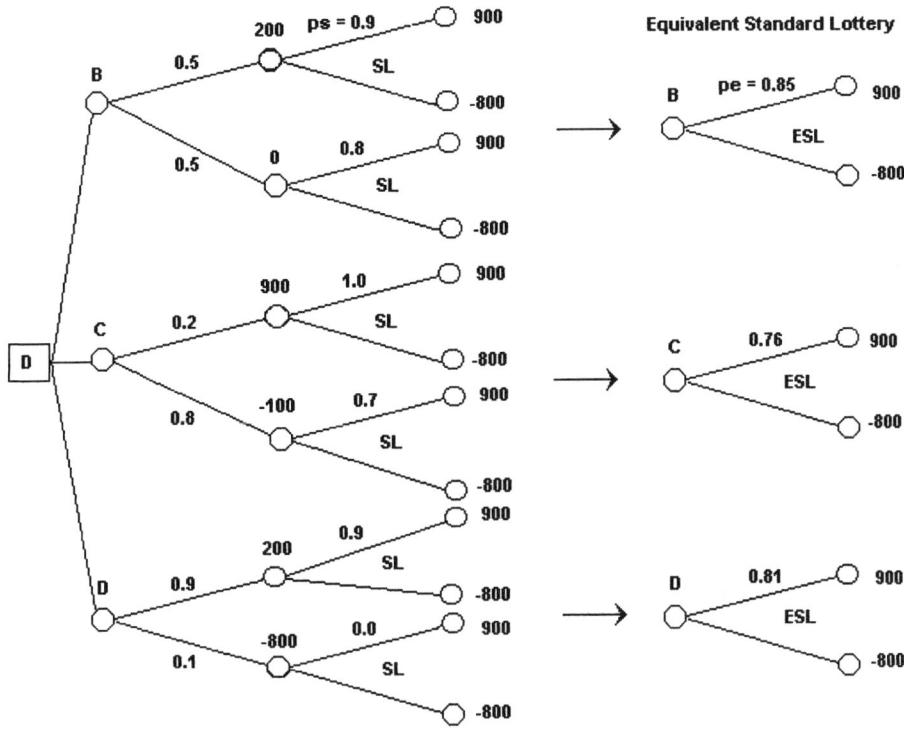

Figure 6-12: Comparing decision alternatives B, C, and D by calculating the indifference probabilities of their equivalent standard lotteries, which are also the utilities of these alternatives.

(3) A decision has three alternatives B, C, and D, with two chance outcomes each, as illustrated in Figure 6-12. Compare the risk inherent in these alternatives. Solution: The decision alternatives B, C, and D are evaluated by first constructing the SL that covers the entire range of outcomes of all alternatives, here $-800 \le x \le$ 900. The SL with branches S_1 and S_2 is attached to the chance outcomes, x_i, of each alternative. Indifference probabilities are then elicited from the DM for each x_i and attached to the favorable branch of the SL. For example, suppose that for alternative B, $p_{s1} S_1 = 0.9$. An ESL is constructed for each alternative by computing their p_e's as the total probability of the favorable outcomes. For alternative B: $p_e = 0.5 \cdot 0.9 + 0.5 \cdot 0.8 = 0.85$, as shown on the right-hand side of Figure 6-12. The ESLs for C and D produce 0.76 and 0.81, respectively. Alternative B has the highest p_e (utility), and C has the lowest. This result is not entirely intuitive. One might think that D should have the lowest p_e, but the high probability on the favorable branch

places it second before C. This ranking is consistent with the one based on the variance in Figure 6-10.

6.5.3 Utility Function

The values-for-certain of Figure 6-11, with a range of $-4 \le x \le 8$, and the corresponding p_{si}, as developed in Example (3), are tabulated in Table 6-11, columns 1 and 2. When one plots the p_{si} against the x_i, they produce a function $p_s = f(x)$. With p_s replaced by the symbol u one gets

$$u = u(x) \tag{6.5-2}$$

where $u(x)$ is called a *utility function*. Usually, but not necessarily, a utility function is a monotonically increasing or decreasing function. This means its gradient does not change sign over its range of validity.

Table 6-11: Elicited Values p_s and Empirical Utility Function Values $u(x)$ for the Chance Outcomes x of the Lottery of Figure 6-11

x	p_s	$u(x)$
(1)	(2)	(3)
-4	0	0
0	0.75	0.77
4	0.95	0.95
8	1.00	0.99

For numerical calculations, it is convenient to fit an analytical function to the elicited utility values, p_s. For the values of Table 6-11, columns 1 and 2 one obtains

$$u(x) = 1 - e^{-0.37(x+4)} . \tag{6.5-3}$$

This function is shown in Figure 6-13. Calculated values of u are juxtaposed to the elicited values p_s in column 3 of Table 6-11. Columns 2 and 3 show reasonably good agreement. Suppose x is a benefit. Then a large benefit has a high utility, whereas a negative benefit or cost has a low utility. The utility being derived from the indifference probability has a range $0 \le u \le 1$. The shape of the curve reveals that as x gets large there is a diminishing incremental utility.

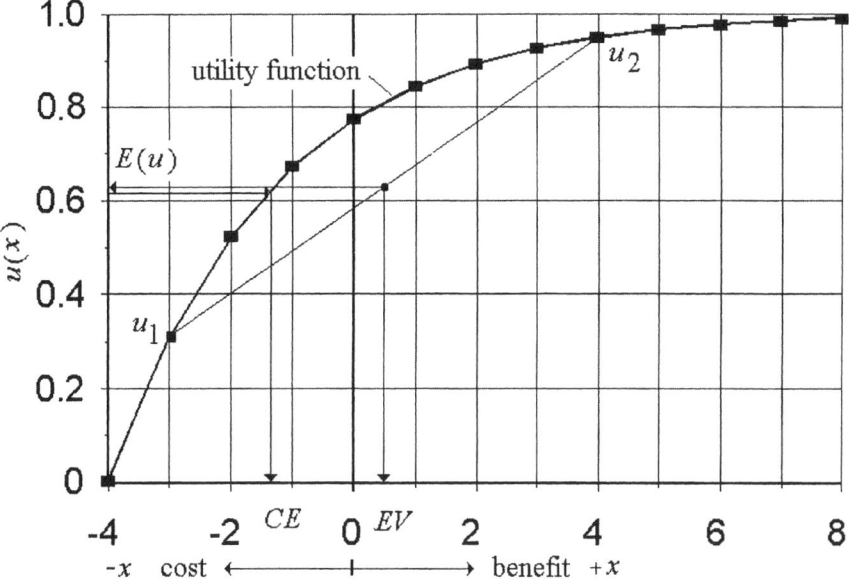

Figure 6-13: The empirical utility function $u(x)$ is assumed to have been elicited from a risk-averse DM for a range of benefits ($+x$) and losses ($-x$). The curve is fitted to experimental points given in Table 6-11, column 2. The expected utility, $E(u)$, is calculated as the probability-weighted sum of utilities u_1 and u_2. For $p_1 = p_2 = 0.5$, the midpoint of the straight line connecting u_1 and u_2 has the ordinate $E(u)$. The abscissa of $E(u)$ is the certainty equivalent CE, which is here located to the left of the expectation of x_1 and x_2, $EV = E(x)$, as a result of the risk-averse shape of the utility function.

Once a utility function is formulated, it allows calculating $u(x)$ as well as the inverse $x(u)$ that is also used in the analysis. The inverse of Equation (6.5-3) is

$$x = -2.7\ln(1 - u) - 4 \qquad \text{for } 0 \le u < 1. \tag{6.5-4}$$

Because of the particular analytical form that was chosen for $u(x)$, the range of the inverse $x(u)$ is limited to $0 \le u < 1$. For $u \to 1$, Equation (6.5-4) is not a good fit. Here, $u = 1$ is simply converted to $x = 8$. This problem can be avoided by choosing another functional form for $u(x)$.

The utility function has a specific shape that is typical of the attitude toward risk of the person or organization whose p_s values it represents. The utility function in Figure 6-13 is concave. This means that its tangent is always above the curve, whereas the straight line connection of two points on the curve is always below the curve. This shape is characteristic of a risk-averse DM from whom the p_{si} were solicited here (Section 6.5.2). A risk-prone DM would be satisfied with much lower p_{si} for small x so that his utility function would take on a convex shape (tangent always below the curve) for which the straight-line connection of two points always lies above the curve (Keeney and Raiffa, 1976, p. 148).

Similar to the expectation of x, an expectation of u can be calculated as

$$E[u(x)] = \sum_{i=1}^{n} p_i u(x_i)$$ (6.5-5)

where the p_i sum to 1 for the n terms $u(x_i)$ that are derived from the n x_i. If an alternative j has $i = 1, ..., n$ chance outcomes x_i with probabilities p_i for which utilities $u(x_i)$ are developed, then the expected utility for alternative j is

$$E_j(u) = \sum_{i=1}^{n} p_{ij} u(x_{ij}) = \sum_{i=1}^{n} p_{ij} p_{sij} = p_{ej}.$$ (6.5-5a)

Equation (6.5-5a) shows that the expected utility of an alternative j also is the indifference probability of the ESL of Equation (6.5-1).

Example: Apply Equation (6.5-5a) to the alternatives 1 and 2 of Figure 6-5. Solution: The outcomes are spanned by $S_1 = 8$ and $S_2 = -4$. Suppose the utility function given by Equation (6.5-3) applies. Instead of going through appending SLs and finding utilities for the NB_{ij}, the $u(x_{ij})$ are calculated by the available utility function for each $x_{ij} = NB_{ij}$. The expected utility of each alternative is obtained from Equation (6.5-5a) as

$$E_1(u) = \sum_{i=1}^{n} p_{i1} u(x_{i1}) = p_{11} u(x_{11}) + p_{12} u(x_{12})$$

and

$$E_2(u) = p_{21} u(x_{21}) + p_{22} u(x_{22})$$

where the p_{ij} are the probabilities of the probabilistic outcomes x_{ij} or NB_{ij} of the alternatives j; and $u(x_{ij})$ are the respective utilities.

6.5.4 Utility Function–Related Risk Measures

The expected utility is the probability-weighted sum of individual utilities. Suppose the sum consists of two probability-weighted utilities, u_1 and u_2, then the expected utility is a linear combination of the two utilities that lies somewhere on the straight line connecting u_1 and u_2. Assuming the probabilities are $p_1 = p_2 = 0.5$, then the ordinate of the middle point is $E(u)$ that can be marked off on the u-axis, as shown in Figure 6-13. If one moves from $E(u)$ back to the right, to the intersection with the utility curve, and down to the x-axis one finds the abscissa x of $E(u)$, which is the *certainty equivalent CE* of $E(u)$. Analytically this inversion is calculated by the inverse of the utility function, Equation (6.5-4). This functional relationship can be written as

$$CE = x[E(u)] \tag{6.5-6}$$

where $x[E(u)]$ is the value x that corresponds to the ordinate value $E(u)$ via the utility function $u(x)$.

For Alternatives 1 and 2 of Figure 6-5 (see also example of Section 6.5.3), one obtains the following utilities:

Alternative 1: $u_1(8) = 1$ $u_2(-4) = 0$
Alternative 2: $u_1(4) = 0.95$ $u_2(0) = 0.75$.

The expected utilities for Alternatives 1 and 2 are

$$E_1(u) = 0.5 \cdot 1 + 0.5 \cdot 0 = 0.5$$

and

$$E_2(u) = 0.5 \cdot 0.95 + 0.5 \cdot 0.75 = 0.85$$

The respective certainty equivalents are

$$CE_1 = -2.7 \ln[1 - E_1(u)] - 4 = -2.13,$$

and

$$CE_2 = -2.7 \ln[1 - E_2(u)] - 4 = 1.12$$

where *CE* can be interpreted as the *risk-adjusted* expected value of a decision alternative.

The certainty equivalent has the units of the original values, x, whereas u is a dimensionless weight. The comparison of the alternatives can be stated in the form of a risk-adjusted expected value as

$$CE_2 > CE_1,$$

which means Alternative 2 is found to be superior to Alternative 1. The risk adjustment for the expectation of Alternative 2 from the $EV = 2$ down to $CE = 1.12$ is small compared to that of Alternative 1, which is reduced from a positive $EV = 2$ to a negative $CE = -2.13$. How this can happen is illustrated in Figure 6-13. A relatively small positive benefit turns into a cost when adjusted for risk attitude.

The difference between the expectation, EV, and the risk-adjusted expectation, CE, is the *risk premium*:

$$RP = EV - CE. \tag{6.5-7}$$

For Alternative 1,

$$RP_1 = EV_1 - CE_1 = 2 - (-2.13) = 4.13.$$

For Alternative 2,

$$RP_2 = EV_2 - CE_2 = 2 - 1.12 = 0.88.$$

The risk premium is a correction of the expected value that reflects the risk attitude of the DM (Keeney and Raiffa, 1976, p. 151). For Alternative 1, RP_1 is more than twice the expected value and wipes out the benefit; for Alternative 2, RP_2 is less than half the expected value and reduces the expected value to less than half. The expected value is also called *risk-neutral*, as it does not reflect risk attitude.

A third risk measure is the *insurance premium*. It is the value for certain that the DM would accept or pay for being relieved of the risky outcomes of the alternatives (lotteries). As was discussed in connection with the derivation of the utility function, the DM becomes indifferent at a level of probability on the favorable branch of an SL (Figure 6-12) between a value-for-certain and the possible real outcomes of the SL. It also was shown that the indifference probability is equal to the utility of the lottery. When the expected utility of all SLs of an alternative is re-transformed into a real value by using the inverse of the utility function, the certainty equivalent is obtained as the risk-modified expectation of the alternative. Keeney and Raiffa (1976, p. 153) define the insurance premium as

$$IP = -CE \tag{6.5-8}$$

where the minus sign reverses the calculated sign of CE. If the alternative has a positive CE (gain), the DM may want to sell the alternative and its possible outcomes for this amount, similar to selling a risky stock for a price. If the CE is negative (cost), the DM may want to pay an amount CE for having the risky outcomes transferred to an insurer. In both cases, the DM either pays or receives an amount for certain, CE, to rid himself of the risky outcomes.

In the example, the insurance premiums are:

for Alternative 1: $IP_1 = 2.13$, for $CE < 0$ (a loss)

for Alternative 2: $IP_2 = -1.12$, for $CE > 0$ (a gain).

For Alternative 1, the insurer may demand a premium $IP_1 = 2.13$ from the client for exchanging the risk of getting either 8 or -4 by a certain payment of 2.13 for a contract period. For Alternative 2, the insurer sees a gain and may *buy the lottery* for $IP_2 = 1.12$ from the risk-averse client. By selling the lottery, the client gets $CE_2 = 1.12$ for sure, whereas the insurer takes the risk of getting 4 or 0. If both insurer and client conduct their own risk analysis, each with his subjective utility function, based on personal (or institutional) risk perception and risk policy, they may come up with different premiums. Through negotiation, or dictated by insurance market conditions, insurer and client may finally settle on an insurance premium that is acceptable to both. If the insurance market is soft, which means capital is chasing risk, the insurer may accept the client's proposal of a low premium. If the market is hard, which means risk is chasing insurers, the client may have to settle for the insurer's premium offer or not get insurance at all.

In principle, the expected value of an alternative represents the insurance premium, as was discussed in Section 6.2.1. The utility function only adds the effect of personal risk attitude or risk perception to the expectation and modifies it into a risk-weighted certainty equivalent. If the alternative (lottery) with uncertain outcomes produces an expected gain that holds up after being adjusted for risk, as the previous Alternative 2, the DM (owner of the lottery) may want to sell the alternative with its uncertain outcomes at a risk-reduced price (premium or lottery ticket) equivalent to CE instead of at EV, like a risky stock. If the lotteries deal with costs, the uncertain outcomes are losses or costs. This case is illustrated in the left hand part of Figure 6-13, but a clear case of exclusive losses or costs is illustrated in Figure 6-14. In this case, $CE > EV$, which means that the risk-averse DM sees a larger expected cost for the alternative than the expected cost. Suppose the utility curve of Figure 6-14 represents the insurers utility function, then the DM has to pay a premium CE that exceeds the risk-neutral expected value EV by the risk premium $RP = CE - EV$, which means an increase in cost over the expected cost EV.

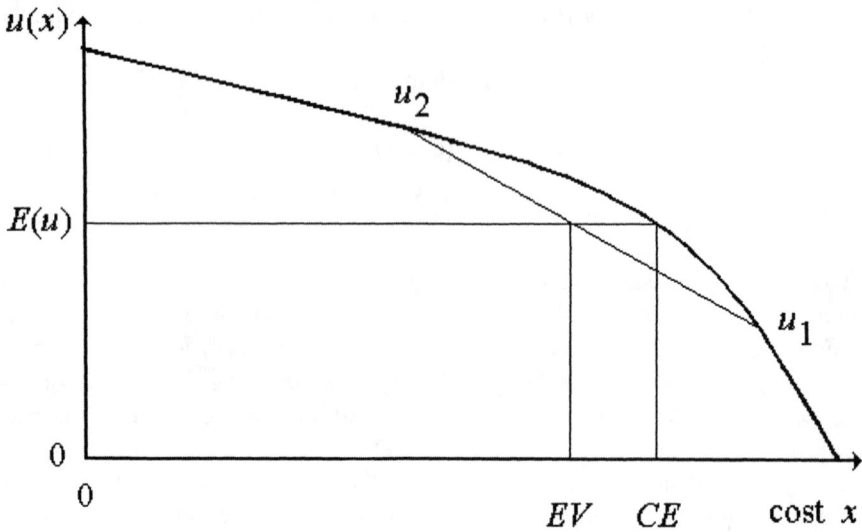

Figure 6-14: Decreasing utility for increasing losses or costs. *EV* is an expected loss and *CE* is the certainty equivalent that exceeds *EV* to reflect the attitude toward risk expressed by the risk-averse shape of the utility function.

The elicitation of the utility function was conducted for a risk-averse DM. If the questions that led to the establishment of the utility function had been posed to a risk-prone DM, a risk-prone utility function would have resulted with a convex shape, a curve that is above its tangent. The analysis of a benefit producing alternative would produce $CE > EV$, because the risk-prone DM would go for the possible high benefits offered by the lottery, as shown by the lower portion of the utility function in Figure 6-15. Also, more complex utility functions may be conceived that include a change in risk attitude from risk-prone to risk-averse, or vice versa, as a function of x. Such a curve is shown in Figure 6-15. A risk-prone approach is non-conservative because it tends to underestimate the exposure to loss (Keeney and Raiffa, 1976, p.157).

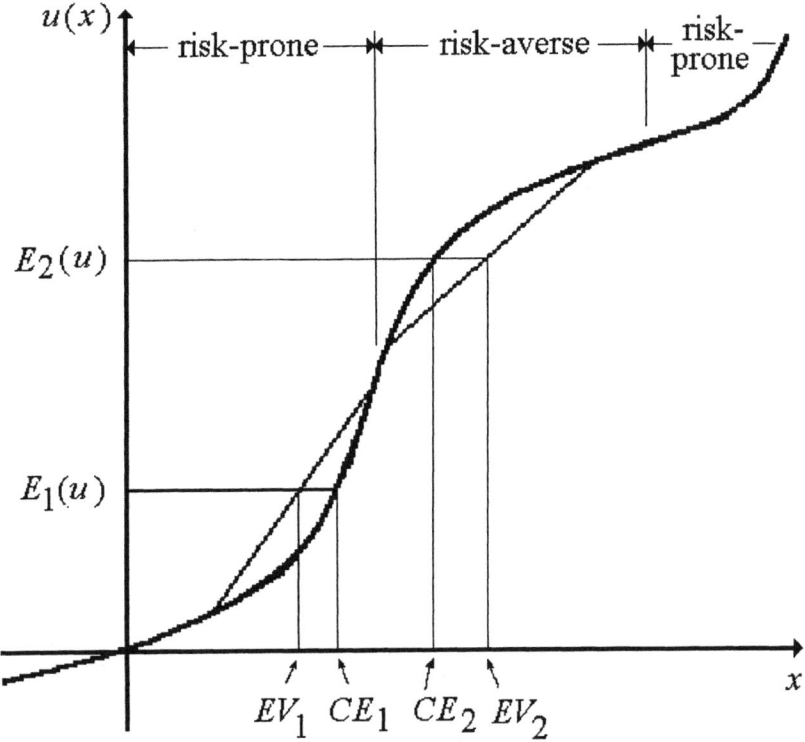

Figure 6-15: Changing attitude toward risk that may be related to changing affordability of risk. The DM's attitude may change from being risk-prone for small risks to being risk-averse when the stakes get bigger, then revert to a more risk-prone attitude for very high stakes.

The utility function can "freeze" the risk attitude of a DM and put it to use when needed. It is comparable to an *expert system* that can conserve an expert's knowledge and make it available when decisions have to be made or when there is no time for extensive deliberations to arrive at a decision. A utility function must be carefully designed, extensively tested, and updated from time to time to reflect the attitude of the DM or that of the institution he represents. Applications of the method can be found in the vast literature that has accumulated on the subject. Only a few examples are cited here (Grayson, 1960; Benjamin and Cornell, 1970, pp. 531– 541; Savage, 1972, pp. 69–104; Keeney and Raiffa, 1976; Lapin, 1978, pp. 740–766; Wunderlich and Giles,

1983, pp. 480–485; Watson and Buede, 1989, pp. 35–59).

Utility theory has received criticism and is not universally accepted. There is definitely ground for improvement of methods for measuring risk. At present there is no method that offers a measure of the subjective dimension of risk that would be superior to the utility method presented here. Perhaps the greatest value of the utility approach lies in bringing a structured approach to risk analysis (Watson and Buede, 1989, pp. 31 and 50).

6.6 Risk Evaluation Example

6.6.1 Dam Rehabilitation Analysis

A dam rehabilitation study of the U.S. Bureau of Reclamation (ACER, 1986) is used as a basis to demonstrate the risk evaluation described in this chapter. Table 6-12 lists the alternatives that were identified as providing various levels of dam rehabilitation. The problem is construed as a decision with a choice of five alternatives, each of which except the "no action" alternative contains two or three sublevels. Thus, a total of 10 alternatives is evaluated. The problem is illustrated by the decision tree in Figure 6-16. It shows the five main alternatives and their sublevels. Only Alternative 1 is shown in detail. The squares mark decisions that result in the choice of an alternative. Once the alternative is chosen and implemented, the resulting project is exposed to the random events as illustrated by the chance branches that point to chance events, such as floods and earthquakes. Their interaction with the structure may activate failure modes that cause damages or costs. There may be many failure modes, such as insufficient spillway capacity, limited reservoir capacity, embankment breaching by overtopping, dam embankment slide, seepage, foundation failure, and so on. Only a few are selected as representative here.

Figure 6-16 shows that Alternatives 2 and 4 can be implemented in steps. Alternative 2.1, for example, proposes raising the dam to eliminate the likelihood of overtopping for the 80 % PMF (probable maximum flood), and Alternative 2.2 proposes additionally raising the dam to further reduce or eliminate the likelihood of overtopping by the 100 % PMF. The branches emanating from F and E represent universes or sample spaces of external events: four floods and three earthquakes, with the numbers on the branches representing their probabilities. External events, such as floods and earthquakes, are independent of the structure or its operation and uncontrollable in magnitude. Floods are to some extent predictable as they are require precursor events in the form of rain or snowmelt. This enables project operation to be accommodating to the heightened seasonal probability of occurrence in the form of an annual operation cycle. Earthquakes are totally unpredictable and can be accommodated only by permanently allocating project capability, such as freeboard. The events within each universe of events are assumed to be unrelated; in other words, a damaging, or more likely, a non-damaging flood and earthquake can happen, most likely at different times during a year. A flood does not favor or preclude an earthquake, and an earthquake does not trigger a flood except for a dam break. A dam can suffer damage from a flood and from an earthquake (spillway damage, overtopping, crack, embankment slide). It is assumed that after a damaging event has occurred the damage is repaired and the next event will encounter a repaired project and cause damage again.

Table 6-12: Dam Rehabilitation Alternatives (ACER, 1986)

Item No.	Alternative No.	Description of Alternative
1	1	No action–leave as is
2	2.1	Raise the dam so that the structure can handle the 80 % PMF (probable maximum flood)
3	2.2	Raise the dam so that the structure can handle the 100 % PMF
4	3.1	Restrict operation so that the structure can handle the 80 % PMF by keeping more flood storage
5	3.2	Restrict operation so that the structure can handle the 100 % PMF by keeping more flood storage
6	4.1	Modify auxiliary spillway for controlling the 60 % PMF
7	4.2	Modify auxiliary spillway for controlling the 80 % PMF
8	4.3	Modify auxiliary spillway for controlling the 100 % PMF
9	5.1	Combine alternatives 2 and 3 for the 80 % PMF
10	5.2	Combine alternatives 2 and 3 for the 100 % PMF

The event branches split into failure mode branches. Each of these branches represents a failure mode that has some probability, given the cause-event occurs. The numbers on the branches are the probabilities that events and associated failure mode occur. The resulting damages have the probabilities of these combined cause-event-failure mode scenarios.

The failure modes that were identified for the study are (ACER, 1986):

- DO: dam overtopping
- ASF: auxiliary spillway failure
- SSF: service spillway failure.

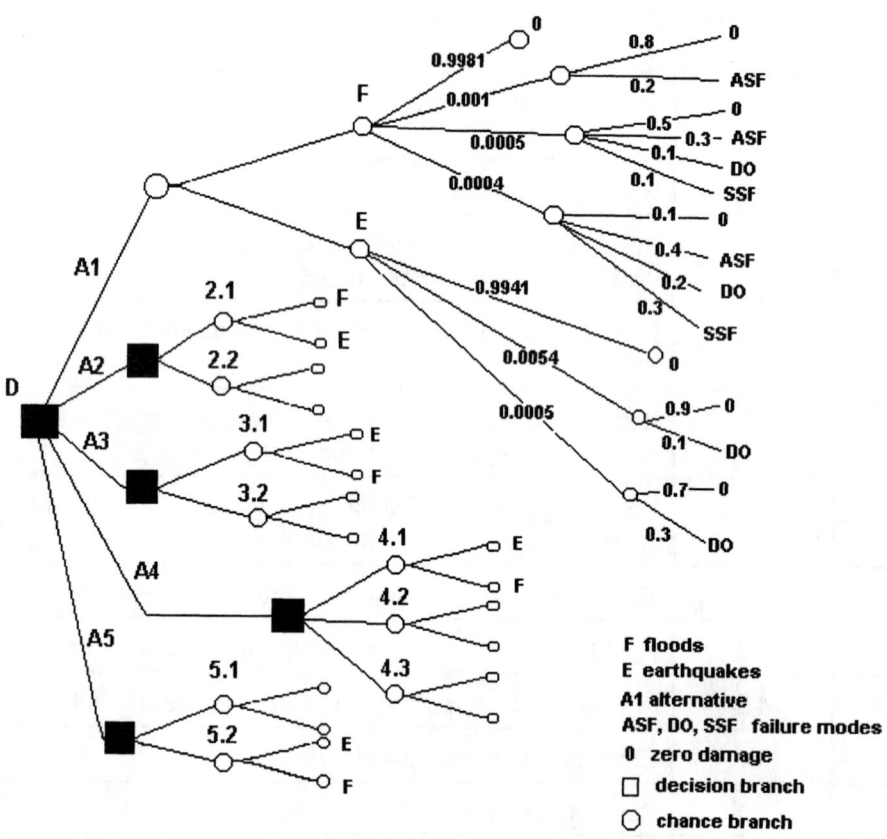

Figure 6-16: Decision tree with alternatives A1 through A5 and their subalternatives. From the decisions emanate the chance branches for floods and earthquakes. From the chance branches emanate the event branches, and from the event branches emanate the failure mode branches and their probabilities. Each branch end point has a cost associated with it (based on data from ACER, 1986). The detailed portion of the decision tree is for Alternative 1, the "leave as is" or "no action" alternative.

The complementary set to the failure mode sets is "normal operation." In the normal operation mode, the structure controls normal events and no damage (at least no structural failure damage) occurs. This mode prevails most of the time and has by far the highest probability of occurrence. Figure 6-16 shows details for Alternative 1. This alternative, as all the others, must cope with the two event universes. For Alternative 1 this produces 16 possible outcomes, eleven for floods and five for earthquakes.

A major part of a risk evaluation is the event-exposure analysis. It determines the damages caused by a particular failure mode. Here, the results of this analysis are taken from ACER (1986). The flood events are expressed in percentages of the PMF. The event-exposure analysis shows that flood events can be controlled without a failure mode up to 0.4 PMF. For larger floods, the probability of no damage decreases. Also, the structure can safely resist earthquakes less than M6.5.

Each damage event, triggered by either flood or earthquake, is the joint occurrence of two events: the natural event E and the structural failure event it may cause. The event probabilities, $P(E)$, are annual frequencies. A return period of "1 in N years" can be construed as an annual frequency of $1/N$ (see also Section 3.7.2). The structural failure F is an event conditioned on a flood or earthquake with probability $P(F|E)$. The rehabilitation alternatives are aimed at reducing or eliminating the probability of failure. A summary of these probabilities for the status quo (Alternative 1) and nine rehabilitation alternatives is given in Table 6-13. The two external event sets (universes) of floods and earthquakes are listed in column 1. Many other events could be specified. If more events are specified, some or all of the events will have smaller probabilities. Whatever the number of events in the event set, the probabilities of each event set, including the normal event, E_0, must add to 1, $\Sigma P(E_i) = 1$, where i are the indices of the events in the set. They run from 0 to $f = 4$ for floods and from 0 to $e = 3$ for earthquakes. Column 2 lists the assumed failure modes that are possible with each event. For small floods and small earthquakes, for which no failure is possible, the failure mode is "None" and is given a probability 1 for all alternatives. For the larger events, various probabilities are assigned to the failure modes, depending on how the structural alternatives cope with these larger events. These failure probabilities for the various alternatives are given in columns 3 though 12 and are the results of a separate structural analysis (ACER, 1986). The failure modes have two mutually exclusive outcomes, failure or no failure. For the larger events, failure and damage is increasingly likely. The exposures in terms of damages in million dollars for each event are given in Table 6-14, column 6. For example, if Alternative 1 is implemented, and a 40 % to 60 % PMF occurs, it is estimated to cause zero damage with 80 % probability, and $700 million damage with 20 % probability.

Table 6-13: Event–Exposure Analysis Results for Dam Rehabilitation Alternatives (from ACER, 1986)

(a) Cause-Event Probabilities

Floods	Probabilities	Earthquakes	Probabilities
0 to 40 % PMF	0.9981	<M 6.5	0.9941
>40 to 60 % PMF	0.0010	>M 6.5 to M 7.5	0.0054
>60 to 80 % PMF	0.0005	>M 7.5	0.0005
>80 to 100 % PMF	0.0004	—	—
Flood Sample Space	1.0000	Earthq. Sample Space	1.0000

(b) Failure Mode Probabilities for Alternatives

Floods, Equ's	Failure Mode	Alt. 1	Alt 2.1	Alt 2.2	Alt 3.1	Alt 3.2	Alt 4.1	Alt 4.2	Alt 4.3	Alt 5.1	Alt 5.2
(1)	(2)	(3)	(4)	(5)	(6)	(7)	(8)	(9)	(10)	(11)	(12)
0 to 40 %	none	1.0	1.0	1.0	1.0	1.0	1.0	1.0	1.0	1.0	1.0
>40 to 60 %	none	0.80	0.80	0.80	0.80	0.80	1.0	1.00	1.00	0.80	0.80
- " -	ASF	0.20	0.20	0.20	0.20	0.20	0.0	0.00	0.00	0.20	0.20
>60 to 80 %	none	0.50	0.60	0.60	0.60	0.60	0.5	0.92	0.95	0.60	0.60
- " -	ASF	0.30	0.30	0.30	0.30	0.30	0.3	0.00	0.00	0.30	0.30
- " -	DO	0.10	0.00	0.00	0.00	0.00	0.1	0.08	0.05	0.00	0.00
- " -	SSF	0.10	0.10	0.10	0.10	0.10	0.1	0.00	0.00	0.10	0.10
>80 to 100 %	none	0.10	0.20	0.30	0.20	0.30	0.1	0.20	0.75	0.20	0.30

- " -	ASF	0.40	0.40	0.40	0.40	0.40	0.4	0.30	0.00	0.40	0.40
- " -	DO	0.20	0.10	0.00	0.10	0.00	0.2	0.20	0.10	0.10	0.00
- " -	SSF	0.30	0.30	0.30	0.30	0.30	0.3	0.30	0.15	0.30	0.30
6.5<M	none	1.00	1.00	1.00	1.00	1.00	1.0	1.00	1.00	1.00	1.00
6.5≤ M <7.5	none	0.90	0.95	1.00	0.95	1.00	0.9	0.92	0.95	0.95	1.00
- " -	DO	0.10	0.05	0.00	0.05	0.00	0.1	0.08	0.05	0.05	0.00
M 7.5	none	0.70	0.85	0.95	0.85	0.95	0.7	0.70	0.80	0.85	0.95
- " -	DO	0.30	0.15	0.05	0.15	0.05	0.3	0.30	0.20	0.15	0.05

Notes for Table 6-13(b): column 1: the first eleven rows refer to floods in % of PMF: 0–40 % means floods from 0 to 40 % of PMF; last five rows refer to earthquakes; M is magnitude on the Richter scale.

column 2: failure modes: DO—dam overtopping; ASF—auxiliary spillway failure; SSF—service spillway failure.

columns 3 to 12: probability of failure mode for event in column 2 for Alternative 1 to 5.2; the sum of probabilities over all failure modes for a particular event including the null event is 1.00; for example, rows 4 to 7 are probabilities of failure according to the failure mode in column 2 for the event "flood greater than 60 % and up to 80 % of PMF": with probability 50 % there is no damage, with probability 30 % ASF occurs, and so on. A probability 1.00 means that one of the failure modes has to occur, usually the no-failure event.

6.6.2 Evaluation of Alternatives

The decision tree of Figure 6-16 is evaluated in Table 6-14. Column 1 identifies the status quo and nine other alternatives of the decision to be evaluated. Column 2 identifies the flood and earthquake events, and column 3 gives the event probabilities, $P(E)$, which are the same for all alternatives. Column 4 contains the possible failure modes. Column 5 contains the conditional probabilities of the failures, given the occurrence of a particular flood or earthquake event, $P(F|E)$, where F designates the failure event, and E is the cause-event. The relation between F and E is probabilistic. Damage is not certain, but it is possible and increasingly likely the bigger the event becomes. The probability of failure is the joint probability of the cause event and the probability that the failure mode is triggered (see Section 6.3.2)

$$P(F) = P(F|E)\, P(E) \tag{6.6-1}$$

where $P(F)$ is the failure probability that could actually be triggered by several cause events but here only one is assumed at a time; $P(F|E)$ is the conditional probability (Table 6-13, columns 3 to 12; Table 6-14, column 5); $P(E)$ is obtained from an analysis of flood and earthquake events (Table 6-14, column 3). If $P(F|E)$ is interpreted as the ratio of k failures per n events E; and $P(E)$ is the number of n events E per total number of exposures, N, then Equation (6.6-1) becomes

$$P(F) = \frac{k}{n}\frac{n}{N} = \frac{k}{N}. \tag{6.6-2}$$

The probability of damage times the damage cost is the probabilistic damage cost,

$$P(F)D = \frac{kD}{N}. \tag{6.6-3}$$

Equation (6.6-3) was encountered in Section 6.2.1 as Equation (6.2-4), where it was used to explain the insurance principle. There the cost of k lost ships was spread over the owners of a fleet of N ships for one event or one trip, or perhaps one year. Here k damage occurrences during the project's life are spread over the N years of the project's life. Hence, Equation (6.6-3) represents an average annual damage cost, or an annual insurance premium. Column 6 of Table 6-14 lists the actual damage costs for all specified events that may happen during a year. These actual costs, multiplied by their annual frequencies of occurrence, give the probabilistic annual costs in column 7. The annual sum of these probabilistic costs represents the annual insurance premium that would have to be collected from the beneficiaries. Equation (6.6-3) still must be adapted to reflect the subjective attitude toward risk of the DM. This is done in Section 6.6.4.

The total probabilistic annual flood cost of alternative c is

$$Df_c = \sum_{j=1}^{m}[P(E_j)\sum_{i=1}^{n}D_i P(F_i|E_j)] \tag{6.6-4}$$

where Df_c is the total annual flood cost for alternative c. The inner sum on the right-hand side of Equation (6.6-4) represents the total damage caused by the mutually exclusive failure modes; each of these sums is then multiplied by the probability of their cause-event and summed again by the outer sum, so that Df_c becomes the total probabilistic cost attributable to flood events for alternative c. For Alternative 1, the "no action" alternative, the total probabilistic annual flood damage cost is:

$$Df_1 = 0.9981 \cdot 0 + 0.001 \cdot (0.8 \cdot 0 + 0.2 \cdot 700)$$

$$+ 0.0005 \, (0.5 \cdot 0 + 0.3 \cdot 700 + 0.1 \cdot 1800 + 0.1 \cdot 1800)$$

$$+ 0.0004 \, (0.1 \cdot 0 + 0.4 \cdot 700 + 0.2 \cdot 3200 + 0.3 \cdot 3200)$$

$$= \$1.18 \text{ million/a}$$

where "a" is the unit symbol for "year" (annus); for completeness, the null event that has the highest annual probability and causes zero damage has been included as the first right-hand side term. Similarly, the total annual damage cost from earthquake damage is the sum over all earthquake branches of Alternative 1:

$$De_1 = 0.9941 \cdot 0 + 0.0054 \, (0.9 \cdot 0 + 0.1 \cdot 1800)$$

$$+ 0.0005 \cdot (0.7 \cdot 0 + 0.3 \cdot 1800) = \$1.24 \text{ million/a}.$$

The total expected annual damage cost from floods and earthquakes is the sum of flood and earthquake damages. For Alternative 1,

$$D_1 = Df_1 + De_1 \tag{6.6-5}$$

$$= 1.18 + 1.24 = \$2.42 \text{ million/a}.$$

These costs are given on the summary line of Alternative 1 in Table 6-14, column 7. In a similar way all other alternatives are evaluated in Table 6-14.

The total expected cost of Alternative 1, D_1, can be construed as a risk-neutral annual insurance premium for Alternative 1 (see Section 6.5.4). The part of the evaluation that deals with the subjective evaluation of the insurance premium is discussed in Section 6.6.4. Before proceeding to this section the benefit-cost analysis is discussed which is carried out with and without risk-adjusted benefits and costs.

Table 6-14: Evaluation of Rehabilitation Alternatives for Protection of Dam Project Against Floods and Earthquakes.

Alt.	Events: Floods & Equ's	Probab. of Floods & Equ's	Failure Mode	Failure Mode Probab.	Damage Cost $M	Probable Dam. Cost $M	Utility Values 1	Probable Utility 1	Expect. Utility 1	Certain. Equival. $M	Expect. Cost $M	Risk Prem. $M
(1)	(2)	(3)	(4)	(5)	(6)	(7)	(8)	(9)	(10)	(11)	(12)	(13)
Alt. 1	0–40 %	0.9981	None	1.00	0	0.000	−1.000	−0.998				
Status	40–60%	0.0010	None	0.80	0	0.000	−1.000	−0.001				
quo		0.0010	ASF	0.20	700	0.140	−2.014	−0.000				
	60–80%	0.0005	None	0.50	0	0.000	−1.000	−0.000				
		0.0005	ASF	0.30	700	0.105	−2.014	−0.000				
		0.0005	DO	0.10	1800	0.090	−6.050	−0.000				
		0.0005	SSF	0.10	1800	0.090	−6.050	−0.000				
	80–100%	0.0004	None	0.10	0	0.000	−1.000	−0.000				
		0.0004	ASF	0.40	700	0.112	−2.014	−0.000				
		0.0004	DO	0.20	3200	0.256	−24.533	−0.002				
		0.0004	SSF	0.30	3200	0.384	−24.533	−0.003	−1.01	5.71	1.18	4.54
	<M 6.5	0.9941	None	1.00	0	0.000	−1.000	−0.994				
	M6.5–7.5	0.0054	None	0.90	0	0.000	−1.000	−0.005				
		0.0054	DO	0.10	1800	0.972	−6.050	−0.003				
	M7.5	0.0005	None	0.70	0	0.000	−1.000	−0.000				
		0.0005	DO	0.30	1800	0.270	−6.050	−0.001	−1.00	3.48	1.24	2.24
Summary 1				0.9992		0.9993	2.42			9.19	2.42	6.77
Alt 2.1	0–40 %	0.9981	None	1.00	0	0.000	−1.000	−0.998				
Raise	40–60 %	0.0010	None	0.80	0	0.000	−1.000	−0.001				
dam for		0.0010	ASF	0.20	700	0.140	−2.014	−0.000				
80 %	60–80 %	0.0005	None	0.60	0	0.000	−1.000	−0.000				
PMF		0.0005	ASF	0.30	700	0.105	−2.014	−0.000				
		0.0005	DO	0.00	1800	0.000	−6.050	0.000				
		0.0005	SSF	0.10	1800	0.090	−6.050	−0.000				
	80–100 %	0.0004	None	0.20	0	0.000	−1.000	−0.000				
		0.0004	ASF	0.40	700	0.112	−2.014	−0.000				
		0.0004	DO	0.10	3200	0.128	−24.533	−0.001				
		0.0004	SSF	0.30	3200	0.384	−24.533	−0.003	−1.005	4.52	0.96	3.57
	<M 6.5	0.9941	None	1.00	0	0.000	−1.000	−0.994				
	M 6.5–7.5	0.0054	None	0.95	0	0.000	−1.000	−0.005				
		0.0054	DO	0.05	1800	0.486	−6.050	−0.002				
	M 7.5	0.0005	None	0.85	0	0.000	−1.000	−0.000				
		0.0005	DO	0.15	1800	0.135	−6.050	−0.000	−1.002	1.74	0.62	1.12
Summary 2.1				0.9993		0.9997	1.580			6.27	1.58	4.69
Alt. 2.2	0–40 %	0.9981	None	1.00	0	0.000	−1.000	−0.998				
Raise	40–60 %	0.0010	None	0.80	0	0.000	−1.000	−0.001				
dam		0.0010	ASF	0.20	700	0.140	−2.014	−0.000				
for	60–80 %	0.0005	None	0.60	0	0.000	−1.000	−0.000				
100 %		0.0005	ASF	0.30	700	0.105	−2.014	−0.000				
PMF		0.0005	DO	0.00	1800	0.000	−6.050	0.000				
		0.0005	SSF	0.10	1800	0.090	−6.050	−0.000				
	80–100 %	0.0004	None	0.30	0	0.000	−1.000	−0.000				
		0.0004	ASF	0.40	700	0.112	−2.014	−0.000				
		0.0004	DO	0.00	3200	0.000	−24.533	0.000				
		0.0004	SSF	0.30	3200	0.384	−24.533	−0.003	−1.004	3.59	0.83	2.76
	<M 6.5	0.9941	None	1.00	0	0.000	−1.000	−0.994				
	M 6.5–7.5	0.0054	None	1.00	0	0.000	−1.000	−0.005				
		0.0054	DO	0.00	1800	0.000	−6.050	0.000				
	M 7.5	0.0005	None	0.95	0	0.000	−1.000	−0.000				
		0.0005	DO	0.05	1800	0.045	−6.050	−0.000	−1.000	0.13	0.05	0.08

Summary 2.2		0.9993	0.99998	0.876						3.71	0.88	2.8
Alt 3.1	0–40 %	0.9981	None	1.00	0	0.000	−1.000	−0.998				
Restrict	40–60 %	0.0010	None	0.80	0	0.000	−1.000	−0.001				
Operation		0.0010	ASF	0.20	700	0.140	−2.014	−0.000				
for	60–80 %	0.0005	None	0.60	0	0.000	−1.000	−0.000				
80 %		0.0005	ASF	0.30	700	0.105	−2.014	−0.000				
PMF		0.0005	DO	0.00	1800	0.000	−6.050	0.000				
		0.0005	SSF	0.10	1800	0.090	−6.050	−0.000				
	80–100 %	0.0004	None	0.20	0	0.000	−1.000	−0.000				
		0.0004	ASF	0.40	700	0.112	−2.014	−0.000				
		0.0004	DO	0.10	3200	0.128	−24.533	−0.001				
		0.0004	SSF	0.30	3200	0.384	−24.533	−0.003	−1.005	4.52	0.96	3.5
	<M 6.5	0.9941	None	1.00	0	0.000	−1.000	−0.994				
	M 6.5–7.5	0.0054	None	0.95	0	0.000	−1.000	−0.005				
		0.0054	DO	0.05	1800	0.486	−6.050	−0.002				
	M 7.5	0.0005	None	0.85	0	0.000	−1.000	−0.000				
		0.0005	DO	0.15	1800	0.135	−6.050	−0.000	−1.002	1.74	0.62	1.1
Summary 3.1		0.99928	0.99966	1.580						6.27	1.58	4.6
Alt 3.2	0–40 %	0.9981	None	1.00	0	0.000	−1.000	−0.998				
Restrict	40–60 %	0.0010	None	0.80	0	0.000	−1.000	−0.001				
Operation		0.0010	ASF	0.20	700	0.140	−2.014	−0.000				
for	60–80 %	0.0005	None	0.60	0	0.000	−1.000	−0.000				
100 %		0.0005	ASF	0.30	700	0.105	−2.014	−0.000				
PMF		0.0005	DO	0.00	1800	0.000	−6.050	0.000				
		0.0005	SSF	0.10	1800	0.090	−6.050	−0.000				
	80–100 %	0.0004	None	0.30	0	0.000	−1.000	−0.000				
		0.0004	ASF	0.40	700	0.112	−2.014	−0.000				
		0.0004	DO	0.00	3200	0.000	−24.533	0.000				
		0.0004	SSF	0.30	3200	0.384	−24.533	−0.003	−1.004	3.59	0.83	2.7
	<M 6.5	0.9941	None	1.00	0	0.000	−1.000	−0.994				
	M 6.5–7.5	0.0054	None	1.00	0	0.000	−1.000	−0.005				
		0.0054	DO	0.00	1800	0.000	−6.050	0.000				
	M 7.5	0.0005	None	0.95	0	0.000	−1.000	−0.000				
		0.0005	DO	0.05	1800	0.045	−6.050	−0.000	−1.000	0.13	0.05	0.0
Summary 3.2		0.9993	0.99998	0.876						3.71	0.88	2.8
Alt 4.1	0–40 %	0.9981	None	1.00	0	0.000	−1.000	−0.998				
Modify	40–60 %	0.0010	None	1.00	0	0.000	−1.000	−0.001				
Auxiliary		0.0010	ASF	0.00	700	0.000	−2.014	0.000				
Spillway	60–80 %	0.0005	None	0.50	0	0.000	−1.000	−0.000				
for 60 %		0.0005	ASF	0.30	700	0.105	−2.014	−0.000				
PMF		0.0005	DO	0.10	1800	0.090	−6.050	−0.000				
		0.0005	SSF	0.10	1800	0.090	−6.050	−0.000				
	80–100 %	0.0004	None	0.10	0	0.000	−1.000	−0.000				
		0.0004	ASF	0.40	700	0.112	−2.014	−0.000				
		0.0004	DO	0.20	3200	0.256	−24.533	−0.002				
		0.0004	SSF	0.30	3200	0.384	−24.533	−0.003	−1.006	5.51	1.04	4.4'
	<M 6.5	0.9941	None	1.00	0	0.000	−1.000	−0.994				
	M 6.5–7.5	0.0054	None	0.90	0	0.000	−1.000	−0.005				
		0.0054	DO	0.10	1800	0.972	−6.050	−0.003				
	M 7.5	0.0005	None	0.70	0	0.000	−1.000	−0.000				
		0.0005	DO	0.30	1800	0.270	−6.050	−0.001	−1.003	3.48	1.24	2.2·
Summary 4.1		0.9994	0.99931	2.279						8.99	2.28	6.7
Alt 4.2	0–40 %	0.9981	None	1.00	0	0.000	−1.000	−0.998				
Modify	40–60 %	0.0010	None	1.00	0	0.000	−1.000	−0.001				
Auxiliary		0.0010	ASF	0.00	700	0.000	−2.014	0.000				
Spillway	60–80 %	0.0005	None	0.92	0	0.000	−1.000	−0.000				
for 80 %		0.0005	ASF	0.00	700	0.000	−2.014	0.000				
PMF		0.0005	DO	0.08	1800	0.072	−6.050	−0.000				

		0.0005	SSF	0.00	1800	0.000	-6.050	0.000				
	80–100 %	0.0004	None	0.20	0	0.000	-1.000	-0.000				
		0.0004	ASF	0.30	700	0.084	-2.014	-0.000				
		0.0004	DO	0.20	3200	0.256	-24.533	-0.002				
		0.0004	SSF	0.30	3200	0.384	-24.533	-0.003	-1.005	5.02	0.80	4.22
	<M 6.5	0.9941	None	1.00	0	0.000	-1.000	-0.994				
	M 6.5–7.5	0.0054	None	0.92	0	0.000	-1.000	-0.005				
		0.0054	DO	0.08	1800	0.778	-6.050	-0.003				
	M 7.5	0.0005	None	0.70	0	0.000	-1.000	-0.000				
		0.0005	DO	0.30	1800	0.270	-6.050	-0.001	-1.003	2.93	1.05	1.89
Summary 4.2				0.9996	0.99942	1.844				7.95	1.84	6.11
Alt 4.3	0–40 %	0.9981	None	1.00	0	0.000	-1.000	-0.998				
Modify	40–60 %	0.0010	None	1.00	0	0.000	-1.000	-0.001				
Auxiliary		0.0010	ASF	0.00	700	0.000	-2.014	0.000				
Spillway	60–80 %	0.0005	None	0.95	0	0.000	-1.000	-0.000				
for 100 %		0.0005	ASF	0.00	700	0.000	-2.014	0.000				
PMF		0.0005	DO	0.05	1800	0.045	-6.050	-0.000				
		0.0005	SSF	0.00	1800	0.000	-6.050	0.000				
	80–100 %	0.0004	None	0.75	0	0.000	-1.000	-0.000				
		0.0004	ASF	0.00	700	0.000	-2.014	0.000				
		0.0004	DO	0.10	3200	0.128	-24.533	-0.001				
		0.0004	SSF	0.15	3200	0.192	-24.533	-0.001	-1.002	2.48	0.37	2.11
	<M 6.5	0.9941	None	1.00	0	0.000	-1.000	-0.994				
	M 6.5–7.5	0.0054	None	0.95	0	0.000	-1.000	-0.005				
		0.0054	DO	0.05	1800	0.486	-6.050	-0.002				
	M 7.5	0.0005	None	0.80	0	0.000	-1.000	-0.000				
		0.0005	DO	0.20	1800	0.180	-6.050	-0.001	-1.002	1.87	0.67	1.20
Summary 4.3				0.9999	0.99963	1.031				4.34	1.03	3.31
Alt 5.1	0–40 %	0.9981	None	1.00	0	0.000	-1.000	-0.998				
Combine	40–60 %	0.0010	None	0.80	0	0.000	-1.000	-0.001				
2 & 3		0.0010	ASF	0.20	700	0.140	-2.014	-0.000				
for	60–80 %	0.0005	None	0.60	0	0.000	-1.000	-0.000				
80 %		0.0005	ASF	0.30	700	0.105	-2.014	-0.000				
PMF		0.0005	DO	0.00	1800	0.000	-6.050	0.000				
		0.0005	SSF	0.10	1800	0.090	-6.050	-0.000				
	80–100 %	0.0004	None	0.20	0	0.000	-1.000	-0.000				
		0.0004	ASF	0.40	700	0.112	-2.014	-0.000				
		0.0004	DO	0.10	3200	0.128	-24.533	-0.001				
		0.0004	SSF	0.30	3200	0.384	-24.533	-0.003	-1.005	4.52	0.96	3.57
	<M 6.5	0.9941	None	1.00	0	0.000	-1.000	-0.994				
	M 6.5–7.5	0.0054	None	0.95	0	0.000	-1.000	-0.005				
		0.0054	DO	0.05	1800	0.486	-6.050	-0.002				
	M 7.5	0.0005	None	0.85	0	0.000	-1.000	-0.000				
		0.0005	DO	0.15	1800	0.135	-6.050	-0.000	-1.002	1.74	0.62	1.12
Summary 5.1				0.9993	0.99966	1.580				6.27	1.58	4.69
Alt 5.2	0–40 %	0.9981	None	1.00	0	0.000	-1.000	-0.998				
Combine	40–60 %	0.0010	None	0.80	0	0.000	-1.000	-0.001				
2 & 3		0.0010	ASF	0.20	700	0.140	-2.014	-0.000				
for	60–80 %	0.0005	None	0.60	0	0.000	-1.000	-0.000				
100 %		0.0005	ASF	0.30	700	0.105	-2.014	-0.000				
PMF		0.0005	DO	0.00	1800	0.000	-6.050	0.000				
		0.0005	SSF	0.10	1800	0.090	-6.050	-0.000				
	80–100 %	0.0004	None	0.30	0	0.000	-1.000	-0.000				
		0.0004	ASF	0.40	700	0.112	-2.014	-0.000				
		0.0004	DO	0.00	3200	0.000	-24.533	0.000				
		0.0004	SSF	0.30	3200	0.384	-24.533	-0.003	-1.004	3.59	0.83	2.76
	<M 6.5	0.9941	None	1.00	0	0.000	-1.000	-0.994				
	M 6.5–7.5	0.0054	None	1.00	0	0.000	-1.000	-0.005				
		0.0054	DO	0.00	1800	0.000	-6.050	0.000				

(1)	(2)	(3)	(4)	(5)	(6)	(7)	(8)	(9)	(10)	(11)	(12)	(13)
	M 7.5	0.0005	None	0.95	0	0.000	−1.000	−0.000				
		0.0005	DO	0.05	1800	0.045	−6.050	−0.000	−1.000	0.13	0.05	0.0$
Summary 5.2				0.9993	0.99998	0.876				3.71	0.88	2.8₄

Notes: column 1: alternatives, as explained in Table 6-12.
column 2: cause-events in the form of floods and earthquakes, as explained in Table 6-13.
column 3: probability of cause events including no damage events.
column 4: failure modes as given in Table 6-13.
column 5: probability of failure mode conditioned on cause-event.
column 6: damage cost caused by failure mode under cause event, million dollars.
column 7: probable damage, million dollars.
column 8: utilities of damage costs; developed in Section 6.5.3; dimensionless.
column 9: utility multiplied by probability of cause event; dimensionless.
column 10: expected utility of damage cost; dimensionless.
column 11: certainty equivalent deduced from expected utility; million dollars.
column 12: expected cost of alternative; million dollars.
column 13: risk premium as difference between certainty equivalent and expected utility; million dollars.
On the summary line of each alternative:
column 5: total probability of no damage from floods.
column 6: total probability of no damage from earthquakes.
column 7: total expected damage (column sum; same as column 12).
column 11: total certainty equivalent.
column 12: total expected cost of alternative.
column 13: total risk premium.

6.6.3 Subjective Risk Measures

The attitude toward risk is introduced into the cost analysis by eliciting a utility function from the DM, as explained in Sections 6.5.2 and 6.5.3. The DM can also directly "shape" a utility function so that it represents his attitude toward risk (Keeney and Raiffa, 1976). For brevity, this approach will be used. Utility functions, because of their arbitrariness, may produce very different "utility" weights. But the essential characteristic of the adopted function is its shape, which must display "risk-aversion"; in other words, it must have a concave shape (see Section 6.5.3). The utility function adopted here is

$$u(x) = -e^{\alpha \cdot x}. \tag{6.6-6}$$

This function, with $\alpha = 0.001$, is displayed in Figure 6-17. It has a concave shape and becomes increasingly negative with increasing damage x.

The gradient of Equation (6.6-6) is

$$\frac{d[u(x)]}{dx} = -\alpha\, u(x) \,. \tag{6.6-7}$$

Equation (6.6-7) states that the utility increment, $d[u(x)]$, for a damage increment, dx, is a fraction α of the present utility, $u(x)$. The minus sign in front of the right-hand side indicates a damage growth rate. The coefficient α is the choice of the DM or his organization that develops and uses the utility function.

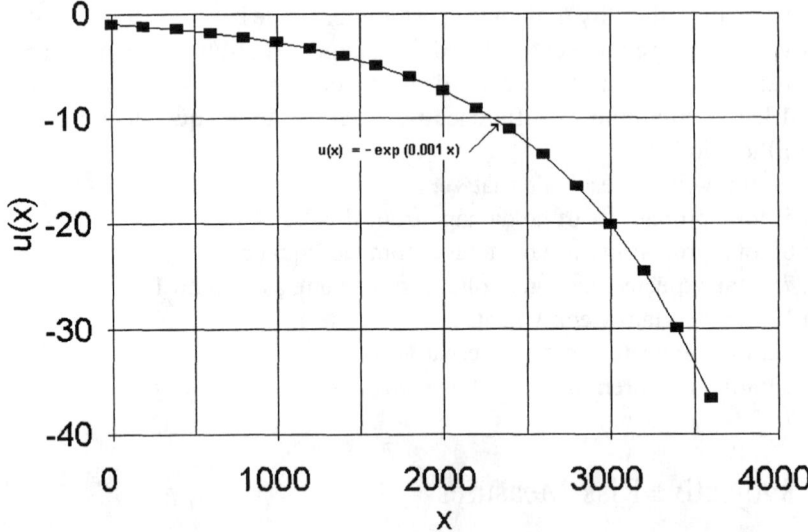

Figure 6-17: Decreasing utility $u(x)$ for increasing damage x for a risk-averse DM.

The utility function is used to convert the damages in column 7 of Table 6-14 into utilities. Expected utilities then are calculated for the flood and earthquake damages. The same utility function is used here for both types of damages. The expected utilities for flood damage and earthquake damage are then reconverted into x-values by using the inverse of the utility function,

$$x = \frac{1}{\alpha}\ln(-u) \qquad\qquad (6.6\text{-}8)$$

where x are damage costs ($million/a), and u is a utility, here a negative number. The inversion of the expected utility $E(u)$ by Equation (6.6-8) produces the *certainty equivalent CE*:

$$CE = \frac{1}{\alpha}\ln[-E(u)] \qquad\qquad (6.6\text{-}9)$$

where CE is a cost for certain in lieu of the chance outcomes of the alternative; $E(u)$ is the expected utility, here a negative number. CE is also the risk-adjusted expected damage, as perceived by the risk-averse DM. A risk-averse utility function produces a certainty equivalent for damage that exceeds the expected damage (see also Figure 6-17):

$$RP = CE - EV > 0 \qquad\qquad (6.6\text{-}10)$$

where RP is the *risk premium*, the surcharge that the risk-averse DM (or the insurer) adds to the expected damage to reflect (perceived) risk. The greater the risk, the greater is the risk premium demanded by the insurer.

The risk premium is a surcharge that changes the expected value into a risk-sensitive value, the certainty equivalent, CE. This risk-adjusted expected value is the *insurance premium* represented as

$$IP = -CE. \qquad\qquad (6.6\text{-}11)$$

The reduction of expected damage EV is a benefit to the project owner. In the form of CE, the risk adjusted expectation, if paid as an insurance premium for the transfer of the risk to an insurer, it becomes a cost (see also Section 6.5.4). Only if the adopted utility function also suits the insurer, the risk owner's CE also represents the insurance premium of the professional risk taker, the insurer. The insurer may have a different utility function or some other way of estimating the risk premium to be added to the expectation.

A probabilistic criterion for judging the alternatives is the total probability of no damage. It is obtained as the sum of mutually exclusive probabilities of no damage of the outcomes of an alternative:

$$P(D = 0) = \sum_{j} P(S_i|E_j)(P(E_j) \qquad\qquad (6.6\text{-}12)$$

where $P(S_i|E_j)$ is the probability of the "no damage" structural state S_i, given event E_j occurs, and $P(E_j)$ is the probability of the cause-event E_j. The sum is taken over all no-damage events. For example, for alternative 1, the total probability of no damage from floods is

$$P(D_f = 0) = 0.9981 \cdot 1.0 + 0.001 \cdot 0.8 + 0.0005 \cdot 0.5 + 0.0004 \cdot 0.1 = 0.99919$$

where the first factor of each product term is from Table 6-14, column 3, and the second factor is from column 5. Each additive term is the joint probability that the cause-event happens and no damage occurs. For each alternative, these events are mutually exclusive. For earthquakes, the no-damage probability is:

$$P(D_e = 0) = 0.9941 \cdot 1.0 + 0.0054 \cdot 0.9 + 0.0005 \cdot 0.7 = 0.99931$$

where the two factors of each additive term are found in the lower portions of columns 3 and 5 of Table 6-14, respectively.

   Another measure of the no-damage probability is obtained by conversion of the no-damage probabilities into return periods by

$$T = \frac{1}{1 - P(D = 0)}. \tag{6.6-13}$$

For Alternative 1, the return period of flood damage is $T_f = 1{,}235$ years, and the return period for earthquake damage is $T_e = 1{,}450$ years.

6.6.4 Benefit-Cost Analysis

   The benefit-cost analysis requires the identification of all benefits and costs that are sensitive to the structural and operational changes proposed by the alternatives. For example, an alternative may advocate operating at lower reservoir levels to provide more flood storage and dam safety, thereby avoiding construction cost, but incurring loss of hydropower and recreation benefits. Benefits include revenues from operations, damage averted, and reduction of harmful exposure. Costs include construction costs, rehabilitation costs, repair and replacement costs, operation costs, and benefits foregone.

   The decision criteria used for comparing alternatives are the following:
(a) *Net benefit*:

$$NB_a = dD_{a1} - C_a - BL_a \tag{6.6-14}$$

where NB_a is the net benefit of alternative a; dD_{a1} is the difference in expected damage between the "no-action" alternative and alternative a; C_a is the rehabilitation cost to avert the failure mode; BL_a is a loss of benefit incurred by alternative a, for example, by requiring operation at lower water levels.

(b) *Benefit-cost ratio*:

$$B_{Ca} = \frac{dD_{ao}}{C_a + BL_a} \, . \tag{6.6-15}$$

Here the benefit-cost ratio is formulated with the damage averted as benefits and the structural costs and benefits forgone as costs.

(c) *Net benefit-cost ratio*: The previous benefit-cost ratio makes the somewhat arbitrary choice of designating benefits forgone as costs. An alternative formulation is to treat the benefits forgone as negative benefits:

$$NB_{Ca} = \frac{dD_{a1} - BL_a}{C_a} \, . \tag{6.6-16}$$

Equations (6.6-15) and (6.6-16) give different benefit-cost ratios. Both ratios are the same only for $BL_a = 0$ and for $BL_a = dD_{a1} - C_a$; in the latter case, both ratios are 1. Equation (6.6-16) is clearly zero when the benefits forgone reduce the benefits to zero, as it should be.

All benefits and costs are annual costs. A one-time expense, such as the front-end cost, is converted into an annual project cost by the capital recovery factor:

$$C_a = (A/P)C \tag{6.6-17}$$

where C_a is the annualized capital cost of the alternative; A/P is the capital recovery factor, a function of the amortization period and of the interest rate; C is the present worth front end investment. The capital recovery factor converges from a maximum of $A/P = 1 + i$, for $n = 1$ year, to $A/P = i$ for large n. For $i = 0.06$, this plateau is reached at 85 years. For short periods Equation (5.8-3) must be used or C_a will be consistently underestimated.

An aspect of the probabilistic analysis that needs consideration is the relationship between benefits and costs. The benefits and costs used to calculate NB and B_C can be related or unrelated to one another. In the case of an existing relationship, a low rehabilitation (rehab) cost may be associated with a small damage reduction and a high loss of benefits; and a high rehab cost may be associated with high damage reduction and a low loss of benefits. Dependence among variables makes a probabilistic analysis more complicated but not infeasible. Here, for demonstration purposes, independence among variables is assumed.

The expected net benefits and costs are calculated as the sums and differences of expected benefits and costs, respectively:

$$E(NB_a) = E(dD_{a1}) - E(C_a) - E(BL_a) \ .$$ (6.6-18)

Since C_a and BL_a are single values, their expectations are equal to these single values. The difference between the damage costs of the "no action" Alternative 1 and alternative a is

$$E(dD_{a1}) = E(D_1) - E(D_a),$$ (6.6-18a)

so that the expected net benefit becomes

$$E(NB_a) = E(D_1) - E(D_a) - C_a - BL_a.$$ (6.6-19)

The net benefit expectations for all alternatives, based on CE and on EV, are given in Table 6-15, columns 7 and 10 respectively. The benefit-cost analysis is carried out with both risk-adjusted CE and risk-neutral EV in the calculation of NB and B_C. The results of the calculations are summarized in Table 6-15. Alternative 1 is the baseline against which the other alternatives are compared. The decision criteria summarized in Table 6-15 include CE and EV of averted damage costs in columns 2 and 3, respectively; RP for flood and earthquake damages in column 4; the annual cost of rehabilitation in column 5; the benefit losses (cumulative for sublevels of alternatives) due to changes in reservoir operation in column 6; the values of both columns 5 and 6 are single costs for each alternative given as input; the net benefit NB based on CE, column 7, is averted damage minus the cost to achieve it; the averted damage is the difference between Alternative 1 and Alternative a; the benefit-cost ratio, B_C, column 8, is obtained by dividing the benefit in terms of damages averted, dD_{a1}, by the sum of costs of the alternative, C_a, plus the benefit foregone, BL_a; the net benefit NB based on EV and the (total) benefit cost ratio, B_C, are given in columns 9 and 10, respectively; the standard deviation and the coefficient of variation of flood and earthquake damages, based on EV, are in columns 11 and 12.

The expected benefit-cost ratio is the expectation of a quotient (Section 3.3.4). Here the costs are single values so that the calculation simplifies to dividing the expected averted damage by the single-valued cost,

$$E(B_{Ca}) = E(\frac{dD_{a1}}{C_a + BL_a}) = \frac{E(dD_{a1})}{C_a + BL_a}.$$ (6.6-20)

CE-based and *EV*-based expected benefit-cost ratios are shown in Table 6-15, columns 8 and 10, respectively.

An objective risk criterion is the standard deviation. It is calculated as the square root of the sum of flood and earthquake damage variances. The standard deviation is chosen as a criterion because it has the same units as the expected value. Equation (2.3-7) was used to calculate the variance:

$$\sigma^2 = \sum_j [P(E_j)P(F_j|E_j)D_j^2] - E(D)^2 \qquad (6.6-21)$$

where the first term under the sum is obtained from the items of Table 6-14, column 7 which are

$$\sum_j P(E_j)P(F_j|E_j)D_j.$$

Each of the column items is again multiplied by D_j (Table 6-14, column 6); flood and earthquake columns are summed separately, the respective EV^2 is subtracted from each sum; the resulting variances for flood and earthquake damages are added and the root of this sum produces the standard deviation for the alternative given in Table 6-15, column 11. The coefficient of variation is the standard deviation divided by the total *EV* of the alternative given in Table 6-15, column 3.

Table 6-15: Summary of Objective and Subjective Decision Criteria

Alt.	CE	EV	RP	C_a	BL_a	NB (CE)	B_C (CE)	NB (EV)	B_C (EV)	σ (EV)	C_v (EV)
	$M	$M	$M	$M	$M	$M	$M	$M	1	$M	1
(1)	(2)	(3)	(4)	(5)	(6)	(7)	(8)	(9)	(10)	(11)	(12)
1-F	5.71	1.18	4.54								
1-E	3.48	1.24	2.24								
1	9.19	2.42	6.77	0.00	0.00	0.00	—	—	—	69.7	28.8
2.1-F	4.52	0.96	3.57								
2.1-E	1.74	0.62	1.12								
2.1	6.27	1.58	4.69	0.82	0.00	2.11	3.58	0.02	1.03	56.3	35.6
2.2-F	3.59	0.83	2.76								
2.2-E	0.13	0.05	0.08								
2.2	3.71	0.88	2.84	1.55	0.00	3.93	3.53	−0.01	0.99	41.5	47.4
3.1-F	4.52	0.96	3.57								
3.1-E	1.74	0.62	1.12								

3.1	6.27	1.58	4.69	0.00	0.33	2.60	8.95	0.51	2.57	56.3	35.6
3.2-F	3.59	0.83	2.76								
3.2-E	0.13	0.05	0.08								
3.2	3.71	0.88	2.84	0.00	0.57	4.91	9.58	0.97	2.70	41.5	47.4
4.1-F	5.51	1.04	4.47								
4.1-E	3.48	1.24	2.24								
4.1	8.99	2.28	6.71	0.82	0.00	-0.62	0.25	-0.68	0.17	69.0	40.3
4.2-F	5.02	0.80	4.22								
4.2-E	2.93	1.05	1.89								
4.2	7.95	1.84	6.11	1.23	0.00	0.01	1.01	-0.65	0.47	64.2	34.8
4.3-F	2.48	0.37	2.11								
4.3-E	1.87	0.67	1.20								
4.3	4.34	1.03	3.31	2.45	0.00	2.40	1.98	-1.06	0.57	48.0	46.6
5.1-F	4.52	0.96	3.57								
5.1-E	1.74	0.62	1.12								
5.1	6.27	1.58	4.69	0.41	0.16	2.35	5.12	0.27	1.47	56.3	35.6
5.2-F	3.59	0.83	2.76								
5.2-E	0.13	0.05	0.08								
5.2	3.71	0.88	2.84	0.82	0.29	4.37	4.95	0.44	1.39	41.5	47.4

Notes: Table headings: row 1: variable names are explained with each column; row 2: (*CE*) and (*EV*) mean that the quantities are based either on certainty equivalent or expected value; row 3: units are in million dollars; ratios of the same units are dimensionless and are designated as 1; row 4: column numbers.

column 1: alternatives: 1-F refers to Alternative 1— flood-related results; 1-E refers to Alternative 1— earthquake-related results; each alternative occupies three consecutive rows, the third row being the summary of the alternative.

column 2: certainty equivalent of expected damage costs.

column 3: expected damage costs.

column 4: risk premium = column 2 – column 3.

column 5: annualized cost of rehabilitation, C_a.

column 6: annual loss of benefits due to operational changes.

column 7: net benefit (*CE*): column 2 (Alt. 1) – column 2 (Alt. a) – column 5 – column 6.

column 8: total benefit-cost ratio (*CE*): (column 2 (Alt. 1) – column 2 (Alt. a))/(column 5 + column 6); / means "divided by".

column 9: net benefit (*EV*): column 3 (Alt. 1) – column 3 (Alt. a) – column 5 – column 6.

column 10: total benefit-cost ratio: (column 3 (Alt. 1) – column 3 (Alt. a))/(column 5 + column 6).

column 11: standard deviation of the alternative obtained as root of the sum of the

variances of flood and earthquake damages. Each variance is obtained by Equation (6.6-21) from values in Table 6-14, as previously explained.
column 12: coefficient of variation: column 11/column 3.

The probabilistic risk-adjusted damage reduction, dD_{a1}, can be construed as an *insurance premium reduction* for Alternative a over what would have to be paid for the "no action" Alternative 1:

$$dD_{a1} = CE_1 - CE_a \qquad (6.6\text{-}22)$$

where dD_{a1} is the insurance premium reduction as a reward for risk control undertaken in the form of the rehabilitation alternative a. Hence, as long as the net benefit of the alternative,

$$NB \geq 0, \qquad (6.6\text{-}23)$$

the insurance premium reduction equals or exceeds the cost to achieve it. If

$$NB < 0, \qquad (6.6\text{-}23a)$$

the insurance premium reduction does not pay off the cost of risk control and the alternative is judged cost-ineffective because of its shortfall in either damage reduction or the creation of disbenefits or both. An alternative must then be found. The cost effectiveness of damage reduction must be increased in this case. Hence, if the insurer is willing to continue insuring Alternative 1, then this would be the economic solution. It would not necessarily be a desirable solution. As was discussed in Section 6.3.3, a soft insurance market that offers low premiums should not be a justification for a risky or dangerous operation.

6.6.5 Discussion of Results

The decision criteria and their preferred values (large or small) are summarized in Table 6-16. The most preferred rank number of a criterion is 1 and the least preferred rank number is 9. The rank sums are calculated separately for objective and subjective criteria. Therefore, the same alternative may have different rank sums. The smallest rank sum signals the preferred alternative.

Table 6-16: Ranking Criteria and Preferred Sizes

Criterion	Symbol	Preferred Size
Net benefit	NB	Large
Benefit-cost ratio	BC	Large
Standard deviation of damages	σ	Small
Coefficient of variation of damages	C_v	Small
Risk premium	RP	Small
Insurance premium	IP	Small
Return period of loss	T	Large
Rank sum	—	Small

The 10 alternatives including the "no action" alternative, are ranked in Table 6-17 by subjective CE-based criteria. Each row of the table represents the criteria for one alternative. The rank numbers are given to the values of a column and are shown as numbers in parentheses under each criterion. When ties occur in the ranking of the column values, they are given the same rank so that the lowest rank (9) may not occur in a column. Summing the rank numbers along each row gives the total rank sum for the ranked criteria of an alternative in column 6. This procedure gives each criterion equal weight. The ranking by subjective (CE-based) risk criteria gives Alternative 3.2 the lowest rank sum of 4. First runner-up is Alternative 5.2 with rank sum 8; and second runner-up are Alternatives 2.2 with rank sum 11.

Objective risk measures, such as the dispersion measures and the cost criteria based on EV are summarized in Table 6-18. The same ranking procedure is used as in Table 6-17. Again Alternative 3.2 has the lowest rank number 15. First runner-up are Alternatives 3.1 and 5.2, tied at 20, and second runner up is Alternative 2.2 and 5.1, tied at 22. The reason for the low rank sum of Alternative 3.2 is the lack of any structural rehabilitation cost. The costs of this alternative are caused only by loss of benefits due to operational changes. The highest ranking Alternative 3.2 has also the lowest σ, whereas the lowest ranking Alternative 4.1 has the second highest σ, which indicates that this alternative is risky.

Table 6-17: Ranking of Alternatives by *CE*-based Criteria

Alt.	*NB* (*CE*)	B_C (*CE*)	*IP* (*CE*)	*RP* (*CE*)	Rank Sum
(1)	(2)	(3)	(4)	(5)	(6)
1	—	—	9.19 (6)	6.77 (6)	(12)*
2.1	2.11 (7)	3.58 (5)	6.27 (3)	4.69 (3)	(18)
2.2	3.93 (3)	3.53 (6)	3.71 (1)	2.84 (1)	(11)
3.1	2.60 (4)	8.95 (2)	6.27 (3)	4.69 (3)	(12)
3.2	4.91 (1)	9.58 (1)	3.71 (1)	2.84 (1)	(4)
4.1	−0.62 (9)	0.25 (9)	8.99 (5)	6.71 (5)	(28)
4.2	0.01 (8)	1.01 (8)	7.95 (4)	6.11 (4)	(24)
4.3	2.40 (5)	1.98 (7)	4.34 (2)	3.31 (2)	(16)
5.1	2.35 (6)	5.12 (3)	6.27 (3)	4.69 (3)	(15)
5.2	4.37 (2)	4.95 (4)	3.71 (1)	2.84 (1)	(8)

Notes: Table headings: *CE*-based decision criteria; units are million dollars.
column 1: decision alternatives.
column 2: net benefit in the form of averted flood and earthquake damage.
column 3: total benefit-cost ratio gives dollar of benefit per dollar of cost.
column 4: insurance premium, *IP*.
column 5: risk premium, *RP*.
column 6: rank sum: the row sum of the numbers in parentheses in columns 2 through 5. *alternative 1 cannot be compared in this way as items are missing.

Table 6-18: Ranking of Alternatives Based on Objective Risk Criteria

Alt.	T_f	T_e	σ (EV)	Cv (EV)	NB (EV)	B_C (EV)	Rank Sum
(1)	(2)	(3)	(4)	(5)	(6)	(7)	(8)
1	1,250 (7)	1,450 (7)	69.7 (6)	28.8 (1)	—	—	(21)*
2.1	1,430 (5)	3,300 (2)	56.3 (3)	35.6 (4)	0.02 (5)	1.03 (5)	(24)
2.2	3,300 (2)	50,000 (1)	41.5 (1)	47.4 (6)	−0.01 (6)	0.99 (6)	(22)
3.1	1,400 (6)	3,000 (3)	56.3 (3)	35.6 (4)	0.51 (2)	2.57 (2)	(20)
3.2	1,430 (5)	50,000 (1)	41.5 (1)	47.4 (6)	0.97 (1)	2.70 (1)	(15)
4.1	1,670 (4)	1,450 (6)	69.0 (5)	30.3 (2)	−0.68 (8)	0.17 (9)	(34)
4.2	2,500 (3)	1,720 (5)	64.2 (4)	34.8 (3)	−0.65 (7)	0.47 (8)	(30)
4.3	10,000 (1)	2,700 (4)	48.0 (2)	46.6 (5)	−1.06 (9)	0.57 (7)	(28)
5.1	1,430 (5)	3,000 (3)	56.3 (3)	35.6 (4)	0.27 (4)	1.47 (3)	(22)
5.2	1,430 (5)	50,000 (1)	41.5 (1)	47.4 (6)	0.44 (3)	1.40 (4)	(20)

Notes: column 1: decision alternatives.
column 2: return period of flood damage, years.
column 3: return period of earthquake damage, years.
 column 4: standard deviation of damages caused by flood and earthquakes (from Table 6-15, column 11).
column 5: coefficient of variation (from Table 6-15, column 12).
column 6: net benefit based on EV.
column 7: total benefit-cost ratio based on EV.
column 8: rank sum: row sum of numbers in parentheses. *alternative 1 cannot be

compared in this way as items are missing.

A comparison of the rankings by *CE*-based and *EV*-based criteria is given in Table 6-19. Both sets of criteria identify Alternative 3.2 as the referred alternative. Alternative 5.2 is first runner-up in both cases but Alternative 3.1 does not appear as first runner-up for the *CE*-based criteria. Similarly, Alternative 2.2 is second runner-up in both cases but Alternative 5.1 does not appear as second runner-up for the *CE*-based criteria. Overall there is not much difference in the rankings by *CE*-based and *EV*-based criteria. The reason for the almost equal ranking is the monotone (continuously decreasing or increasing) utility function that once selected does not change the relative position of *EV* and *CE* to each other.

Table 6-19: Alternatives Occupying the Three Highest Ranks According to *CE*- based and *EV*-based Decision Criteria

Ranking	*CE*-Based	*EV*-Based
Preferred	Alt. 3.2 (4)	Alt. 3.2 (15)
1. Runner-up	Alt. 5.2 (8)	Alt. 3.1 and 5.2 (20)*
2. Runner-up	Alt. 2.2 (11)	Alt. 2.2 and 5.1 (22)

*Alt 1 would range here but was excluded.

The relationship between *CE* and *EV* for monotone increasing or decreasing utility functions is given in Table 6-20. A monotonously increasing risk-averse utility function is given in Figures 6-13 and a monotonously decreasing risk-averse utility function is given in Figure 6-14. The inequalities of Table 6-20 illustrate the locations of *CE* and *EV* on the abscissa of each curve. The positions of *CE* and *EV* for the risk-prone case in the last column of Table 6-20 would be found if the curves were replaced by convex curves, which are bowl shaped as indicated by the lower part of Figure 6-15. The inequalities indicate that the risk-prone DM would see greater benefits than the risk-neutral benefits and smaller losses than the risk-neutral losses, the risk-neutral values in both cases being represented by *EV*.

Table 6-20: Relationship between *EV* and *CE* as a Function of Attitude toward Risk

Utility Function	Risk-averse	Risk-prone
Benefits	$EV > CE$	$CE > EV$
Damage or Loss	$CE > EV$	$EV > CE$

A comparison of the CE-based NB and B_C of Table 6-17 with the EV-based values of Table 6-18 shows that the former are substantially larger than the latter. The CE-based values are "inflated" by higher costs produced by the risk-averse utility function. This also makes the CE-based cost reductions (averted costs), dD_{a1}, larger than the EV-based cost reduction. While the utility function used in the example does not change the ranking compared to objective (risk-neutral) criteria, it changes the weights on the benefits (damage forgone). The EV-based B_C for Alternative 3.2 is 2.7, whereas the CE-based B_C is 9.58. This means that a risk-averse DM sees a considerably higher cost effectiveness for this alternative than a risk-neutral DM. The risk-neutral approach may be considered a non-conservative approach as it plays down risk in the face of uncertainty. Risk control measures that are economically attractive for a risk-averse DM may be considered economically unjustified by a risk-neutral DM. This may lead to postponing or entirely disregarding the correction of a risky condition. For the risk-averse DM (Table 6-17), 8 out of 9 alternatives have a $B_C > 1$, whereas for the risk-neutral DM (Table 6-18) 5 out of 9 have a $B_C > 1$. If the owner of a risky project has to pay an insurance premium that corresponds to CE, according to Equation 6.6-11, an incentive exists to reduce the insurance premium by carrying out risk control measures.

The difference between the expected flood damage of Alternative 1 (as is) and Alternative a, dD_{a1}, when weighted by a risk-averse utility function becomes an inflated loss reduction, but it also is the risk-adjusted insurance premium reduction. Of course, the premium reduction ultimately granted for the risk control measures undertaken by the risk owner depends on the insurer's perception of the risk reduction, in other words, on the insurer's, not the owner's, utility function.

In principle, many different decision criteria should be examined, because they may give hints about strengths and weaknesses of the alternatives in different areas of concern. There may be additional criteria to those tested here that could shed additional light on how to identify the preferred alternative. Compromises may have to be made because most likely not all objectives can be satisfied at the same level, as the rank numbers of the rows in Tables 6-17 and 6-18 imply. Probabilistic methods can only be *decision aids*, as is true for all analysis methods. The selection of an alternative for implementation, remains the responsibility of the DM.

6.7 Concluding Remarks

A decision analysis was presented in the form of an examination of several mutually exclusive alternatives, each with probabilistic outcomes, with the intention of improving risk control. The choice of an alternative is similar to picking a lottery, which, for a front end investment in structural improvement or operational change, returns random outcomes. The DM selects what seems to him the most favorable of these lotteries. The real outcomes are not expected damages but actual damages associated with the various mutually exclusive events, of which one will occur in each period. Most likely, of course, it will be the zero-damage event, because it has by far the highest probability. But as long as damaging events are not impossible they also can occur. The situation is not unlike a horse race. The owner does everything within his power to prepare the horse for the win, but still he cannot be assured that his horse will win.

The owner/operator of a potentially harmful process may want to or may be mandated to transfer the exposure to these infrequent but large damages to a professional risk taker, an insurer, by making a *payment for certain* in the form of an insurance premium. Both the DM and the insurer may come up with an estimate of this insurance premium, but their estimates may not be the same, because they may not have the same subjective risk perception. If a premium can be agreed upon, it is payable at contract conclusion time and it relieves the DM of possible damages in excess of any deductibles or other exclusions that the contract may specify. An insurer does not save for a specific client and each new contract must stand on its own merit. In a so-called hard insurance market, following a period of exceptional losses, insurers need to analyze and select their exposures carefully. The evidence of risk control measures in place may induce the insurer to extend a contract to a perceived low risk new client or refuse a contract renewal with a perceived high risk old client. The alternative with the largest expected damage reduction may be the most desirable for the insurer, whereas the most cost-effective alternative, the one with the highest net benefit or the highest benefit-cost ratio may be the preferred alternative for the project (risk) owner.

Risk is a burden or a threat that is perceived differently by different people. Therefore, it is natural that risk assessment, if done by different people with different interests produces different answers. The two parties that most likely differ in their risk assessment are on the one hand those who create the exposure and on the other hand those who are exposed to the consequences. Exposure to possible harm is assumed here to result from externalities of projects, here dangers created by a dam in the event of large floods and earthquakes (internalities would be lack of maintenance, operator error, structural failure under operational stress, and so on, which are not considered here). The dam (or risk) owner may have a tendency of belittling the exposure, whereas the insurer who is called upon to shoulder the risk may have the tendency to exaggerate it. A conflict may arise over the size of an insurance premium as a consequence of differences in risk perception. It requires that the parties develop an understanding of their subjective risk perceptions and then come to some compromise between willingness to pay on the part of the risk owner and demand for compensation on the part of the risk taker (insurer). If no compromise is achievable, other ways of exposure reduction must be found, such as self-insurance, group insurance, more effective risk control, or a combination of measures.

The methods of risk evaluation described in this chapter have been around for a long time but they are still underused and not accepted or recognized for their potential of being effective engineering tools. The excuse is usually arbitrariness and subjectivity, which are actually the attributes of the traditionally used deterministic methods. More engineers need to familiarize themselves with probabilistic concepts and use them to advantage in their engineering work. As with all unfamiliar subjects, a threshold of knowledge must be acquired before the potential of the methods becomes obvious. Maintenance of hydraulic structures was singled out here as an application area for these rather generic methods, because water infrastructure maintenance planning and decision making in the face of uncertainty constitutes an almost classic application opportunity for these methods. Aside from actually using the methods, simply acquiring a basic understanding of the concepts of probabilistic methods should have a mind-expanding effect on an engineer's perception of the usually probabilistic real world problems he or she may be called upon to address. More comprehensive methods and more comprehensive thinking in turn will lead with a high probability to better solutions for many engineering problems.

****

Glossary

acceptance region. Also nonrejection region; used in hypothesis testing.

accessibility. The probability of being able to access a repair site according to plan.

adverse selection. Selection by an insurer of too many high-risk clients.

affordability. In connection with risk, the availability of financial assets that allow a risk taker to be aggressive in taking risk for perceived gains.

aggregation. Assembling the system from its components. The opposite of disaggregation.

annual maximum series. A selection of annual maximum events, e.g., floods or earthquakes, for design purposes.

annuity. An annual payment that reduces an initial capital to zero over a specified number of years at a specified interest rate.

arithmetic mean. The sum of the values of all elements in a population divided by the number of elements. The same as the expectation, with each element having the same probability.

autocorrelation. In a time series, when the next element is correlated to the previous element, or to several previous elements. Accordingly, one distinguishes lag-1, lag-2, and so on, correlations.

availability. The probability of being in operation when scheduled for operation.

back-of-the-head decision. A decision made after some subjective deliberation of evidence by a person, without the benefit of a decision analysis and not necessarily completely explainable.

bathtub curve. A failure rate distribution that exhibits relatively high failure rates at the beginning of the life of a component or system and with increasing age, with low failure rates in between.

Bayes' formula. Also called Bayes' theorem. A formula for the updating of prior probabilities, which are marginal probabilities, to posterior probabilities, which are conditional probabilities based on additional information obtained by an experiment affecting the sample space.

beta function. A probability density function that is adaptable to very different shapes, ranging from the bell shape to the bathtub shape.

binomial distribution. The probability distribution for a random process that has two possible outcomes, for example, success or failure, with constant success probability p and complementary failure probability $q = 1 - p$.

birth defect. A flaw introduced into a project or structure during the design or construction phase. As a hidden hazard it increases the probability of failure in the operation phase.

birth-death process. A process that has the characteristic of being on or off, with time intervals between on and off having random length.

black box model. A statistical correlation between input and output that does not require in-depth knowledge of the system.

Boolean algebra. Algebra of sets.

capital recovery factor. Ratio of an annuity (annual down payment) to a present worth capital investment; a function of interest rate i and number of payment years, n.

cause event. An event that is the cause of one or more follow-up events.

cdf. Cumulative distribution function; the integral of the pdf. Represents the probability of non-exceedance of some upper limiting value x.

central limit theorem. Also CLT. The tendency of sums of random variables or of sample means to approach the normal pdf as the sample size becomes large. This trend makes the normal pdf or Gaussian the limiting distribution of sample means, which   may be drawn from any distribution. This property makes the normal distribution the most important distribution in statistics and probability calculus.

central moment. The moment of the random variable with respect to its expected value. The central moment arm is $x - \mu$; the first central moment of a random variable is zero. The first ordinary moment (around the origin) is the expected value. The second central moment is the variance.

certainty equivalent. The true value equivalent to an expected utility.

chance branch. One of two or more possible outcomes of a random event illustrated by branches coming from an origin or root which marks the event itself.

chance event. An event that has a chance of occurring; also random event.

Chebychev inequality. A statistical measure that gives an approximate estimate of the probability of a random realization falling into a specified band around the mean.

chi-square test. An index, chi-square, is computed that represents a comparison of the normal pdf with the experimental pdf. For a selected small exceedance probability, say 5 %, or $F(z) = 0.95$, and a given degree of freedom (number of element classes – number of parameters – 1), a parameter c is picked from a table. If chi-square is less than or equal to c, the probability that chi-square is less than or equal to c is 95 %, and the hypothesis that the experimental distribution is normal is not rejected.

coefficient of variation. The ratio of the standard deviation to the mean, a measure of dispersion.

collective. The total of all elements; also the universe of elements or the sample space.

combination. Grouping of elements without regard to their sequence.

complementary probability. The difference between the probability of the sample space, $P(S) = 1$, and the probability of the union of all sets.

complementary set. A set that together with the other sets fill the sample space.

component redundancy. The backing up of components of a primary system.

concave. A shape of a curve that imitates a cave roof, sometimes also called "convex from above"; the curve remains below its tangent in the considered interval.

conceptual model. One or more related mathematical formulas that describe in some detail the working of a physical, industrial, or other process.

conditional probability. Probability of an event whose occurrence depends on another event.

confidence band. A range centered on the population mean in which most values of the sampled random variable or test statistic are found. The bandwidth is a function of the specified confidence level or significance level.

confidence estimate. The width of an interval that contains the sample mean with a desired probability; the confidence bandwidth is a function of the sample standard deviation, the chosen confidence level, and the sample size.

confidence interval. *See* confidence band.

confidence level. The probability that the sampling point falls inside the confidence band. The significance level is the complementary probability of the point falling outside the confidence band.

contingency. An unexpected situation that may arise and must be considered in advance planning.

continuous probability. A probability associated with a continuous random variable.

continuous random variable. A variable that varies continuously over a range.

control tunnel. An passage in a concrete dam or in the foundation of the core of an embankment dam that runs across the valley. Used to inspect for cracks, movements, leakage, and for remedial foundation work.

convex. A curve is convex from below, if it remains above its tangent in a given interval.; a curve is convex between two points if a straight line connecting the two points does not cut the curve.

corrective maintenance. Corrective action that is taken after a breakdown has occurred.

cost-effectiveness factor. The benefit-cost ratio is the ratio of benefit dollars per dollar of cost.

CPM. Critical path method.

crash time. Ultrashort completion time of an activity requiring overtime and/or overstaffing.

critical path time. The shortest possible completion time for a series of activities.

cumulative distribution function. Integral of the probability density function (pdf) from its lower limit, x_l, to an upper limit, x_u; the integral, abbreviated cdf, represents the non-exceedance probability, $F(x_u)$.

cycle time. Period that includes operation time and repair time, between two consecutive start-ups.

decision. Making a choice among two or more alternative courses of action.

decision analysis. Establishing criteria that quantify the pros and cons of possible alternative solutions for a problem.

decision branch. Choices a decision maker has that are illustrated as branches emanating from a point that represents the decision.

decision criterion. A measure used to judge competing alternatives, such as net benefit and benefit- cost ratio; usually several decision criteria are compared.

decision evaluation . Attaches values on alternatives that are available for implementation.

decision maker. The person in charge of making a choice, such as selecting an alternative for implementation; abbreviated DM.

decision tree. The various alternatives available to a decision maker to accomplish a complex project illustrated in the form of root branches emanating from a point that marks the decision and secondary chance branches that mark possible events the alternative must cope with.

decision under uncertainty. A deliberate decision to do or not to do something is combined with the probabilistic events and failure modes possibly caused by these events; the final outcome of alternatives is not under the control of the decision maker but can only be assessed through a probabilistic approach.

delay time. Time that passes with inactivity when an unexpected event occurs and defensive measures must be activated.

design flaw. An error committed during the design stage of a project, such as inadequate resistance of a component.

design flood. A major flood used for the layout of the spillways and the height of a dam, for the heights of levees, and for the cross section of canal or river channels.

detection. The effort of finding flaws that are not obvious.

deterioration process. Over time, a structure and the materials of which it is constructed undergo a slow decline in resistance due to exposure to the elements, vibrations, temperature changes, freeze-thaw cycles, corrosion, abrasion, and other wear.

deterministic method . A mathematical method that assumes all relationships between variables are well-defined by mathematical formulas, and all input variables and formula parameters have precisely known single values (e.g., means).

diagnosis. A finding on the underlying causes of a problem based on detection and analysis of symptoms.

dimensionless. Actually a quantity with dimension 1; for such a quantity the dimensional exponents of all component quantities add to zero. For example, the ratio of two lengths, $r = x_1 x_2^{-1}$ consists of two quantities each having the dimension of length, L, with the dimensional exponents 1 and -1, respectively, so that the dimension of r is dim $r = L^{1-1} = L^0 = 1$. The units of r are [m/m] $=$ [1].

disaggregation. Subdividing a complex system into subsystems and the subsystems into lower level subsystems and, finally, into basic components. The opposite to aggregation or synthesis.

disaster. An event, natural or human-caused, that causes great harm to many persons and considerable property and/or environmental damage.

discount rate. Rate used to reduce compound amounts of net cash flow to present worth for comparison with front end investment cost. It is comparable to the long-range bond rate.

discrete. Something that is separate and countable in contrast to being continuous.

discrete probability. The probability of a discrete event.

discrete random variable. A random variable with a countable number of realizations, for example, the states of a structure (new, deteriorated, failed), or the heads or tails in coin tosses.

disjoint events. Events that cannot occur jointly.

dispersion. The spread of individual outcomes around an expected value.

distribution. The function values $f(x)$ associated with x. If $f(x)$ is the frequency of x then the distribution is a frequency distribution or probability density function (pdf).

DM. Decision maker.

dot product. The inner product or scalar product of two vectors; it is the sum of the products of corresponding components that produces one number.

downtime. The time a project or component spends in planned or unexpected outage.

durability. The property of resisting deterioration and breakdown.

dynamic maintenance. Maintenance that is carried out according to need, not in response to breakdown or according to a plan.

economy of scale. The often experienced fact that a large project can produce a product at lower cost than many small ones.

economy-centered maintenance. Maintenance that is primarily guided by economic considerations for maintaining operations.

efficiency. The ratio of output produced to input provided; in compatible units, a dimensionless number.

emergency. A rapidly emerging dangerous condition that threatens the survival of a project or major components, and lives and property of people.

environmental exposure. The effects of sun, wind, rain, heat, cold, ice, snow, air, water, and aggressive pollutants in air and water on structures, components, and equipment.

EPP. Emergency Preparedness Plan.

equity. Value of property beyond the mortgage owed on it.

equivalent standard lottery. Abbreviated ESL. An alternative has appended to each of its possible real outcomes a two-pronged lottery, called the standard lottery, whose outcomes span the real outcomes of the alternative. The expectation of the indifference probability of all standard lotteries is the indifference probability of the ESL. The ESL makes the possible outcomes of different alternatives comparable with each other.

Euler formula. A formula for numerically integrating functions by calculating the areas of vertical strips as product of width Δx and average ordinate $[f(x_1) + f(x_2)]/2$ and adding them.

EVA. Expected value analysis, an analysis that determines the expected value of possible outcomes.

event tree. The sequence of events as they lead from one event to others including all possible follow-up events depicted in the form of branches of a tree emanating from a cause event.

exceedance probability. The probability that an event is greater than a specified event and all lesser events.

expectation. The probability-weighted mean of a random variable; also expected value.

expected life. The expectation computed from a pdf of life length.

expected utility. The expected value calculated from utility values.

expert system. A mathematical model that simulates the decision process that an expert might use to proceed from the discovery of a flaw to diagnosis of the cause, and on to benefit-cost analysis of repair alternatives. Usually an interactive mathematical model for the user to input values or make a decision while running the model.

exponential decay function. A function that represents the rate of reduction or decay of a quantity as a fraction of the amount on hand.

exponential distribution function. A function that describes probabilities which decline with increasing size of the random variable according to an exponential decay function whereby small sizes of the variable have high probability of occurrence, while large sizes have low probability; times between failures may be exponentially distributed.

exposure. Placing lives and/or property into the reach of harmful events that may damage or destroy life and property should they occur.

external event. An event that occurs by natural or human causes unrelated to the structure or its operation, such as floods or earthquakes.

extreme value. A minimum or maximum of a function.

failure rate. Rate at which members of a surviving population or age group discontinue functioning in the upcoming time interval.

fault tree. A graphical display of system failure traced back to all possible cause events. Or in reverse, the display of cause events and the succession of possible actions or omissions by humans or sensors that can lead to failure of the system (failure paths). May be represented as an inverted tree that starts with possible cause events and shows one or more failure paths that if not blocked result in one or more possible failure events.

FEMA. Federal Emergency Management Agency.

FERC. Federal Energy Regulatory Commission.

first moment. Integral over $x f(x)$ for a continuous function, or summation of $x f(x)$ for a discrete function. If extended over the entire distribution it amounts to calculating the probability-weighted mean of a population.

frequency. The absolute frequency is the number of elements (set) with the same property in a population. The relative frequency is the absolute frequency divided by the total number of elements in the population, a number less than or equal to 1.

fuzzy calculus. A calculation method that deals with variables that have number ranges, like probabilistic variables, but instead of probabilities a degree of belief from 0 to 1 is allocated to the numbers.

gamma distribution. A probability distribution with two parameters that can simulate many different shapes and that makes use of the gamma function.

gamma function. Also factorial function that is used to evaluate factorials. Appears in the integration of the normal distribution and other mathematical functions.

Gaussian. the normal pdf.

harmful event. An event that causes loss or damage to life, health, and property.

hazard. A condition that increases the components of risk, such as event size and exposure size, and their probabilities; thus, it increases risk itself.

hazard function. A mathematical function of instantaneous failure rate distribution; the number of failures per surviving elements during a time increment Δt that starts at a time t.

histogram. A graphical display of a partitioned sample space in which the bar widths are class sizes (subranges or sets) of the random variable and the bar heights are absolute or relative frequencies of elements per class; usually assembled from observations.

hydraulic structure. A structure or a set of structures required for the management of water for various uses; it usually includes also the mechanical, electrical, and other equipment needed for its operation.

ICOLD. International Commission on Large Dams.

indifference probability. A probability elicited from a decision maker at which he is truly indifferent (with no preference) between a value he may obtain at that probability or some smaller value he would get for certain.

infrastructure. The technological basis on which a modern society functions. It includes all public engineering works, such as drinking water supply systems, waste water drainage and treatment systems, energy production and distribution systems, solid and hazardous waste management systems, water management facilities, navigable waterways, irrigation systems, land and air transportation systems, and others.

insurance. A protection against exposure to severe loss by various kinds of events, natural or human-caused; by a time-limited, usually annual contract, the client pays a relatively small premium to an insurer who in turn pays contractually agreed upon monetary compensation if loss occurs.

insurance premium. A fixed payment made by a client to the insurer for the promise by the insurer to compensate the client for specified damages should they occur during the contract period.

insurer. Person or business that takes on the risk of large losses in return for an insurance premium (professional risk taker).

intangible. Something that cannot be touched or easily measured by money.

interarrival time. Time between two successive arrivals or incidents.

interference. An action that has the connotation of a hindrance or obstruction that slows a process or throws it off track, here the slowing down or halting the deterioration of infrastructure by maintenance.

internal rate of return. The interest rate that makes the present value of the net benefit stream (net cash flow) equal to the present value of the investment cost; in other words, the interest required to reach the break-even point between a capital expenditure and the benefits derived from it.

intersection. Parts of sets that are common to two or more sets.

joint probability. The probability of events occurring jointly.

law of large numbers. A limit approached as the number of observations becomes very large. For example, if the number of elements used in the calculation of the sample mean goes to infinity, the probability that the sample mean becomes identical to the population mean tends toward 1. In a fair coin toss, the process probability (probability of head or tail in each toss) is $p = 0.5$. If ten coins are tossed, simultaneously or in sequence, there may be 8 heads and 2 tails. If the number of coins tossed is only large enough, the probability $p = 0.5$ will be approached with probability 1.

level of significance. A selected probability α of the exceedance of an acceptance region. A null hypothesis cannot be rejected as long as the test statistic (e.g., the sample mean) falls into the acceptance region. If the test statistic falls into the rejection region, the null hypothesis is rejected and the test result is called statistically significant at level α. The probability that the null hypothesis is rejected, given it is true, is the probability of the type I error, which is equal to α.

liability. The responsibility for loss caused to others.

life cycle. The time from start-up to final shutdown, replacement, or demise of a system.

life-cycle benefit-cost analysis: Economic calculation that includes the time value of all benefit streams and cost streams (net cash flow); the net cash flow is compounded over the life of the project and discounted to present worth for comparison with the front end investment.

life-cycle costing. An economic analysis that compares all costs of competing alternatives over a project's economic life discounted to present worth.

limiting distribution. The distribution of sample means (the Gaussian pdf) that is approached when the number of elements in the samples approaches infinity.

liquefaction. A soil mass (embankment dam) taking on a liquid consistency by loss of intergrain friction due to excess internal pore water pressure; causes for liquefaction are shaking by an earthquake or sudden external pressure relief by water level drop.

logarithmic transformation. Representing a variable by its logarithm to find a suitable pdf.

LOL/PAR. Ratio of the number of people who loose their life and the total population at risk.

lottery. A construct by which two chance outcomes are compared with an outcome for certain; a probability is assigned by the owner of the chance outcomes (DM) at which he is exactly indifferent between accepting the chance outcomes or the outcome for certain. *See* indifference probability.

maintainability. The probability of carrying out maintenance within the planned scope of time and cost.

maintenance. Support provided to an operating project which ensures its performance as expected.

Markov chain. A stochastic (random) process that moves from state to state but the probability of a future state depends only on the present state without regard of how it arrived in the present state.

Markov process. A special case of a stochastic process for which the probability of transition to the next state is conditioned on the present state; the process can be discrete or continuous.

matrix. A rectangular array of numbers or elements, such as the coefficients of a set of linear equations, with each element being assigned to a proper location. Each element a has two indices, a_{ij}, with i referring to its row and j to its column. Matrices with one row are row matrices and matrices with one column are column matrices. Determinants are matrices used for solving linear equations.

maximum possible event (MPE). A synthetic event that is constructed by assuming the coincidence of many factors that contribute to increasing the size of the event to the point that this is meaningful. Preferably, an MPE is obtained probabilistically by random selection of the contributing factors, the construction of an event pdf, and finally by the selection of an event near or at the upper limit of the event range.

mean. The sum of the values of the elements of a sample divided by the number of sample elements.

metric units. Units of the International System of Units (SI).

minimization. Procedure of finding the smallest variable value of the range.

mixed system. A system that is made up of serial and parallel subsystems.

model. One or more mathematical relationships that simulate a process, also *mathematical model*, that are evaluated in sequence or simultaneously for the unkown output variables. *See* black box model, and conceptual model.

monitoring. A continuous measurement-taking process.

monotonical. A function whose gradient does not change sign within a considered range.

Monte Carlo simulation. A method that uses random sampling of inputs and parameters of relationships, as required, to create an input set for a mathematical model; the model is then used to produce an output. Many such selections and model outputs produce a random number collective that is examined for its probabilistic information content.

mortality rates. Failure rates for human beings, or systems and components; the number of deaths per thousand of the population segment at a given age.

multiobjective. Several purposes considered together.

mutually exclusive. Cannot occur together.

negligence. Failure to exercise reasonable care.

net benefit. The surplus of benefits over costs; also net cash flow.

non-exceedance probability. The probability of all events equal to or less than some limiting event; computed by the cdf. The complementary probability of the exceedance probability.

normal distribution. A symmetric bell shaped pdf with a range from minus infinity to plus infinity. Its cdf is an S-shaped monotonically rising curve; also called Gaussian.

null-hypothesis. A first hypothesis made on the range of a test statistic, such as the sample mean. *See* significance level.

numerical integration. The adding of small area increments under a function between specified integration limits.

O & M. Operation and maintenance.

objective risk. Risk evaluation based on statistical parameters, which are based on data, such as mean, standard deviation, and confidence bands; leads to the same result regardless of who evaluates the data.

odds. Used in the sense "the odds are..." is a way of assigning numbers usually to two subsets of a sample space to express their probability. For example, "the odds are 3 to 1 that the accident will happen" means the sample space of 4 is divided into two mutually exclusive sets, *accident* and *no-accident*; 3 is assigned to *accident* and 1 to *no-accident*. This gives the sample space 4 elements. The probability of *accident* is then 3/4, and the probability of *no-accident* is 1/4.

optimization. A mathematical method for finding the extremum (maximum or minimum) of a function, usually constrained by bounds on the variables. Sometimes several functions need to be optimized. The resulting compromise solution is known as Pareto optimality.

outage. State of being out of operation.

overmaintaining. Doing maintenance when it is not required; may be cost-ineffective and wasteful of resources.

parallel system. Arrangement of components that allows one or more alternative functional paths through the system, in contrast to the serial system.

partial-duration series. A time series that includes annual and subannual events.

partitions. Mutually exclusive and disjoint (non-intersecting) sets that exhaustively fill the sample space; in other words, the sum of all set probabilities is 1.

pdf. Probability density function; assigns a frequency to each value of the range of a random variable.

perception. Individual view of a problem.

performance. Workings of a system to achieve its purpose.

permutation. Change of a set by changing the order of elements. The set (1, 2) is permutated to (2,1). One combination of n elements has $n!$ permutations. There are permutations with and without repetition of elements.

PERT. Program Evaluation and Review Technique, a project planning technique.

piping. Erosion of fine material by seeping water that forms an upstream progressing pipe or tunnel into the core of an embankment dam or its foundations; usually a consequence of a defective earth filter in the dam; if it connects to the reservoir, it may breach the dam.

population. A collective or universe of elements; refers to the total number of elements of the sample space in contrast to a sample.

possibility. A probability greater than zero.

posterior probability. A probability that has been modified by additional information on which it becomes conditioned. Calculated by Bayes' formula.

predictive model. A mathematical model that simulates the essential workings of a process so that it can make estimates of what is going to happen if changes in input variables and model parameters or constraints occur.

premature demise. A failure that happens before it is expected to happen.

preventive maintenance. Maintenance undertaken before a forced outage occurs.

prior probability. The probability of an event before its updating by additional information.

probabilistic approach . A strategy that permits variables, parameters, and coefficients to be random variables; see also probabilistic method.

probabilistic decision analysis. An evaluation of alternatives whose costs and benefits are influenced by chance events.

probabilistic method. A mathematical method or model that uses random variables as input and produces random variables as output. It may use deterministic or probabilistic relationships, the latter having random variables as parameters and/or coefficients.

probabilistic model. A mathematical model that simulates a probabilistic process by a probabilistic method.

probabilistic process. A sequence of realizations of a random variable; a process by which a system accesses different states by transition probabilities; see also stochastic process.

probabilistic variable. A variable that has a range of values (realizations) and a pdf over the range. Also called random variable.

probability. The ratio of the number of elements of a specific kind to the total number of elements in a large number of elements. Must always be defined on a collective or sample space. Also a degree of possibility from 0 (impossible) to 1 (certain). A probability based on data (statistical probability) is an objective probability; a probability based on judgment is a subjective probability.

probability density function. Also pdf. An analytical or empirical function that assigns a frequency to each realization of a random variable over its range. A necessary condition of a pdf is that its integral from the lower bound of the range to the upper bound must be 1.

probability of survival. Probability of nonfailure; also called reliability.

probable maximum event. A synthesized event; also PME; for a flood, PMF.

product rule. The formula for the probability of simultaneous, independent events.

prognosis. A prediction of future events based on detected symptoms.

queuing theory. Theory that deals with waiting lines that form when random arrivals are serviced by one or more servers with random service time.

random. Without aim or purpose.

random process. Also stochastic process. A process that produces random variables.

random variable. A variable that has a range of values (realizations) each of which has a frequency of occurrence given by a pdf.

range. A set of values of a variable between a lower bound and an upper bound; may be finite or infinite; discrete or continuous.

ranking. Sorting a set of elements in descending order, with the largest element being given rank 1.

rate. A number of occurrences per time and/or per total population; the flow of a quantity per time.

redundancy. The presence of duplicate components or of a duplicate system that can take over operation in case the primary components or system fails.

regression. Determination of a relationship between a dependent variable and one or more independent variables using data fitting, such as the least squares method.

rehabilitation. Overhaul of a project that has deteriorated from wear and tear, or reconstruction of a project that is not in compliance with present construction standards.

reinsurance. Transfer of excess risk from a primary insurer to a secondary insurer or reinsurer; the insurer's insurance.

reliability. The probability of a component or system of not failing during a specified period; also survival probability.

reliability function. A function that quantifies the reliability of a system of components.

reliability-centered maintenance. Maintenance that emphasizes reliability.

remaining life. Average length of life that is left at a given age.

REMR. Repair-Evaluation-Maintenance-Rehabilitation, a U.S. Army Corps of Engineers project.

renewal process. Also renewal counting process. A collection of time periods between failures or between repair jobs in a queue, or interarrival times; the interarrival times are independent and identically distributed random variables with an arbitrary distribution. An alternating on-off process is a renewal process with two independent random variables "on-time" and "off-time."

repair. Rebuilding or replacing parts that are malfunctioning or have ceased functioning so that the system is restored to operation as expected.

response time. Time until action can be taken to cope with an unexpected outage or emergency.

return period. The average time span that passes until an event of a given magnitude is expected to recur. The "time" is actually the average number of trials between occurrences of the event. For one trial per year, the return period is said to be in years.

Richter scale. Logarithm of the ratio of the maximum oscillation amplitude read from a specific seismograph to a standard amplitude. Used as an indicator of earthquake magnitude. Extends from 0 to 9. Has been empirically related to energy released, to horizontal and vertical ground accelerations, and others. Damage generally occurs from magnitude 5.5 on up. Each magnitude represents a 31.5-fold increase in energy released. Two magnitudes represent an approximate 1,000-fold increase ($31.5 \cdot 31.5$) in energy released.

risk. Exposure to a harmful event with at least one of the two being probabilistic; the event may or may not occur during a specified period and the scope of the exposure may or may not be precisely known. If nothing is known about event and the extent of the exposure, then the exposure represents risk under uncertainty.

risk acceptance. A willingness to face the risk, usually when there is a perceived reward.

risk analysis. Assembling the components and circumstances that produce risk and evaluating the associated objective and subjective risk measures, such as expectation and standard deviation, and risk premium and insurance premium, respectively.

risk assessment. A study that attempts to describe and to the extent possible to quantify risk.

risk control. Measures that limit or reduce the severity of the components of risk, such as event; event probability, if human-caused; extent of exposure and exposure probability.

risk evaluation. Quantitative assessment of risk in the form of an insurance premium or other monetary value.

risk factor. A condition or activity that increases risk; a hazard.

risk imposition. Some entrepreneur forcing others to accept risk emanating from his or her planned or ongoing activity, often without knowledge of and approval by the risk owners.

risk management. A strategy that keeps exposure, events, and associated likelihood of occurrences of harmful events under control, by hazard recognition, reduction, and elimination, either before or after risk has been recognized or created.

risk owner. Someone under risk.

risk perception. Subjective assessment of the exposure to harm; the risk assessor, the risk owner, and the risk imposer may have a different risk perception.

risk premium. Amount of money assessed for an activity over and above the cost of a risk-neutral activity; a reward for a risky activity.

risk sharing. A group of people who pool assets to compensate a group member in the case the member incurs damage; the principle of mutual insurance.

risk taker. A person or business that accepts risk usually for a fee (reward); an insurer.

risk-averse. An attitude toward risk that tries to minimize or avoid risk.

risk-neutral. An attitude that acts as if there is no risk; insensitivity to risk.

risk-prone. An attitude that seeks or attracts risk, usually for monetary gain, personal thrill, lack of information, or other reason.

RO&M: Review of Operation and Maintenance Program.

routine maintenance. Maintenance activities that are carried out on a fixed schedule to meet basic requirements.

safety. Condition of being protected from exposure to harm.

safety factor. The ratio of resistance to load. An analogy is the economic safety factor which is the benefit-cost ratio.

safety margin. The amount by which resistance exceeds the load. An economic analogy is the net benefit.

safety-centered maintenance. Maintenance policy that emphasizes the safety of existing structures and equipment in contrast to emphasis on other objectives, such as economy or service reliability.

sample. A number of elements taken at random from a much larger population as a representation of the population, when the population itself is too large or incomplete so that it cannot be analyzed as a whole.

sample parameters. Parameters, such as mean and standard deviation calculated from a sample in contrast to those calculated from the total population.

sample space. The ensemble of elements that represent a data population; also a collective, a universal set, or a population.

sampling distribution. The pdf of a test statistic, usually the sample mean. According to the central limit theorem, sample means tend to be normally distributed, if only the number of sample elements is large enough.

seat-of-the-pants decision. A decision that is made by informal subjective weighing of information and experience. The flaws of this decision method are its lack of repeatability, and the impossibility of making a sensitivity analysis of critical information items.

second moment. For a continuous function, the integral (or sum) over the product of area increments of the pdf times the square of their distance from the origin; for a discrete function, the pdf function value times the square of the distance from the origin; if x is replaced by the distance from the mean, $x - \mu$, then result is the"second central moment" or variance.

SEED: Safety Evaluation of Existing Dams.

self-insurance. A company or a group of companies with similar objectives create a reserve fund for the protection from unexpected, potentially ruinous loss that may affect one of the group's members at a time; self-insurance is necessary if insurance companies refuse coverage. Self-insurance may be economical if it can be accomplished at substantial savings in insurance premiums. A deductible on an insurance policy is self-insurance that is clearly equivalent to "no insurance"; self-insurance is usually limited to small losses that can be borne by the policy holder. Group insurance is self-insurance by individuals or industries that cannot obtain regular insurance.

sensitivity analysis. A mathematical model can be used to change a parameter or variable by a small amount and calculate the resulting change in the outcome.

serial correlation. In a time series, a future outcome may be influenced by a previous outcome. A typical example is auto-correlation where the next outcome is a function of the present outcome.

service life. The useful life of a component from start-up to final shutdown or failure.

service rate. The number of repair jobs accomplished in a time unit.

set. A group of elements that have similar properties. Venn diagrams are graphical illustrations of sets as parts of a universe, or a collective, or a sample space. The probability of a set is defined on the sample space, as the ratio of the set's number of elements to the total number of elements of the sample space. Ordinary sets may intersect with each other. A set and its complementary set exhaustively fill the sample space.

shutdown. The state of being out of operation. Also the action of transferring a system into a shutdown state. A shutdown can be planned, as in preventive maintenance, or it can occur as a random failure resulting in a forced shutdown.

SI. International System of Units, the modernized metric system. Public Law 100-418, section 5164, of August 1988, designates "the metric system of measurement as the preferred system of weights and measures for United States trade and commerce ...". The SI is based on seven independently defined base units from which all other units are derived. Internal coherence, decimality, and one quantity-one unit are the hallmarks of the metric system.

simulation. The imitation of the functioning of a system or process by simultaneous or quasi-simultaneous evaluation of numerous and often complicated relationships by computer.

SL. Standard lottery, an expression from game theory used in decision analysis. *See* also ESL (equivalent standard lottery).

slack time. A period that an activity can be delayed past its earliest start time without exceeding its latest completion time.

smart system. A system that signals need for maintenance.

SOP: Guide for Preparation of Standing Operating Procedures.

speculative risk. An exposure to financial loss a risk-prone gambler or money manager accepts in the hope of making money quickly; delaying maintenance for monetary gain falls into this risk category.

standard deviation. A statistical parameter that indicates the dispersion of elements of a sample or population around the mean; a shape parameter in probability distributions. It is calculated as the sum of the squared departures of elements from the mean divided by the number of elements reduced by 1, because one degree of freedom is lost to the mean, and taking the root of this average; Its square is the variance.

standard error of the estimate. The standard deviation from a regression line; here, the mean is represented by the regression line and the standard deviation with respect to the regression line is computed as the sum of the squared departures from the regression line divided by the number of elements used in the regression minus 2, as two degrees of freedom are lost to the coefficients of the regression line; then the root is taken of this averaged sum.

standard error of the sample mean. The standard deviation of the sample mean; it is calculated as the population standard deviation divided by the square root of the sample size.

standard normal distribution. The subtraction of the mean from the random realization x, and the division of the difference by the standard deviation produces a dimensionless variable, z , that is called the *standard normal variable* , also sometimes the standard deviate. This variable is used in a quadratic exponential decay function to produce the normal pdf or standard normal distribution, $f(z)$, which has a bell shape with a symmetry axis at $z = 0$, and a range $-\infty \leq z \leq +\infty$. The normal probability density function (pdf) is usually tabulated for $z \geq 0$, but can be easily calculated for use in spreadsheet calculations or computer programs. The probability of z is defined only for an infinitesimally narrow strip around z as $f(z)dz$.

standard of care. The exercise of state-of-the-art professional circumspection in professional work.

standby. A component or system held in non-operating reserve in case of a breakdown and that is ready for immediate use subject to a successful switch.

state. The condition a project is in, e.g., operating state, shutdown state, or other.

stationarity Invariance with time of a process.

stochastic decision process. A stochastic process that is under a DM's policy that favors desirable outcomes by increasing transition probabilities toward desirable states or by decreasing transition probabilities toward failure states without being under full control of such a policy.

stochastic process. A process governed by probabilistic laws that produces random outcomes or moves among possible states in a random fashion; may be discrete or continuous; an example of a discrete process is a deterioration process that moves in a random fashion from a new state to a deteriorated state and finally to a failure state; a continuous process is the generation of (white) random noise where the increments of the continuous output are the random variables.

structural integrity. Soundness of the structure based on all components performing as expected.

structure function. An index of a system that marks it as functioning or not functioning.

structure reliability function. A relation to assess the reliability of an assembly of components.

structured decision process. A method that analyzes all available information in an analytical way in contrast to seat-of-the-pants decision making. The structured decision process is repeatable, can be documented, and can be subjected to sensitivity analysis. **subjective**. Influenced by an individual's personal judgment.

subjective risk. Possible gains or losses as perceived by a person (DM) through a method that adjusts a risk-neutral result according to risk attitude and risk perception; quantitative assessment by the utility function concept.

subset. A set that is part of another set.

subsystem. A part of a larger system or a smaller system within a system.

surprise effect. Delayed action that occurs when an unexpected event happens and the response needs some time to become effective.

survival probability. Probability that a failure will not occur during a period; same as reliability.

symptom. A detectable condition that points at an underlying cause.

syndicate. In insurance, a group of people who invest money for risk coverage.

system. An assembly of components that operates on input to produce an output.

system redundancy. A system that has a second system in reserve.

system response. Reaction of a system's output to a change in input or a change in a system component.

system state. The condition a system is in, for example, operating state, shutdown state, failure state, and so on.

tangible. Something whose value can be expressed in money.

time series. A sequence of states or values over time, usually totals or averages for consecutive time increments, such as days, weeks, months; for example, the number of pipe failures per month over a ten year period.

TQM. Total quality management.

transition probability. Probability of a stochastic process moving to a new state, given its present state.

trigger event. Event that causes other events to follow; also cause event.

unbiased estimator. An estimator that does not depend on any other estimator, e.g., the mean of a population.

uncertainty. Lack of knowledge that extends to probabilities; mild uncertainty is characterized by data-based probabilities; severe uncertainty exists when there are no data-based probabilities.

underwriting. Making a contract for taking over a risk from a client for an insurance premium paid by the client for a specified period and assuming liability for specified damages should they occur.

uniform distribution. A pdf of a random variable whose realizations all have the same frequency ordinates.

uninsurable risk. An exposure to harm that is in progress after the probabilistic aspects have been removed.

union. The joining of two sets.

universal set. The collective of all elements; also called "universe."

unreliability. The probability of failure.

utility. A subjective weight that is substituted for the cost of a risky project; this weight reflects the attitude toward risk of the decision maker for whom or by whom the utility is developed.

utility function. The functional relationship that can be used to convert costs into utilities and vice versa in subjective risk assessments. The shape of the curve reflects the attitude toward risk of those for whom the utility function was developed.

value of life. A controversial measure that attempts to put a value on a person's life. It can be used in risk analysis to steer the decision maker away from selecting high risk, life threatening alternatives.

variance. The average of the squared departures from the mean. A measure of dispersion that is not dimensionally compatible with the units of the random variable or the mean (see standard deviation).

vector. A quantity that requires more than one number for its definition; for example, a force is a vector that is defined by magnitude and direction; more generally, any array of numbers, such as a row in a matrix or a column in a matrix, which are referred to as row vector and column vector, respectively. Arrays that can be construed as row and column vectors arise as transition probabilities from and to states of a Markov chain. Dot products or matrix products arise as probabilities of mutually exclusive paths in Markov chains.

Venn diagram. Illustration of sets, their relationships with each other, and of the sample space that contains all sets.

waiting line. In queuing theory, the accumulation of random arrivals at a server.

warning time. The period that is available for alerting people of an impending disaster. Earthquakes and explosions have zero warning time; fair-weather dam breaks, floods, hurricanes, and tornadoes have varying warning time based on observations and prediction methods.

warranty. The assurance given by a supplier that the item will perform as expected for some limited period beginning with start-up.

wear-out failures. Failures that are caused over time by operational wear and tear on the system and its components.

****

References

Abramowitz, M. and I. A. Stegun, edit. (1970). <u>Handbook of Mathematical Functions with Formulas, Graphs and Mathematical Tables</u>. U.S. Department of Commerce, National Bureau of Standards, Applied Mathematics Series 55. Issued June 1964. Ninth Printing, June 1970. Superintendent of Documents. U.S. Government Printing Office, Washington, DC 20402.

ACER (1986). Guidelines to Decision Analysis. ACER (Assistant Commissioner— Engineering and Research) Technical Memorandum No. 7, Bureau of Reclamation, Denver, CO.

ASCE (1998). ASCE 1998 Report Card for America's Infrastructure. Internet: http://www. asce.org. Copyright 1996, 1998.

ASCE (2000). Corps Research Reduces Maintenance Costs. <u>Civil Engineering</u>, November 2000, p. 27. American Society of Civil Engineers, 1801 Alexander Bell Drive, Reston, VA 20191-4400.

ASCE (2001). ASCE—1998 Report Card for America's Infrastructure and Issue Briefs. Internet: http://www.asce.org/reportcard. May 1, 2001. <u>See also</u>: Report card a great success, local assessments now needed, by R. W. Bein, <u>ASCE News</u>, May 2001.

ASCE (2002). Dam Safety Bill Advances in the House, by Jay Landers. <u>Civil Engineering</u>, September 2002, p. 12. American Society of Civil Engineers, 1801 Alexander Bell Drive, Reston, VA 20191-4400.

ASCE/USCOLD (1975). Lessons from Dam Incidents—USA. Committee on Failures and Accidents to Large Dams of the United States Committee on Large Dams, J. F. Redlinger, Chairman. ASCE, New York.

ASDSO (1996). <u>ASDSO Newsletter</u>, Nov/Dec 1996, Vol. 11, No. 6, p. 11. Association of Dam Safety Officials, 450 Old Vine, Lexington, KY 40507.

Bath, M. (1979). Introduction to Seismology. Second Revised Edition. Birkhäuser Verlag, Basel, Boston, Stuttgart.

Beard, L. R. (1974). Probabilities of Rare Floods. From Inspection, Maintenance and Rehabilitation of Old Dams. Proceedings of the Engineering Foundation Conference, Asilomar Conference Grounds, Pacific Grove, CA, September 23–28, 1973, pp. 314–319, ASCE, New York.

Benjamin, J. R. (1974). Making Decisions in the Face of Uncertainty. From Inspection, Maintenance and Rehabilitation of Old Dams. Proceedings of the Engineering Foundation Conference, Asilomar Conference Grounds, Pacific Grove, California, September 23-28, 1973, pp. 356–381, ASCE, New York.

Benjamin, J. R. and C. A. Cornell (1970). Probability, Statistics, and Decision for Civil Engineers. McGraw-Hill, New York.

Biswas, A. K. and S. Chatterjee (1971). Dam Disasters—An Assessment. Engineering Journal (Canada). Vol. 54, No. 3, pp. 3–8. March 1971.

Blind, H. (1983). The Safety of Dams. Water Power & Dam Construction, Vol. 35, No. 5, pp. 17–21, May 1983.

Bloch, H. P. and F. K. Geitner (1990). An Introduction to Machinery Reliability Assessment. Van Nostrand Reinhold, New York.

Bloomberg, R. (1990). "Seepage cutoff wall is deepest yet." Engineering News Record, April 12, 1990, pp. 31–32, McGraw-Hill, New York.

Bock, P. K. (1974). Bureau of Reclamation Examination Program. Inspection, Maintenance and Rehabilitation of Old Dams. Proceedings of the Engineering Foundation Conference, Asilomar Conference Grounds, Pacific Grove, CA, September 23–28, 1973, pp. 50–60. ASCE, New York.

Boggs, H. L., G. S. Tarbox, and R. B. Jansen (1988). Arch Dam Design and Analysis. Chapter 17, p. 493 - 539. Advanced Dam Engineering for Design, Construction and Rehabilitation , R. B. Jansen, ed. Van Nostrand Reinhold, New York.

Bolt, B.A. (1978). Earthquake Hazards. EOS–Transactions of the American Geophysical Union, Vol. 59, No. 11, November 1978, pp. 946–962. Washington, DC.

Bolt, B. A., W. L. Horn, G. A. MacDonald, and R. F. Scott (1975). Geological Hazards. Springer-Verlag, New York, Heidelberg, Berlin.

Brigham, E. O. (1974). The Fast Fourier Transform. Prentice-Hall, Englewood Cliffs, NJ.

Britannica (1996). Britannica Book of the Year (events of 1995), p. 47. Encyclopaedia Britannica, Inc., Chicago.

Britannica (1998). Britannica Book of the Year (events of 1997). Insurance: pp.164–165. Dams, by T. W. Mermel, pp. 139–141. Encyclopaedia Britannica, Inc., Chicago.

Britannica (1999). Britannica Book of the Year (events of 1998), pp.142–144. Encyclopaedia Britannica, Inc., Chicago.

Bronstein, I. N. and K. A. Semendjajew (1984). Taschenbuch der Mathematik (Handbook of Mathematics). G. Grosche, V. Ziegler, and D. Ziegler, eds., 21. edition (English edition available). Verlag Harri Deutsch, Thun and Frankfurt/Main.

Brown, D. (1974). Dam Inspection Program of the Federal Power Commission. Inspection, Maintenance and Rehabilitation of Old Dams . Proceedings of the Engineering Foundation Conference, Asilomar Conference Grounds, Pacific Grove, CA, September 23–28, 1973, pp. 23–38. ASCE, New York.

Brown, L. F. (1994). Principles for Specifying Risk Acceptance Guidelines. Los Alamos National Laboratory, Los Alamos, NM. Presentation at WATTec Conference, February 23, 1994, Knoxville, TN.

Brouwer, G. (2002). Up into the sky. Civil Engineering, ASCE, January 2002, pp. 50–57.

Bryant, L. M. and Mlakar, P. F. (1987). Evaluation of Civil Works Steel Structures. Report J650-87-003/1377, U.S. Army Construction Engineering Research Laboratory, Champaign, IL 61820-1305.

BUREC (1975). Concrete Manual: A Water Resources Technical Publication. 8th Edition. U.S. Department of the Interior, Bureau of Reclamation. U.S. Government Printing Office, Washington, D.C. 20402, and the Bureau of Reclamation Engineering and Research Center, P.O. Box 25007, Denver Federal Center, Colorado 80225.

BUREC (1980). Safety Evaluation of Existing Dams: A Manual for the Safety Evaluation of Embankment and Concrete Dams (SEED). U. S. Department of the Interior, Water and Power Resources Service. Denver Federal Center, P.O. Box 25007,

Denver, CO 80225, Attn 922.

BUREC (1986). Guide for Preparation of Standing Operating Procedures for Dams and Reservoirs (SOP), by Neil J. Gillis. Bureau of Reclamation Engineering and Research Center. Library of Congress Catalog Card No. 85-600635. U.S. Government Printing Office, Denver, CO, January 1986.

BUREC (1991). Review of O&M Program: Field Examination Guidelines (RO&M). U.S. Dept. of the Interior, Bureau of Reclamation, Engineering Division, Facilities Engineering Branch, Denver Office, Denver, CO.

BUREC (1991a). Hydropower 2000: Reclamation's Energy Initiative. U.S. Department of the Interior, Bureau of Reclamation, November 1991.

Cohen, G. (1998). Nightmares and Hope in Bhopal. Corporate Watch. Internet. http://www.igc.apc.org/trac/bhopal/nightmare.html

Cornell, C. A. (1972). First-Order Analysis of Model and Parameter Uncertainty. Proceedings, International Symposium on Uncertainties in Hydrologic and Water Resources Systems, Tucson, AZ, Vol. 2, pp. 1245–1272.

CSO (1958). Commissioners Standard Ordinary Mortality Table. See Greene, M. R. (1977), p. 412.

Daus, G. (2001). Feasibility Study Proposes Inflatable Dam. Water Engineering and Management, November 2001 (www.waterinfocenter.com).

Davis, M. L. (1993). Rivers in the Desert: William Mulholland and the Inventing of Los Angeles. HarperCollins, New York.

De Camp, L. S. (1993). The Ancient Engineers. Barnes & Noble Books, New York.

Dell'Isola, A. J. and S. J. Kirk (1981). Life Cycle Costing for Design Professionals. McGraw-Hill, New York.

Dell'Isola, A. J. and S. J. Kirk (1983). Life Cycle Cost Data with Educational Supplement. McGraw-Hill, New York.

DOE/BUREC (1989). Replacements, Units, Service Lives, Factors. Contract No. DE-AC65-87WAO2032, May 1989. Prepared for USDOE and USBUREC by Stone and Webster Management Consultants, Inc., 5500 South Quebec Street, Englewood, CO

80111-1914.

Dolcimascolo, A. (1980). Safety inspection of dams. <u>Water Power and Dam Construction</u>, October 1980.

Draper, N. R. and H. Smith (1966). <u>Applied Regression Analysis</u>. John Wiley & Sons, New York.

Electro-Watt (1959). Erfahrungen beim Betrieb der Kraftwerke Mauvoisin (experiences with the operation of Mauvoisin hydropower project). <u>Schweizerische Bauzeitung</u>, Vol. 77, No. 39, pp. 645–654.

ENR (1970/1). Vaiont Dam trial ends. <u>Engineering News Record</u>, January 1, 1970, p. 14. McGraw-Hill, New York.

ENR (1985/9). Some dams don't need safety frills: Committee looks at risk procedures, finds diversity and lack of knowledge. <u>Engineering News Record</u>, September 5, 1985, pp. 32–33. McGraw-Hill, New York.

ENR (1989). There is progress but the threat remains. <u>Engineering News Record</u>, April 27, 1989, pp. 34–35. McGraw-Hill, New York.

ENR (1990). BuRec forks over cash. <u>Engineering News Record</u>, August 16, 1990, pp. 12–13. McGraw-Hill, New York.

ENR (1997). <u>Engineering News Record</u>, August 16, 1990, p. I-20 and p. I-24. McGraw-Hill, New York.

ENR (2000/10/9). U.S. safety officials seek funds for more inspections and repairs. <u>Engineering News Record</u>, October 9, 2000, p. 18. McGraw-Hill, New York.

ENR (2001/2/12). Red Hills' red ink shows big risks, by R. Korman. <u>Engineering News Record</u>, February 12, 2001, p. 12. McGraw-Hill, New York.

ENR (2001/10/22). Market braces for terror shock, by W. G. Krizan and R. Korman, <u>Engineering News Record</u>, October 22, 2001, pp. 10–11. McGraw-Hill, New York.

Farmer, F. R. (1967). Proceedings of the Symposium on Containment and Siting of Nuclear Power Plants, International Atomic Energy Agency, Vienna, Austria, April

3–7, 1967 (as cited by L. F. Brown).

FEMA (1996). National Dam Safety—Mitigation Directorate, Federal Emergency Management Agency. January 1996. FEMA Dam Safety Office, 500 C Street S.W., Washington, DC 20472.

FERC (1987). Engineering Guidelines for the Evaluation of Hydropower Projects. Federal Energy Regulatory Commission, Office of Hydropower Licensing. July 1987 (Reprint December 1989)
http://www.ferc.fed.us/hydro/docs/engguide/guidelines.htm

FERC (1989). Update of FERC (1987). In January 2002, latest update was November 1998.

FERC (1998). Engineering Guidelines for the Evaluation of Hydropower Projects. Section 11-5.6.1: Sliding on abutment contact with flat abutment slopes (web site); see also Section 11-8.3.2.

Fiering, M. B. and B. B. Jackson (1971). Synthetic Streamflows. Water Resources Monograph 1, American Geophysical Union, Washington, DC.

Fireman's Fund (1997). Personal communication with R. D. Farnsworth, Fireman's Fund Insurance Co., a subsidiary of Allianz, Novato, CA.

Franz, D. D., B. A. Kraeger, and R. K. Linsley (1991). Estimating the Frequency of Extreme Flood Events. EOS, June 25, 1991, American Geophysical Union, Washington, DC.

Freund, J. E. (1993). Introduction to Probability. First published in 1973. Republication by Dover Publications, New York.

Galambos, T. V. (1981). Load and resistance factor design. Engineering Journal, American Institute of Steel Construction, Third Quarter, 1981, pp. 74–84.

Goldberg, S. (1986). Probability: An Introduction. Dover Publications, New York (unabridged republication of the work first published in 1960).

Goldman, A. S. and T. B. Slattery (1964). Maintainability: A Major Element of System Effectiveness. John Wiley & Sons, New York. Reprinted 1967. Reprinted by R. E. Krieger Publishing Company, Huntington, NY, 1977.

Grayson, C. J., Jr. (1960). Decisions Under Uncertainty: Drilling Decisions by Oil and Gas Operators. Cambridge, Harvard University Press.

Grant, E. L. and W. G. Ireson (1970). Principles of Engineering Economy. 5th ed. The Ronald Press Company, New York.

Greene, M. R. (1977). Risk and Insurance. 4 th ed., South-Western Publishing Co., Cincinnati, OH.

Gross, D. and C. M. Harris (1974). Fundamentals of Queuing Theory. John Wiley & Sons, New York.

Gruetter, F. and N. J. Schnitter (1982). Analytical Risk Assessment for Dams. Commission International des Grand Barrages, Quatorzieme Congres des Grandes Barrages, Rio de Janeiro, 1982. Q. 52, R. 39. pp. 611–625.

Habibian, A. (1988). Washington Suburban Sanitary Commission—Final Report on The Water Main Condition Analysis Program Pilot Study. Volume II: Technical Supplement.

Hall, W. D. (1993). Dams and hydropower in the Tennessee Valley. Water Power and Dam Construction, August 1991, pp. 47–49.

Hammersley, J. M. and D. C. Hanscomb (1964 and 1967). Monte Carlo Methods. Methuen & Co LtD, London, U.K.

Hammond, J. D. (1980). Risk-spreading through underwriting and the insurance institution. In Societal Risk Assessment: How Safe Is Safe Enough? R. C. Schwing and W. A. Albers, Jr., eds. Plenum Press, New York and London, pp. 147–180.

Heffron, R. E. (1990). The Use of Submersible ROV's for the Inspection and Repair of Water Conveyance Tunnels. Water Resources Infrastructure: Needs, Economic and Financing, pp. 35–40. John F. Scott and Reza M. Khanbilvardi, eds., ASCE, New York,

Henley, E. J. and H. Kumamoto (1985). Designing for Reliability and Safety Control. Prentice Hall, Englewood Cliff, N.J. 07632.

Hillier, F. S. and G. J. Lieberman (1974). Operations Research. 2 nd ed., Holden-Day, Inc., San Francisco.

Hirsh, J. (2001). Sticker Shock. Special Advertising Section— Insurance, pp. I3 - I23. Engineering News Record, February 12, 2001. McGraw-Hill, New York.

Hoel, P. G. (1971). Introduction to Mathematical Statistics. 4 th ed. John Wiley & Sons, New York.

ICOLD (1984). Deterioration of Dams and Reservoirs: Examples and their analysis. ICOLD Committee on Deterioration of Dams and Reservoirs, M. Rocha, chair. Imprimerie Louis-Jean, Publications scientifiques et litteraires, 05002 Gap. Janvier 1984. ISBN 90 6191 546 5. Distributed by A. A. Balkema, P.O. Box 1675, Rotterdam, Netherlands. For U.S. and Canada: A. A. Balkema Publishers, Old Post Road, Brookfield, VT 05036.

ICOLD (1991). Q65: Ageing of dams and remedial measures. Report of the 17 th ICOLD Congress. Water Power & Dam Construction, 43:10, p. 65-70.

I.I.H.S. (1992). Crash Problem On A Per Mile Basis. Special Issue: Crashes, Fatal Crashes per Mile. Insurance Institute for Highway Safety. Status Report, Vol. 27, No. 11, p. 3. 1005 North Glebe Road. Arlington, VA 22201.

I. I. I. (2003). The Insurance Information Institute Fact Book 2003. Insurance Information Institute, 110 William Street, New York, NY 10038.

Ireson, W. G., ed. (1982). Reliability Handbook. Reissue of 1966 Edition. McGraw-Hill, New York.

Ivey, D. L. (1965). Splitting tensile tests on structural light weight aggregate concrete. Texas Transportation Institute, College Station, Texas (as quoted by E. Kreyszig, Advanced Engineering Mathematics, p. 840).

IWPDC (1995). ICOLD reports on dam failures. International Water Power and Dam Construction, Vol. 47, No. 5, p. 2, May 1995.

Jaffee, D. M. (1989). Money, Banking, and Credit. Worth Publishers, 33 Irving Place, New York, NY 10003.

Jansen, R. B. (1980). Dams and Public Safety. U.S. Department of the Interior, Water and Power Resources Service. U.S. Government Printing Office, Engineering Research Center, Denver Federal Center. P.O. Box 25007, Denver, CO 80225.

Jansen, R. B., ed. (1988). Advanced Dam Engineering for Design, Construction, and

Rehabilitation. Van Nostrand Reinhold, New York.

Kanamori, H. and E. E. Brodsky (2001). The Physics of Earthquakes. Physics Today, June 2001, pp. 34–40. American Institute of Physics, 2 Huntington Quadrangle, Suite 1N01, Melville, NY 11747-4502.

Katz, A. (1967). Principles of Statistical Mechanics. W. H. Freeman and Company, San Francisco and London.

Keeney, R. L. and H. Raiffa (1976). Decisions with Multiple Objectives: Preferences and Value Tradeoffs. John Wiley & Sons, New York.

Klemes, V. (2000). Common Sense and other Heresies: Selected Papers on Hydrology and Water Resources Engineering. C. D. Sellars. ed. Canadian Water Resources Association. P.O. Box 1329, Cambridge, Ontario, N1 R7 G6 Canada.
KNS (1989). Coolidge Dam likely to break? In Briefs–Nation and World, Knoxville News Sentinel, Tuesday, May 2, 1989.

KNS (1995/2). Sinking insurance companies: Many consumers slip through states' insurance safety nets, by D. Sword. Knoxville News Sentinel, February 19, 1995. Knoxville, TN,

KNS (1995/12). This year's disaster bill reaches record $180 billion, p. A9, December 31, 1995. The Financial Times for Scripps Howard News Service. Knoxville News Sentinel, Knoxville, TN.

KNS (1997/8). Muskrats making Swiss cheese of Holland's protective levees. By The Associated Press. Knoxville News Sentinel, August 1997. Knoxville, TN.

KNS(2001/9/13). World Trade Center attack likely to cost insurers billions. By T. Agovino, Associated Press. Knoxville News Sentinel, September 13, 2001. Knoxville, TN.

KNS(2001/9/27). Lloyds set to ante up on $1.9 billion in claims from terrorist attack. Knoxville News Sentinel, September 27, 2001. Knoxville, TN.

KNS(2001/10). Tennessee set to crack down on all insurance fraud with new law. By R. Locker, The Commercial Appeal. Knoxville News Sentinel, October 1, 2001. Knoxville, TN.

Kreyszig, E. (1979). Advanced Engineering Mathematics, 4 th ed. Chapter 20:

Probability and Statistics, pp. 838–939. Sections 20.2: Tabular and Graphical Representations of Samples, pp. 840–846. Section 20.9: Binomial, Poisson, and Hypergeometric Distributions, pp. 873–879. John Wiley & Sons, New York.

LaPay, W. S. (1990). Probabilistic Basis for Managing Maintenance. Water Resources Infrastructure: Needs, Economic and Financing. John F. Scott and Reza M. Khanbilvardi, eds., pp. 101–104. ASCE, New York.

Lapin, L. L. (1978). Statistics for Modern Business Decisions. 2nd edition. Harcourt Brace Jovanovich, New York.

Lau, T. and T. Sohre (1992). An Insurer's Experience with Small Hydroelectric Powerplants. Proc. American Power Conference, Vol. 54-I, p. 173. 54th Annual Meeting, 1992, Chicago. Sponsored by Illinois Institute of Technology, IIT Center, Chicago, IL.

Linsley, R. K., M. A. Kohler, and J. L. H. Paulhus (1958). Hydrology for Engineers. McGraw-Hill, New York.

Lloyds (1996). Lloyds Heritage Historical Data 1990–1996. Corporate Communications Department, World wide Web: www.lloydsoflondon.co.uk.

Locks, M. O. (1973). Reliability, Maintainability and Availability Assessment. Spartan Books, Hayden Book Company, Rochelle Park, NJ.

Lundin, L. (1974). TVA Dam Inspection Activities. Inspection, Maintenance and Rehabilitation of Old Dams. Proceedings of the Engineering Foundation Conference, Asilomar Conference Grounds, Pacific Grove, CA, September 23–28, 1973, pp. 39–49. ASCE, New York.

MacDonald, T. C. and J. Langridge-Monopolis (1984). Breaching characteristics of dam failures. ASCE, Journal of Hydraulic Engineering, Vol. 110, No. 5, pp. 567–586, May 1984.

Maher, M. L. (1987). Expert Systems for Civil Engineers. ASCE, 345 East 47th Street, New York, NY 10017-2398.

Marzulla, R. J. (1990). Behind bars: prosecutors sting corporate executives. Environmental Protection, October 1990.

Matthews, R. A. J. (1997). The science of Murphy's Law. Scientific American,

April 1997, pp. 88–91. See also Letters to the Editor, <u>Scientific American</u>, August 1997, p. 8.

Mays, L. W. and Y-K. Tung (1992). <u>Hydrosystems Engineering and Management</u>. McGraw-Hill, New York.

McManamy, R. (1992). Chicago shifts to finger pointing. <u>Engineering News Record</u>, May 4, 1992, pp. 10–11. McGraw-Hill, New York.

Merck (1982). Generalized Cardiovascular Disorders, chapter 24, p. 386 ff. <u>The Merck Manual</u>, 14th ed. Merck Sharp & Dohme Research Laboratories, Division of Merck & Co., Inc., Rahway, NJ.

Melching, C. S. and C. G. Yoon (1996). Key sources of uncertainty in QUAL2E Model of Passaic River. <u>Journal of Water Resources Planning and Management</u>, March/April 1996.

Mermel, T. W. (1989). The World's Major Dams and Hydroplants. Water Power and Dam Construction, July, 1989. Quadrant House, The Quadrant, Sutton, Surrey SM2 5AS, UK.

Miles, S. H. (1977). Status Report on Public Law 92-367. <u>The Evaluation of Dam Safety</u>. Engineering Foundation Conference Proceedings, pp. 3–19. ASCE, New York.

Moan, O. B. (1982). Application of Mathematics and Statistics to Reliability and Life Studies. <u>Reliability Handbook</u>, Chapter 4. W. G. Ireson, ed., McGraw-Hill, New York. Reissue of the 1966 edition.

Morgan, M. (1954). The Dam. <u>Reader's Digest</u>, 1954, pp. 155–168. A condensation from the book <u>The Dam</u> published by Viking Press (copyright by Murray Morgan). 18 East 48th Street, New York, N.Y.

Morris, J. W. (1974). National Program for Inspection of Dams.   <u>Inspection, Maintenance and Rehabilitation of Old Dams</u>   . Proceedings of the Engineering Foundation Conference, Asilomar Conference Grounds, Pacific Grove, CA, September 23–28, 1973, pp. 14–22. ASCE, New York.

Morrow, L. C., ed. (1957). <u>Maintenance Engineering Handbook</u>, chapter 12, Maintenance of Electrical Equipment, Sec. 7, p. 84. 2nd ed. 1966. McGraw-Hill, New York.

Muckenthaler, P.(1989). Hydraulische Sicherheit von Staudämmen (Hydraulic Safety of Earth Dams), Report 61, Technical University Munich, Civil Engineering IV, D-8000, München 2, Arcis Strasse 21, Germany.

Myers, R. and W. O. Wunderlich (1983). A Program to Compute Quarterly Flow Histograms. Report No. WR28-2-500-159, pp. 18–19. Tennessee Valley Authority, Norris, TN.

Navord (1970). Maintainability Engineering Handbook, published by Direction of Commander, Naval Ordnance Systems Commands, Superintendent of Documents, Washington, DC 20402.

NCPWI (1987). National Council on Public Works Improvements. The Nations Public Works: Report on Water Resources. K. Schilling, C. Copeland, J. Dixon, J. Smythe, M. Vincent, and J. Peterson. 1111 18th St., N.W., Suite 716, Washington, DC 20036.

NCPWI (1988). Fragile Foundations: A Report on America's Public Works. J. M. Giglio, P. C. Goldmark, Jr., F. Holmer, L. Jackson, and F. K. Wallison. 1111 18th St., N.W., Suite 716, Washington, DC 20036.

Nielsen, K. L. (1962). Differential Equations. College Outline Series. Barnes & Noble, New York.

NRC (1983). Safety of Existing Dams: Evaluation and Improvement. Committee on the Safety of Existing Dams, U.S. National Research Council, National Academy Press, Washington, DC.

NRC (1984). Risk Assessment in the Federal Government: Managing the Process. Committee on Institutional Means for Assessment of Risk to Public Health. Commission on Life Sciences. U.S. National Research Council. National Academy Press, Washington, DC, 1983. Third Printing 1984.

O'Donovan, T. M. (1983). Short Term Forecasting: An Introduction to the Box-Jenkins Approach. John Wiley & Sons, New York.

OMB (1992). Office of Management and Budget. Circular No. A-94 Revised (Transmittal Memo No. 64), October 29, 1992. Memorandum for Heads of Executive Departments and Establishments. Subject: Guidelines and Discount Rates for Benefit-Cost Analysis of Federal Programs (for update see OMB web site).

Parzen, E. (1965). Stochastic Processes. Third printing. Holden-Day, San Francisco.

Post, N. M., and H. B. Stussman (1989). Inflatable rubber weir makes comeback. Engineering News Record, September 14, 1989, pp. 30–31. McGraw-Hill, New York.

Prichard, B. A. (1977). Bureau of Reclamation's Assessment of Safety of Old Dams. ASCE (1977). The Evaluation of Dam Safety. Engineering Foundation Conference Proceedings, pp. 163–172. Asilomar Conference Grounds, Pacific Grove, CA. November 28–December 3, 1976. Homer B. Willis, Organizing Committee Chair. ASCE, New York,

Raiffa, H. (1970). Decision Analysis: Introductory Lectures on Choices Under Uncertainty. Addison-Wesley, Reading, MA.

Remenieras, G. (1960). L'Hydrologie de L'Engenieur. Eyrolles, Paris.

Reynolds, G. M. (1976). Liability of Consultants in Dam Investigations. Responsibility and Liability of Public and Private Interests on Dams. Engineering Foundation Conference, Asilomar Conference Grounds, Pacific Grove, CA. September 28–October 3, 1975. Joseph J. Ellam, chairman. ASCE, New York.

Rissler, P. (1988). Effects of Measuring and Control Systems on the Reliability of Dams. Probabilistic Safety Investigations of Dams, part VIII, Vol. B, pp.195–201, and Figure VIII-2, p. 200. K. H. Idel, ed. (Original: Sicherheitsuntersuchungen von Staudämmen auf probabilistischer Grundlage: Auswirkungen von Mess- und Kontrolleinrichtungen auf die Zuverlässigkeit von Dämmen, Teilband B, Teil VIII). Ruhrtalsperrenverein–Ruhrverband, Essen.

Rizzo, P. C., W. R. Argentieri, J. M. Bair, K.L. Massey, and S. H. Moxley (2002). Saving Saluda. Civil Engineering, October 2002, pp. 56–61. ASCE, Reston, VA 20191.

Robison, R. (1992). Smart structures. Civil Engineering, November 1992, Vol. 62, No. 11, pp. 66–68. ASCE, New York.

Ross, S. M. (1992). Applied Probability Models with Optimization Applications. Unabridged and unaltered republication of the work published by Holden-Day, San Francisco, in 1970, by Dover Publications, Mineola, NY 11501.

Samson, D. (1988). Managerial Decision Analysis. Richard D. Irwin, Inc., Homewood, IL 60430.

Sasser, M. (1996). The Role of Risk Communication. Lockwood Greene Technologies, Inc., Oak Ridge, TN (presentation at WATTec Conference, Knoxville, TN, February

22, 1996).

Savage, L. J. (1972). The Foundations of Statistics. 2nd revised ed. Dover Publications, New York.

Scanlon, J. M., J. E. McDonald, C. L. McAnear, E. D. Hart, R. W. Whalin, G. R. Williamson, and J. L. Mahloch (1983). REMR Research Program Development Report, Final Report, U.S. Army Engineer Waterways Experiment Station, P.O. Box 631, Vicksburg, MS 39181.

Sherard, J., L. Richard, J. Woodward, S. F. Gizienski, and W. A. Clevenger (1963). Earth and Rock Dams: Engineering Problems of Design and Construction. John Wiley & Sons, New York.

Shooman, M. L. (1968). Probabilistic Reliability: An Engineering Approach. McGraw-Hill, New York.

Siddall, J. N. (1972). Analytical Decision-Making in Engineering Design. Prentice-Hall, Englewood Cliffs, New Jersey.

Sieh, K., and S. LeVay (1998). The Earth in Turmoil—Earthquakes, Volcanoes, and their Impact on Humankind. W. H. Freeman, New York.

Simmons, M. (1981). Minimizing Risk of Flood Loss in the National Flood Insurance Program. In Risk/Benefit Analysis in Water Resources Planning and Managment, pp. 41–52. Y. Y. Haimes, ed. Plenum Press, New York and London.

Slovic, P., B. Fischhoff, and S. Lichtenstein (1981). Rating the Risks. In Risk/Benefit Analysis in Water Resources Planning and Management, pp. 193–217. Y.Y. Haimes, ed. Plenum Press, New York and London.

Stamatis, D. H. (1995). Failure Mode and Effect Analysis: FMEA from Theory to Execution. ASQC Quality Press, Milwaukee, WI 53202.

Stedinger, J.R., R. M. Vogel, and E. Foufoula-Georgiou (1993). Frequency Analysis of Extreme Events. Chapter 18, Handbook of Hydrology, D. Maidment, ed., McGraw-Hill, New York.

Swets, J. A., R. M. Dawes, and J. Monahan (2000). Better Decisions Through Science. Scientific American, October 2000, pp. 82–87. 415 Madison Ave., New York, NY 10017.

Takasaki, M. (1989). The Omata inflatable weir, at the Kawarabi hydro scheme, Japan. International Water Power and Dam Construction, November 1989.

Thorndike, K. E. (1994). FuzziCalc®, Version 1.51. FuziWare, Inc., P.O. Box 11287, Knoxville, TN 37939-1287.

Tschantz, B. A. (1977). Status of States' Dam Safety Program. The Evaluation of Dam Safety. Engineering Foundation Conference Proceedings, pp. 20–38. ASCE, New York.

UN (1992). United Nations Department of Humanitarian Affairs Glossary: Internationally Agreed Glossary of Basic Terms Related to Disaster Management, p. 93. Geneva, Switzerland.

USACE (1987). Accident at Lower St. Anthony Falls drains navigation pool. Crosscurrent, Vol. 10, No. 12, December 1987 (obtained by personal communication with Stuart V. Dobberpuhl, U.S. Army Corps of Engineeris, St. Pauls District).

U.S.WRC (1980). Principles and Standards for Water and Related Land Resources Planning. U.S.Water Resources Council. Federal Register, September 28, 1980.

VDI (1995/11). Energiesparlampen legen ihre Kinderkrankheiten ab (Energy saving lamps overcome their childhood diseases). VDI-Nachrichten, 17. November 1995. Düsseldorf, Germany.

VDI (1996/4). Russisch-Roulette in Tschernobyl (Russian roulette in Chernobyl). VDI Nachrichten, No. 17, 26. April 1996. Düsseldorf, Germany.

Viessman, W. J., G. L. Lewis, and J. W. Knapp (1989). Introduction to Hydrology. Harper & Row, New York.

Von Mises, R. (1981). Probability, Statistics and Truth. 2nd revised English edition prepared by Hilda Geiringer based on the third (1951) German edition by Springer-Verlag. Dover Publications, New York.

Von Neumann, J. and O. Morgenstern (1953). Theory of Games and Economic Behavior. 3rd ed. (1st ed. 1944, 2nd ed. 1947). Princeton University Press, Princeton, NJ.

Vrijling, J. K. and A. de Graaf (1987). Some Considerations on an Acceptable Level of Risk in the Netherlands. Ministry of Transport and Public Works, Rijkswaterstaat, Rijkskantorengebouw Westraven, 3502 LA Utrecht, The Netherlands.

Vrijling, J. K. (1993). Development in Probabilistic Design of Flood Defenses in the Netherlands. In Reliability and Uncertainty Analyses in Hydraulic Design, pp. 133–178. B. C. Yen and Y. K. Tung, eds. ASCE, New York.

Walpole, R. E. and R. H. Myers (1989). Probability and Statistics for Engineers and Scientists. 4th ed. Macmillan, New York.

Watson, S. R. and D. M. Buede (1989). Decision Synthesis: The Principles and Practice of Decision Analysis. Cambridge University Press, Cambridge, UK. In U.S.A.: 40 West 20th Street, New York, NY 10011.

Webster (1993). Merriam-Webster's Collegiate Dictionary. 10th ed. Merriam-Webster, Springfield, MA.

Willis, H. B. (1976). Federal Legislation and Activities for Dam Safety. From "Responsibility and Liability of Public and Private Interests on Dams," pp. 5–16. An Engineering Foundation Conference," September 28–October 3, 1975, J. J. Ellam, Chair. ASCE, New York.

World Almanac 1999. World Almanac Books, One International Boulevard, Suite 444, Mahwah, NJ 07495-0017.

Wunderlich, W. O. and J. Giles (1983). Risk Evaluation in Reservoir Operations. Proceedings of the Conference on Frontiers in Hydraulic Engineering, pp. 480–485. Hung Tao Shen, ed. ASCE, New York.

Wunderlich, W. O. and J. E. Prins (1984). Water for the Future: Water Resource Developments in Perspective. Part 1: History of Water Resource Developments, pp. 3–176. A. A. Balkema, Rotterdam and Boston.

Wunderlich, W. O. (1991). Probabilistic Aspects of Hydroproject Maintenance. Hydraulic Engineering, Proceedings of the 1991 National Conference, pp. 978–983, R. M. Shane, ed. ASCE, New York.

Yen, B. C., S. Cheng, and C. S. Melching (1986). First order reliability analysis. Stochastic and Risk Analysis in Hydraulic Engineering, B. C. Yen, ed. Water Resources Publications, Littleton, CO.

Yen, B. C. (1986). Stochastic and Risk Analysis in Hydraulic Engineering. Preface, p. 1. Water Resources Publications, Littleton, CO.

Yen, B. C. and Y-K. Tung (1993). Reliability and Uncertainty Analysis in Hydraulic Design. ASCE, New York.

Yevjevich, V. (1972). Probability and Statistics in Hydrology. Water Resources Publications, Fort Collins, CO.

****

Index